Astronomy

For Beginners

Frederick A. Ringwald

California State University, Fresno

Kendall Hunt
publishing company

Cover image courtesy of Frederick Ringwald © 2013.

Kendall Hunt
publishing company

www.kendallhunt.com
Send all inquiries to:
4050 Westmark Drive
Dubuque, IA 52004-1840

ISBN 978-1-4652-2844-4

Printed in the United States of America
10 9 8 7 6 5 4 3 2 1

CONTENTS

PREFACE

This book is primarily for university students in one-semester, introductory astronomy classes, mainly for students not majoring in science. Because of the information-rich nature of astronomy, the book may also be useful for students who are majoring in science, high-school students, amateur astronomers, and anyone else interested in learning about the Universe.

Too many textbooks leave the students gasping, "Do I need to know all this?" This book isn't intended to be an encyclopedia. It's intended to be an introduction, so that it's short, but it covers the essentials. It's also written to accommodate how today's students learn about the world around them, and to be user-friendly for their instructors. Above all, it stresses what everyone needs to know about science: that the Universe follows orderly, predictable laws, and that we human beings can learn these laws, with careful observation, logic, and reason, although we will always still need to test our understanding by experiment—or in other words, by trying things out.

This course is designed for college undergraduates with no prior knowledge at all of science. Too many texts for courses like this assume a quite extended background, and indeed, are largely books for majors with the math excised. A fresh approach is what's really needed here: the point is to help our students learn science, not to impress them with how clever we scientists think we are.

If you are a new instructor and want to get higher teaching evaluations by both your students and colleagues, this is the text for you. This book has been designed to accommodate both traditional "teacher-centered" and "student-centered," "active" learning teaching methods. Provisions designed to be useful especially for new teachers include the PowerPoints, test bank, and plans for what to cut when time is short (see the Class Schedules for 50– and 75–minute classes).

When this author was young, every kid in America knew quite a bit about space because of Project Apollo, but that was just a wee while ago. The point of a science text for students not majoring in science is not to leave them feel like they're drinking from a fire hose, nor to dazzle them with how clever we scientists think we are, nor to teach them that science is incomprehensible. The point is to teach the students some science, in ways they can understand and remember.

I have avoided extensive end-of-the-book tables of numbers. They pose a hazard for students who haven't yet realized that one can't learn science by trying to replace reasoning with rote memorization. Again, this book isn't meant to be an encyclopedia. Anyone who does want a periodic table or extensive tables of data on the planets or the Greek alphabet can find them easily enough online.

CHAPTER

1

Why Study Science, If You're Interested in Something Else?

This book is an introduction to astronomy, and to science in general, primarily for students who are not majoring in science, or for students who are just beginning to study science. But why study science in the first place? It is because we live in a society in which science is important—too important to be left only to the scientists. Science affects *everyone*.

> Science is the only news . . . all the human interest stuff is the same old he-said-she-said, the politics and economics the same sorry cyclic dramas . . . and even the technology is predictable if you know the science. Human nature doesn't change much; science does, and the change accrues, altering the world irreversibly.
>
> –Steward Brand, in *The Next Whole Earth Catalog*

Scientific innovation is the leading driver of American economic growth.

> . . . More than 50% of American economic growth for the past 50 years has come from scientific knowledge and innovation. . . .
>
> –John Holdren

> Almost everything that distinguishes the modern world from earlier centuries is attributable to science. . . .
>
> –Bertrand Russell

> It's still almost unbelievable: Men Walk on Moon. Next to that headline, everything else seems provincial. The biggest events since July 20, 1969 [the Apollo 11 landing] have indeed involved wars, scandals, terrorism, disasters. Maybe we should give a nod to the invention of the Internet, and the decoding of the human genome . . . But nothing tops the Apollo program.
>
> –Joe Achenbach

The only event of the past century that will be remembered a thousand years from now may well be the first Moon landing. Science is the human activity that, more than any other, defines our times.

Do you really want to be left out?

More to the point: Do you trust scientists to do all your thinking for you?

Remember, as free people, it is your *right,* and your *responsibility,* to be well-informed, critical-thinking citizens. You have a practical need to get informed. I care, because I want to live in a free society!

> An enlightened citizenry is indispensable for the proper functioning of a republic.
>
> –Thomas Jefferson

> Some people ask: 'Why should I care what happens in space? Don't we have enough to deal with right here on Earth?' Well, consider the dinosaur.
>
> –Bruce Betts (*The Planetary Society*)

(Large, non-avian dinosaurs are thought to have become extinct because Earth was hit by a large asteroid, or rock from space, at the time when they did become extinct.)

The good news is that the public is overwhelmingly positive about science:

- 85% say science and technology will make our lives better and our work more interesting.
- 72% say the benefits of research outweigh its harmful results.

Culturally, science matters too:

- One-third of the top 50 highest-grossing films of all time have been science fiction.
- Eight of the 100 all-time "best" movies (rated by the American Film Institute) were science fiction.

Still, it *hasn't* been all good. Science and technology *can* be used to do great harm; for example, with nuclear weapons. Snap judgments, as usual, aren't helpful. We need to be able to think rationally about science so we can increase the good and decrease or eliminate the harm.

A practical reason to study science is to answer questions from children, who are naturally curious about the world around them.

Some Questions that Kids Ask (*not* a complete list):

- What is taste?

  Acid = Sour
  Base = Bitter
  Sugar = Sweet
  Salt = Salty

  Things aren't always as they seem, at first!

▶ **Figure 1-1** People are naturally curious about the world around them, especially when they're kids. A major practical reason for a course like this is to answer questions, for kids of all ages. (Image courtesy of Frederick Ringwald)

- Why is grass green?
  Before you answer "chlorophyll," remember that simply repeating a funny word isn't an explanation.
- Why is the sky blue? Why is the ocean blue?
  Parents often lie to their children about this, because they don't know the real answer. Never lie to your children: admit you don't know.
- Why is Earth round?
- How old is Earth?
- How did the Solar System form?
- How did the Universe begin?

Unanswered Questions—and Please, Be Honest about Saying "I Don't Know":

- What's outside the Universe?
- What came before the beginning of the Universe?
  This is equivalent to "What's outside the Observable Universe?"
- If the Universe is expanding, what is it expanding into?
- How exactly did life on Earth originate?
- Are humans alone in the Universe?

CHAPTER 2

Powers of Ten and Scientific Notation

William Thomson was a 24-year-old engineer when he helped to install the first successful transatlantic telegraph cable. The project succeeded largely because he insisted on making measurements, and carefully recording them. Because of this success, he was given the title "Lord Kelvin," and we will meet him again later in this book. He said:

> When you can measure what you are speaking about,
> and express it in numbers, you know something about it;
> but when you cannot measure it,
> when you cannot express it in numbers,
> your knowledge of it is of a meager and unsatisfactory kind:
> It may be the beginning of knowledge, but you have scarcely,
> in your thoughts, advanced to the stage of science.
>
> –William Thompson (later Lord Kelvin)

> The great book of nature is written in mathematical symbols.
>
> –Galileo

The Universe follows mathematical laws. This is the key to how humans can understand it, because everyone can understand at least some mathematics.

Nevertheless, "**Math is hard,**" says Thomas Garrity, a mathematician. He knows of no serious mathematician who finds the subject easy. In fact, he quips, "most, after a few beers, will confess to how slow they are." *You* don't need to be intimidated by it!

> Do not worry about your difficulties in mathematics. I can assure you that mine are still greater.
>
> –Albert Einstein, in a letter to a 12-year-old.

Therefore, unless you're smarter than Einstein, you'll need to practice.

Math Terms for Teachers

1 thousand = 10^3 = 1,000
1 million = 10^6 = 1,000,000
1 billion = 10^9 = 1,000,000,000
1 trillion = 10^{12} = 1,000,000,000,000

These are the largest numbers in common use. Kids are often intrigued by larger ones:

1 quadrillion = 10^{15} = 1,000,000,000,000,000
1 quintillion = 10^{18} = 1,000,000,000,000,000,000
1 sextillion = 10^{21} = 1,000,000,000,000,000,000,000
1 septillion = 10^{24} = 1,000,000,000,000,000,000,000,000
1 octillion = 10^{27} = 1,000,000,000,000,000,000,000,000,000
1 nontillion = 10^{30} = 1,000,000,000,000,000,000,000,000,000,000
1 decillion = 10^{33} = 1,000,000,000,000,000,000,000,000,000,000,000

Larger numbers exist, but they're rarely used because there aren't enough things in the Universe to count:

- 1 googol = 10^{100} = 10,000,000,000,000,000,000,000,000,000,000,000,000,000,000,000,000,000, 000,000,000,000,000,000,000,000,000,000,000,000,000,000,000,000,000, or a 1 followed by 100 zeroes.
- 1 googolplex = $10^{1 \text{ googol}}$ = a 1 followed by a googol zeroes. There isn't enough paper on Earth to write this.

Here are some examples of these numbers, **but don't memorize these for exams:**

- Half a million people live in Fresno.
- Your grandparents may be over 2 billion seconds old.
- There are over 7 billion people on Earth.
- The U.S. government spends about $4 trillion a year.
- The U.S. economy is worth over $14 trillion.
- There are on the order of 100 quintillion grains of sand on all the beaches on Earth.
- There are over 5 sextillion stars in the Universe.
- There are over one nontillion air molecules in this classroom.
- There are over 10 trillion decillion decillion (10^{80}) particles in the observable Universe.

Here are some distances, expressed in the large numbers, **but don't memorize these for exams:**

- The Moon is about quarter of a million miles from Earth.
- Mars is over 50 million kilometers (35 million miles) from Earth.
- The Sun is 150 million kilometers (93 million miles), or 1 Astronomical Unit (1 AU), from Earth.
- Pluto, at the outer edge of the Solar System, is about 40 AU from Earth and the Sun, or about 6 billion kilometers (4 billion miles).
- Proxima Centauri, the nearest star outside the Solar System, is 4.2 light-years away. One light-year is about 10 trillion kilometers, or 6 trillion miles, so Proxima Centauri is over 40 trillion km (25 trillion miles) away.
- The center of the Milky Way, our galaxy, is 250 quadrillion km away.
- Andromeda, our nearest large neighbor galaxy, is 25 quintillion km (2.5 million light-years) from Earth.
- The observable Universe is over 100 sextillion kilometers across.

► **Figure 2-1** It can be good for the human spirit to look at a million of something, and know it. There may be a million stars here, but this picture shows less than one-tenth of them. We would need a much larger and more detailed picture to see them all. (NASA/Space Telescope Science Institute)

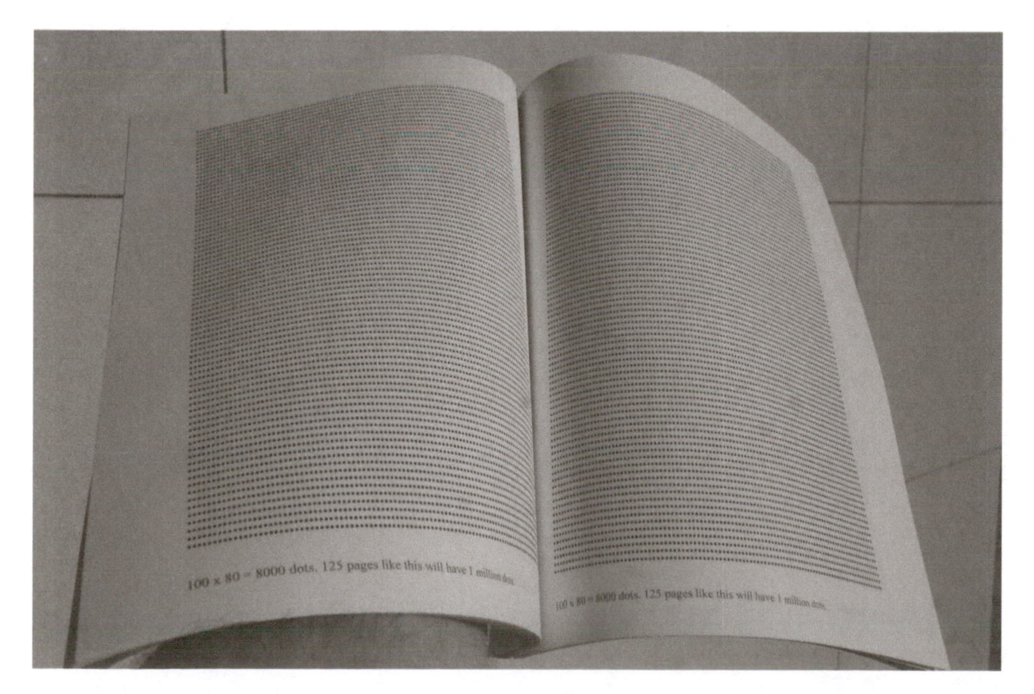

► **Figure 2-2** How big is one million? Each page shown here has 100 rows of 80 dots. A book of 125 of these pages will have 1 million dots. One billion dots would be a stack of these books over 6 meters (20 feet) high. One trillion dots would be a large classroom filled with these books. (Image courtesy of Frederick Ringwald)

► **Figure 2-3** There are more stars in the Observable Universe than there are grains of sand on all the beaches of Earth. (Image courtesy of Frederick Ringwald)

Powers of Ten and Scientific Notation

In astronomy, large numbers are common. We use scientific notation to express them and to understand them.

10^6 = 1,000,000 = 1 million. It's OK to spell out the word *million* here. In fact, it's better, since many zeros confuse the eye.

10^7 = 10 million
10^8 = 100 million
10^9 = 1 billion
10^{12} = 1 trillion

There are larger numbers (e.g., 10^{15} = 1 quadrillion) but, generally, trillions are the largest numbers in common use.

With scientific notation, the general rule is that: 10^x = a 1 with *x* zeroes after it.

Scientific notation is sometimes called exponential notation, since the x in this equation is called the *exponent*. The "powers of ten" are also sometimes called *orders of magnitude*, with each factor of ten equal to one order of magnitude.

Example: the average distance from the Earth to the Sun = 1 Astronomical Unit = 1 AU

= 93000000 miles	Writing it this way is confusing to the eye.
= 93,000,000 miles	This is slightly better.
= 93 million miles	This is better.
= 9.3×10^7 miles	This is even better.
= 1.5×10^8 km (kilometers)	This is best, since we like metric units.

So 1 AU = 1.5×10^8 km = 150,000,000 km = 150 million km.

We'll use the following symbols often:

= equals
≡ is defined by
≈ is about equal to (within 2%)
~ is roughly equal to (within 10%)
≠ is not equal to

We often need small numbers in astronomy, too. Stars get their power from nuclear energy. A typical nucleus of an atom has a diameter of:

0.000 000 000 000 001 m = 10^{-15} m

or a 1 preceded by 14 zeroes (or 15 zeroes, if you count the one before the decimal).

Small numbers are therefore expressed in this way:

10^1 = 10
10^0 = 1
10^{-1} = 0.1 = 1/10
10^{-2} = 0.01 = 1/100
10^{-3} = 0.001 = 1/1000, etc.

Standard scientific notation (for example, 1.5×10^8) is always preferable for any number larger than five decimal places long, or in other words, 1 million or larger. This works because most people have serious trouble visualizing more than six of anything. Likewise for any number smaller than 1/1000.

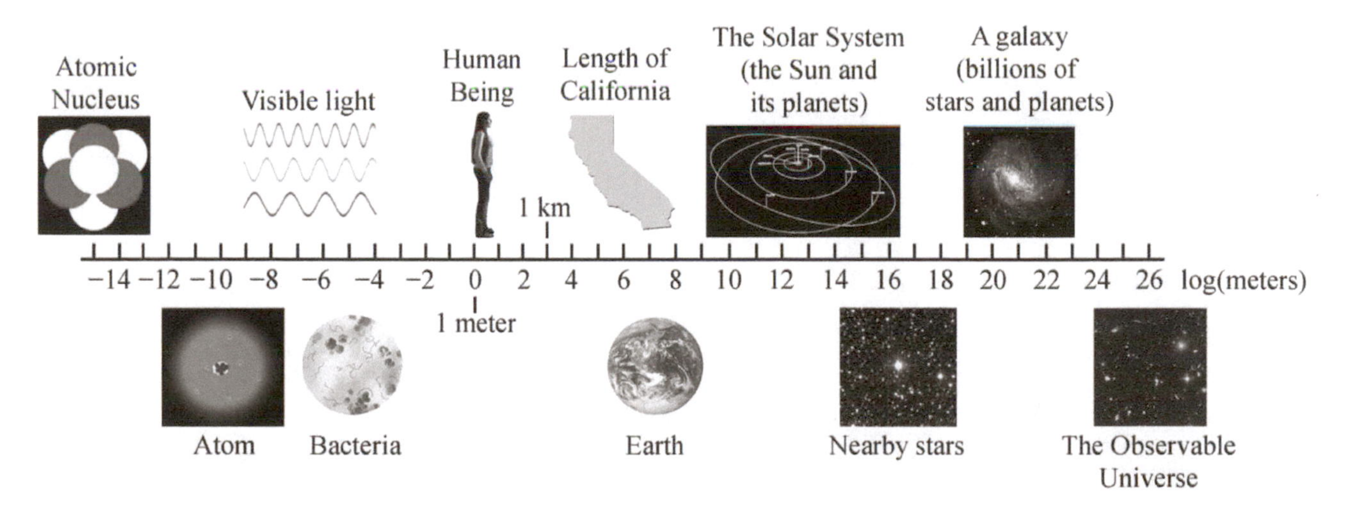

► **Figure 2-4** This is a **logarithmic** scale, which counts by *multiplying* powers of ten. This contrasts with a **linear** scale (used in everyday life), which counts by *adding* ones. Often, the smallest examples of anything are by far the most common. Examples include stars, galaxies, atoms, and animals. Shakespeare called this "the modesty of nature." Logarithmic (or log) scales are useful for counting things that can be both very large and very small and are equally important regardless of size. Examples include earthquakes, stars, and meteorite impacts. (All images courtesy of Frederick Ringwald except for the bacteria, by the Centers for Disease Control and Prevention; California, by NOAA; Earth, by NASA; The Solar System, by NASA/Lunar and Planetary Institute; Nearby stars, by 2MASS/IPAC-Caltech/University of Massachusetts; the Observable Universe, by NASA/Space Telescope Science Institute.)

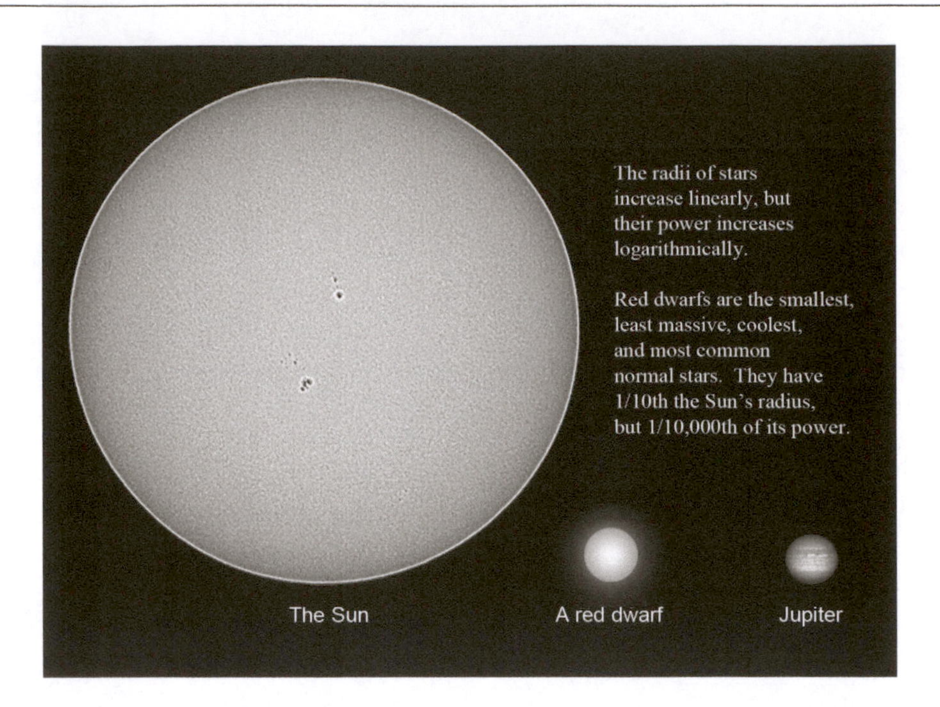

▶ **Figure 2-5** Stars increase in radius linearly but increase in power output logarithmically. This means that the smallest stars have one-tenth of the Sun's radius, but only one-ten-thousandth of its power. (Image courtesy of Frederick Ringwald)

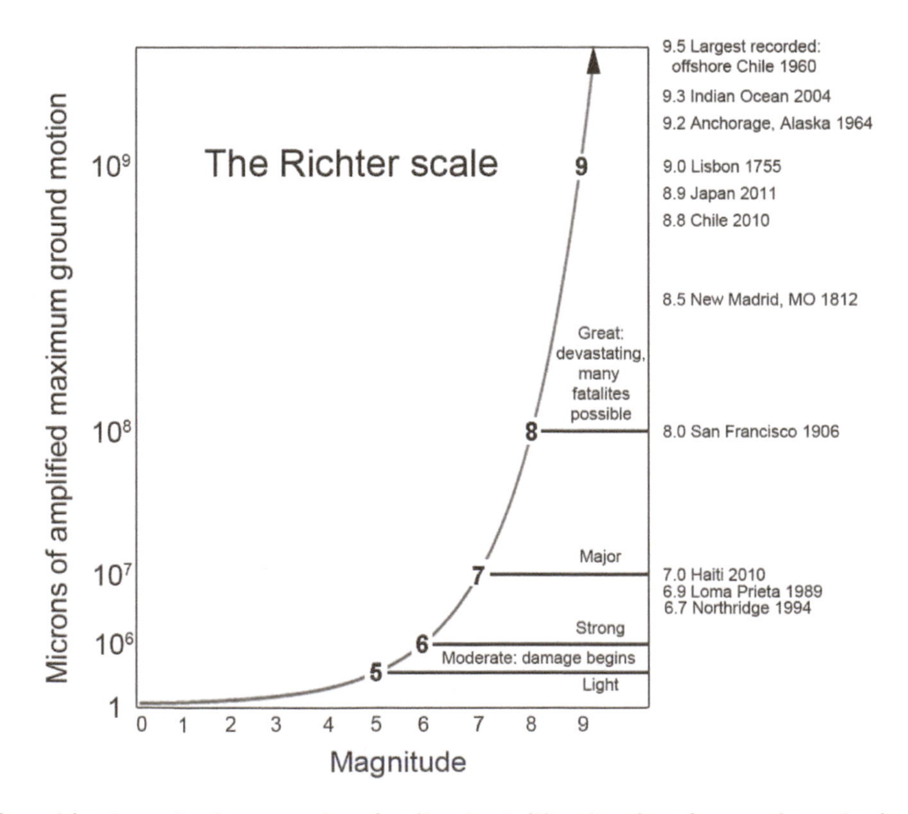

▶ **Figure 2-6** A logarithmic scale that people who live in California often know about is the Richter scale. An earthquake that rates an 8 on the Richter scale causes ten times more shaking than one that rates a 7. A logarithmic scale is useful here because small earthquakes can come before big ones, so it's important to count them all. (Image courtesy of Frederick Ringwald)

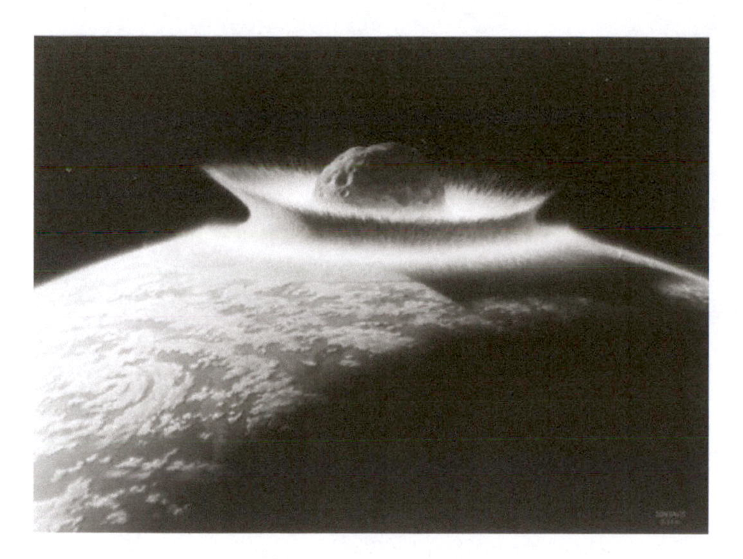

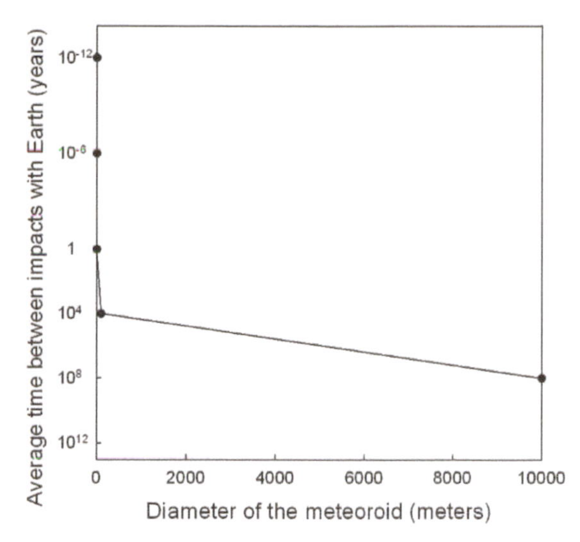

▶ **Figure 2-7** The average time between impacts of rocks in space, also called asteroids, are plotted as powers of ten on only one axis (the vertical axis, which is the y-axis). This is like the Richter scale plot, but steeper, so the points bunch up at the left. This makes this plot hard to read. (*Left:* NASA/Don Davis; *Right:* Image courtesy of Frederick Ringwald)

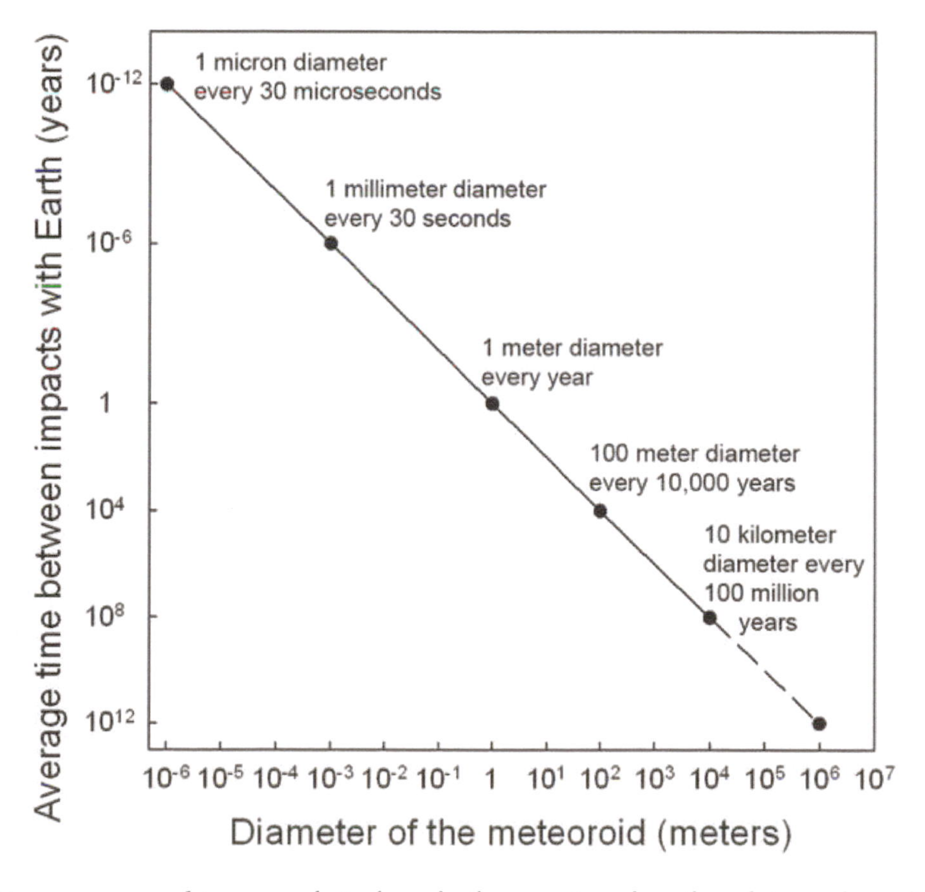

▶ **Figure 2-8** Here, powers of ten are plotted on *both* axes. It makes the plot much easier to read and understand: it shows that small impacts (dust hits) are, by far, the most common. It also shows that large impacts are rare. (Image courtesy of Frederick Ringwald)

CHAPTER 3

Units, Light-Years, and Look-Back Time

Americans often use inches, feet, and miles to measure distances. There are 12 inches in 1 foot, and 5,280 feet in 1 mile. So, how many inches are in a mile? It's not so easy to figure out, is it?

This is why American scientists, and most other people in nearly all other countries, prefer to use the metric system. The metric system is easier to use because it's based on powers of ten.

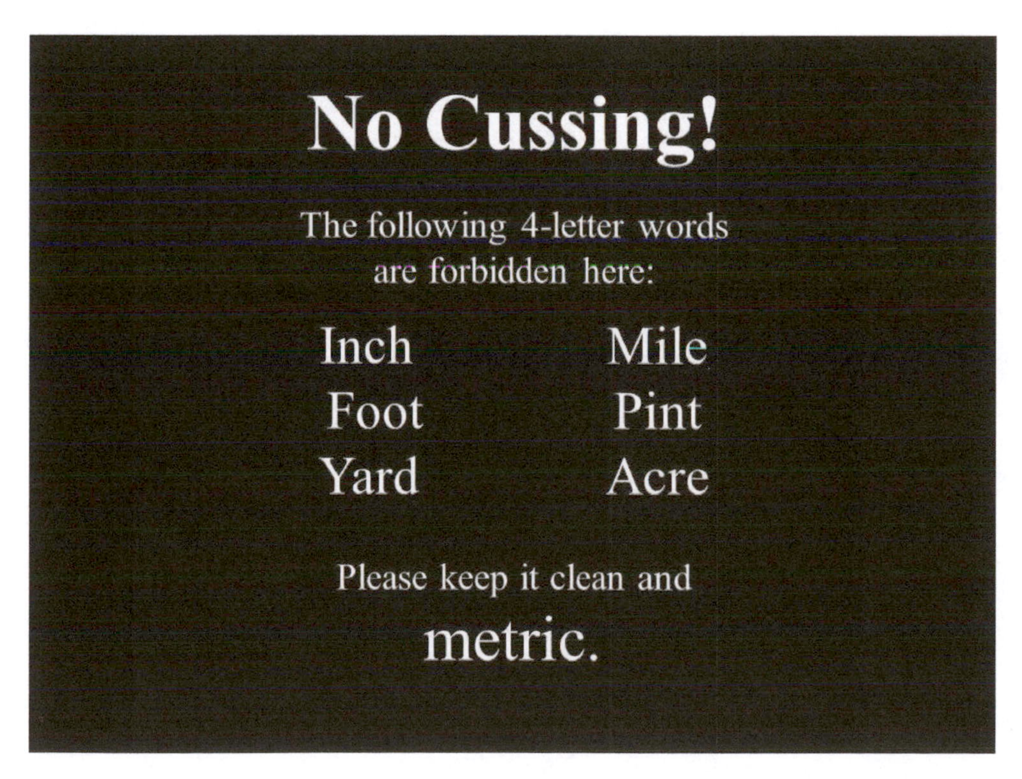

► **Figure 3-1** Avoid using the English (or Imperial) system of units—it's hard to use. Use the metric system instead. (Image courtesy of Frederick Ringwald)

The Metric System

The metric system uses meters for length, and grams for mass. We can express larger units, too—for example, kilometers (1 km = 1000 m) and kilograms (1 kg = 1000 grams).

1 inch ≡ 2.54 cm (centimeters), exactly, with 100 cm = 1 m (meter).
1 kg (kilogram) = 2.2 pounds, so 454 grams = 1 pound, on Earth. (Work it out yourself!)
1 metric ton = 1 tonne = 1 T = 1000 kg = 2200 pounds (1 U.S. ton = 2000 pounds)

The metric system uses prefixes in front of units, such as meters or grams, to show *powers of ten:*

Prefix	Meaning	Abbreviation	Example
kilo	$\times 10^3$ (thousand)	k	1 kg = 1000 g (grams)
mega	$\times 10^6$ (million)	M	1 MT = 1 megaton
giga	$\times 10^9$ (billion)	G	1 Gb = 1 gigabyte
tera	$\times 10^{12}$ (trillion)	T	

Using small units in the metric system, we have:

centi	$\times 10^{-2}$ (hundredth)	c	100 cm = 1 m
milli	$\times 10^{-3}$ (thousandth)	m	10 mm = 1 cm
micro	$\times 10^{-6}$ (millionth)	micro, or μ	1 micron = 1/1000 mm
nano	$\times 10^{-9}$ (billionth)	n	
pico	$\times 10^{-12}$ (trillionth)	p	

Some further examples of this are:

Light travels at a speed of about 186,000 miles per second.
This is about equal to 670 million miles per hour, or 300,000 kilometers per second.
At this speed, light travels 30 cm (about one foot) in **1 ns = 1 nanosecond = 10^{-9} s.**
1 micron is one-millionth of a meter, or 1/1000 of one millimeter (mm).
A human hair is about 0.1 mm = 100 microns thick.
Bacteria are one-tenth this size, or about 10 microns long.
Atoms are 1/100,000th this size, or 0.1 nm = 10^{-10} m across.

Working with Scientific Notation and the Metric System

One can add, subtract, multiply, and divide in scientific notation, just as with regular numbers. To do this, remember the general rules for how exponents work:

$a^m a^n = a^{m+n}$ For example: $10^2 \times 10^3 = 10^5$ or alternatively, $100 \times 1000 = 100{,}000$.
$a^{mn} = (a^m)^n$ For example: $(10^3)^2 = 10^6$ or alternatively, $1000 \times 1000 = 1$ million.

Here's an example of how to multiply large numbers in scientific notation:

$$(5 \times 10^3) \times (3 \times 10^2) = 15 \times 10^{3+2} = 15 \times 10^5 = 1.5 \times 10^6$$

Here's an example of how to multiply small numbers in scientific notation:

$$(0.05)\,(0.003) = (5 \times 10^{-2})\,(3 \times 10^{-3}) = 15 \times 10^{-5} = 1.5 \times 10^1 \times 10^{-5} = 1.5 \times 10^{-4}$$

Here's an example of how to divide in scientific notation:

$$5 \times 10^3 / 2 \times 10^2 = 5/2 \times 10^{3-2} = 2.5 \times 10^1 \text{ or, as we can see, } 5000/200 = 25.$$

How to work with scientific notation on a calculator:

To enter 1.5×10^{-8} on your calculator, use the [EE] or [EXP] button.

1. Punch in [1.5]
2. Punch in [EE] or [EXP]
3. Punch in [8]
4. Punch in [+/–]

How to raise any number to a power on your calculator:

To calculate $(0.5)^{12}$ on your calculator, use the [x^y] button.

1. Punch in [0.5]
2. Punch in [x^y]
3. Punch in [12]
4. Punch in {=}

If your calculator doesn't work this way, bring it to your instructor during office hours, and ask your instructor to show you how to use it. It will help if you can bring the manual.

Units Conversions

How many inches are in one mile?

$$\left(\frac{12\text{ inches}}{1\text{ foot}}\right)\left(\frac{5280\text{ feet}}{1\text{ mile}}\right)=\frac{63{,}360\text{ inches}}{1\text{ mile}}$$

So there are 63,360 inches in one mile.

To do this, we chose to write each term inside the parentheses to be equal to 1. By doing this, and by choosing factors of 1 for which units we don't need to cancel, we can change units in any way we like, to solve whatever problems we have.

For example, how many kilometers are in one mile?

Recall that 1 inch = 2.54 cm, exactly.

From above:

$$\left(\frac{63{,}360\text{ inches}}{1\text{ mile}}\right)\left(\frac{2.54\text{ cm}}{1\text{ inch}}\right)\left(\frac{1\text{ m}}{100\text{ cm}}\right)\left(\frac{1\text{ km}}{1000\text{ m}}\right)=\left(\frac{63{,}360\times 2.54}{10^2\times 10^3}\right)\left(\frac{\text{km}}{\text{mile}}\right)$$

$$=\frac{160934.4}{10^5}\left(\frac{\text{km}}{\text{mile}}\right)=\frac{1.609344\times 10^5}{10^5}\left(\frac{\text{km}}{\text{mile}}\right)=\frac{1.609344\text{ km}}{1\text{ mile}}$$

So, 1.609344 km = 1 mile.

How many miles are in 1 km?

This is easy if you use the previous equation, but write it upside down now, as it's still equal to 1:

$$1\text{ km}=\left(\frac{1}{1.609344}\right)\text{miles}=0.62137\text{ miles.}$$

Similarly, 1 m = 39.37 inches, since:

$$\left(\frac{1\text{ inch}}{2.54\text{ cm}}\right)\left(\frac{100\text{ cm}}{1\text{ m}}\right)=\frac{100}{2.54}\frac{\text{inches}}{\text{m}}=\frac{39.37\text{ inches}}{1\text{ m}}$$

We can also find that:

1 m = 3.2808 feet, since 12 inches = 1 foot, so 1 m = (39.37/12) feet = 3.2808 feet

and that:

$$1\text{ foot}=0.3048\text{ m, since }3.2808=\frac{1}{0.3048}.$$

Doing this last calculation is easy: just use the 1/x key on your calculator.

Significant Digits, Precision, and Accuracy

Here's a slightly different way to find how many kilometers (km) are in one mile. Again, it uses the fact that 1 inch = 2.54 cm (centimeters), *exactly*.

$$\left(\frac{2.54\text{ cm}}{1\text{ inch}}\right)\left(\frac{12\text{ inches}}{1\text{ foot}}\right)\left(\frac{5280\text{ feet}}{1\text{ mile}}\right)=\frac{(2.54)(12)5280\text{ cm}}{1\text{ mile}}$$

$$=\frac{160934.4\text{ cm}}{1\text{ mile}}\left(\frac{1\text{ m}}{100\text{ cm}}\right)\left(\frac{1\text{ km}}{1000\text{ m}}\right)=\frac{1.609344\text{ km}}{1\text{ mile}}$$

So 1.609344 km = 1 mile, exactly.

Here, we may include all the digits to the right of the decimal, since the relation is *exact.*

We can't always do this, because our ability to measure is never infinitely precise.

Here's an example to illustrate this:

How many km are in one AU?

1 Astronomical Unit = 1 AU = 93 million miles, so:

$$93\text{ million miles}\left(\frac{1.609344\text{ km}}{1\text{ mile}}\right)=150\text{ million km.}$$

Note that this is 150 million km, and *not* the exact result of 149,668,992 km. Those extra digits aren't *significant*, and are probably wrong. This is because the output can't be more precise than the most precise number in the input.

Another example of significant digits is with money. All financial transactions are calculated to the dollar and cent—but we ignore fractions of a cent, because they're too small to be significant.

Be careful: calculators will still show these fractions of a cent. This is because calculators are built to assume that all digits are significant. They may not be: it takes human intelligence to check whether they are.

This illustrates the difference between **precision** and **accuracy**.

Precision is the *detail* of a measurement
Accuracy is the *correctness* of a measurement

They're quite different! Another issue in measurement is **uncertainty**, sometimes also called *measurement error*. Here the words *uncertainty* and *error* are used differently than in everyday speech. They don't mean the measurement is inaccurate—or in other words, is just plain wrong. They mean *any measurement has limited precision*—or in other words, *there is always a limit to how precisely one can measure*.

For example, the age of the Universe is now known to be:

13.80 ± 0.04 billion years.

This means that the Universe is between 13.76 and 13.84 billion years old. It's based on measurements done with modern spacecraft (and described in the section on cosmology later in this book). Before these spacecraft, the uncertainty in the measurements was much greater, at:

15 ± 5 billion years.

Notice how the term after the "±" is larger. This means the measured age was somewhere between 10 and 20 billion years old: not as precise as 13.80 ± 0.04 billion. It's still accurate, though, since 13.80 ± 0.04 billion and 15 ± 5 billion do overlap.

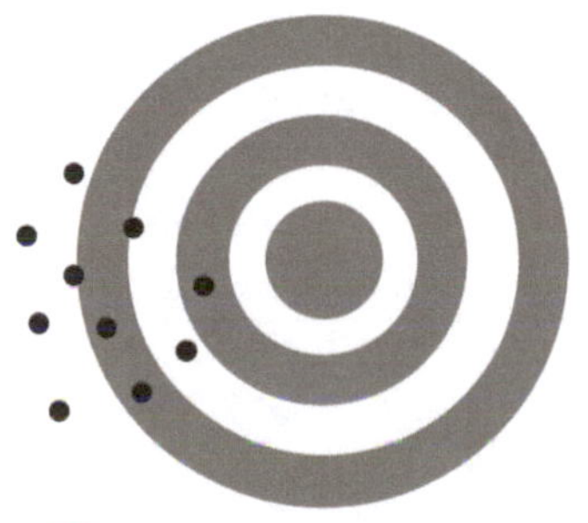

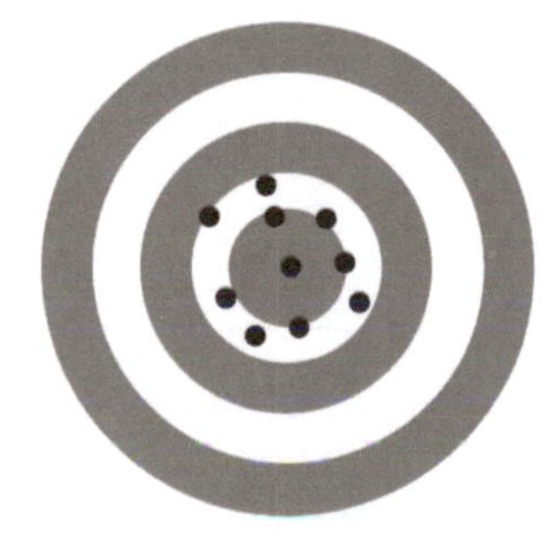

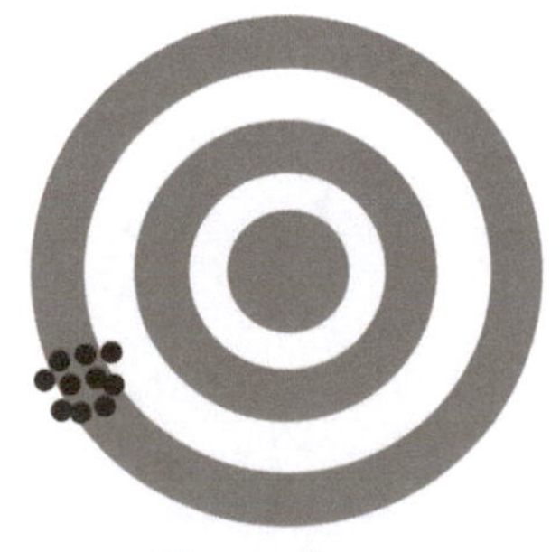

▶ **Figure 3-2** Accuracy and precision are different, subtly. (Image courtesy of Frederick Ringwald)

Uncertainty ***doesn't indicate science that is wrong, or in other words, inaccurate.*** It shows only that all measurements have limits—but then, most people realize this when they think about it.

The important question **isn't:** "Is there uncertainty?"
The important question is: "Is the uncertainty *significant?*"

It is *vital* that scientists be careful and *honest* about uncertainty. Only by doing so does science have any chance of progressing, because an essential part of science is *correcting errors.* Unfortunately, many people crave certainty ever since they were children.

Television and radio, in particular, have *many* confident-sounding voices that have *no* idea what they're talking about. Many people still believe the confident-sounding voices, however. *Beware of this!*

Do not confuse *confidence* with *certainty*—they are very different. Notice how anyone who is trying to lie to you, such as astrologers, psychics, and some politicians, always sound so confident, even when they know what they say is wrong. They know that many people crave certainty, and often confuse confidence with certainty. Good scientists are honest about uncertainty. Sadly, they are sometimes made fun of because of it.

> The demand for certainty is one which is natural . . . but is nevertheless an intellectual vice. If you take your children for a picnic on a doubtful day, they will demand a dogmatic answer as to whether it will be fine or wet, and be disappointed in you when you cannot be sure. The same sort of assurance is demanded, in later life, of those who undertake to lead populations...
>
> –Bertrand Russell (*Philosophy for Laymen,* 1946)

Let's return to units conversion problems and do an important one:

The Light-Year

- A light-year is **a unit of distance,** *not time.*
- One light-year is *the distance light travels in one year.*
- Light has a finite speed of 186,000 miles/second = 300,000 km/s.

From riding in cars, you may already know that:

$$\text{speed} = \frac{\text{distance}}{\text{time}}, \text{ for example, when your speed is } 60\,\frac{\text{miles}}{\text{hour}}.$$

We can rearrange this equation to find that:

$$\begin{aligned}
&\text{distance} = \text{speed} \times \text{time, so that:}\\
&1 \text{ light-year} = \text{the distance light travels in one year} = (\text{the speed of light}) \times (1 \text{ year})\\
&= (300{,}000 \text{ km/s}) \times (1 \text{ year})\\
&= \left(3.00 \times 10^5 \frac{\text{km}}{\text{s}}\right)(365.2422 \text{ days})\left(\frac{24 \text{ hours}}{1 \text{ day}}\right)\left(\frac{60 \text{ minutes}}{1 \text{ hour}}\right)\left(\frac{60 \text{ s}}{1 \text{ minute}}\right)\\
&= (3.00 \times 10^5)(365.2422)(24)(60)(60) \text{ km}\\
&= 9.47 \times 10^{12} \text{ km.}
\end{aligned}$$

Therefore, **1 light-year = 9.47 × 10^{12} km,** about 10 trillion km, or about **6 trillion miles.**

Look-Back Time

It takes time for light to travel. This is useful for astronomers, since when we look deep into space, it's like looking into a time machine. We can literally *see* the way the Universe was, in the distant past.

We see the Moon because it reflects light from the Sun. The Moon is 239,000 miles from Earth. The speed of light in empty space is 186,000 miles per second. It therefore takes 1.3 seconds for light to travel from the Moon to the Earth. We therefore say that the Moon is at a *distance* of 1.3 *light-seconds* from Earth. When we look at the Moon, we therefore see it the way it was 1.3 seconds ago. The *look-back time* from Earth to the Moon is therefore 1.3 seconds.

The Sun is 400 times farther away from Earth than the Moon. The Sun is 93 million miles from Earth. It takes light 8.3 minutes to travel from the Sun to Earth. This means that we see the Sun the way it was 8.3 minutes ago. If a flare goes off on the Sun, we won't see it until 8.3 minutes after it happened.

Looking deep into space, this effect becomes profound.

The nearest star (other than the Sun) is Proxima Centauri. It's 4.2 *light-years* from Earth. We therefore see it the way it was 4.2 years ago. If there were a flare on Proxima Centauri, we wouldn't see it until 4.2 years after it happened.

Let's pause for a moment to ponder a very big thought: *the Sun is a star.*

All the stars we see at night are Suns. About 10%, or 1 in 10 of them, are similar enough to the Sun to be hospitable for life as we know it, if these stars have planets like Earth.

Most stars are in vast islands of hundreds of billions of stars, called *galaxies*. Our galaxy is called *the Milky Way* because it is visible as a faint band across the night sky—but only from a dark, country sky.

Look-back time is useful for learning the history of the Universe. *Hubble Space Telescope* has taken images so deep, one can see galaxies that are over 12 billion light-years away. Because the speed of light is finite, this means we're seeing the galaxies the way they were 12 billion years ago—when the Universe was not even one billion years old.

▶ **Figure 3-3** The Moon is 384,000 km (239,000 miles) from Earth. Light takes 1.3 seconds to travel from the Moon to Earth. Therefore, the Moon is at a distance of 1.3 *light-seconds* from Earth. We see the Moon the way it was 1.3 seconds ago. (Image courtesy of Frederick Ringwald at Fresno State's Campus Observatory)

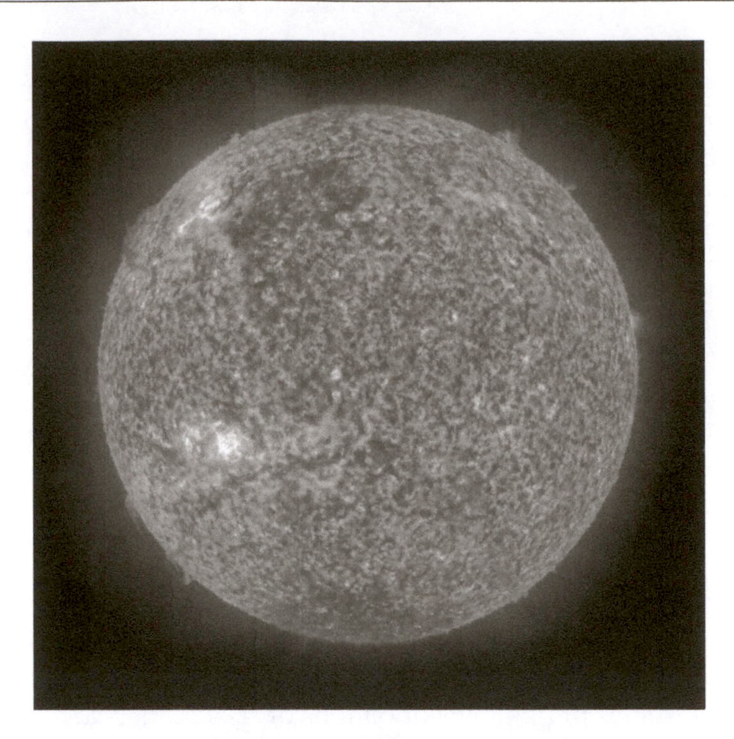

► **Figure 3-4** **The Sun is 150 million km (93 million miles) from Earth. Light takes 8.3 minutes to travel from the Sun to Earth. Therefore, the Sun is at a distance of 8.3 *light-minutes* from Earth. We see the Sun the way it was 8.3 minutes ago.** (NASA Solar Dynamics Observatory)

► **Figure 3-5** **Proxima Centauri is the closest star to Earth, other than the Sun. It is 4.2 light-years from Earth. We therefore see it the way it was 4.2 years ago.** (Atlas Image (or Atlas Image mosaic) obtained as part of the Two Micron All Sky Survey (2MASS), a joint project of the University of Massachusetts and the Infrared Processing and Analysis Center/California Institute of Technology, funded by the National Aeronautics and Space Administration and the National Science Foundation.)

▶ **Figure 3-6** On a dark night away from city lights, the Milky Way looks like a faint band of light, stretching across the sky. The Milky Way is our galaxy, or island of hundreds of billions of stars. It's not easy to get a good picture of the Milky Way, since we're in it. The stars we can see in it with just our unaided eyes are up to about 1,000 light-years away. (Image courtesy of Frederick Ringwald at Fresno State's station at Sierra Remote Observatories)

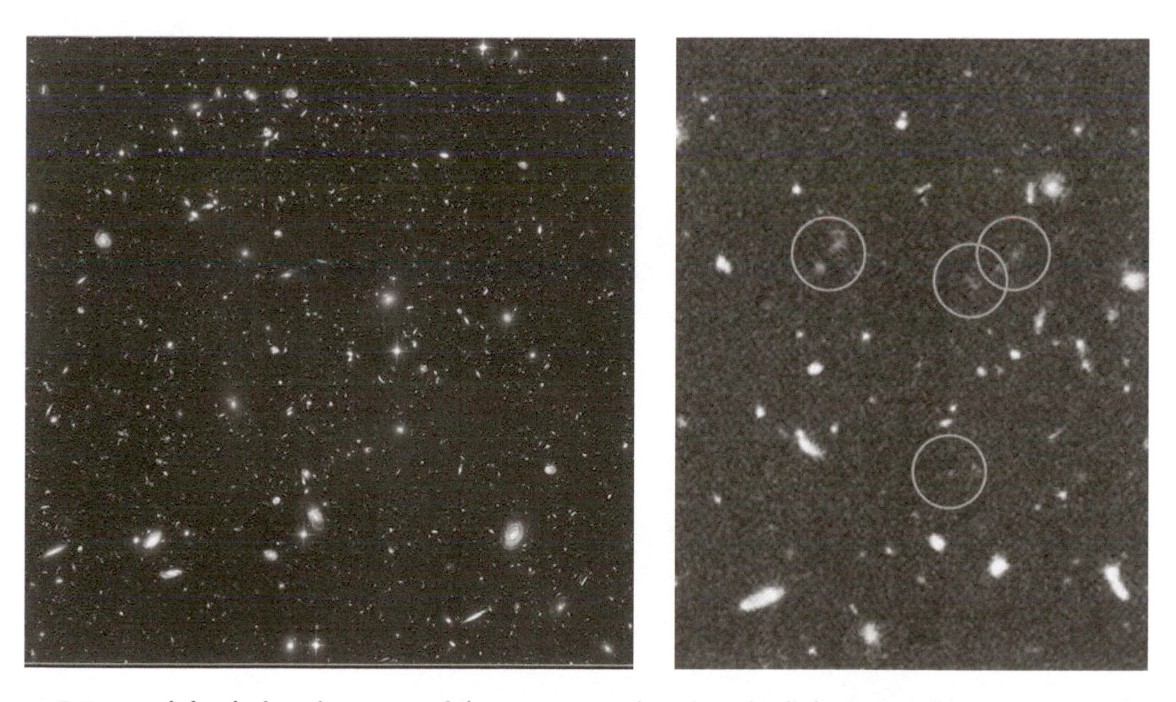

▶ **Figure 3-7** Look-back time is a powerful concept. It takes time for light to travel, so looking deeply into the Universe is like looking into a time machine. By looking deep, we can *see* the history of the Universe. *Left:* Galaxies are islands of hundreds of billions of stars. The brighter ones in this deep field are about 100 light-years away: we see these galaxies the way they were when dinosaurs walked the Earth. (NASA/STScI) *Right:* The little faint galaxies in the background, shown in this close-up, are over 12 billion light-years away. (NASA/STScI)

CHAPTER

4

Proportions

A Brief Tour of Space and Time

One light-year is awfully big. How can the human mind comprehend such enormous distances? The answer is that it's like comprehending anything—it just depends on what one compares it to. Notice that there are about 60,000 Astronomical Units in one light-year. We can verify this, since we know that:

1 AU = 1.5×10^8 km, and that 1 light-year = 9.47 trillion kilometers;

therefore;

$$\frac{\text{1 light-year}}{\text{1 AU}} = \frac{9.47 \times 10^{12} \text{ km}}{1.5 \times 10^{8} \text{ km}} = 63{,}100$$

Remember how we determined how many inches there are in one mile? Let's redo this calculation to make it look like the previous equation:

$$\frac{\text{1 mile}}{\text{1 inch}} = \left(\frac{\text{5280 feet}}{\text{1/12 feet}}\right) = (12)(5280) = 63{,}360$$

In other words, there are about as many Astronomical Units (AU) in one light-year as there are inches in one mile, about 63,000. An equation expressing this would be:

$$\frac{\text{1 light-year}}{\text{1 AU}} \approx \frac{\text{1 mile}}{\text{1 inch}}.$$

You know how long an inch is. You can easily imagine how long a mile is. It's therefore not hard to imagine 1 AU, since one sees the Sun often enough. Now imagine this distance shrunk down to one inch. On this scale, *one light-year* would be a mile.

A Scale Model of the Local Universe

(Don't memorize this, but do understand the principle of proportions).

James Irwin, who walked on the Moon during the Apollo 15 mission in 1971, said that from the Moon, the Earth looked like a little blue marble:

> "the most beautiful marble you can imagine."

Let Earth be represented by a little blue marble 1.50 cm in diameter (0.75 cm radius). The radius of the real Earth is about 6380 km, or 4000 miles. On this scale, the Moon would be about the size of the head of a small finishing nail, since the Moon's radius is about 1/4 that of Earth's.

The Moon is about 60 Earth radii away from the Earth. On our scale, this would put it 44 centimeters away from Earth. On this scale, the Sun would be 174 meters away from Earth. The Sun's radius is 108 times that of Earth's, so on our scale it would be 1.6 m in diameter—about as tall as the average woman, at about 5′4″.

At closest approach to Earth (also known as opposition), Mars is about 55 million km from Earth. On our scale, Mars would be 64 meters away. The radius of Mars is about half that of Earth's, so on our scale this would make it about the size of an orange seed.

The diameter of Jupiter is about 11 times that of Earth. On our scale, this would make it 16.5 cm in diameter—about the size of a cantaloupe. At 5.2 times the average distance between the Sun and Earth (an Astronomical Unit or AU, 1.5×10^8 km) from the Sun, at closest approach to Earth, Jupiter would be about 1 kilometer away on our scale.

Pluto, at 39 AU from the Sun, would be 6.6 kilometers away. The nearest star (other than the Sun) is 4.2 light-years away. On our scale, this would be 460,000 kilometers away—greater than the distance to the (real) Moon.

This reveals something basic about the nature of the Universe: the Universe is *very* spread out. The stars are Suns, but they are *very* far away.

▶ **Figure 4-1** James Irwin walked on the Moon during the Apollo 15 mission in 1971. (NASA/Apollo 15 crew/David R. Scott)

► **Figure 4-2** Colonel Irwin said that from the Moon, Earth looked like a little blue marble... (NASA/Apollo 17 crew/Captain Eugene A. Cernan)

► **Figure 4-3** "...the most beautiful marble you can imagine." (NASA/Apollo 8 crew/William A. Anders)

► **Figure 4-4** The Solar System is the Sun and its planets. It is our local neighborhood in space. This artist's concept is slightly misleading, since the planets never line up in a straight line like this. Also, while the planets are shown with correct size relative to each other (11 Earths could stretch across Jupiter, and 10 Jupiters could stretch across the Sun), the planets are never this close to each other. (Image courtesy of Frederick Ringwald, using images by NASA/Jet Propulsion Laboratory)

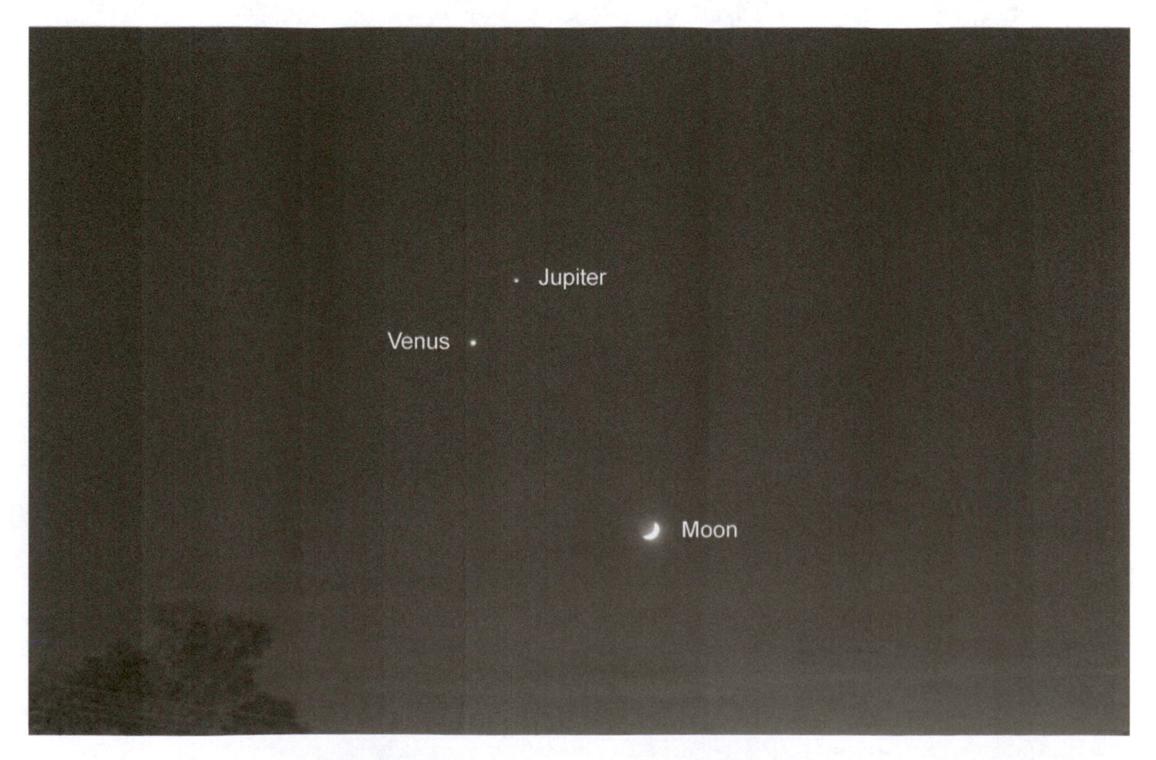

► **Figure 4-5** The planets in the Solar System are so far away that, to the unaided eye, they look like dots, just like bright stars do. The Moon is close enough to Earth for the unaided eye to see some detail. (Image courtesy of Frederick Ringwald)

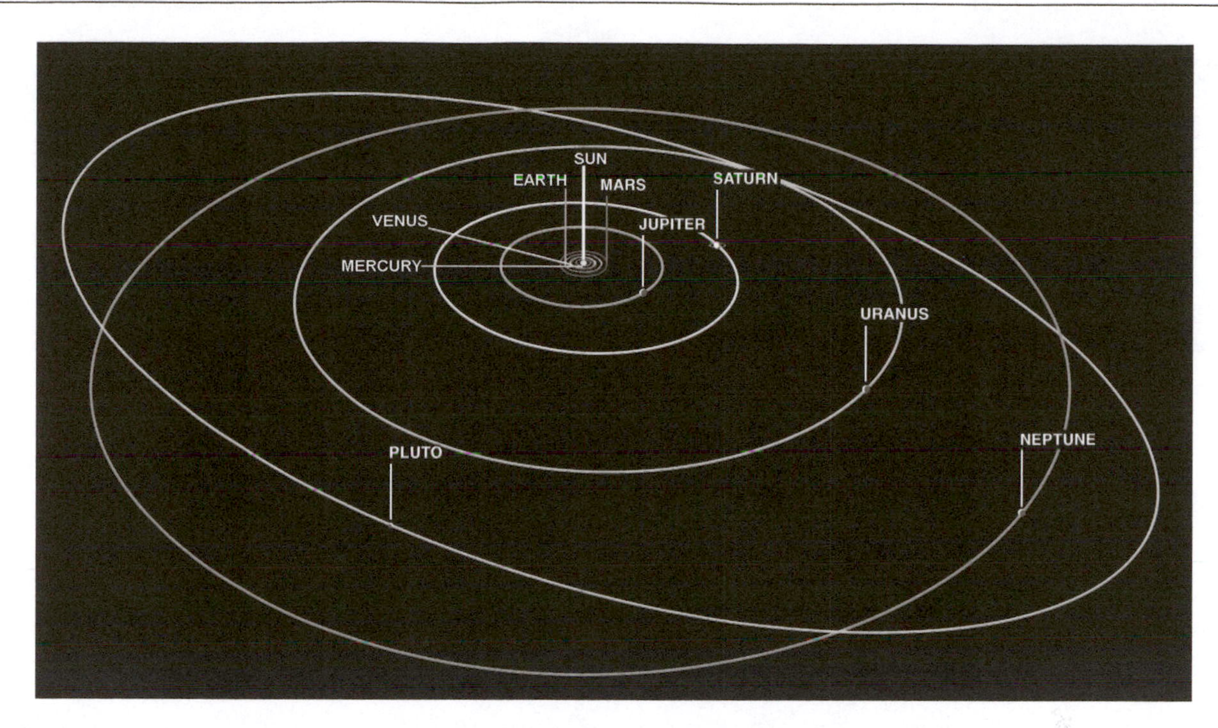

► **Figure 4-6** The true nature of the Solar System is more closely shown here, since the planets are *very* spread out. (NASA/Lunar and Planetary Institute)

► **Figure 4-7** Proxima Centauri is 4.2 light-years from Earth. It is one star in a system that has its own planets. (Atlas Image (or Atlas Image mosaic) obtained as part of the Two Micron All Sky Survey (2MASS), a joint project of the University of Massachusetts and the Infrared Processing and Analysis Center/California Institute of Technology, funded by the National Aeronautics and Space Administration and the National Science Foundation.)

The Basics of Astronomy

To learn science, students may need to learn many unfamiliar words. If the instructor isn't careful, this can be a problem, since there can be more new words in a science course than there are in a foreign-language course.

In this course, we want to keep the number of unfamiliar terms manageable. To help make them understandable, we will often say "in other words," include an alternative meaning in parentheses (immediately after the new word), or just say "or." We will do this below, with the terms *orbits*, *Solar System*, and *nebula*.

The Sun is a star. About 1 in 10 of them, or 10%, are just like the Sun.
The stars appear to be pinpoints of light because they are so far away.
The closest star (other than the Sun) is 4.2 light-years away.
Recall that one light-year is about 6 trillion miles. The stars are trillions of miles away.

We don't see stars in the daytime because the Sun's light scatters in Earth's atmosphere, making it blue and bright. Sadly, this is also why people who live in cities can't see many stars, because city lights drown out nearly all the faint stars.

Earth's atmosphere ends only about 60–100 miles above Earth's surface. Above Earth's atmosphere is outer space. Outer space is **a vacuum**, or empty space, with no air. (A *vacuum cleaner* works by sucking air. If a window in a spacecraft broke, an astronaut could be sucked out into space like a cork popping out of a bottle. This is because the air in the spacecraft will expand into the empty space, and it can take an astronaut with it.)

The Moon moves around, or **orbits,** Earth, about once **a month.**
The Moon is about 250,000 (quarter of a million) miles from Earth.

For comparison, Earth is about 8,000 miles in diameter.

The Sun and the planets make up **the Solar System.**
("Sol" means Sun in Latin and in Spanish.)
The planets of the Solar System move around, or **orbit**, the Sun.

Other planets have moons, too. **A moon orbits a planet.** These "moons" used to be called "satellites," but in 1957, the first artificial satellite, called Sputnik, was launched into orbit around Earth. Ever since, when people say "satellite," they usually mean "artificial satellite," or in other words, spacecraft orbiting Earth. Maybe it'll take another century for the language to straighten itself out, with the word *moon* meaning "natural satellite," the word *satellite* meaning "artificial satellite," and with Earth's Moon being called by a proper name, such as its Latin name, "Luna." If the first human settlement on Earth's Moon is named anything other than "Luna City," this author is going to be mad.

Earth orbits the Sun once **a year.** The Sun is 1 Astronomical Unit, or about 93 million miles, from Earth, or about 400 times farther away from Earth than the Moon.

Earth spins, or rotates, once every 24 hours. This is what causes **day and night.** From our home on the surface of Earth, the Sun appears to rise in the east in the morning, and set in the west in the evening. It's Earth that is moving, however.

The Solar System is just one small part of the Universe. In 1991, astronomers discovered that **other stars have planets, too.** Hundreds of these "extrasolar planets," or "exoplanets," are now known. There are probably billions more to be found. So far, we have neither discovered life in space, nor (reliably) proven that life from space has visited Earth. Many claims of this have been made, such as with UFOs, but none have held up to careful examination.

Nebulae are clouds of gas and dust in the space between the stars. Nebula means "cloud" in Latin. "Nebula" is singular, and "nebulae" is plural. Some nebulae, such as the Orion Nebula, are where we can see new stars forming. Others, such as the Crab Nebula, are the remnants of stars that have died.

Galaxies are vast islands of hundreds of billions of stars. The one we live in is called the Milky Way. It contains about 600 billion stars. The Milky Way is visible as a faint band of light across a dark sky. The Milky Way is faint so it can't be seen in the city because of city lights. Our nearest neighbor is the Andromeda Galaxy: it's 2.5 million light-years away.

► **Figure 4-8** The only place in space humans have walked on is the Moon. Neil Armstrong was the first human to walk on the Moon, during the first landing by humans, the Apollo 11 mission in 1969. He took this picture of Buzz Aldrin, the second human to walk on the Moon. Only 12 astronauts have so far walked on the Moon, all American men, all between 1969 and 1972. (NASA/Apollo 11 crew/Neil A. Armstrong)

► **Figure 4-9** Some dream of sending astronauts to Mars. Whether it will happen in your lifetime remains to be seen. One reason is that Mars is 200 times farther from Earth than the Moon, so sending astronauts there is harder. (NASA/Pat Rawlings)

► **Figure 4-10** The *International Space Station* orbits Earth. Low-Earth orbit is only about 400 km (250 miles) above Earth's surface. It is the only other place in space where humans have been, aside from the Moon. The *International Space Station* has been continuously occupied since 2000 by crews of 3 to 13. (NASA/STS 132 crew)

► **Figure 4-11** Earth's atmosphere is only about 100 km (60 miles) deep. Outer Space, usually just called "space," begins outside of this. Outer Space is empty space, also called a *vacuum*. (A vacuum *cleaner* works by pumping air.) (NASA/STS 117 crew)

► **Figure 4-12** More than one astronaut has been surprised by how thin Earth's atmosphere looks, from space. Ulf Merbold noted, "Obviously this was not the ocean of air I had been told it was so many times in my life. I was terrified by its fragile appearance." (NASA/International Space Station 7 crew)

► **Figure 4-13** Robot spacecraft have visited all the planets of the Solar System. As Neil deGrasse Tyson observes, "Robots don't have to eat, they don't have to breathe, and they don't get upset if they can't come home." Robot spacecraft are therefore often much less expensive than human spaceflight, but many people just don't get excited by robots.

Don't call spacecraft with astronauts "manned spacecraft." That hasn't made sense since women started flying in space in 1963. Call them "human" or "piloted" spacecraft. Likewise, don't call robots "unmanned" spacecraft. Call them robots, or robotic spacecraft.

The robot shown in this artist's concept is *Pioneer 10*. It crossed the orbit of Neptune in 1983, and is now drifting out into interstellar space, because of inertia. (NASA/Don Davis)

► **Figure 4-14** Telescopes show us what's outside the Solar System because there is information in starlight. This is Kitt Peak in Arizona, at sunset. (Image courtesy of Frederick Ringwald)

CHAPTER

5

The Cosmic Calendar

Scientists now know the history of the Universe in a fair amount of detail. Sometimes someone might ask, "How do they know? Were they there?" The surprising answer is: YES, because there is here. The remnants of ancient times are still among us today, and scientists have devised a variety of methods of measuring how old they are.

Another way to say this is: if an elephant had helped itself to the contents of your refrigerator last night, would you know it, even though no one was there to see the elephant? *Of course* you would know it: the elephant would have left behind ***evidence*** of its presence. (Any footprints in the cheesecake would be a dead giveaway.) The following *isn't* "just a theory": every item on this list is based on evidence, such as fossils, and has been carefully *checked*—and being checkable is what defines information as *facts*.

Still, it can be difficult for the human mind to grasp the billion-year timescales of natural history. We can make it easier to understand if we use simple direct proportions. The Universe has been measured to be 13.8 billion years old. Let's represent this as one year.

In other words, imagine that the origin of the Universe occurred at the stroke of midnight, 2014 December 31. Now will be the stroke of midnight exactly one year later, on 2015 December 31. With time scaled in this way, each day on the Cosmic Calendar represents about 40 million years of the age of the real Universe.

The following events in cosmic history would therefore occur at the following times on the Cosmic Calendar:

January 1, 12:00 a.m. (13.80 ± 0.04 billion years ago)

The Universe began in a hot, dense fireball called the Big Bang. It expanded rapidly, which it continues to this day. There are now indications of an early time of even more rapid expansion, or inflation, in the first 10^{-35} of a second.

The First Three Minutes (*real* minutes, not scaled)

The whole Universe was as hot and dense as the center of the Sun. Nuclear fusion reactions, as which power the Sun, were occurring everywhere. They turned hydrogen nuclei into helium nuclei, again as in the Sun. This left the Universe about 3/4 hydrogen and 1/4 helium, by mass—which we observe today.

January 1, 12:12 a.m. (380,000 years later)

The Universe cooled enough to become transparent to light. Thanks to look-back time, we can still detect this fireball, called the Cosmic Background Radiation—quite literally, the echo of the beginning of the Universe. After this, with no other sources of light, the Universe became dark, until the first stars formed.

January 6 (200 million years later)

The first stars formed. So far, most of the evidence of this is indirect, from how they affected the Cosmic Background Radiation. We plan to observe the first stars directly with the successor to *Hubble Space Telescope*, called the *James Webb Space Telescope (JWST)*, which is scheduled for launch in 2018.

The first stars probably had brief lives and explosive deaths. This is because the nuclear reactions that generate stellar energy also make chemical elements heavier than helium, including carbon, nitrogen, and oxygen—the material that makes up most of living animals and plants on Earth. By exploding, the first stars seeded the Universe with these elements, and we do see these elements everywhere in the Universe. We can also see short-lived stars exploding and making heavy elements today: they're called *supernovae*.

January 18–22 (500–700 million years later)

The most distant galaxies (which are islands of hundreds of billion stars, which are often disk or spiral-shaped) that *Hubble Space Telescope* can see are this old. We still don't know many of the details of how the galaxies formed: observing this will be a project for *James Webb Space Telescope*, which will be much larger and more sensitive than *Hubble Space Telescope*.

May 1 (8 billion years ago)

The disk of our Galaxy—notice it's capitalized here, for our Milky Way Galaxy—was formed. We know this because the oldest stars in the Galaxy are this old.

Star formation, throughout the Universe, was probably more active at early times than now. We know this because of how galaxies have changed, or evolved, over the age of the Universe, which we can see because of look-back time. The Sun formed *after* the time of maximum star formation.

August 23 (4.57 ± 0.02 billion years ago)

The Solar System formed in less than 100 million years. To appreciate what a *short* time this really was, consider:

- The Appalachian Mountains are 300–500 million years old.
- The Rocky Mountains are 60–80 million years old.
- The Sierras and the Klamaths are 50–60 million years old, and are still going up.
- The Cascades are less than 2 million years old and are still going up.

The planets formed within 10–100 million years after the Sun formed. Note the precision of the known age of the Solar System, 4.57 ± 0.02 billion years: this was measured from the oldest meteorites and Moon rocks.

These dates, like most that follow, were found by radioactive dating. Radioactive atoms split at precise rates over time. They therefore can be used as clocks. Some are long-lived, with decay rates in the billions of years. A rock with much long-lived radioactive material and almost no short-lived radioactive material is older than one with nearly the same amount of long-lived material, and relatively plentiful short-lived radioactive material. (You can work this out yourself.)

Radioactive dating is supported by other methods used by geologists to find out how old rocks are. These methods, including erosion and sedimentation rates, are also used to do useful things, such as searching for oil and minerals.

August 26 (100 million years later)

The Sun and the planets were mostly formed. Much debris was left over. The planets therefore had much planetesimal bombardment—in other words, they were often hit by *big* meteorites, or rocks from space.

Toward the end of this process, large leftover pieces were hitting other large leftover pieces. A planetesimal about the size of Mars is thought to have hit the early Earth, and some of the debris spattered from Earth's interior settled into orbit around Earth, to clump together by gravity to form the Earth's Moon. This idea is called the "giant impact theory" of the origin of the Moon. It is sometimes referred to as "The Big Splat."

September 16 (3.8 billion years ago)

Planetesimal bombardment mostly ended in the Late Heavy Bombardment. Most of the largest features on the Earth's Moon are at least this old. After this, Earth's surface cooled and solidified.

Before September 25 (3.5 billion years ago)

The earliest definite signs of life on Earth appeared. These are stromatolites, which are rocks carved by bacteria. Notice how early this was: Earth has had life for over 90% of the time the planet has been capable of supporting life.

October 27 (2.4 billion years ago)

The Great Oxydation Event: blue-green algae made Earth's atmosphere rich in oxygen. Before this, it had been composed of chemically reducing gases, mainly from outgassing from Earth's interior.

November 21 (1.5 billion years ago)

The first known multicellular life appeared. No one knows why. Before this, all life had been single-celled, and so was relatively simple.

December 17 (545 million years ago)

The Cambrian Explosion, of diversity of species of life, occurred. The fossil record got much richer, beginning in rocks this old. No one knows why. Included were the first hard-shelled animals.

December 18 (510 million years ago)

First fish.

December 19 (470 million years ago)

First plants on land.

December 20 (430 million years ago)

First animals on land (millipedes).

December 22 (370 million years ago)

First amphibians; emergence of vertebrates onto land; first insects.

December 23 (251 million years ago)

"The Great Dying," the most extensive mass extinction known in the history of life on Earth. Over 90% of marine species and 70% of land species became extinct. No one knows why, but there was extensive volcanism at the time.

December 25 (230 million years ago)

First dinosaurs.

December 25 (210 million years ago)

First mammals.

December 27 (140 million years ago)

First flowering plants.

December 30 (65 million years ago)

Non-avian dinosaurs became extinct, thought to be because of a giant meteorite impact on Earth. That an impact happened is clear, and it happened at the time of the extinction. This is shown by microfossils abruptly becoming less common in a worldwide 65-million-year-old rock layer that is rich in iridium, which is common in meteorites but uncommon in Earth rocks. Whether the impact was from an asteroid or comet is under debate. How exactly the impact caused the extinctions is also unclear: it might have thrown so much dust into the air that Earth became dark and cold for a long time.

December 31, 9:18 p.m. (4.2 million years ago)

The first known hominid (human-like) ancestor, *Australopithecus*, appeared in Africa. It was essentially an ape that walked upright.

December 31, 10:51 p.m. (1.9 million years ago)

Hominids (*Homo Erectus*) used stone tools and fire, but maybe not articulate speech. Their fossils and tools have been found from Africa, China, and Java.

December 31, 11:52 p.m. (190,000 years ago)

The first known modern *Homo Sapiens*, our species, appeared in East Africa. All living humans are descended from one woman who lived about 100,000 years ago in South Africa, as shown by DNA taken from people around the world.

In 1998, a fossil trackway (footprints) just this old was discovered in South Africa. The footprints were of a modern human, probably a woman, since the footprints and the stride were small. It could have been her!

December 31, 11:57 p.m. (70,000 years ago)

Homo Sapiens almost became extinct. No one knows why: perhaps this was because of climate change caused by Mount Toba, a supervolcano that erupted at this time. Gene differences in humans today indicate a population bottleneck at this time, to possibly fewer than 1,000 individual humans worldwide.

December 31, 11:59 p.m. (30,000 years ago)

Our nearest relatives, *Homo Neanderthalensis*, became extinct. Why them and not us? They were certainly stronger, and had larger brains—but their tool use (or in other words, their technology) was not as advanced as ours.

December 31, 11:59:32 p.m. (10,000 years B.C.)

The first known agriculture began in Mesopotamia (now called Iraq). Soon after were the earliest known human settlements: towns and cities.

December 31, 11:59:48 p.m. (3,300 years B.C.)

The first recorded history began in Sumeria. The Bronze Age began in the Middle East.

December 31, 11:59:49 p.m. (2,600 years B.C.)

The pyramids were built in Egypt. The stone parts of Stonehenge were built in Britain.

December 31, 11:59:53 p.m. (1,100 years B.C.)

Iron Age in Europe and Asia.

December 31, 11:59:59.5 p.m. (about 1800 A.D.)

The Industrial Revolution began in England.

December 31, 11:59:59.9 p.m. (1957 October 4)

The Space Age began with the launch of Sputnik 1 from the U.S.S.R.

AND NOW FOR THE PUNCHLINE

We occupy only the very TINIEST sliver of cosmic history! This is obvious, perhaps, but worth making clear.

OTHER IDEAS TO TAKE HOME WITH YOU

On Life

Life has existed on Earth for over 90% of the time the planet has been cool enough to have a solid surface. The exact origin of life, however, is still poorly understood and is a lively topic of research.

Most of this time, the only life on Earth was single-celled. We know this because of the fossil record. Complex, multicellular life is relatively recent.

On the Universe

The Universe had a definite origin. We can measure how long ago it was, to within 2%. We know this because we can see it, because of look-back time.

We don't know the ultimate fate of the Universe. Current evidence suggests it will expand forever.

On the Stars

The stars are SUNS. This is a very big thought.

There are *very* many stars, with over 100 billion per galaxy, and over 50 billion galaxies in the Universe. We know this because we can see them.

Stars, like the Sun, are quite common, comprising 5–10% of all stars. Again, we know this because we can see these stars.

Stars are powered by nuclear energy, which also makes all the chemical elements heavier than helium. We know this because we see the stars doing it, and we understand the process well enough to reproduce it ourselves.

YOU are literally made of star dust—and so are connected to the Universe in a *very* basic way.

On Human Understanding of the Universe

Pseudoscience, which is making up stories and pretending to know the answer, is not necessary. Science provides a way to learn about the real Universe, which is far larger and more intricate than any fiction ever imagined.

► **Figure 5-1** *Top left:* Earth was hit by a large asteroid or comet, or rock or snowball from space, respectively, 65 million years ago. It is estimated to have been 10 kilometers (6 miles) in diameter. (NASA/Don Davis) *Top right:* The impact left a layer of asteroidal material worldwide. This layer coincides with an abrupt change in the fossils. (Image courtesy of Frederick Ringwald) *Bottom left*: The crater from this impact is in the Yucatan in Mexico and is 200 km in diameter. (U.S. Geological Survey) *Bottom right:* The impact is thought to have caused the extinction of the dinosaurs shortly afterward. How exactly this happened is still unclear: one idea is that the impact threw so much dust into Earth's atmosphere, the lack of sunlight shut down plant life, which affected animal life. (Image courtesy of Mark Garlick)

January

Sunday	Monday	Tuesday	Wednesday	Thursday	Friday	Saturday
		1	2	3	4	5
6	7	8	9	10	11	12
13	14	15	16	17	18	19
20	21	22	23	24	25	26
27	28	29	30	31		

February

Sunday	Monday	Tuesday	Wednesday	Thursday	Friday	Saturday
					1	2
3	4	5	6	7	8	9
10	11	12	13	14	15	16
17	18	19	20	21	22	23
24	25	26	27	28		

March

Sunday	Monday	Tuesday	Wednesday	Thursday	Friday	Saturday
					1	2
3	4	5	6	7	8	9
10	11	12	13	14	15	16
17	18	19	20	21	22	23
24	25	26	27	28	29	30
31						

April

Sunday	Monday	Tuesday	Wednesday	Thursday	Friday	Saturday
	1	2	3	4	5	6
7	8	9	10	11	12	13
14	15	16	17	18	19	20
21	22	23	24	25	26	27
28	29	30				

May

Sunday	Monday	Tuesday	Wednesday	Thursday	Friday	Saturday
			1	2	3	4
5	6	7	8	9	10	11
12	13	14	15	16	17	18
19	20	21	22	23	24	25
26	27	28	29	30	31	

June

Sunday	Monday	Tuesday	Wednesday	Thursday	Friday	Saturday
						1
2	3	4	5	6	7	8
9	10	11	12	13	14	15
16	17	18	19	20	21	22
23	24	25	26	27	28	29
30						

July

Sunday	Monday	Tuesday	Wednesday	Thursday	Friday	Saturday
	1	2	3	4	5	6
7	8	9	10	11	12	13
14	15	16	17	18	19	20
21	22	23	24	25	26	27
28	29	30	31			

August

Sunday	Monday	Tuesday	Wednesday	Thursday	Friday	Saturday
				1	2	3
4	5	6	7	8	9	10
11	12	13	14	15	16	17
18	19	20	21	22	23	24
25	26	27	28	29	30	31

September

Sunday	Monday	Tuesday	Wednesday	Thursday	Friday	Saturday
1	2	3	4	5	6	7
8	9	10	11	12	13	14
15	16	17	18	19	20	21
22	23	24	25	26	27	28
29	30					

October

Sunday	Monday	Tuesday	Wednesday	Thursday	Friday	Saturday
		1	2	3	4	5
6	7	8	9	10	11	12
13	14	15	16	17	18	19
20	21	22	23	24	25	26
27	28	29	30	31		

November

Sunday	Monday	Tuesday	Wednesday	Thursday	Friday	Saturday
					1	2
3	4	5	6	7	8	9
10	11	12	13	14	15	16
17	18	19	20	21	22	23
24	25	26	27	28	29	30

December

Sunday	Monday	Tuesday	Wednesday	Thursday	Friday	Saturday
1	2	3	4	5	6	7
8	9	10	11	12	13	14
15	16	17	18	19	20	21
22	23	24	25	26	27	28
29	30	31				

► **Figure 5-2** Imagine that the Universe began on January 1 at midnight. Imagine also that December 31 just before midnight is now. Then each "day" stands for 38 million years. On this scale, dinosaurs appeared on December 25, and became extinct on December 30. Dinosaurs are supposed to be so ancient, yet they lived only during the most recent 2% of the history of the Universe.

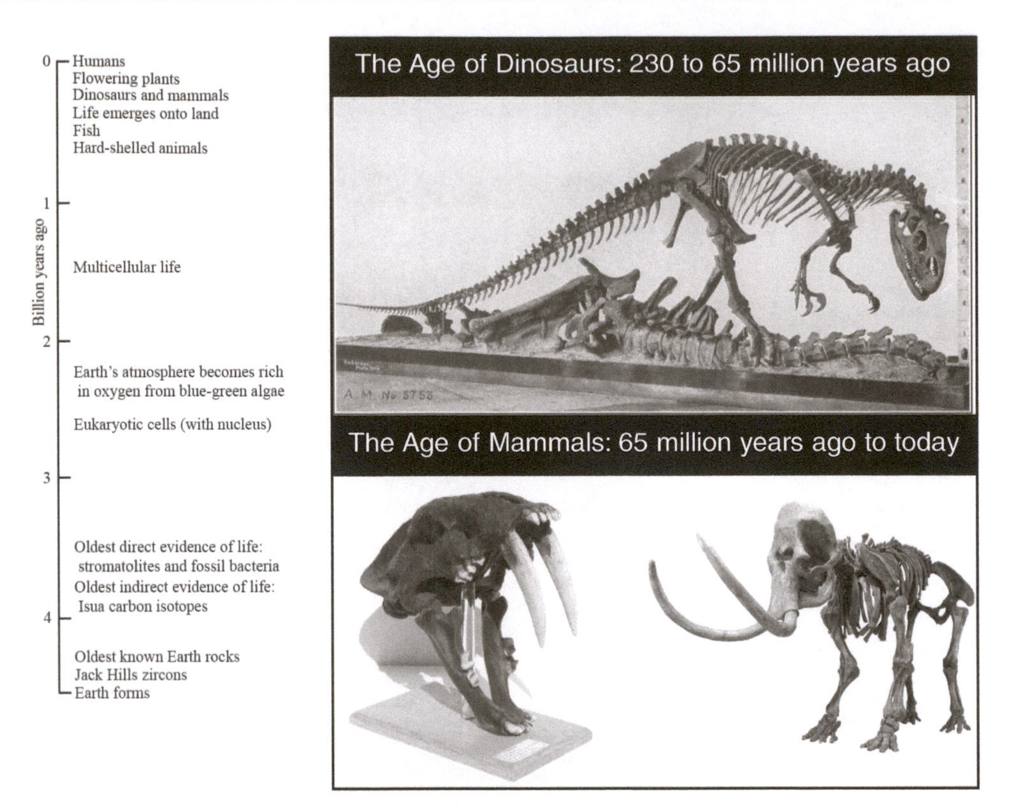

► **Figure 5-3** *Left:* Again, the common perception of dinosaurs is that they are old, but dinosaurs lived only during the last 2-5% of Earth's history. (Image courtesy of Frederick Ringwald) *Top right:* Since dinosaurs lived between 230 to 65 million years ago, this time is sometimes called the Age of Dinosaurs. This is misleading, since mammals lived during most of this time, since 210 million years ago. (1909 image by the American Museum of Natural History) *Bottom center:* A skull of a saber-tooth cat, in the author's collection. (Image courtesy of Frederick Ringwald) *Bottom right:* A skeleton of a mammoth. The time, from 65 million years ago to now, is sometimes called the Age of Mammals. This is also misleading, since one family of dinosaurs did not become extinct: the birds. We may just as well call today the Age of Birds, or even the Age of Bacteria, since one might argue that bacteria are the dominant life forms on Earth. (Image © Eduard Kyslynskyy, 2013. Used under license from Shutterstock, Inc.)

► **Figure 5-4** Bones, stone tools, genetics, and footprints indicate that anatomically modern humans (*Homo sapiens*) originated in Africa about 190,000 years ago. *Everything* directly related to humans and their direct ancestors happened only after 9 p.m. on December 31 of the Cosmic Calendar, mentioned in this chapter. (Public Domain)

▶ **Figure 5-5** Trilobites ruled Earth in their time, over 500 million years ago, with over 10 thousand species in North America alone. They became extinct 50 million years before the dinosaurs. The one at lower left is a horseshoe crab, their nearest living relative. (Image courtesy of Frederick Ringwald, from the author's collection)

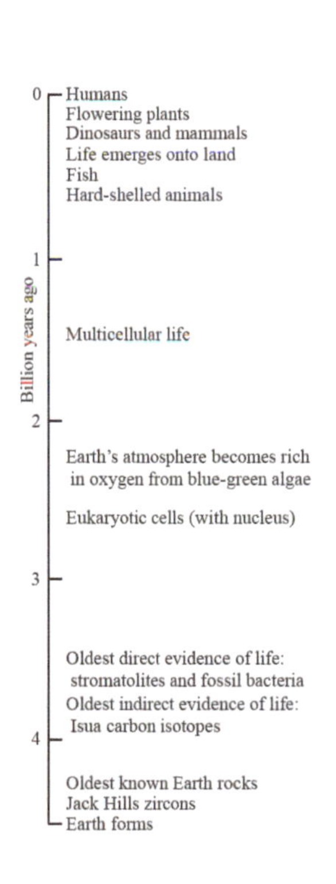

▶ **Figure 5-6** *Left:* Earth has had simple life for most of its history. (Image courtesy of Frederick Ringwald) *Top right:* Living stromatolites, in Shark Bay, Australia (Image © Rob Bayer, 2013. Used under license from Shutterstock, Inc.) *Bottom right:* These stromatolites, where bacteria ate into rock, are 2.3 billion years old. (Image courtesy of Frederick Ringwald, from the author's collection)

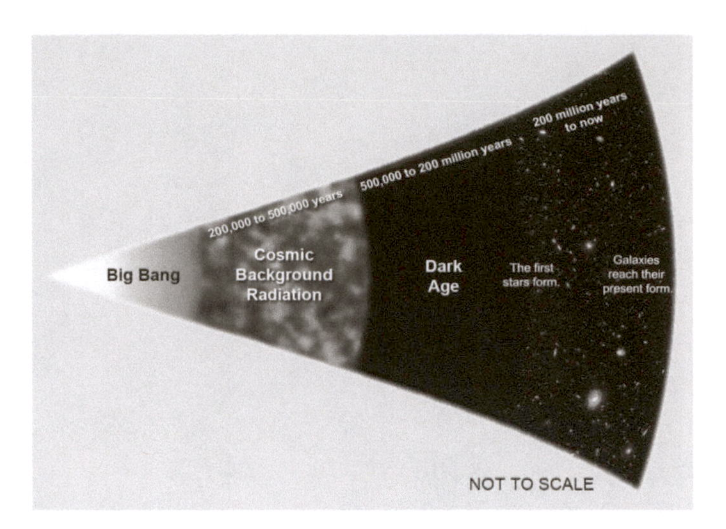

► **Figure 5-7** Recall *look-back time.* Because it takes time for light to get around, when one looks deeply into the Universe, it's like looking into a time machine. What does one see, when one looks back all the way? (Image courtesy of Frederick Ringwald, using NASA images)

► **Figure 5-8** What do you see, when you look back all the way? You see the origin of the Universe. This is the cosmic background radiation. It is an actual picture of the hot origin of the Universe, widely referred to as the Big Bang. (Adapted from NASA/WMAP)

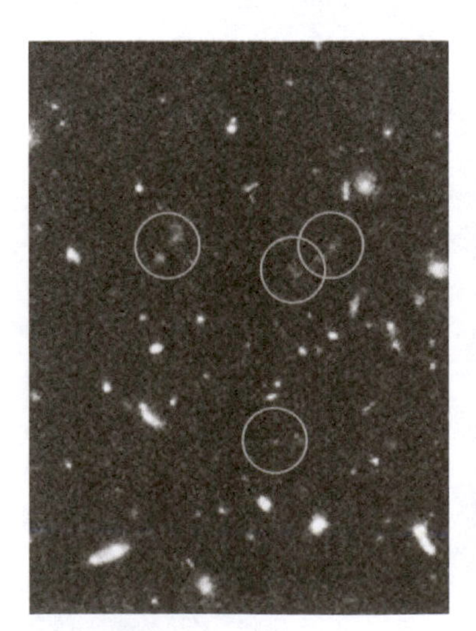

► **Figure 5-9** *Left:* The little faint red galaxies circled here are over 12 billion years old. (NASA/STScI) *Right:* Our galaxy is called the Milky Way. It was pulled together by gravity 9 ± 2 billion years ago. We know the age from how the oldest burned-out cinders in it (which used to be stars) cool down. When it was forming, it may have looked like this, the Sombrero Galaxy. (NASA/STScI)

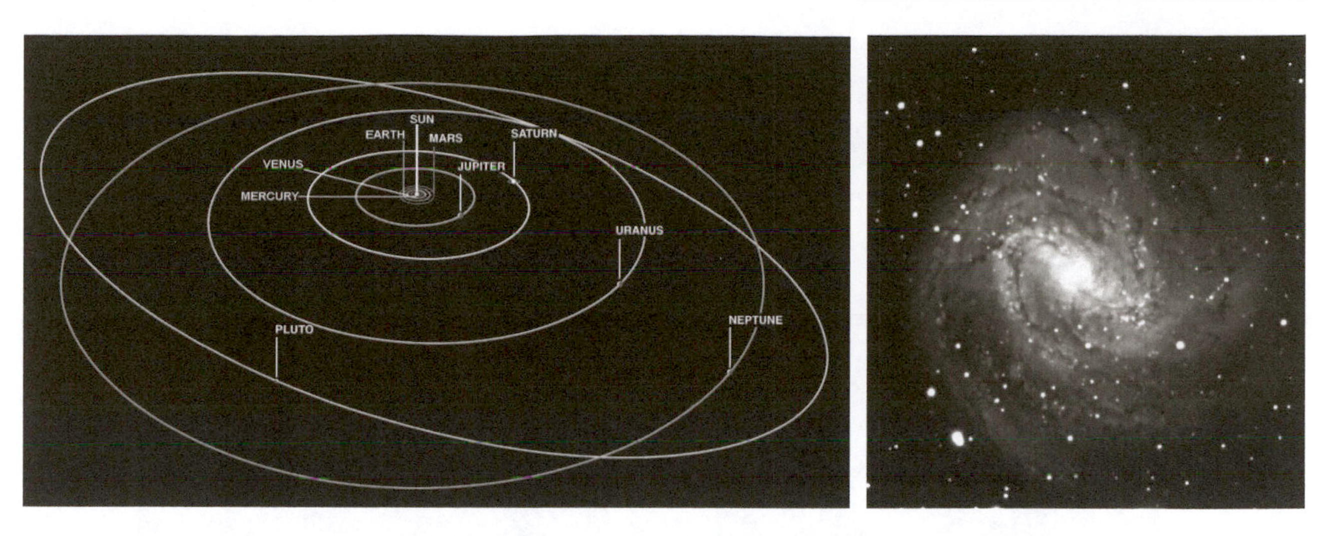

► **Figure 5-10** Don't confuse *galaxies* with *planetary systems,* or *systems of planets*. *Left:* The Solar System is our planetary system. It has one star (the Sun) and its system of planets. (NASA/LPI) *Right:* This is a whole galaxy, with hundreds of billions of stars. (Image courtesy of Frederick Ringwald at Mount Laguna Observatory)

► **Figure 5-11** *Every star* in this galaxy may have its own system of planets. Each red object shown here is a cloud of gas, or a *nebula.* (NASA/STScI)

▶ **Figure 5-12** Stars are forming by gravity in this nebula. "Nebula" means cloud. "Nebulae" is plural for "nebula," since it's a Latin word. Nebulae aren't water vapor, like clouds in Earth's atmosphere. They're mostly hydrogen gas, like what makes up stars. (NASA/STScI/Robert Gendler)

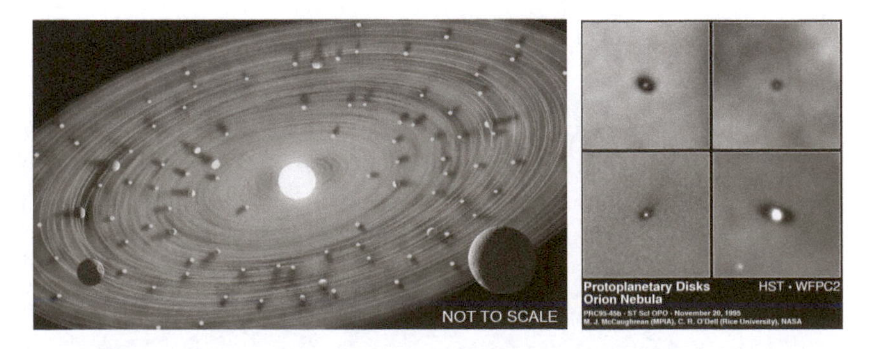

▶ **Figure 5-13** *Left:* In the Solar Nebula, 4.57 ± 0.02 billion years ago, the Sun and the planets formed by gravity. We know the age because we have leftover pieces: ***meteorites***. (Image courtesy of Frederick Ringwald) *Right:* One reason to think that the Sun and planets formed in this way is that we can see it happening elsewhere, with other stars and planets forming now. (NASA/STScI)

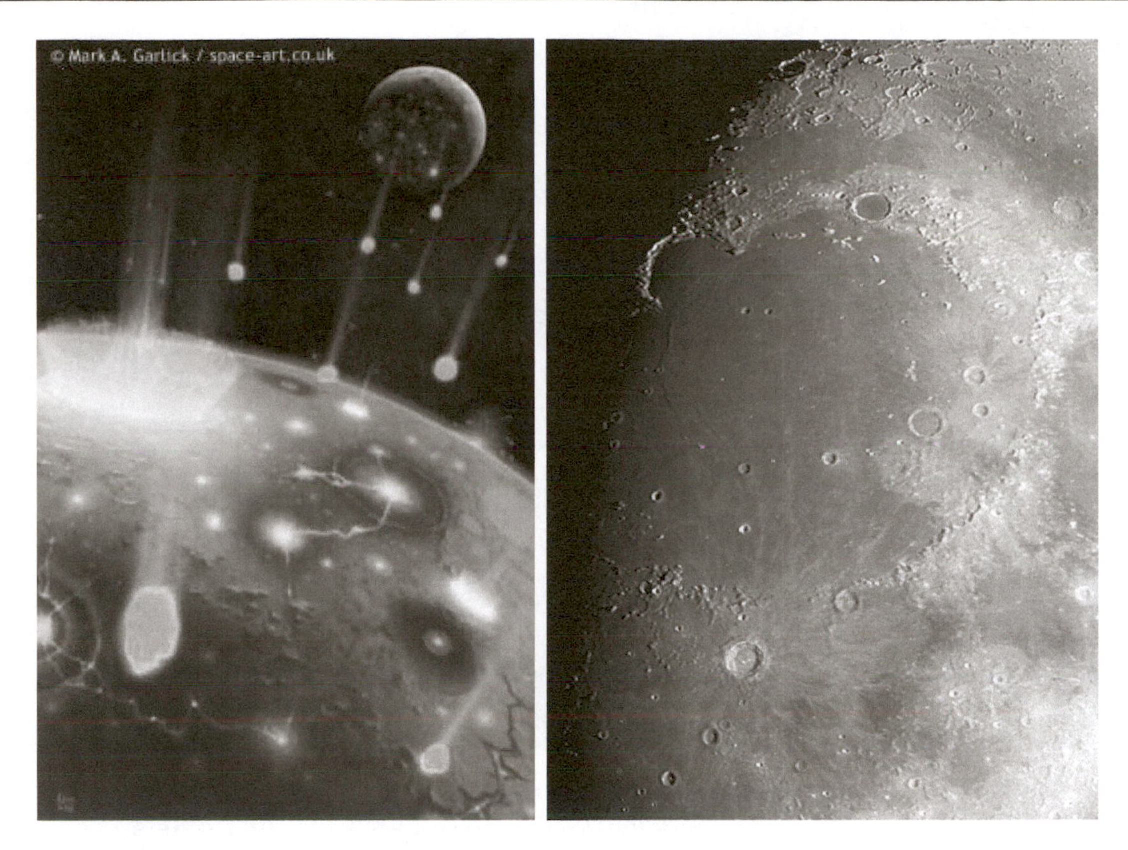

► **Figure 5-14** Another reason to think that the planets formed by gravity is that we see the remnants, or craters, left by it. *Left:* This artist's concept shows how planet formation may have looked when it was happening. (Image courtesy of Mark A. Garlick) *Right:* This photo of the Moon shows the craters. (Image courtesy of Frederick Ringwald at Fresno State's Campus Observatory)

► **Figure 5-15** The early Earth wasn't at *all like* the home we know and love—it had almost no free oxygen in its atmosphere. This artist's concept shows how it may have looked. (Image courtesy of Frederick Ringwald)

▶ **Figure 5-16** *Top left:* Traces of ancient events still exist today. Radioactive decay can be used to tell how old things are. (Image courtesy of Frederick Ringwald) *Top right:* Rock layers and how they were deposited by water can also show how old rocks are. (Image courtesy of Frederick Ringwald) *Bottom left:* Rain traps carbon dioxide in the mineral siderite. Earth's early atmosphere was rich in CO_2. (NOAA) Inset: Siderite (USGS) *Bottom center:* Zircons from the Jack Hills in Australia are 4.404 ± 0.008 billion years old. They form only in liquid water, evidence of oceans and greenhouse heating. (Image courtesy of Frederick Ringwald) Inset: Zircon (Image © Morgenstjerne, 2013. Used under license from Shutterstock, Inc.) *Bottom right:* Iron oxide bands show that Earth's atmosphere first became oxygen-rich 2–2.5 billion years ago. Fossils of blue-green algae support this. (Image © Paulo Afonso, 2013. Used under license from Shutterstock, Inc.)

CHAPTER

6

Classical Astronomy

The Positions and Motions of Objects in the Sky

A constant challenge to science education is that we carry a strong internal model of how the world works—and if this model is wrong, it can be difficult to correct it. (The video *A Private Universe,* at the Harvard commencement, shows this.)

> It's not ignorance does so much damage; it's knowing so darned much that ain't so.
>
> –Josh Billings (often misattributed to Mark Twain or Will Rogers)

Here are some common misconceptions in astronomy and in science in general:

- *It is a myth* that Earth has seasons because it's closer to the Sun during summer.
- *It is a myth* that the Moon's phases are caused by Earth's shadow.
- *It is a myth* that accidents and crime are worse during Full Moon—even though many smart, reliable people such as doctors, nurses, and police believe this.
- *It is a myth* that astrology makes predictions that are more reliable than random chance. In fact, in any fair, objective test, astrology just plain *does not work*.
- Notice that the world *didn't* end in 2012, contrary to the prediction that the Mayans *didn't* make. There is also *no validity* to prophecies of Nostradamus.
- *It is a myth* that Mars will ever look "as big as the Moon." This is an Internet hoax: the unaided eye can see the planets (except for Neptune), but to the eye they look like bright stars (except for Uranus), unless one uses a telescope.
- *It is a myth* that a planetarium is the same as an observatory. A planetarium is a special theater that shows what the sky looks like. An observatory is a housing for a telescope.
- *It is a myth* that astronauts on the Moon could see the Great Wall of China. (They couldn't.) It is also a myth that the Great Wall of China is the most visible human-made object from orbit. (Farms and city lights are.)
- *It is a myth* that objects float weightless in spacecraft because "there is no gravity" in space. If this were true, the spacecraft wouldn't orbit Earth at all.

- *It is a myth* that global warming is a hoax. Climate change is real, and it is caused by humans. Vaccines don't cause autism. Cell phones don't cause cancer.
- *It is a myth* that any high-energy physics experiment could create a black hole that could swallow Earth. If one could, natural high-energy cosmic rays would have done this long ago. Black holes are real, but it is a myth that they "suck."
- *It is a myth* that there is any *reliable, physical* evidence that alien life is visiting Earth, or that there were ever "ancient astronauts," or that any human was ever abducted by a flying saucer, or that the so-called "lost" continents of Atlantis or Lemuria ever existed, or that crop circles are extraterrestrial activity. Don't get me wrong: life in space may exist, but I don't think we've found it, yet.
- It is *highly implausible* the U.S. (or any other) government or military has a crashed flying saucer that it's hiding from the public.

Classical Astronomy is the study of the positions and motions of objects in the sky.

The unaided eye: One can learn a great deal about the Universe just by looking up at the sky with one's unaided eyes. Nowadays, we say *unaided* eye, not *naked* eye. This is because kids couldn't look at *Sky & Telescope's* web page on "Naked Eye Astronomy," because parental-control software was flagging the word *naked*. Another astronomy term to avoid because it makes people giggle is "heavenly bodies."

The ecliptic is *the path of Earth's orbit around the Sun*. The ecliptic looks like the line in Earth's sky where the Sun appears to move throughout the year. The other planets orbit the Sun near the ecliptic plane, too. (Pluto doesn't because it isn't a planet.)

The constellations of the zodiac are the twelve constellations through which the ecliptic passed during the time of the ancient Greeks, who made up many of these constellations. The Sun, Moon, and planets are usually in these constellations.

Seasons do NOT occur because Earth's orbit is closer to the Sun during summer. In fact, Earth is closest to the Sun on January 3—directly contradicting this. Notice also that the seasons are reversed in the Northern and Southern Hemispheres. In July, it is summer in California, but it is winter in Chile and Australia.

boreal = northern   austral = southern

The seasons happen because *the Earth's equator is tilted by 23.5 degrees to the ecliptic.*

Because of this:

In summer: the Sun rises north of east, and sets north of west.
It passes high in the sky,
but not directly overhead unless you're in the Tropics,
and is above the horizon for a long time, so the days are long.

In winter: the Sun rises south of east, and sets south of west.
It doesn't pass very high in the sky,
and is above the horizon for a short time, so the days are short.

All of this is reversed in the Southern Hemisphere.

Stonehenge and other ancient monuments worldwide used this to tell the time of year.

Archaeoastronomy is the astronomy of extinct ancient cultures—for example, the Mayans.
Ethnoastronomy is the astronomy of extant ancient cultures—for example, the Chinese.

("Extant" means "still-existing.")

Architecture for solar energy still uses all this today.

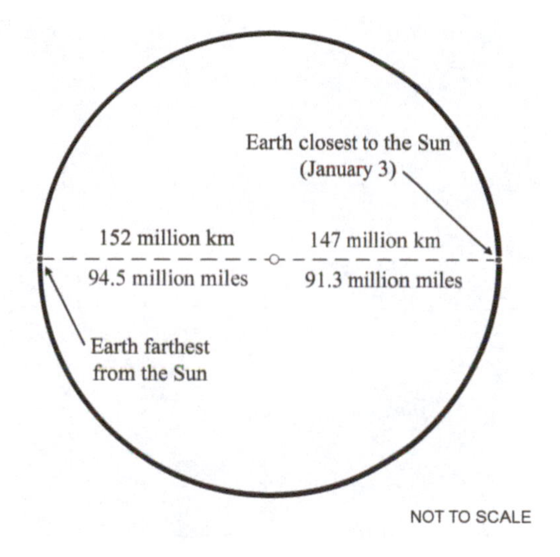

► **Figure 6-1** This is a plot of Earth's orbit around the Sun. Notice how it is nearly a circle. Although it is in fact an ellipse, it's not very elongated. (In other words, it has an eccentricity of less than 2%.) Its being an ellipse, therefore, has almost no effect on the seasons. (Image courtesy of Frederick Ringwald)

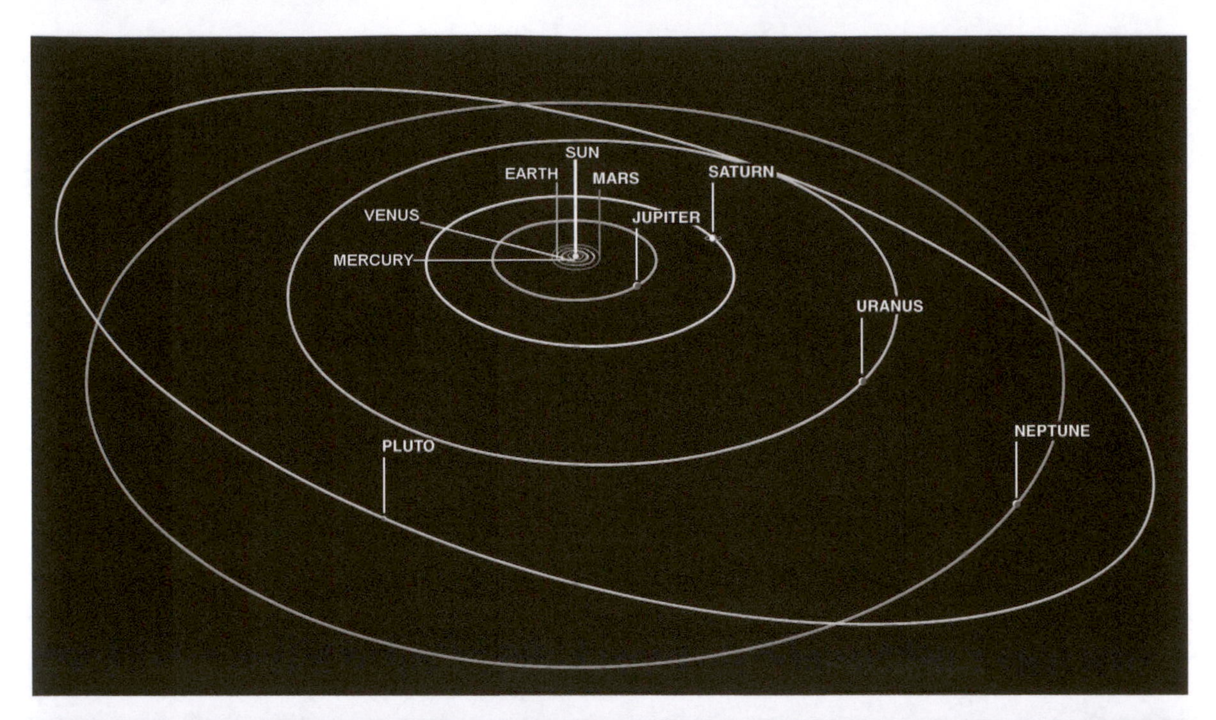

► **Figure 6-2** *Top:* The ecliptic is the plane of Earth's orbit around the Sun. Notice how the planets all orbit within a few degrees of the ecliptic. (Pluto does not because it isn't a planet.) (NASA/LPI) *Bottom:* The ecliptic is shown drawn onto a photo of the sky. Notice how three planets (Saturn, Mars, and Venus) and the Moon are always within a few degrees of the ecliptic. (Image courtesy of Frederick Ringwald)

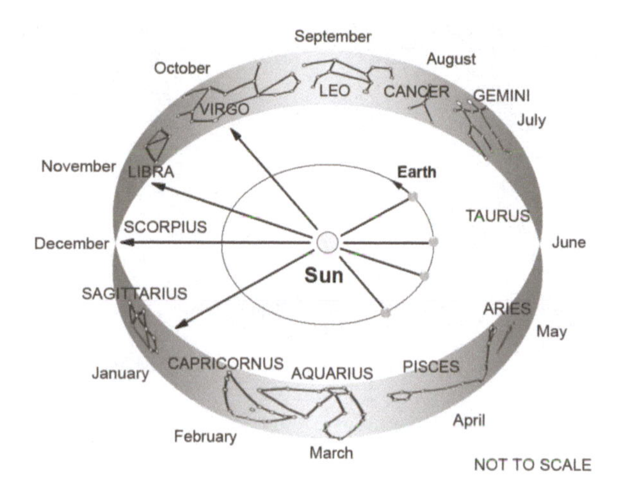

▶ **Figure 6-3** **The constellations of the zodiac are the twelve constellations through which the ecliptic passed during the time of the ancient Greeks, who made up many of these constellations. The Sun, Moon, and planets are usually in these constellations, as seen from Earth.** (Image courtesy of Frederick Ringwald)

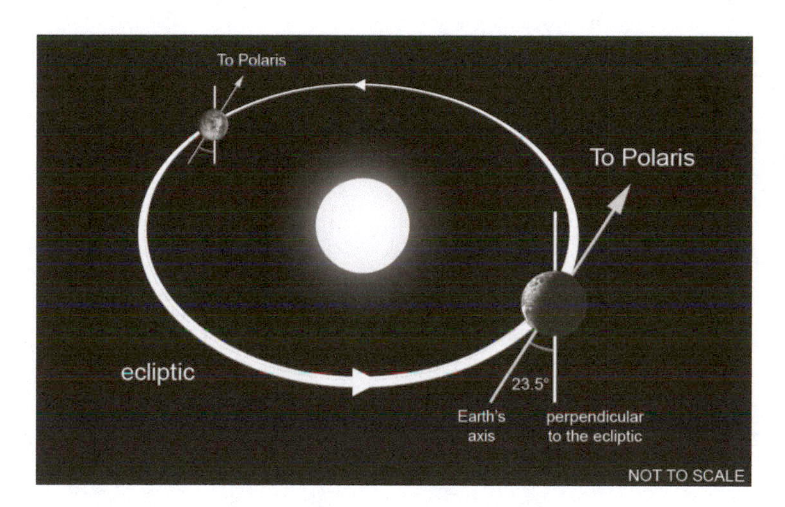

▶ **Figure 6-4** **As Earth orbits the Sun, the direction its axis points stays the same (roughly). It points toward the north celestial pole, which is within 1° of Polaris, the North Star.** (Image courtesy of Frederick Ringwald, using NASA images of Earth)

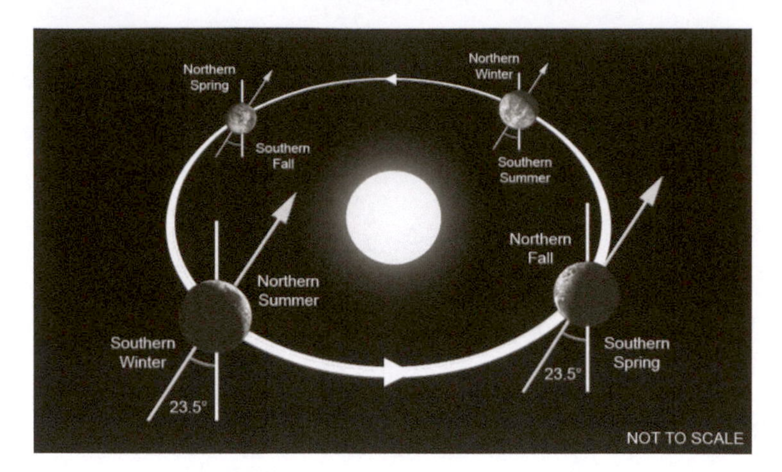

► **Figure 6-5** Why Earth has seasons: Earth's axis is tilted by 23.5° to perpendicular to the ecliptic plane. *In other words:* Earth's equator is tilted by 23.5° to the ecliptic. This means that, as the Northern Hemisphere tilts toward the Sun, it has summer. At the same time, the Southern Hemisphere tilts away from the Sun, so it has winter. Likewise, six months later, the seasons are reversed. At this time, the Southern Hemisphere has summer, also called austral summer. This is at the same time as northern winter, also called boreal winter. (Image courtesy of Frederick Ringwald, using NASA images of Earth)

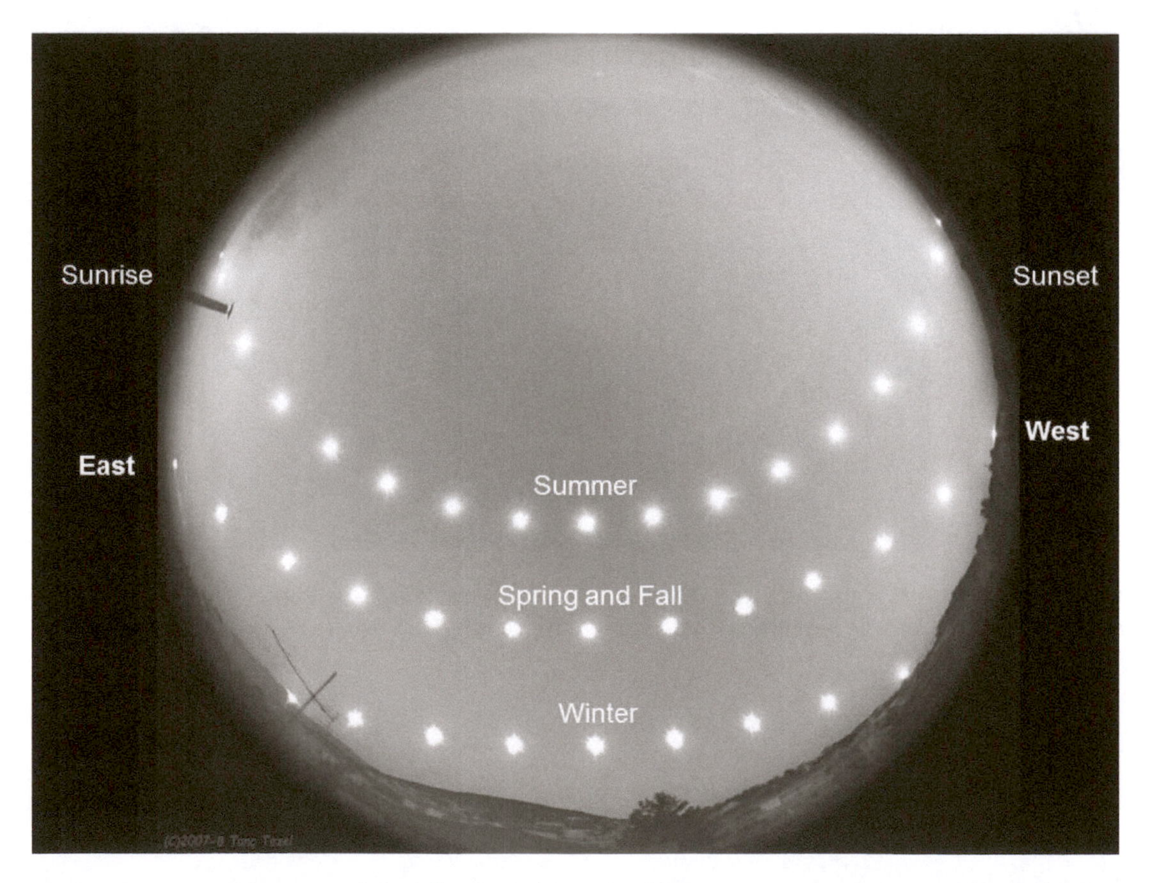

► **Figure 6-6** *Top trace:* Because of the tilt of Earth's axis, the Sun appears to rise north of east in summer, and set south of west. This makes for long, warm days, with the Sun high in the sky for more of the day. *Middle trace:* The Sun rises due east and sets due west during the vernal equinox, the first day of spring. The only other time of year it does this is during the autumnal equinox, the first day of fall. *Bottom trace:* In winter, the Sun appears to rise south of east. It sets north of west, making for short, cold days with the Sun low in the sky all the time. This is shown and described here the way it happens in the Northern Hemisphere. It is reversed in the Southern Hemisphere. (Image courtesy of Tunc Tezel)

► **Figure 6-7** Stonehenge in England and other ancient monuments worldwide used where the Sun rises to tell the time of year. The Heelstone, shown at right in both figures, casts a shadow into the center of the circle at sunrise at summer solstice. Ancient people *weren't* stupid: they used their eyes and their brains. (Images courtesy of Frederick Ringwald)

► Figure 6-8 **El Caracol ("the Snail" in Spanish) in México was a building used by the ancient Mayans to observe the rising of Venus. This allowed the Mayans to develop a calendar that was nearly 20 times more accurate than the Gregorian calendar, used today.** (Image © Ms Deborah Waters, 2013. Used under license from Shutterstock, Inc.)

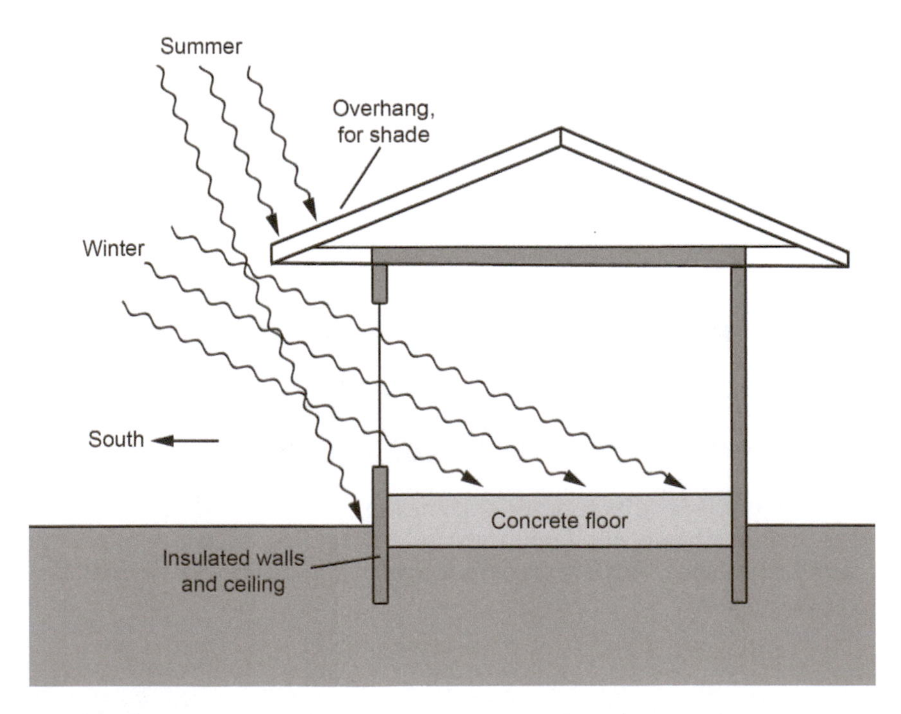

► Figure 6-9 **Solar energy can be used for home heating and cooling by designing for the Sun's different positions in the sky throughout the year.** (Image courtesy of Frederick Ringwald, based on a figure from Hinrichs, Roger A., & Kleinbach, Merlin H. (2006). *Energy: Its Use and the Environment,* 4th ed., pg. 178)

Annual Motion: We see different stars during different times of year because of Earth's motion around the Sun. A star wheel shows this, including both the daily (diurnal) motion of Earth's rotation, and the annual motion of Earth's orbit around the Sun.

The celestial sphere is the imaginary sphere of the sky, centered on the Earth. Ancient people thought it was real. It isn't, but it still helps us map the sky.

Ancient people worldwide needed to know the sky because they needed some way to keep time. (Clocks weren't invented until the 1300s.) Knowing the time of year was crucial for agriculture, when getting the dates of planting and harvesting wrong could result in starvation. The ancients also used the stars for navigation at sea.

Constellations are patterns of stars in the sky, created in ancient times to make it easier to find one's way around the sky. In 1930, the International Astronomical Union divided up the sky into 88 official constellations. Many of the ones in the northern sky date from ancient times and are named for myths and legends—for example, Hercules the Hero, Orion the Hunter, Taurus the Bull, Scorpius the Scorpion, and Sagittarius the Archer.

(*D'Aulaire's Book of Greek Myths* is a wonderful book on Greek myths. So is *The Metamorphoses* of Ovid. See also the pages at the end of this section of the Class Notes.)

Many constellations in the Southern Hemisphere were named in the 1600s and 1700s and reflect scientific interests of the time. These include Telescopium the Telescope, Microscopium the Microscope, Fornax the Furnace, and Antlia the Air Pump.

Asterisms are unofficial constellations; you may make them up yourself. Widely used ones are the Big Dipper (which is part of the constellation Ursa Major the Great Bear), the Little Dipper (which is most of the constellation Ursa Minor the Little Bear), Orion's Belt, the Great Square of Pegasus, and the Summer Triangle.

Star Names: Most stars visible to the unaided eye were given names by the Arabs in 900–1000 A.D. Some examples include these stars visible during Fall semester:

- *Vega*, *Deneb*, and *Altair* are the three bright stars that make up the Summer Triangle.
- *Antares* is the red supergiant star at the heart of the constellation Scorpius the Scorpion.

Stars visible during Spring semester include:

- *Sirius*, the brightest star in the night sky, in the constellation Canis Major the Big Dog.
- *Betelgeuse* is a red supergiant star in the constellation of Orion the Hunter.
- *Rigel* is a blue supergiant star in Orion.
- *Capella* is a yellow giant star in the constellation Auriga the Chariot Driver.
- *Aldebaran* is a red giant star in the constellation Taurus the Bull.

Star Registries: Some people will tell you that if you pay them a fee, they will name a star after you. These names have ***no official recognition*** by the scientific community.

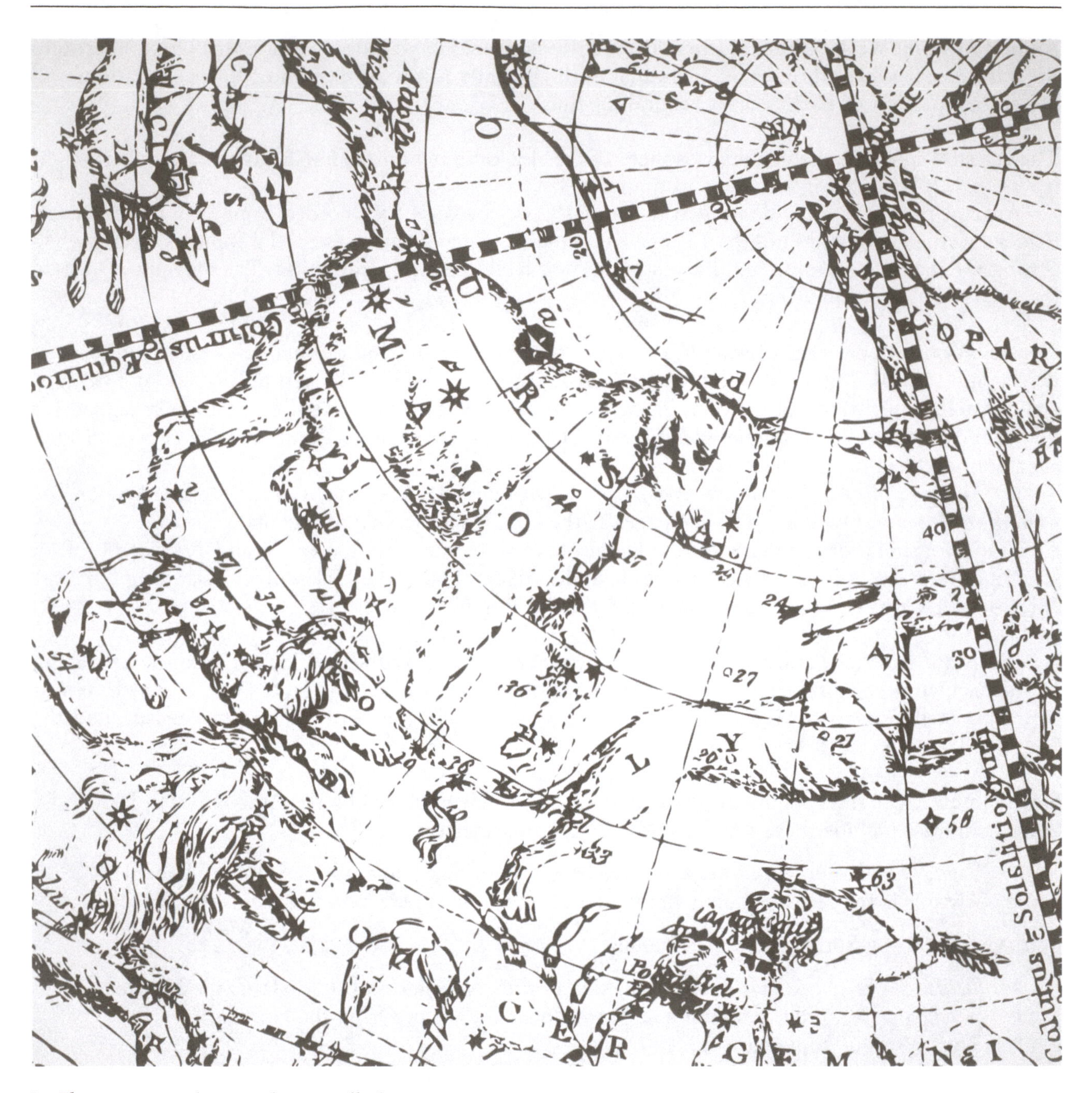

► **Figure 6-10** The Greek constellations (Image © alicedaniel, 2013. Used under license from Shutterstock, Inc.)

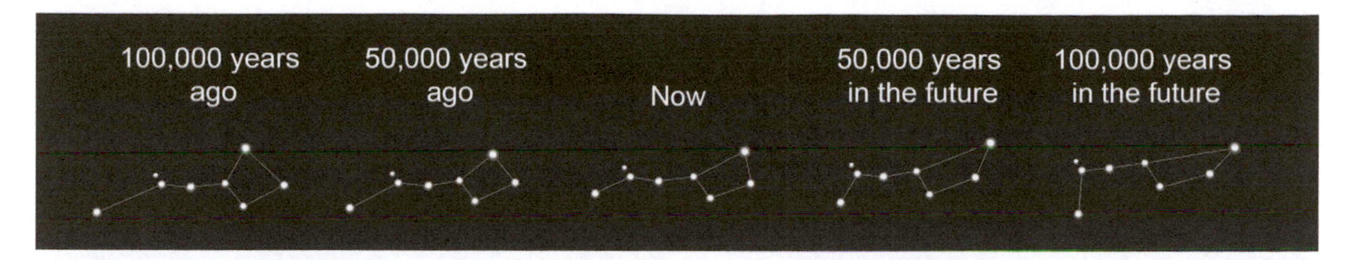

▶ **Figure 6-11** The Big Dipper changes over time because of its stars' proper motion through space. This happens slowly by human reckoning, though, because the stars are so far apart. Because of this, the same constellations used in ancient times can be used today. Also, the yellow star is about twice as far away as most of the other stars shown here. This shows that most constellations aren't physical groups of stars: most are just arbitrary patterns picked out by the human eye and brain. (Image courtesy of Frederick Ringwald)

▶ **Figure 6-12** Asterisms are "unofficial" patterns of stars in the sky: you may make them up yourself. (FOXTROT © 2005 Bill Amend. Reprinted with permission of UNIVERSAL UCLICK. All rights reserved.)

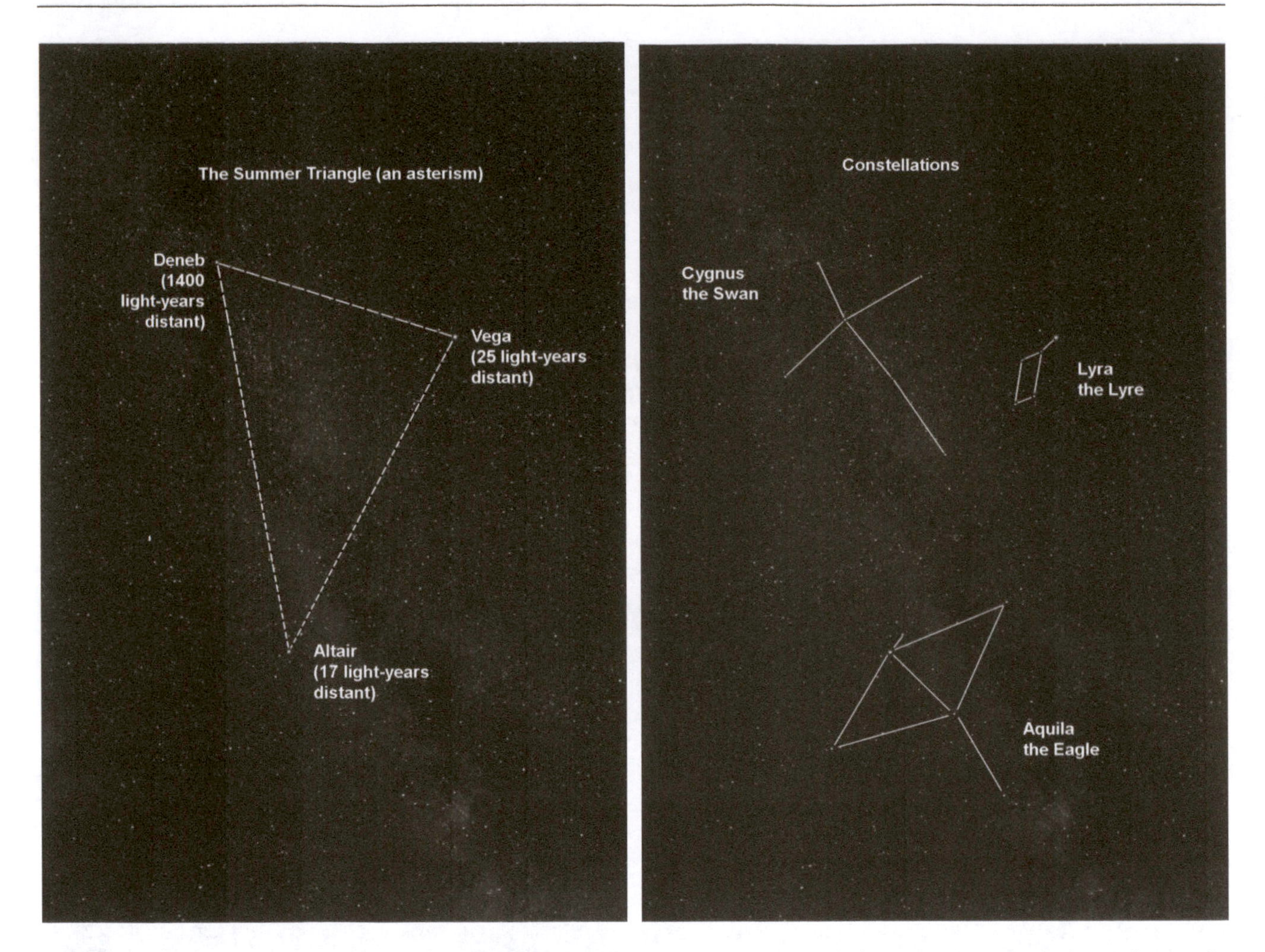

▶ **Figure 6-13** *Left:* The Summer Triangle is an asterism, consisting of the stars Vega, Altair, and Deneb. Deneb is much farther away than Vega and Altair. These stars are not a physical group, but make a triangle to human eyes. (Image courtesy of Frederick Ringwald) ***Right:*** The Summer Triangle is an asterism, and not one of the official 88 constellations. The stars of the Summer Triangle are in the constellations Lyra the Lyre, Aquila the Eagle, and Cygnus the Swan. (Image courtesy of Frederick Ringwald)

Bayer notation: In 1603, Johannes Bayer gave the brightest star in a constellation the first letter of the Greek alphabet, "alpha" (or α), and the second brightest star in a constellation the next Greek letter, "beta" (or β), and the third brightest "gamma" (or γ), etc. Because of this:

- α Centauri = the brightest star in the constellation Centaurus the Centaur,
- β Centauri = the 2nd brightest star in the constellation Centaurus,
- γ Centauri = the 3rd brightest star in the constellation Centaurus, etc.

The problem with this is that the Greek alphabet has only 24 letters, and there are many more than 24 stars in each constellation. Since then, many other designations have come into use. Most in use now use the coordinates of where stars are in the sky.

North circumpolar stars are so far north, they never set below the horizon as Earth spins.

Fresno is at latitude 37° north. Because of this, stars within 37° of the north celestial pole are always above the horizon when observed from Fresno. There are also stars so far south one can never see them from Fresno—these are called south circumpolar stars.

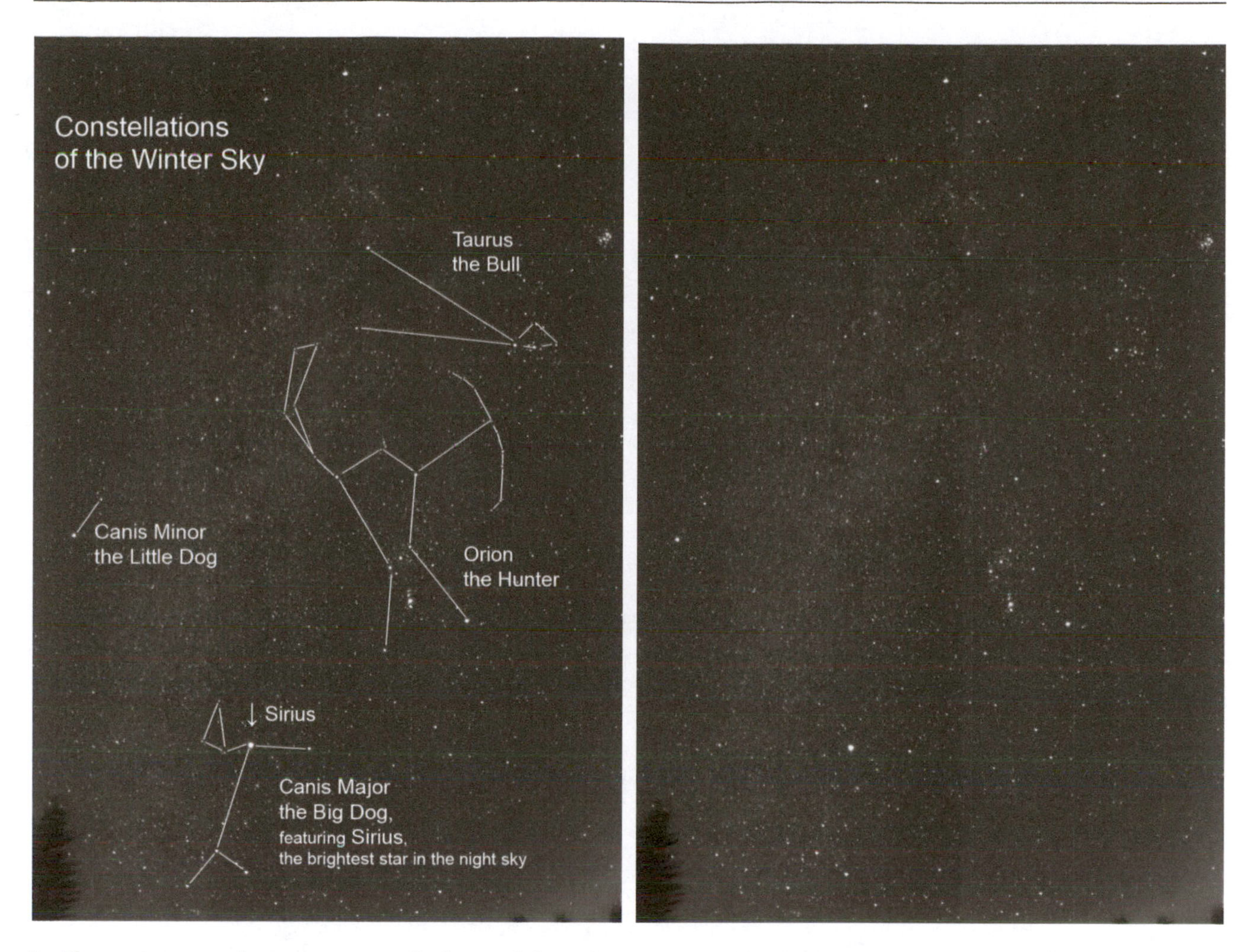

► **Figure 6-14** *Left:* Some constellations of the winter sky. *Right:* The winter sky. (Images courtesy of Frederick Ringwald)

Polaris, also called the North Star, is ***not*** the brightest star in the sky. There are 48 stars brighter than Polaris, including the Sun. Polaris is special because, purely by luck, it is within 1° of the north celestial pole, the place in the sky directly above Earth's North Pole.

This means that Polaris is always in the north. You can use it to find your direction. Pull off the road and stop the car before you do!

The Clouds of Magellan, also known as **the Magellanic Clouds,** are two companion galaxies of the Milky Way. They are *south circumpolar*: one can't see them from Fresno since they're too far south. The Magellanic Clouds do appear high in the sky from places in Earth's Southern Hemisphere, such as Chile, South Africa, or Australia.

The Midnight Sun: The Arctic Circle is at 23.5° north latitude. If an observer is north of this at the summer solstice, the Sun is so far north it becomes a north circumpolar star. This gives 24 hours of perpetual daylight, and even more, farther north. The North Pole gets six months of daylight and six months of darkness.

► **Figure 6-15** As Earth spins, stars appear to move around the north celestial pole. Stars so far north they never set below the horizon are called *north circumpolar* stars. (Image courtesy of Frederick Ringwald)

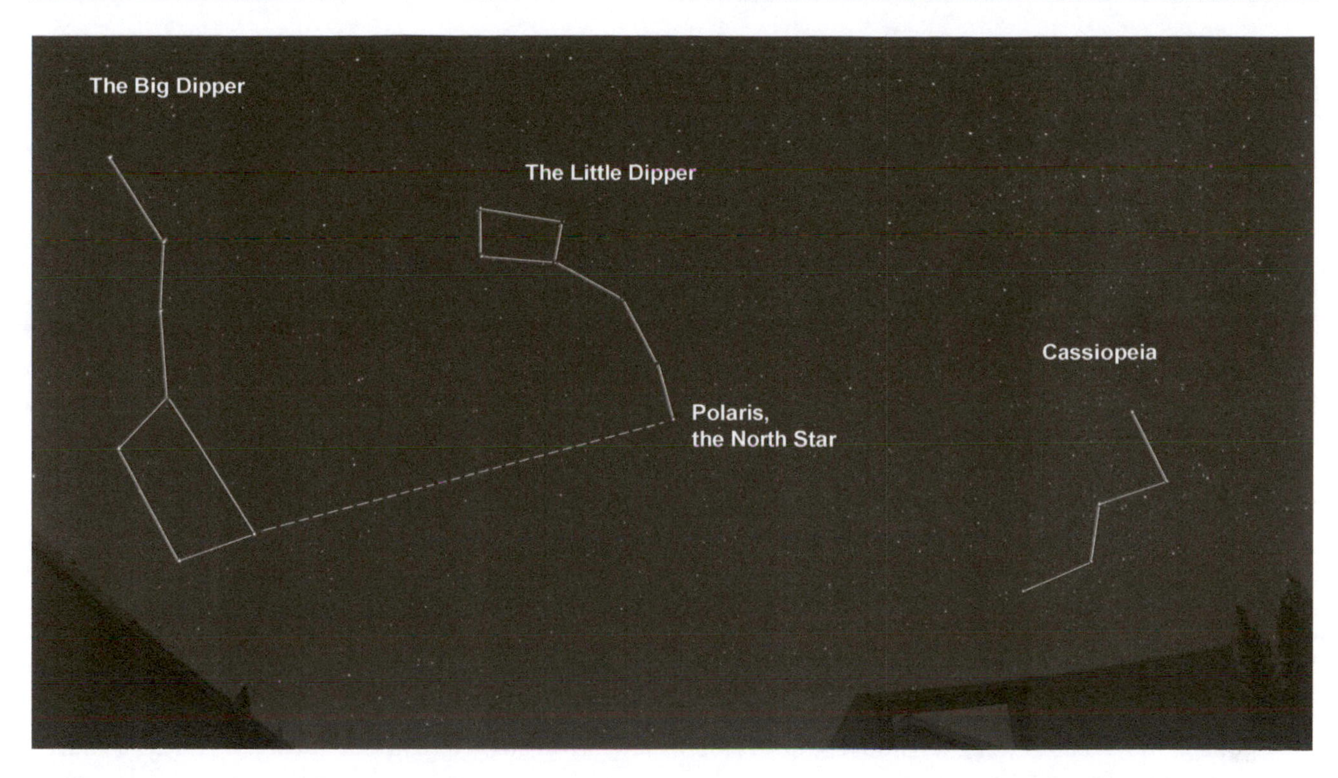

▶ **Figure 6-16** The north circumpolar constellations include Ursa Major the Great Bear (also known as the asterism the Big Dipper), Ursa Minor the Little Bear (also known as the asterism the Little Dipper, which has Polaris the North Star at the top of its handle), and Cassiopeia the Queen (also known as the asterism the Big W, or the Big M, depending on the time of year). The two stars at the end of the bowl of the Big Dipper are called "the pointers": if you follow them for five lengths, they point to Polaris. (Image courtesy of Frederick Ringwald)

▶ **Figure 6-17** This is the north circumpolar sky with the lines removed. (Image courtesy of Frederick Ringwald at Fresno State's station at Sierra Remote Observatories.)

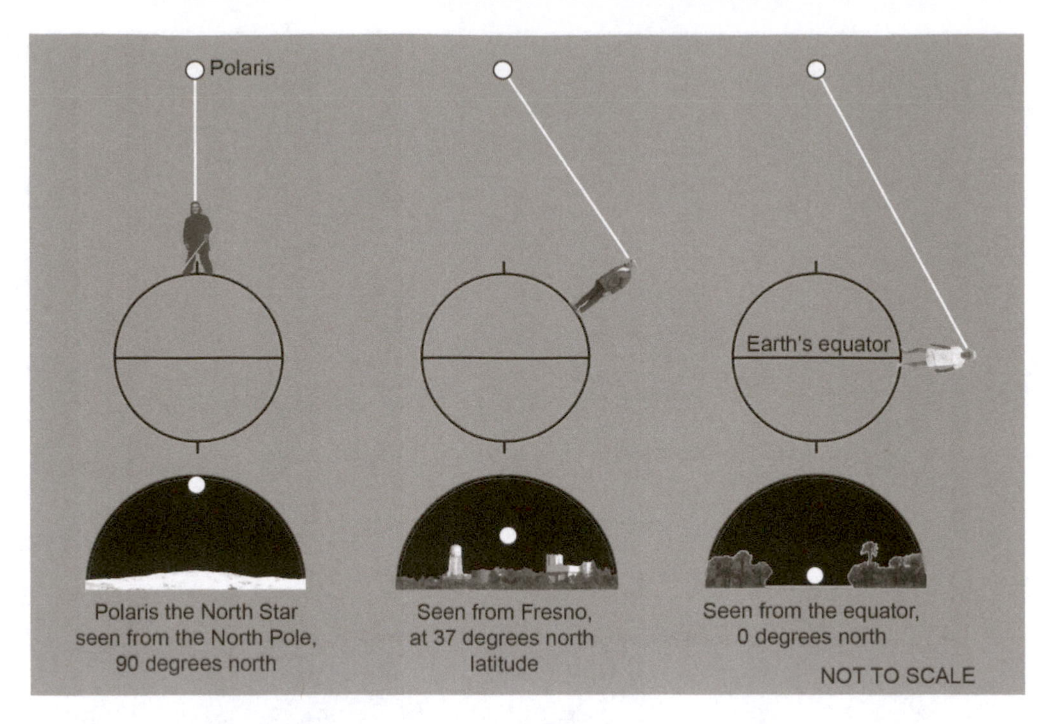

▶ **Figure 6-18** Polaris is higher above the northern horizon at higher latitudes, because Earth is round. (Image courtesy of Frederick Ringwald)

▶ **Figure 6-19** A view from Chile in South America of the Clouds of Magellan, or the Magellanic Clouds (at left). These are two companion galaxies to the Milky Way. They are *south circumpolar,* within 20 degrees of the south celestial pole. The Magellanic Clouds are therefore not visible to anyone farther north than 20 degrees north latitude. The southern Milky Way is visible at right. (National Optical Astronomy Observatories/ Cerro Tololo Interamerican Observatory)

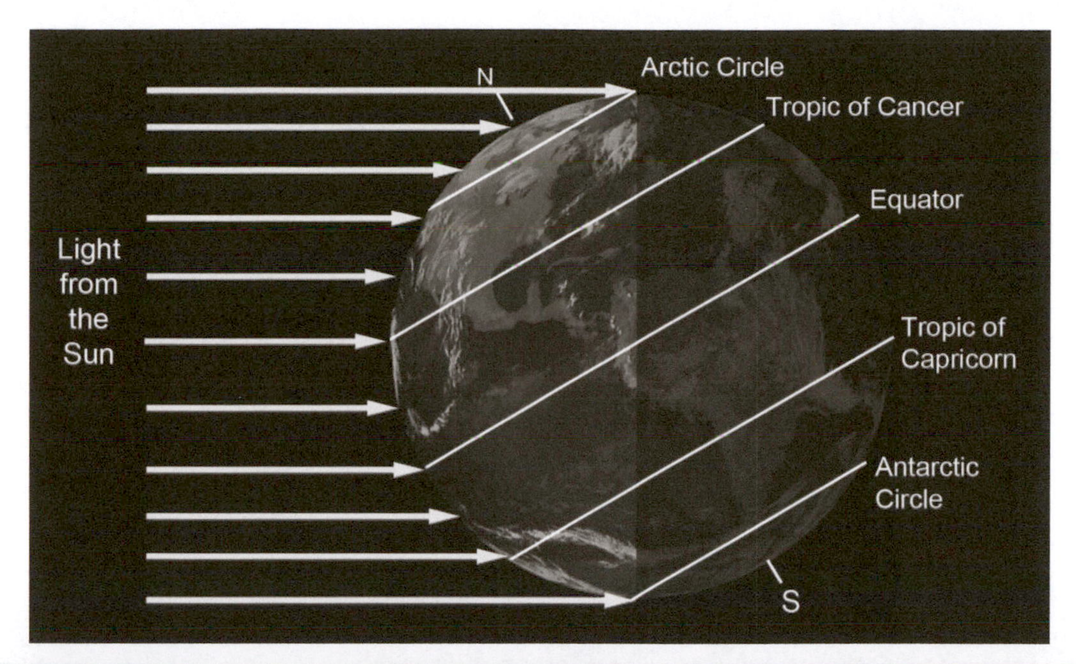

▶ **Figure 6-20** *Top:* Above the Arctic Circle at the summer solstice, the Sun can become a north circumpolar star. (Image courtesy of Frederick Ringwald, using a NASA image) *Bottom:* Midnight sun on island of Vaeroy, Lofoten islands, Norway. (Image © Harvepino, 2013. Used under license from Shutterstock, Inc.)

Equinoxes and Solstices

The celestial equator is the line in the sky directly above Earth's equator. (More technically, the celestial equator is a great circle on the celestial sphere.)

- The *vernal equinox* is when the Sun crosses the celestial equator, going from south to north. It is celebrated as the first day of spring, often on March 21, even though it can take two months for Earth's oceans and atmosphere to warm up.

  The vernal equinox can be from March 19 to 21 because Earth's orbit isn't a perfect circle. It will be on March 20, from 2008 to 2043.
- *The autumnal equinox* is when the Sun crosses the celestial equator, going north to south. It's celebrated as the first day of fall, often on September 21.
- *The summer solstice* is when the Sun is farthest north on the ecliptic.
  It's celebrated as the first day of summer, often on June 21.
  It is the longest day of the year.
- *The winter solstice* is when the Sun is farthest south on the ecliptic.
  It's celebrated as the first day of winter, often on December 21.
  It is the shortest day of the year.

Angles

There are 360° in a circle.
There are 60 arcminutes in one degree (60′ = 1°).
There are 60 arcseconds in one arcminute (60″ = 1′).

The Moon covers about half a degree, or about 30 arcminutes.

One arcsecond is a *tiny* angle. It's only 1/3600th of 1°. If I were 2 miles from you, and I held up a dime, you would see it cover one arcsecond (1″).

Astronomers now routinely measure angles of less than 1″.
The state-of-the-art is now approaching milliarcseconds, or 1/1000th of 1″.
This is the angle covered by a dime that is 2000 miles away (in Chicago, from Fresno).

Equatorial Coordinates

Equatorial coordinates measure where stars are on the celestial sphere.

Right ascension measures east/west, analogous to longitude on Earth.
Declination measures north/south, analogous to latitude on Earth.

The celestial equator has a declination of 0°.
Polaris has a declination of nearly 90° north.

Right ascension goes to the left, as seen from the Northern Hemisphere.
The location of the vernal equinox along the ecliptic defines 0 hours right ascension.

Altazimuth

Altazimuth coordinates measure where stars are in your local sky.

The zenith is the point in the sky directly overhead. It is 90° from the horizon.

Altitude: measures up/down, from the horizon.
0° altitude = the horizon
90° altitude = the zenith

Azimuth: measures left/right, east from north.
0° azimuth = north
90° azimuth = east
180° azimuth = south
270° azimuth = west

Altazimuth coordinates are convenient for describing where a star is at a given time. The problem is that Earth spins, and moves around the Sun, so stars appear to move through the sky. We therefore need both altazimuth coordinates *and time,* to tell a star's equatorial coordinates.

Meridian

The **Meridian** is the line in the sky from the North Point (the point on the northern horizon) to the zenith (overhead) to the South Point (the point on the southern horizon).

a.m. = ante meridian, or before the Meridian: the time before the Sun crosses the Meridian, or morning.
p.m. = post meridian, or after the Meridian: the time after the Sun crosses the Meridian, or afternoon.

Usually the best time to observe any star is when it is crossing the Meridian, since you're looking up through the least amount of Earth's atmosphere.

Daylight saving time, called "summer time" in some countries, is when clocks are deliberately set ahead an hour. This is done to make the hours of daylight better match the time when most people are awake. This is set by the government: it has nothing to do with astronomy. Many countries, and some U.S. states, do not use it. In 2007, the U. S. federal government set daylight saving time in most of the United States to extend from the second Sunday in March to the first Sunday in November.

Please note: it is not "daylight savings time," it is "daylight saving time."

The Calendar

The Babylonians lived in Mesopotamia, in what now is Iraq. Agriculture originated there in about 10,000 B.C. The first towns, the first cities, and the earliest known writing also originated there. Perhaps as early as 3500 B.C., and certainly after 1800 B.C., the Babylonians recorded observations of the Sun, the Moon, and Mercury, Venus, Mars, Jupiter, and Saturn, the planets that are bright enough for the unaided eye to see. (They didn't realize that Earth is a planet, too.)

By 2400 B.C., the ancient Egyptians realized there are 365 days in the year. Since then, the problem has been that there aren't exactly 365.0 days in one year, because it takes the Earth 365.2422 days to orbit the Sun. If one assumes that a year is exactly 365.0 days, one will lose a day every four years.

After 100 years of this, the calendar will be nearly a month behind schedule. Since planting crops too late in the year can result in a poor harvest, many cultures devised calendars that are more accurate.

When Julius Caesar was Emperor of Rome, he instituted the Julian calendar in 45 B.C. In it, every fourth year had an extra day, or a leap year day. We still celebrate this by putting February 29 into the calendar during leap years. The Julian calendar therefore had 365.25 days, which meant it lost a day only every 128 years.

By 1582, the Julian calendar had lost over 11 days. This meant that the growing season was arriving noticeably late. Pope Gregory XIII got the calendar back on time by abolishing 11 days, so that October 4, 1582 was followed by October 15, 1582. He also instituted a new calendar, called the Gregorian calendar. This is the calendar most countries use today. In the Gregorian calendar, leap year days still occur every four years, but not during century years such as 1900 or 2000, unless the year is divisible by 400. The years 1700, 1800, and 1900 were not leap years, but 1600 and 2000 were.

The Gregorian calendar therefore has 365.2425 days per year. This means it will lose a day only about every 3000 years. The ancient Mayans had a calendar nearly 20 times more accurate—it lost a day only once every 57,000 years.

Christmas, Easter, and Halloween

It is now definitely known that Jesus Christ was *not* born on what is now called December 25, 0 A.D. There was no Year 0 A.D., since the number zero was unknown to the Romans. Zero had an ancient origin in India, but the idea wouldn't be brought to Europe until about 900 AD by Arab traders. The birth of Christ is thought to have occurred in 4 B.C., since the records of at least three years were lost because of poor recordkeeping by monks during the Middle Ages. Also, the *New Testament* describes shepherds tending their flocks at night during the birth of Christ. They would much more likely have been doing this in spring, since it would have been too cold for this in December.

In ancient Rome, the winter solstice was celebrated by a holiday season that lasted several days, called the Saturnalia. Surprisingly many pagan customs live on now, in their Christianized forms. The early Christians knew that converting people was much easier with the old and the new coexisting side-by-side.

The pagan celebration of fertility, Easter, was also adopted into the Christian calendar. It comes about nine months before Christmas, the celebration of birth. Easter and Christmas correspond to the vernal equinox and the winter solstice, although Easter is now celebrated by the Roman Catholic Church and most Protestant denominations on the first Sunday after the first Full Moon that happens on or next after the vernal equinox. Fertility symbols such as eggs and rabbits are still common at Easter, even in secular Easter egg hunts in the United States.

The pagan celebration of death lives on as Halloween. Roman Catholics consider All Saints Day on November 1 a holiday. In some Catholic countries, such as México and Poland, the dead are celebrated at that time of year.

The Christmas Tree

The pagan Germanic religion was a curious one. The Christmas tree represents Yggdrasil, the great ash tree believed to hold up the nine worlds. These did not correspond to the planets of the Solar System—remember that Uranus and Neptune weren't discovered until the invention of the telescope—these were mythical worlds, of which Midgard (the Earth) and Asgard (where the gods lived) were but two. Another was Jotunheim, from which the frost giants, the Jotuns, came. Thor, the thunder god, spent much of his time using his great war hammer to defend the other worlds from the Jotuns.

The round ornaments on a Christmas tree represent the worlds. The garlanding represents the pathways between the worlds on which the Valkyrie, maidens on horseback, would take slain warriors into Asgard. (The Valkyrie are the subject of an opera by Wagner, and may be remembered best for their distinctive outfits, but how would you design armor for women?) The lights represent the sparks from Thor's hammer, and the tinsel represents the icicles from the Jotuns. The angel or star on top is a Christianized version of the eagle that would stand guard atop the tree against the dragon of

Nidhogg, which gnarled in its roots in one of the underworlds. In the intervening centuries, the Christmas tree turned from an ash into a pine; and although Martin Luther had one, it's less clear how the Christmas tree got over to England, where the author saw them, since the Vikings weren't exactly good neighbors.

Greek Mythology and Constellations

Greek and Roman myths are not politically correct. Still, if one is to learn from history, it must be honest.

Mythology is not history, of course, but it shows the values of the real, historical people who made up the stories. Athena, the war goddess, was portrayed as much more powerful than her brother Ares, the war god, because Athena was also the goddess of wisdom. This is because the Greeks admired cleverness in battle more than brute strength. On the other hand, from their stories, you can tell that the ancient Greeks were not as kind to their women as we like to think we are. What will people 23 centuries from now think of us, one might wonder?

To learn why the planets, satellites, and features on them have the names they have, see the Gazetteer of Planetary Nomenclature, by the International Astronomical Union, at:

http://wwwflag.wr.usgs.gov/USGSFlag/Space/nomen/nomen.html

The Pantheon

In Greek and Roman mythology, twelve gods and goddesses reigned supreme from Mount Olympus, with minor gods in every river and sprites in every tree.

- *Zeus*, the thunder god, was the chief god. The Romans admired Greek culture, so they incorporated many Greek gods into their religion. Zeus therefore became the Roman god *Jupiter*.
- *Hera* (Roman *Juno*) was wife to Zeus and the goddess of marriage and the home.
- *Poseidon* (Roman *Neptune*) was a brother of Zeus and the god of the sea.
- *Athena* (Roman *Minerva*) was a daughter of Zeus and the goddess of war. She was also the goddess of wisdom, and was therefore much more powerful than her half-brother, Ares.
- *Ares* (Roman *Mars*), a son of Zeus, was the god of war.
- *Apollo*, a son of Zeus, was god of light and music. In some traditions, he was also identified as the Sun god.
- *Artemis* (Roman *Diana*), a daughter of Zeus, was the goddess of hunting.
- *Hermes* (Roman *Mercury*), a son of Zeus, was the messenger of the gods.
- *Aphrodite* (Roman *Venus*), who rose mysteriously from the sea, was the goddess of love and beauty.
- *Hephaistos* (Roman *Vulcan*), a son of Zeus and Hera, was the god of fire.
- *Demeter* (Roman *Ceres*), a sister of Zeus, was the goddess of the harvest.
- *Dionysus* (Roman *Bacchus*), a son of Zeus, was the god of wine.
- Attending the fire was *Hestia* (Roman *Vesta*), a sister of Zeus, the goddess of the hearth, who gave her throne to Dionysus.
- Another major god who rarely spent time in Olympus, preferring his own realm, was *Hades* (Roman *Pluto*), a brother of Zeus, the god of the dead.
- The Titans were a previous generation of gods who had been deposed by Zeus and the other Olympians. One of these was Cronus (Roman *Saturn*), the god of time. Cronus was the father of Zeus. The Titaness Rhea was the mother of Zeus.

These live on as the days of the week, in Spanish and French, respectively:

Sunday	=	domingo	=	dimanche	from dominus, Latin for "lord"
Monday	=	lunes	=	lundi	from luna, Latin for "Moon"
Tuesday	=	martes	=	mardi	the day of Mars
Wednesday	=	miércoles	=	mercredi	the day of Mercury
Thursday	=	jueves	=	jeudi	the day of Jupiter
Friday	=	viernes	=	vendredi	the day of Venus
Saturday	=	sábado	=	samedi	the day of Saturn

English uses the corresponding northern European deities:

Sunday	=	the Sun's day
Monday	=	the Moon's day
Tuesday	=	day of Tyr, the war god, the Norse counterpart of Mars
Wednesday	=	day of Woden, another war god, (loose) counterpart to Mercury
Thursday	=	day of Thor, the thunder god, counterpart to Jupiter
Friday	=	day of Freya, the love goddess, counterpart to Venus
Saturday	=	Saturn's day

The week got seven days because the ancient Babylonians—long before the Greeks, Romans, or Norse—noticed bright objects that moved through the sky, relative to the stars. These seven objects were the Sun, the Moon, and five planets, namely Mercury, Venus, Mars, Jupiter, and Saturn, all of which are bright enough to be seen with the unaided eye. The Babylonians therefore took the number seven to be a lucky number. During the French Revolution, along with adopting the metric system, a new calendar with a ten-day week was proposed. It never caught on because weekends came less often.

The Constellations

In 1930, the International Astronomical Union (IAU) divided the entire sky into 88 constellations. Many constellations in the Northern Hemisphere were named in ancient times, and have names from Greek and Roman mythology. That the IAU adopted Greek and Roman constellations in no way means that the Greeks and the Romans were the only people to make up constellations. Nearly all cultures do, since the patterns make the stars easier to identify.

For example, Ursa Major is the Greek constellation the Great Bear. In America, part of it is known as the Big Dipper. In England, this part is known as the Plough, or Charles's Wain, the wagon of King Charles. In China, it is the Carriage. It is therefore pointless to argue about constellations: for all of them, someone always has "another story." Even the "official" Greek ones, adopted by the IAU, are based mostly on Greek myths, which were told around campfires for centuries before they were written down. There are therefore always alternative ways that many stories come out. In the same way the children's game "telephone" can change stories, when repeated person to person, myths and legends will change, being based on oral tradition.

Orion the Hunter is a striking constellation with many bright stars. Particularly striking are three bright stars in a row—the belt of Orion. Opposite the sky is Scorpius the Scorpion, who stung Orion to death. Next to Scorpius is Ophiuchus the Snake Bearer, or the physician, since Asclepius, the greatest of physicians in Greek mythology, was said to have learned his art and all the secrets of the Earth from snakes because he gave their mother a proper funeral, which the Greeks considered important.

Gemini the Twins, is another winter constellation, next to Orion. In it are two first-magnitude stars, Castor and Pollux. In Greek mythology, Castor and Pollux were twin brothers of Zeus and

Leda, a mortal. Pollux was immortal, but Castor was not. The brothers loved each other so much that Pollux gave Castor half of his immortality—meaning that, when they died, they became the constellation.

To the south and west of Orion are his two hunting dogs, the constellations Canis Major the Big Dog and Canis Minor the Little Dog. Canis Major contains Sirius, the brightest star in the night sky. Canis Minor has first-magnitude Procyon. To the north of Orion is the constellation Auriga the Charioteer, with the first magnitude star, Capella. To the northeast of Orion, south of Auriga, is the constellation Taurus the Bull. Orion is shown fighting him, with his great club raised. His shield is an arc of faint stars, warding off the bull's attack.

Centaurus is too far south to be easily visible from mid-northern latitudes. This is too bad because it is a large and spectacular constellation, rivaling Orion in splendor and in number and diversity of astrophysically interesting objects. It has two first-magnitude stars, Alpha Centauri and Beta Centauri; nowhere else in the sky are two stars this bright so close to each other. In Greek mythology, the Centaurs were wild, uncouth, destructive creatures—except for Chiron, who was kind and wise and who educated young heroes before they would do their famous deeds.

A prominent spring constellation is Ursa Major the Great Bear. In America, part of it is often referred to at the Big Dipper, with seven bright, second-magnitude stars tracing the shape of a dipper. In Greek mythology, the bear was Callisto, a lover of Zeus. Hera, his wife, became so angry she turned Callisto into a bear, and then picked her up by the tail and hung her in the sky, which is why she has such a long tail. (Most bears don't.) Hera put Callisto just in front of the constellation Canes Venatici the Hunting Dogs, which chase her around the Pole Star, Polaris, forever. Ursa Minor the Little Bear is the constellation Polaris is in: it is Callisto's child, Arcas. Arcas was hunting with his dogs and didn't recognize his mother when he encountered her as a bear. Rather than allow the dogs to attack Callisto, Zeus put Callisto and Arcas together into the sky as constellations.

Polaris is at the tip of the tail of Ursa Minor. The two stars at the end of the bowl of the dipper point to Polaris. One can use the dipper to point to other constellations, too—the other two stars in the bowl point down to Regulus, the first magnitude star in Leo the lion. The handle of the Big Dipper points to Arcturus, a first-magnitude star in the constellation Bootes the Herdsman. If one continues following this arc, one comes to another first-magnitude star, Spica, in the constellation Virgo the Maiden. One can remember this by "follow the arc to Arcturus, and speed on to Spica." To the Greeks, Virgo was Persephone, goddess of fertility; when she arrived in the sky each spring the Earth bloomed, and when she left the sky in fall, it withered.

In the summer sky is the Summer Triangle. The triangle is an asterism composed of three first-magnitude stars—Vega, Deneb, and Altair. Vega is in the constellation Lyra the Lyre. It is a distinctive little constellation, with five third- to fourth-magnitude stars within 10 degrees, mostly south of Vega. Lyra the Lyre is the instrument of Orpheus, the greatest of musicians, who could make rocks weep with his playing. Deneb is in the constellation Cygnus the Swan. Cygnus is Zeus, who appeared in the form of a swan to impress Leda, the mother of the Gemini twins, Castor and Pollux. Altair is in the constellation Aquila the Eagle, which depicts another manifestation of Zeus.

The constellation Hercules the Hero is in the spring sky, just west of the Summer Triangle. Hercules slew Hydra the 9-Headed Monster, a constellation of which straddles the equator, and stretches over 100 degrees across the sky. Slaying the Hydra was the second of the twelve labors of Hercules. The first was when he slew Leo the Lion, a lion with a hide so thick, weapons couldn't pierce it. After strangling it with his enormous strength, he skinned it with his own claws and wore the skin as armor. This is why artwork of Hercules often shows him wearing a lion skin. The constellation Hercules is also next to the constellation Leo the Lion. Next to them is the constellation Cancer the Crab, which was sent by Hera to pinch Hercules to distract him.

Hercules is just south of the northern constellation Draco the Dragon. There were many dragons in Greek mythology. This one is probably Ladon, the dragon that Hercules slew while carrying out the eleventh of his twelve labors. Ladon guarded the tree bearing the golden apples at the end of the world. Hercules had to ask Atlas, the Titan who held up the sky, to pick them for him or else he would die. Hercules, with his great strength, was able to hold up the sky, and Atlas, freed of his burden, almost left him there. Only because Hercules tricked him into holding up the sky again, "so I can make a pad of the lion skin on my shoulders," did Hercules manage to get away.

Constellations in the fall sky include Pegasus, Perseus, Andromeda, Cetus, Casseiopeia, and Cepheus. The Great Square of Pegasus is an easily identifiable asterism, again helpful for finding the other constellations. North of the square, along the Milky Way, is Perseus the Swordsman. Perseus cut off the head of Medusa, a monster with snakes growing out of her head instead of hair, so ugly that looking at her would turn a human into stone. Perseus carries the head in his hand, at the location of the eclipsing binary star Algol the Demon's Eye. This star, easily visible to the unaided eye, noticeably changes brightness because it is an eclipsing binary.

Perseus, a mortal son of Zeus, had help from his immortal siblings. He is flying with the winged sandals borrowed from Hermes (or Mercury), the messenger of the gods. He cut off Medusa's head with a sharp knife borrowed from Artemis, the goddess of hunting. He also had the polished shield of Athena, the goddess of wisdom, and was able to look at Medusa without being turned into stone because he looked at the reflection in the shield.

In the constellation, he is shown saving Andromeda the Princess from being eaten by Cetus the Sea Monster (or Whale). She was chained to a rock to appease Poseidon, god of the sea, who was angry because her mother, depicted by the constellation Casseiopeia the Queen, had boasted she was more beautiful than his daughters. Her husband, depicted in the constellation Cepheus the King, looks on. Next to Andromeda is the constellation Pegasus the Flying Horse, which sprang from the severed head of Medusa, much to the surprise of Perseus. It was later tamed by Bellerophon, who slew the Chimera, a monster with three heads—one of a lion, one of a goat, and one of a snake.

The constellation Aries the Ram, is also visible in fall. Aries was a magical, flying ram, with a golden fleece. A band of heroes, including Orpheus and Hercules, and led by Jason, recovered this fleece after a long voyage across the Black Sea on Argo the Ship. Argo is also a constellation, but it is no longer used; it has been broken up into four constellations: Vela the Sail, Carina the Keel, Puppis the Stern, and Pyxis the Compass. Another southern constellation is Sculptor. This sculptor was Pygmalion, who created a statue of a woman so beautiful he fell in love with it. A prayer to Aphrodite, the love goddess, was answered when the statue turned into a woman.

There are many other myths and legends, about many other constellations. Telling them all could fill a book, and indeed, there are several wonderful books in which one can read more about these stories. *D'Aulaires' Book of Greek Myths*, by Ingri and Edgar Parin D'Aulaire, is a children's book, but it's still my favorite on mythology, partly because it's so beautifully illustrated. Nearly all the information provided here was reproduced from memory, from having read this book. Much of the rest came from *The Metamorphoses,* by Ovid, the Roman poet of the 1st century AD. By writing these myths down, Ovid became the authoritative source on them: his book is still in print, in both Latin and English. Another is Plutarch's *Lives*. Finally, *The Stars*, by H. A. Rey, is another children's book that adults can read and enjoy. It's my favorite for learning constellations, since the author was brave enough to redraw many constellations, to make them look more like what they're supposed to look like. An extensive source of names of the major and minor Greek gods and goddesses, and what major and minor planets they are named after, is *The Nine Planets: A Multimedia Tour of the Solar System*, by Bill Arnett, at:

http://nineplanets.org/

CHAPTER

7

What Is Science?

An Introduction to Scientific Method

Delphinus the Dolphin is a distinctive little constellation, between the constellations Cygnus the Swan and Aquila the Eagle. In the myth of Dionysus (Roman *Bacchus*), the god of wine, Dionysus fell asleep on the beach and was kidnapped by pirate sailors who thought that someone so well dressed must be a prince. Dionysus woke up and told them who he was. The sailors didn't believe him. Dionysus then revealed himself in all his glory: he grew to such a large size, he filled the ship, with vines growing up

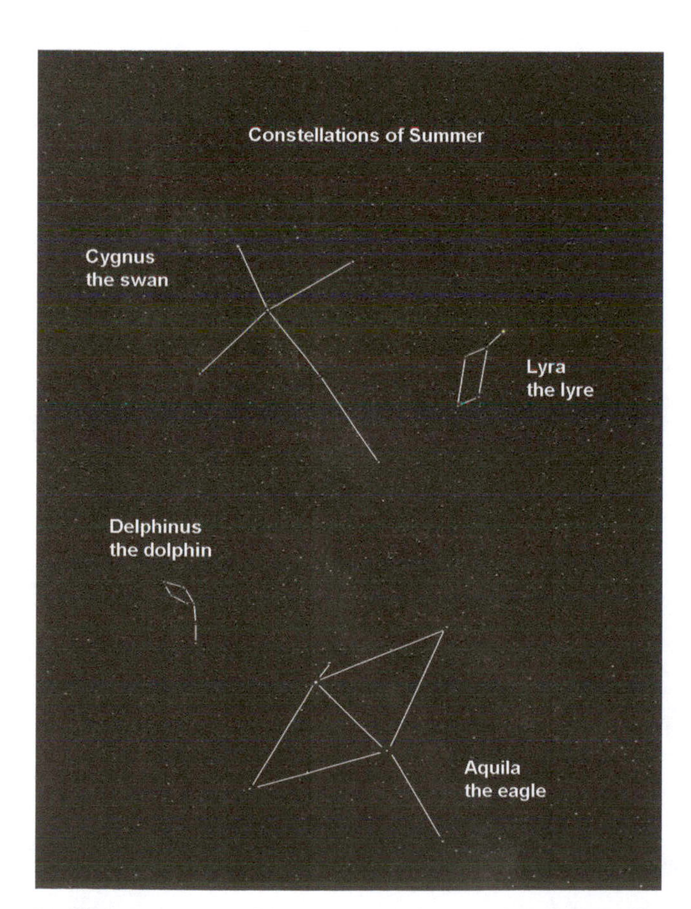

► **Figure 7-1** *Left:* The constellation Dephinus the Dolphin, next to the asterism the Summer Triangle. (Image courtesy of Frederick Ringwald) ***Right:*** The myth of Dionysus is shown on this Greek postage stamp. (Image © Lefteris Papaulakis, 2013. Used under license from Shutterstock, Inc.)

the sails. The sailors panicked and jumped into the sea. Dionysus was a kind god: he didn't want them to drown, so he turned them into dolphins. The ancient Greeks thought this explained why dolphins are so intelligent: they used to be people.

This is an example of an explanation after the fact. It's also called a *post hoc* argument, "*post hoc*" being Latin for "after this." The full saying is "after this, therefore because of this." An expla-

USA TODAY Snapshots

Jobs that rate respect

Top five professions that people in the USA perceive as having the most prestige:

Scientist **57%**
Firefighter **55%**
Doctor **52%**
Teacher **49%**
Nurse **47%**

Source: Harris Interactive Poll

By Elizabeth A. Crowley and Bob Laird, USA TODAY

▶ **Figure 7-2** **There are exceptions, but by and large the general public holds scientists in a great deal of respect.** (From *USA Today,* "Jobs that Rate Respect," January 22, 2004, Gannett-USA Today. All rights reserved. Used by permission and protected by the Copyright Laws of the United States. The printing, copying, redistribution, or retransmission of this Content without express written permission is prohibited.)

▶ **Figure 7-3** *Left:* **This author wishes that the public was better at telling science from the cheap imitation. Here is a sign at the hotel bar during a meeting of the American Astronomical Society. It says, "Welcome, American Astrological Society." The hotel staff didn't know the difference between the science of astronomy and the pseudoscience of astrology.** (Image courtesy of Frederick Ringwald) ***Right:*** **Here is the sign, corrected.** (Image courtesy of Frederick Ringwald)

nation is supposed to explain the facts: one shouldn't make up what one claims to be new facts (such as that dolphins used to be humans, of which there is no evidence) to make up an explanation.

Mythology often tries to explain things this way. Many people who are not scientists think this is what science does, too. It isn't. This is an example of what science *doesn't* do.

Many stereotypes of science abound. A typical one, at upper left of Figure 7-4, is a middle-aged to elderly white male with bad hair, bad eyewear, questionable motives and sanity, and of course the ever-present white lab coat. For the record, gentle reader, your present author has worn a white lab coat only once ever in his life.

Figure 7-4 *Top left:* A stereotypical "mad" scientist. Most genuine scientists in fact do not want to "rule the world." Many aren't terribly enthusiastic about managing their own laboratories. (Image © Louis D. Wiyono , 2013. Used under license from Shutterstock, Inc.) Top right: Some scientists do wear white laboratory coats, particularly in a chemistry lab like this where the chemicals they are working with might ruin their clothes, if spilled. (Photo courtesy of California State University, Fresno) *Bottom left:* This posed picture shows some genuine physicists with a genuine experiment that can reach temperatures within a few degrees of absolute zero. (Photo courtesy of California State University, Fresno) *Bottom center:* This shows some geologists having a class on a glacier. (Photo courtesy of California State University, Fresno) Bottom right: Two computer scientists are shown smiling because something works. Two genuinely stereotypical scientists are sitting behind them: these days, the most typical scientist looks like a person sitting at a computer. (Photo courtesy of California State University, Fresno)

Science is a recent phenomenon. The words *science* and *scientist* weren't in common use until the 1830s. Before this, scientists were called "natural philosophers." For much of history, science was illegal because it could bring change, which many people in power didn't want.

Galileo Galilei is credited as the originator of modern, experimental, evidence-based, inductive science, diagrammed in the flowchart in Figure 7-5. He did this in the early 1600s. Galileo was first to write about scientific method, in a way recognizable as scientific today—ideas are tested by experiment, and are discarded if proved wrong by experiment. Also, science is inductive, meaning that individual examples are used to find out generalizations about what nature is observed to do.

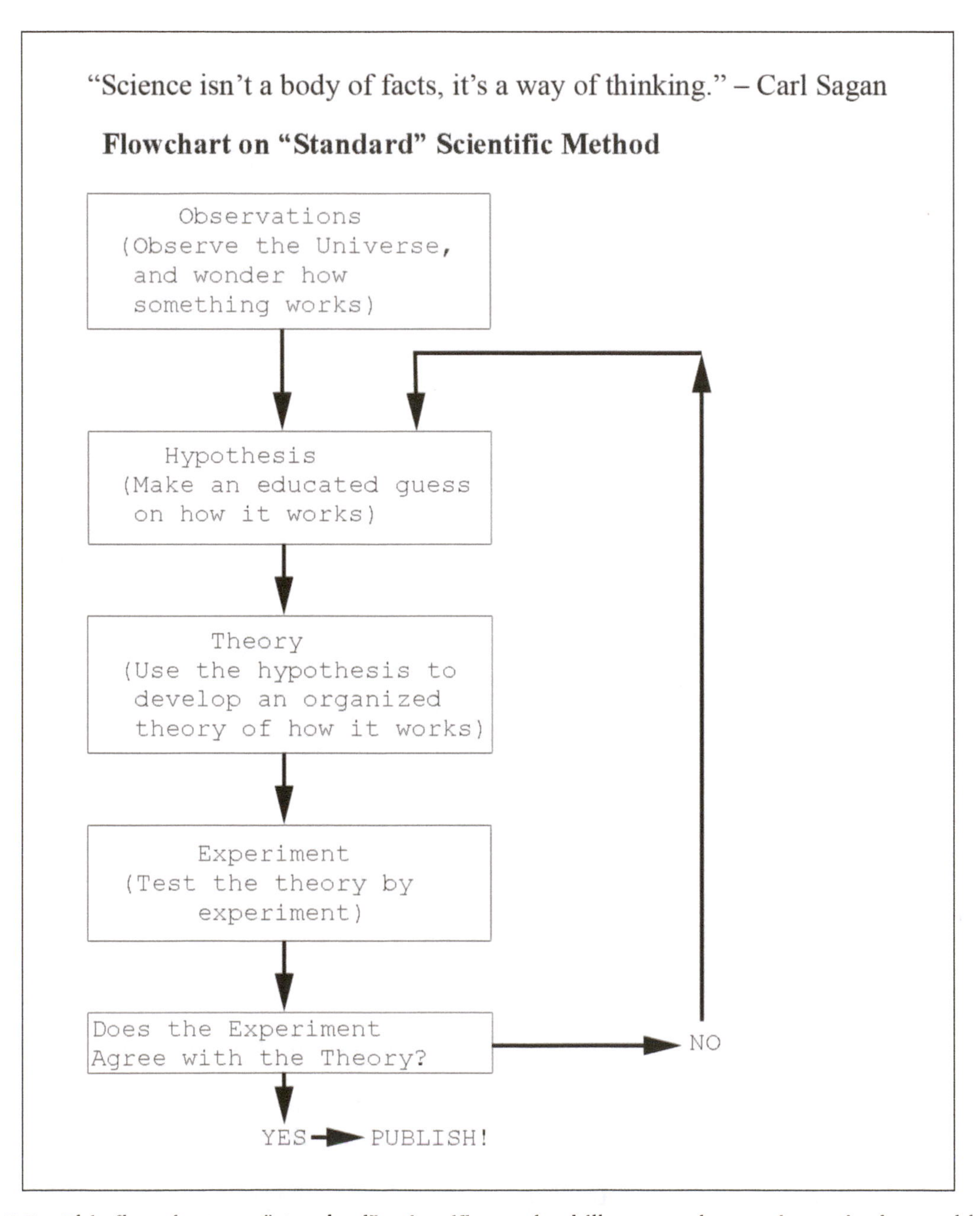

▶ **Figure 7-5** This flowchart on "standard" scientific method illustrates how science is done, although the steps can be done in more ways than are shown here. (Image courtesy of Frederick Ringwald)

Francis Bacon lived at the same time as Galileo (and Shakespeare), in the late 1500s and early 1600s. He wrote extensively about scientific method, but he does not get credit for inventing science because he didn't put enough emphasis on hypothesis and theory—in other words, that it's desirable to understand what one is doing. Bacon thought that if one's observations are orderly, how they worked would be obvious. Unfortunately, this is rarely true: imagining how something works, or in other words forming a hypothesis, requires human imagination. This is why computers don't do science: people do science.

Bacon's emphasis on trying things out experimentally, in the absence of theoretical understanding, strikes many people today as potentially dangerous. Bacon also underestimated the importance of mathematics in science. Galileo had no such trouble.

Nevertheless, Francis Bacon did understand the importance of science in a way few people before him did. As he wrote, "Knowledge is power."

Before this, science was *only* a search for truth. Bacon recognized that science can be used to make life better. It may seem elementary today, but he did some of the first experimenting with refrigeration, by using ice to preserve food by keeping it cold. Only during the Industrial Revolution in the 1830s, about 200 years after Galileo and Bacon, would it be commonly realized that science can be used to make money.

> "One thing you have to say about science: it delivers the goods."
>
> "...If you want to know when the next eclipse of the Sun will be, you might try magicians or mystics, but you'll do much better with scientists...They can routinely predict a solar eclipse, to the minute, a millennium in advance. You can go to the witch doctor to lift the spell that causes your pernicious anemia, or you can take vitamin B12...Try science."
>
> –Carl Sagan

> "Science isn't a body of facts, it's a way of thinking."
>
> –Carl Sagan

> "I think it'll be nice to have gotten close on some of it. But being wrong is part of how science works. I'm looking forward to seeing why I was wrong."
>
> –Ryan Anderson

This last step, *publishing*, is often neglected in science courses that are taught in many schools. This is too bad, because it is essential: scientific results have to become known to the world. Otherwise, why bother?

> "Science is the radical idea, quite foreign to human nature, that an objective reality exists apart from our own internal whims, conceits, and made-up stories–and that this reality, huge beyond imagining, can be revealed and endlessly unfolded by carefully examining evidence."
>
> –Stephen P. Maran and Laurence A. Marschall

> "... Great laws are not divined by flashes of inspiration, whatever you may think. It usually takes the combined work of a world full of scientists over a period of centuries ... Theory after theory was advanced and checked and counterchecked and modified and abandoned and revived and converted to something else. It was a devil of a job."
>
> –Isaac Asimov, in the short story, "Nightfall"

Science is specifically about testing hypotheses and understanding the results rationally in the physical, material world. To be considered *scientific*, an idea must be:

- **Verifiable**: There must be some way to test the idea by experiment.
- **Objective** and **repeatable**: It must work for anyone, in a fair test.
- **Predictive:** It must make predictions that can be tested.
- **Falsifiable**: The idea *can* be shown to be false, by some unambiguous test.

"Falsifiable" means *open to scrutiny and potential invalidation.* There must be, at least in principle, at least one test that can prove the idea wrong, if it's to be considered a scientific idea. Science requires a willingness to follow ideas wherever they might lead.

> "If an experiment does not hold out the possibility of causing one to revise one's views, it is hard to see why it should be done at all."
>
> –Peter Medawar

> "Science is fundamentally something you do with your *hands*."
>
> –Barbara Ehrenbach

Nevertheless, science need *not* involve lots of expensive equipment that goes PING!

It's fashionable among some philosophers of science today to make fun of the flowchart in Figure 7-5 as oversimplified. Since this is an introductory course, we will use it anyway, to help us understand the basics of how scientists work. Be aware that in actual practice by scientists, the steps can be done in nearly any order—sometimes to good effect, sometimes not. (Publishing first, before doing any other work, is not recommended. Neither is yelling "Feyerabend" in a crowded auditorium.)

Notice also: the word ***theory*** is often misused in everyday speech, where it generally means a hypothesis, or a guess. What theory really means is any organized body of knowledge, such as Film Theory, Music Theory, or Einstein's Theory of Relativity.

There's no doubt that Einstein's Theory of Relativity is valid, because it's been tested many times every day in every high-energy physics lab in the world. It's still called Einstein's "Theory" of Relativity because Einstein presented the idea in an organized manner. Another theory is Pasteur's Germ Theory of Disease, the idea that microscopic bacteria and viruses cause disease. Another theory is Maxwell's Electromagnetic Theory, which gave us radio and television. Another is Darwin's Theory of Evolution—without which nothing in the entire science of biology makes sense, and which gave us antibiotics, most else of modern medicine, and modern agriculture. If you doubt this, consider how much antibiotics have changed over the past 50 years. Old antibiotics no longer work because the germs have changed—or in other words, evolved.

Induction is generalization. *Science is inductive*, with reasoning from observed facts to general principles. In other words, in science we start by observing facts, and use them to discover general principles. Importantly, *scientific theories **are** these general principles.* (Deduction is reasoning from general principles to particulars, for example, the logical proofs done in high-school geometry and in higher mathematics.)

Empirical means "derived from observations and measurements." Any empirical science should make *predictions*.

Predictions are essential to science. If a theory doesn't make testable predictions, it often isn't taken seriously, since it can't be tested. If a theory can't be tested, it stops being science, and turns into a game of "let's pretend."

VOLUME 41 ISSUE 03 — AMERICA'S FINEST NEWS SOURCE™ — 19 JANUARY 2005

WHAT DO YOU THINK?

Georgia's Evolution Stickers

Last week, a U.S. district judge ordered a Georgia school district to remove stickers reading, "Evolution is a theory, not a fact" from its textbooks. What do *you* think?

Jered Garza
Driver

"The thing is, they're right. Evolution is nothing more than a well-supported, predictive, scientifically rigorous theory."

Brad Dawson
Novelist

"I hope they replaced the old stickers with new ones that read, 'Do not burn.'"

▶ **Figure 7-6** This is a joke from a humor magazine (*The Onion*). It makes fun of how the word *theory* is often used incorrectly in everyday life. All that "theory" means is any organized body of knowledge, such as film theory or music theory. Some scientific theories that have been thoroughly tested include Maxwell's Electromagnetic Theory, which gave us radio and television, Einstein's Theory of Relativity, and Darwin's Theory of Evolution. (With permission from *The Onion,* January 19, 2005)

Some people (almost none of whom are scientists) might dismiss the dates in "the Cosmic Calendar" (covered in Chapter 5) by saying "it's just a theory" and "who really knows, since no one was there." Remember, though, that this theory is what geologists use to predict where to drill into the ground to find oil.

When proved wrong, scientists are supposed to be good sports, and let go of the old, discredited ideas. This is what makes science so powerful: *Errors are corrected, since scientific ideas are* ***falsifiable.***

Unlike in some other human endeavors, scientists who are quick to "see the light" and change their ways *often get more respect*. This doesn't mean they should abandon old ideas at the first sign of trouble: just that, when shown clearly that the old ideas are *wrong*, they shouldn't hesitate to abandon them.

Nevertheless, *scientists are human*. The human ego has, throughout history, delayed or gotten in the way of much great science. This is particularly so when large amounts of money or political power are involved. Scientists don't like this, and we want to avoid it. Be aware, though, that it sometimes does happen.

Arguments from authority (for example, you must believe a professor's ideas, only because he's a professor) have little or no weight in science. A favorite example was Albert Einstein, who revolutionized science with his Theory of Relativity. At the time, he was just 26 years old and working in obscurity as a patent clerk because he couldn't get a teaching job, because he'd argued with his professors more than they'd liked. The editor of the journal that published Einstein's papers said he "didn't know who this Einstein fellow was," but it was clear that "the world had changed," because Einstein's ideas were so compelling. (It helped that Einstein wrote his early papers well, in ways that other people could understand.)

Michael Shermer says it well whenever he's asked, "Why should we believe you?" He replies, "You shouldn't." You should think for yourself!

> "It doesn't matter how beautiful your theory is, it doesn't matter how smart you are. If it doesn't agree with experiment, it's wrong"
>
> –Richard Feynman

Corroborating evidence is also important. Almost nothing in science is ever an isolated incident: good science *fits together*, like a jigsaw puzzle. Anything important should have multiple lines of evidence, and often does.

But then, only one *correct* line of evidence can be enough.
Multiple incorrect lines of evidence can be *wrong*.

> "The first principle is that you must not fool yourself, and you are the easiest person to fool."
>
> –Richard Feynman

Stringy Holes: Hawking Concedes Defeat

July 23, 2004 by Robert Naeye

Courtesy Caltech / Heidi Aspaturian.

SKY tonight.com

Famed Cambridge University physicist Stephen Hawking has finally come around to accepting what many of his colleagues have thought for decades: black holes preserve information about the material that falls into them ...*Mainstream media outlets have reported this story largely because of Hawking's celebrity status...* ***"I think it is a bit overhyped,"*** says physicist Greg Landsberg.

► **Figure 7-7** Scientists are expected to be gracious when proved wrong. Willingness to let go of ideas that have been discredited is what gives science its great power, since errors are corrected over time. No scientist is ever right all the time—even Newton and Einstein were sometimes wrong. (Text used with permission from Robert Naeye and Sky and Telescope.com. Photo: NASA/Paul Alers)

▶ **Figure 7-8** Arguments from authority carry little weight in science. An example of this is how Albert Einstein did some of his best scientific work while working in obscurity, as a clerk at the patent office in Bern, Switzerland. He might have been a nobody at the time, but this didn't matter: what mattered was whether he was *right,* and he was.

Examples of arguments from authority include: "Because I say so!" or "Because I'm the professor, that's why!" Arguments from authority don't work in science because the only real authority in science is nature itself.

As a friend said to the young Einstein, "You have only one fault: no one can tell you anything!" As Einstein himself said in old age: "To punish me for my contempt for authority, fate has made me an authority myself."
(© Bettmann/Corbis)

Examples of Non-Scientific Ideas

1. The idea that the Universe might be inside an atom that's in another, gigantic Universe, and that this Universe might be inside an atom that's in another, even bigger Universe, and so on, to infinity. This isn't a scientific idea, because it's not possible to prove it, *or to disprove it*, with any conceivable observation.
2. The idea that the Universe came into existence five minutes ago, complete with our memories of what happened before that. Again, one can't prove this, or disprove it. A similar idea is that the Universe was created 6,000 years ago, with fossils put in the ground to make it look older.
3. Bigfoot, also known as the Sasquatch, or Yeti, or Abominable Snowman. How come no one ever catches one of these creatures alive, or finds a dead one? (A gorilla suit in a freezer, as

► **Figure 7-9** An example of a non-scientific idea, because it's not falsifiable, is the idea that the Universe is inside of an atom that is in another, much larger Universe, and that this Universe is inside of an atom that is in another, much larger Universe, and so on to infinity. Unless there is some way to detect or otherwise to show by experiment that any of these larger Universes exist, this isn't a scientific idea, because there's no way to prove it, or to disprove it. (1888 image "Universum," from L'atmosphère: météorologie populaire by Camille Flammarion)

some sellers of Bigfoot merchandise tried to claim in 2008, does not count.) We have 30-million-year-old fossils of primates from remotest Africa, but we don't have a single authentic bone from a creature claimed to be 8 feet tall and said to live in the contiguous 48 United States, of all places.

Related to this is the philosophical principle called **Occam's razor:**

This is the idea that *the most likely explanation is usually the one that requires **the fewest assumptions.***

It was named for William of Occam, who in the 1300s wrote that when constructing an argument, one should not go beyond what is logically required. Occam's "razor" therefore "cuts out" unnecessary assumptions.

UFOs: an example of unnecessary assumptions

Many people assume "UFO" means "spacecraft from another world." ***It doesn't:*** **it means "Unidentified Flying Object." Over 95% of UFOs can be identified, as known phenomena.**

Most scientists are *not* convinced that UFOs are spacecraft from other worlds. The physical evidence is poor, consisting mainly of broken tree branches and burned spots on the ground—not uniquely compelling. More convincing would be if a UFO occupant were to drop something clearly not from Earth, for example, a ray gun that really worked. It would also be convincing if an alien spacecraft were to land at the White House, and its occupants were to open up diplomatic relations. In these cases, the evidence would be so clear, we'd be silly if we weren't convinced.

> Spinoza's Dictum: "I have made a ceaseless effort not to ridicule, not to bewail, not to scorn human actions, but to understand them."
>
> –Baruch Spinoza

> "Ridicule is not part of the scientific method."
>
> –J. Allen Hynek

Most scientists would be *delighted* to be proved wrong about this. So far, though, the evidence that has been claimed hasn't been *compelling*. If one points out the lack of physical evidence and plausibility for UFOs, one may be accused of being part of a conspiracy to cover them up, usually sponsored by the government, or worse, a "government dupe." If one points out that the real government has trouble doing mundane things, such as balancing its budget, one can be told of rumors of a "shadow government." Science doesn't deal in rumors. I've been called "unscientific" when being skeptical about UFOs. I was being perfectly scientific: *I have examined the evidence carefully, and have found it wanting.*

Assuming that UFOs *are* alien spacecraft makes *many* extra assumptions!

> "Many homes in America now have moderately sophisticated burglar alarm systems, including infrared sensors and cameras triggered by motion. An authentic videotape, with time and date denoted, showing an alien incursion–especially as they slip through the walls–might be very good evidence. If millions of Americans have been abducted, isn't it strange that not one lives in such a home?…"
>
> –Carl Sagan

If alien abductions are so common, how come they never wake up the neighbors? A fallacy (or false idea) common among UFO enthusiasts is that, because they don't understand everything about something they've seen, it *cannot* be understood. *Unexplained* doesn't necessarily imply *cannot be explained. Unexplained* also doesn't mean *will never be explained.*

Another idea that does badly by Occam's razor is to suppose that aliens must have visited Earth in the past, solely because one archaeologist in the 1970s couldn't figure out how the Egyptians or the Mayans built their pyramids. Other archaeologists have learned quite a lot about how this was done: we now know that inferring "ancient astronauts" is insulting to the Egyptians and the Mayans. The Egyptians and the Mayans weren't as advanced as we are, but they weren't stupid!

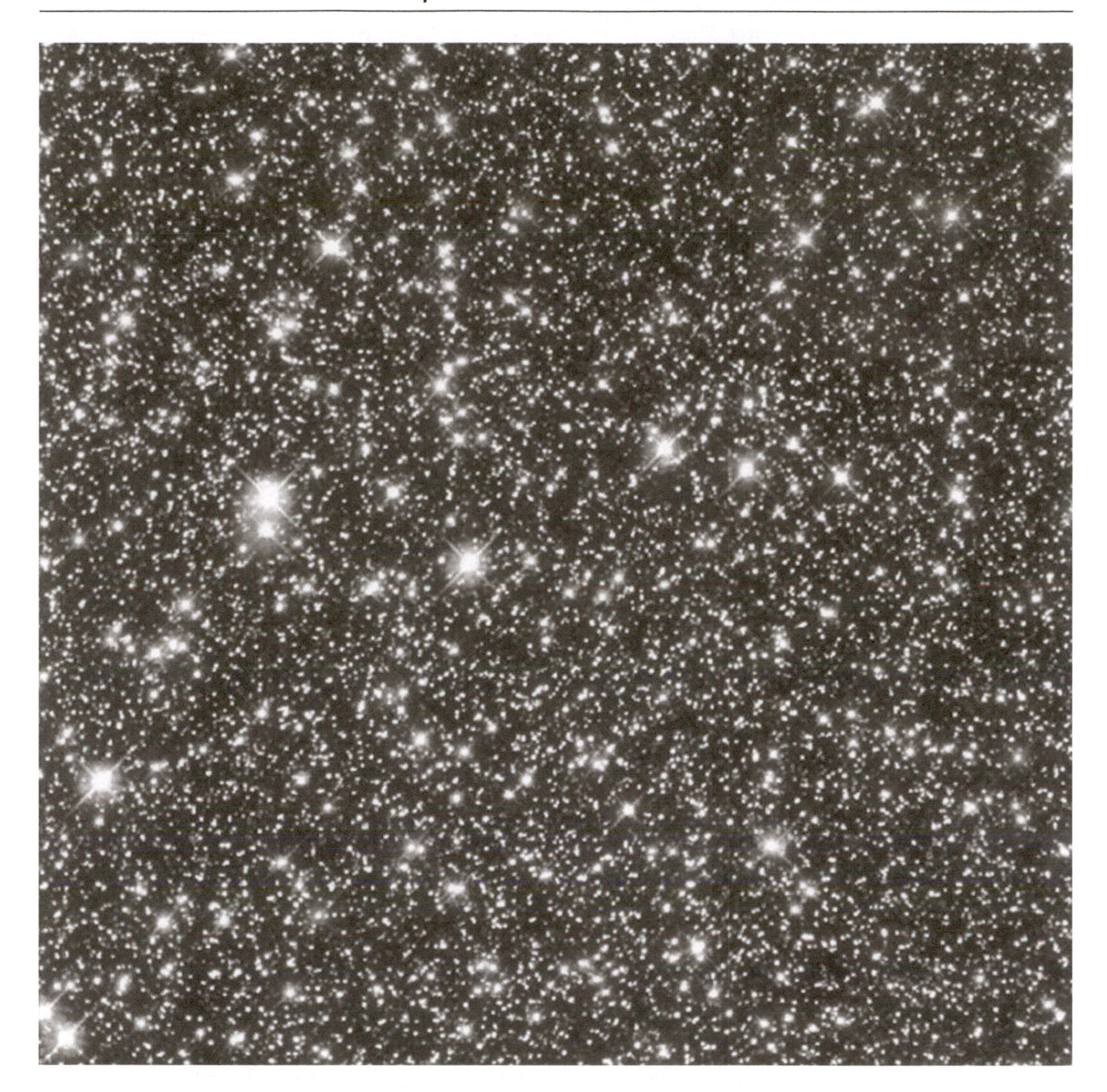

► **Figure 7-10** Don't get the author wrong: I don't want to dismiss the possibility that life may exist elsewhere in the Universe, beyond Earth. As Greek philosopher Metrodorus said in about 500 BC, "Is it reasonable to suppose that in a large field, that only one shaft of wheat should grow, and in an infinite Universe, to have only one living world?" Yet, despite what many UFO enthusiasts would say, the evidence for this so far has not been compelling. Real compelling evidence of life in space would be among the most influential discoveries in the history of science. (Source: NASA/STScl)

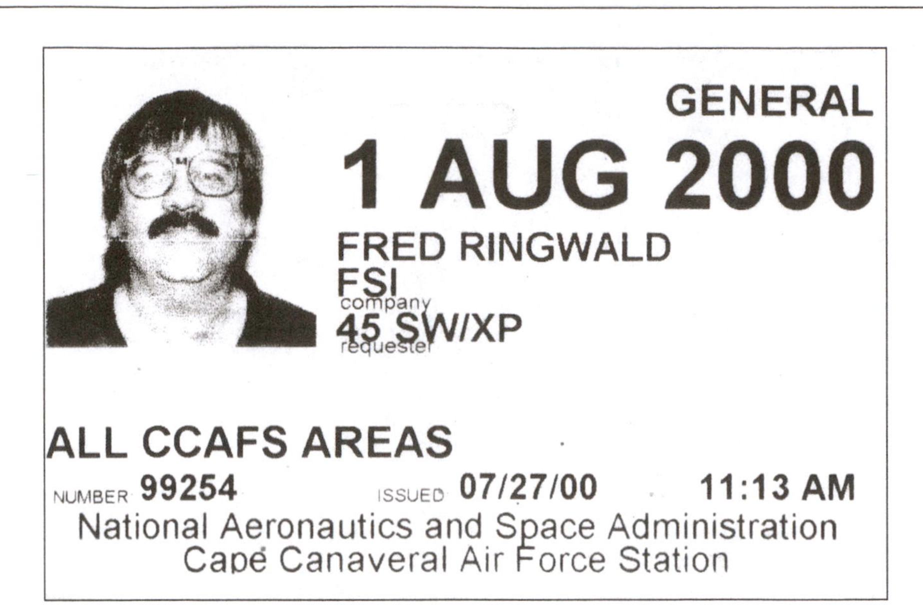

► **Figure 7-11** Because I am skeptical about claims that UFOs are alien spacecraft, I often get accused of being part of some kind of conspiracy. Here is my badge from NASA Kennedy Space Center in Florida. Notice that it authorizes entry into "All CCAFS (Cape Canaveral Air Force Station) Areas." If they're hiding anything at Kennedy Space Center, I didn't see it. (Image courtesy of the 45th Space Wing of the United States Air Force)

Science and the Human Spirit

How do you recognize a genuine miracle, when you see one? As David Hume observed:

> "No testimony is sufficient to establish a miracle, unless the testimony be of such a kind, that its falsehood would be more miraculous than the fact which it endeavors to establish."

On the other hand,

> "There is no quicker way for a scientist to bring discredit upon ...[the] profession than roundly to declare—particularly when no declaration of any kind is called for—that science knows or soon will know the answers to all questions worth asking, and that the questions that do not admit to a scientific answer are in some way nonquestions or 'pseudoquestions'... I am happy to say that however many scientists may think this, very few nowadays are mugs enough or rude enough to say so in public."
>
> —Peter Medawar, from *Advice to a Young Scientist*

Many questions cannot be answered by scientific method, because they don't deal with testable, falsifiable hypotheses in the physical, material world. Many of these are among the most important aspects of life, such as:

ethics, inspiration, morals, justice, beauty, love, aesthetics, or religion.

All students in this course are free to believe in any religion they wish. One doesn't have to be an atheist to be a good scientist, or to get an A in this course.

It is the received wisdom of nearly all human cultures that one's life consists not in the abundance of things one possesses. Nevertheless, it *is* important to put food on the table, and science *can* do that. The attitude that "the encyclopedia knows so you needn't" is little more than an excuse for incompetence. It is dangerous to rely on scientists to do all of our thinking for us.

Anyone who wants to get a good education should spend some time thinking about these questions that science can't answer. It is vital that science, engineering, and especially pre-medical majors do so. I don't like to put that kind of power into the hands of people who don't understand what matters most to most other people.

When I was a boy, some medical researchers tried to develop an artificial heart. (The technology did not progress in this way, though: it turned out that heart transplants work better.) I asked my mother, "Where does the love come from in an artificial heart?" She answered, "It comes from all the hard work of all the doctors and nurses who made the artificial heart possible." Not all bird songs are necessarily just territorial shrieks: to my ears, some birds sound like they're singing for their own enjoyment, because they're happy. Why do some people insist that reason and beauty be mutually exclusive? There really is no reason for it.

What about religion and science? They have long been in conflict, mostly needlessly.

Absolute truth is difficult to find anywhere. Perhaps the only sure place to find it is in pure mathematics. If you want absolute truth, you won't find it in inductive science.

Knowing science doesn't necessarily make one a good person. The Nazis during World War II were dangerous exactly because they had excellent scientists, and horrible morals. The ancient Greeks knew this: in Plato's *Apology*, Socrates commented that artisans, politicians, and poets aren't necessarily "wise," even if they are technically masterful. Although science is an important part of any modern education, it can't be the only part.

None of my students are *required* to believe anything in my courses, nor should they be. But then, science isn't about belief: science is about proof. In particular, please don't deny that evolution exists, or say, "evolution is just a theory."

Evolution is a *fact*, since one can *observe* it happening. Evolution does *not* deny the existence of a deity, nor does it imply that the Universe "just happened," or was "an accident" that occurred "by chance." All the word *evolution* means is change, and we do *see* the Universe change.

Claiming that the Universe was a cosmic "accident," or that we are "simply coils of self-replicating DNA," is like saying the Grand Canyon is a big hole in the ground. It may be correct, but this description is grossly inadequate.

If scientists aren't allowed to teach how we *see* evolution happening, we won't be able to give you any more antibiotics. It isn't like we scientists will be going on strike, or because we're pushing some political agenda: it's because we will be *unable to do our jobs*. This will be trouble, because many germs that cause disease are *observed* to have changed over time—in other words, they *evolved*—and have become resistant to old treatments. Examples of diseases caused by germs that have been observed to have evolved include MRSA, SARS, antibiotic-resistant tuberculosis, and AIDS.

Some politicians might say they don't "believe" in evolution. When they also say they'll make the country independent of foreign oil within ten years, they have *zero* chance of doing this without science. Science, religion, and society need to learn to co-exist peacefully. If they don't, all are in peril. Religion may still be able to stamp out science: it has in the past. This would result in a lowered standard of living that would be noticed and resented, in this modern age.

The word *believe* has two different meanings, and it's important to understand them. The first meaning is to accept something only on faith, without questioning its truth. The second meaning is to accept something because of evidence, understood rationally. These two meanings are therefore *exact opposites* of each other: don't confuse them.

Science is a matter of *evidence*, by experimental tests. No rational person can ignore it or dismiss it out of hand: it has produced too many things that work for anyone, whether they "believe" in it or not, the way that antibiotics, cell phones, airplanes, lightning rods, and a host of others do. There are *no* documented cases of anyone having been cured of any physical ailment by faith healing alone, nor of any crime ever having been solved by a psychic. (See *Why People Believe Weird Things*, by Michael Shermer.)

"Where ignorance is our master, there is no possibility of real peace."

–The Dalai Lama

Even intelligent, reliable people can be misled. Many doctors, nurses, and police officers believe that accidents are nastier and crime is worse during Full Moon. This is wrong: repeated careful looks into "the lunar effect" (also called "Moon madness") show that this is nothing but selective memory, in which people remember the hits, but forget the misses. There are also many cases of well-trained pilots—and your present author!—having been fooled in UFO cases.

It will not work to try to skirt around this, by allowing science if it is kept only among scientists, but not to be revealed to "the common person." Although this author is a scientist, he likes to think he is also a common person. Science is too important to be left exclusively to the scientists. Too much power can corrupt anything, even science.

Just because something isn't scientific doesn't mean that it *must* be useless or bad. *Beware, however, of anything that pretends to be scientific, but* ***isn't***—for example, astrology.

James "The Amazing" Randi and the $1 Million Paranormal Challenge

James "The Amazing" Randi is a professional magician who lives in Florida. He works tirelessly to expose trickery and fraud. So did Harry "The Great" Houdini, the most famous professional magician of all time. A magician is an entertainer who amazes audiences with illusions. Well-performed illusions and tricks are fun to watch, because they look like real magic.

Magicians are in an excellent position to expose fraud, since they know how people make assumptions about what they see. Magicians are often better at detecting fraud than scientists are. Scientists have often been fooled by frauds, because scientists too often assume that what they see is honest.

"I'm in the strange business of telling folks what they should already know," writes Mr. Randi, "I meet audiences who believe in all sorts of impossible things, often despite their education and intelligence. My job is to explain how science differs from the unproven, illogical assumptions of pseudo-science—and why it matters."

A paranormal, supernatural, psychic, or pseudo-scientific phenomenon is something that does not follow the laws of nature, even though at least part of it allegedly takes place in the physical, material world. Examples of paranormal phenomena include astrology, UFOs, alien abductions, psychics, healing by faith alone, extrasensory perception (ESP), telepathy, mind reading, clairvoyance, crystal balls, fortune tellers, tarot, palmistry, ghosts, haunted houses, talking with the dead, séances, and channeling. Other examples include crop circles, reincarnation, levitation, astral projection, homeopathy, acupuncture and "ki" (life force), auras, spirit photography, voodoo, sympathetic magic, the Bermuda Triangle, dowsing, faeries, near-death experiences, telekinetic movement such as spoon bending, spontaneous combustion, the "lost" continents of Atlantis and Lemuria, etc.

For 25 years, Mr. Randi sponsored a $10,000 prize for "the performance of any paranormal event...under proper observing conditions." The James Randi Educational Foundation has since raised the prize to $1 million. To date, no one has ever passed even the preliminary tests.

An application form for the prize is available at: http://www.randi.org/research/index.html

Mr. Randi notes that whenever someone gets a Ph.D. degree, it becomes nearly impossible for that person to say, "I don't know" or "I was wrong." Don't do that, if you ever get a Ph.D. Don't do that, if you *don't* get a Ph.D!

Objections to Astrology

A student once asked the following:

> "Do you believe in horoscopes? The signs? I am a Cancer and I have all the traits, but should I believe the horoscopes I read in magazines and newspapers?"

Here was my reply:

> Absolutely not. I don't think astrology has any validity whatsoever.

For starters, are you sure you have all the traits of a Cancerian? According to which astrologer? Their opinions vary widely, after all. Also, are you sure you aren't subtly fitting the perception of your traits to fit their profile, which is almost always flattering? They rarely list negative character traits in horoscopes. Fortune tellers of all sorts, including palmists, "mind readers," psychics, and other frauds, often do this too.

One objection to astrology is a moral argument. Aren't civilized people supposed to deplore discriminating against groups of people because of accidents of birth? It's so unfair: one can't choose one's parents, nor can one do anything about the circumstances of one's birth. And yet, this is precisely what astrology says: it says that all people with similar circumstances of birth are all alike! Shouldn't we condemn astrology as a form of bigotry? For how is refusing to date a Leo or hire a Cancerian any different from refusing to date a member of a certain ethnic group, or hire a member of a certain religion?

Another objection I have to astrology is a philosophical one: Doesn't astrology contradict the idea that we have free will? After all, if we are destined from birth to do whatever the stars say, doesn't this mean that we have no free will, or ability to control our lives? I wouldn't want to dispense with the concept of free will: it gives me the freedom (and the responsibility) to choose what I do with my life.

A scientific argument against astrology is even more damning. The well-documented, thoroughly tested, reproducible fact is that astrology just plain doesn't work. I always demonstrate this in class, with a randomized survey to see if the class can recognize their own personality type, according to two different astrologers' opinions. They never can, any more than by random chance.

Some astrologers do become wealthy, by preying off superstition and gullibility. Everything they predict is vague and ill-defined. Why is it never precise and exact? Why can't they ever predict what the numbers at the stock market will do? If they could, they'd be richer than Bill Gates.

More to the point: **there were no warnings by astrologers** the day **before September 11, 2001** of the attacks on the World Trade Center and the Pentagon. **There were also no warnings by astrologers** the day before the tsunami on December 26, 2004, that killed nearly 230,000 people. You can check both items easily, by going to the library and looking up the newspaper horoscopes from the previous day. Astrology, psychics, etc., just plain ***do not work!***

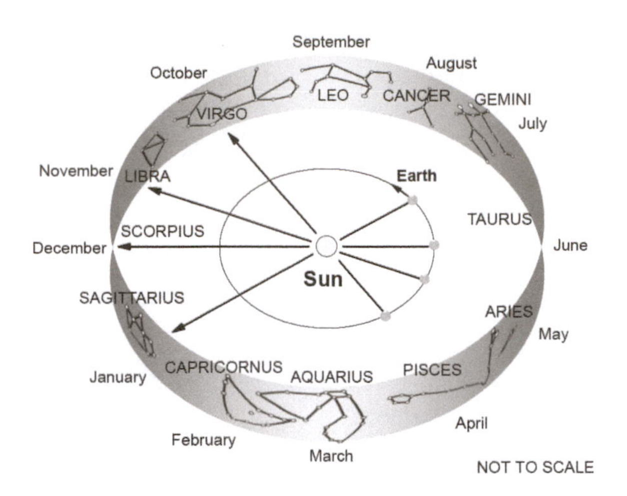

► **Figure 7-12** The constellations of the zodiac are the 12 constellations through which the ecliptics passed, in 140 AD. The ecliptic is now different. If you were born between November 30 and December 17 during any year of the 20th or 21st centuries, you were born under the sign of Ophiuchus, a constellation that is not one of these 12. (Image courtesy of Frederick Ringwald)

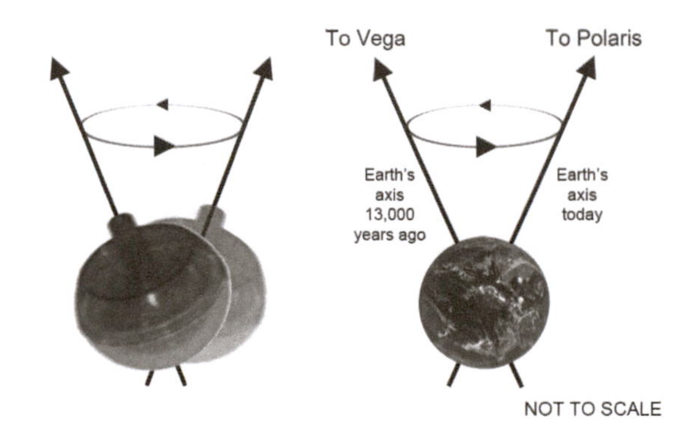

► **Figure 7-13** Precession causes Earth's axis to wobble like a toy top, over a 26,000-year cycle. Because of it, the constellations of the zodiac change over time. (Image courtesy of Frederick Ringwald)

Astrology also has two technical problems that any astronomer should know about

1. As the Earth spins, it wobbles like a top, in a slow (26,000-year) cycle called the Precession of the Equinoxes. This means that the position of everything in the sky, as seen from Earth, is slowly changing, including the constellations of the Zodiac. The ancient Greek astronomer Claudius Ptolemy defined these constellations in the 2nd century AD. They're where the Sun appears in the sky, for a given date, throughout the year. Ptolemy defined them in this way:

Aries	Mar 21 – Apr 19
Taurus	Apr 20 – May 20
Gemini	May 21 – Jun 20
Cancer	Jun 21 – Jul 22

Leo	Jul 23 – Aug 22
Virgo	Aug 23 – Sep 22
Libra	Sep 23 – Oct 22
Scorpio	Oct 23 – Nov 21
Sagittarius	Nov 22 – Dec 21
Capricorn	Dec 22 – Jan 19
Aquarius	Jan 20 – Feb 18
Pisces	Feb 19 – Mar 20

However, because of the precession in the nearly 2000 years since Ptolemy, the Sun today passes through the following constellations, on the following dates:

Aries	Apr 19 – May 13
Taurus	May 14 – Jun 19
Gemini	Jun 20 – Jul 20
Cancer	Jul 21 – Aug 9
Leo	Aug 10 – Sep 15
Virgo	Sep 16 – Oct 30
Libra	Oct 31 – Nov 22
Scorpius	Nov 23 – Nov 29
Ophiuchus	Nov 30 – Dec 17
Sagittarius	Dec 18 – Jan 18
Capricornus	Jan 19 – Feb 15
Aquarius	Feb 16 – Mar 11
Pisces	Mar 12 – Apr 18

That's right, although the present author was born in early June, which according to Ptolemy would make me a Gemini, the Sun was in fact in Taurus when I was born. The Sun signs have no validity whatsoever, because they just plain aren't correct.

Amusingly, if you were born between 30 November and 17 December, the Sun was in the constellation Ophiuchus. Ophiuchus, of course, is *not* one of the twelve constellations of the Zodiac defined by Ptolemy. If you were born between November 30 and December 17, and some lounge lizard asks you "Hey, baby, what's your sign?", you can say "Ophiuchus: crossed serpents!" Somehow, it seems appropriate.

2. New planets have been discovered since astrology was defined by Ptolemy. These include Uranus (discovered in 1781) and Neptune (1846). Over 100,000 asteroids, also called minor planets, have also been discovered, beginning with Ceres in 1801. So have hundreds of Kuiper-Belt Objects, beginning with Pluto in 1930. Does this mean that all horoscopes predating these discoveries were invalid?

Aside from technical problems, yet another objection to astrology is a philosophical one, based on a scientific observation. Astrology made much more sense in a cozy little Universe, purpose-built just for us, the way Ptolemy thought it was. Modern observations, however, show that the Universe is *huge*, much vaster than Ptolemy ever dreamed. We also now know that we are *not* at the center of the Universe, again as Ptolemy incorrectly thought. I therefore find astrology very hard to believe, in the context of what we now know about the Universe. Do you really believe that the Universe, in all its magnificent splendor, revolves around you? This idea would appear to me to be the height of egotism.

For more on problems with astrology, see the following sites:

http://dionysia.org/astrology/sun-signs.html
http://www.astrosociety.org/education/astro/act3/astrology3.html#defense

Different Kinds of Objections to Astrology

A **philosophical** objection:

Astrology says that people don't have free will, because they're subject to whatever fates are predestined in the stars.

A **moral** objection:

Astrology says that all people with the same circumstances of birth are all alike, which civilized people ought to condemn as a form of bigotry.

An ***invalid*** attempt at a scientific objection:

Astrology involves physical laws that are currently unknown to science, so it can't possibly be right.

A **scientific** objection:

Astrology just plain doesn't work, more often than random chance, as shown by the experiment in class.

Science and Technology

Science ***is not*** the same as technology.

Science is the search for knowledge, for its own sake. Technology is the application of knowledge.

It is possible to have highly developed science without much technology. The ancient Greeks valued learning, and were superb mathematicians. They didn't like working with their hands, though, because they saw it as beneath them, because they were a slave society. Their technology didn't get far, because of this.

It is also possible to have highly developed technology without much science. The ancient Romans were mainly concerned with expanding and defending their empire. They were good at building roads and forts that still exist. They didn't encourage science, because science causes change, and they didn't want change.

Science today owes much of its power to being *technology driven*. We are free from prejudices against technology, unlike the ancient Greeks. We also can see the value of learning more about the world around us, unlike the ancient Romans.

Modern science and technology **advance each other**.

Pure Versus Applied Science

The originators of modern science, including Copernicus, Kepler, and Galileo, were interested in science only as a way of discovering truth. It was Francis Bacon who recognized that "Knowledge is power." This illustrates the difference between pure and applied science.

Pure, or basic, science is simply science with no immediate practical application. Applied science is research undertaken to address practical problems. Much confusion with this exists from misunderstanding the word *pure*.

The gap between pure and applied science—or in other words, between the kooky and the commercial—is closing fast.

> "Technology's most important obligation is to get out of the way. Machinery should make life easier. Useless features and bad design make technology a self-important nuisance instead of a help."
>
> –David Gelernter

"Technology always has unforeseen consequences, and it is not always clear, at the beginning, who or what will win, and who or what will lose."

– Neil Postman

"There is a limit to the promise of new technology, and that it cannot be a substitute for human values."

– Neil Postman

Some Fields of Science Mentioned In This Course

(by no means a complete list)

Physics is the study of *how things work*.
In other words, physics is the study of matter and energy. It's difficult to understand much that goes on in this world without it.

Chemistry is the study of *what things are made of*.
In other words, chemistry is the study of substances, and how molecules and atoms combine to make substances.

Astronomy is the study of the sky, particularly the positions and motions of objects in it, including the Sun, Moon, planets, and stars.

Astrophysics is the study of how things work in the sky. Astrophysics = astronomy + physics.
The distinction between astronomy and astrophysics made sense when astrophysics was new, around 1900. It doesn't today: astronomy and astrophysics are now so interconnected, it's hard to tell them apart anymore. This is *good*—astrophysics has made excellent progress.

Cosmology is the study of the whole Universe, or the cosmos, and its origin and history. (This should *not* be confused with cosmetology.)

Astronautics is the study of space flight. This should not be confused with astronomy: most astronomers do not fly in space or become astronauts.

Dynamics is the study of motion, a branch of physics.

Geology is the study of Earth.

Biology is the study of life.

Biochemistry is the study of the chemistry of life.
(Etc.)

The Nobel Prize

Alfred Nobel was a chemist, and he invented dynamite. When he died, he left his fortune to fund prizes awarded annually since 1901. The Nobel Prizes are awarded on December 16, the anniversary of Nobel's death, in Stockholm, Sweden. The Nobel Prizes are awarded for:

1. Physics,
2. Chemistry,
3. Medicine or physiology (how the body works),
4. Literature, and
5. Peace.

In 1970, a prize for economics was added.

The Nobel Prize for physics, chemistry, and medicine or physiology are the most famous prizes in the world given to scientists. This is probably because of the large amount of money awarded, now $1.4 million, tax-free. When Einstein got his in 1921, it was $20,000.

The Nobel Prize probably hasn't affected the course of science much, since science has grown so much since Nobel's day. To award all the deserving science done today, there would need to be more than ten Nobel prizes in each category. For example, Jonas Salk never got one for developing the polio vaccine.

The Nobel Prize also hasn't funded much scientific research, since research today can easily cost many millions of dollars. Marie and Pierre Curie, who won the Nobel Prize in Physics in 1903, were famous examples of when it did: their lab before they won the prize was unheated and famously wretched.

The Nobel Prize *has* raised public awareness of science. There is no Nobel Prize for Astronomy, but the physics prize has been awarded for work in astronomy several times. Much of this work will be mentioned later in this book.

Ethics in Science

Ethics is the study of what is right and what is wrong. When it comes to ethics, science is no different from any other human activity. If scientists can foresee that their work will cause harm, they should *stop doing it*. If they don't, any "beating of the breast" afterward will "make a hollow and unconvincing sound," as Peter Medawar wrote in *Advice to a Young Scientist*.

What makes the study of ethics interesting isn't when right and wrong are clear: it's when there is a gray area between them. Is it ethical for scientists to develop new weapons systems when their country is under attack? Does the greater good really justify any cost that might befall an individual? Does the end really justify the means? Does the promise of benefit give scientists the right to do anything they want?

> ...Few scientists have a burning desire to rule the world; typically, they don't enjoy managing people and research budgets.
>
> –Sidney Perkowitz, in *Hollywood Science*

Real scientists, of course, may *not* do anything they want. Scientific research is governed by law, like any other human activity. It can be wrong to impose a ban on scientific advancements because they might be harmful when misused. We don't let just anyone take X-rays, or do surgery: medical people are licensed to practice, and they can lose their licenses if they are careless or derelict in their duty.

> The heart must mediate between the head and the hand.
>
> – Fritz Lang, in *Metropolis*

A common theme in science fiction is that it's not nice to play God. Human beings don't have the wisdom for it. The search for knowledge is a fine thing, but it must never be put above human concerns. The Hippocratic Oath, which many doctors take, includes: "First, do no harm." Intentionally causing harm to test subjects in scientific experiments, or deceiving them, is wrong. It is also illegal, because there have been abuses of this kind in the past. People don't like to be experimented on without their consent. Even with their consent, people don't like to be treated like numbers, or carelessly, or callously, or disrespectfully. Confidentiality of personal information, such as medical records, requires care. People *especially* don't like anyone messing with their children, their families, or their sex lives!

> Science can tell you how to clone a Tyrannosaurus Rex. Humanities can tell you why this might be a bad idea.
>
> –University of Utah College of Humanities

It is ***not*** true that "you can't stop progress." The Supersonic Transport (SST) was a design for a faster-than-sound airliner, in the early '70s. It was never developed in the United States, because the American public didn't like how noisy and costly it would have been. (Another supersonic airliner, the Concorde, was developed in Britain and France. It did fly, but it was taken out of service because it never made a profit.) In the early 20th century, some people expected that everyone would own their own helicopter by 1980. This never came true because it would have been costly, inefficient, and dangerous. Flying cars are a bad idea, which your present author hopes never comes true. With the way that people drive, I don't want them flying!

As with ethics in general, simplistic "snap" judgments rarely work. What is needed is careful reasoning. Science and technology may put food on the table, but they do *not* provide wisdom by which to live. This is one reason that areas of human thought outside of science are important.

Langmuir's Laws of Science Practiced Badly

Irving Langmuir won the Nobel Prize in Chemistry in 1932. He was the first scientist to use the word *plasma* to mean hot gas, and helped to invent television.

In 1953, Langmuir gave a talk on cases in which science had been practiced badly, because of improper use of scientific method. In this talk, he presented some observations, or "laws," of what can happen when science is done badly.

You know it is science done badly when...

1. The claimed effect is often very faint, barely above the background.
2. Seemingly nothing can be done to make the claimed effect stronger.
3. There are claims of great accuracy.
4. Fantastic theories contrary to experience are suggested.
5. Criticisms are met with excuses apparently thought up on the spur of the moment.
6. The ratio of supporters to critics starts at about 50-50, and gradually falls to zero.

Why are these important? They show how to misuse scientific method. Science, like any human invention, is a tool. As Jay Baldwin noted, "If you don't know at least three ways to misuse a tool, you don't understand how to use that tool."

One of Langmuir's examples was the false discovery of N-rays, the existence of which was claimed by a scientist who was jealous of Wilhelm Roentgen's discovery of X-rays. Other examples of science done badly that illustrate Langmuir's laws were Percival Lowell's claims of having discovered "canals" on Mars (which spacecraft to Mars later showed do not exist), the "cold fusion" claim of 1989, and most pseudoscience.

Scientists are human. Although science can make remarkable progress, be aware that scientists don't always do what's right.

Graphics in Science

Mathematics is important in science, because the Universe follows mathematical laws. Mathematics can be hard to understand, but *drawing a picture* can help enormously. Avoid presenting results as just a table of numbers: use the numbers to *draw a picture*.

Edward Tufte has written several books about this. They include *The Visual Display of Quantitative Information*, *Envisioning Information*, *Visual Explanations*, and *Beautiful Evidence*. Another good book on graphics is *Thinking with a Pencil*, by Henning Nelms. Another is *How to Lie with Statistics*, by Darrell Huff. Not all statistics are lies, and I don't want to teach you to lie: I want you to know how to recognize it when someone is lying to you.

Here are some important principles for graphics:

- Remember that the purpose of any plot is to show the reader the *connections* in the data. A plot should invite *comparison* by the eye.
- Don't make your readers puzzle out your graphics. Avoid using the "Legend" function, even though it is the default when making a scatter plot in spreadsheet software: it forces your readers to have to puzzle out your graphics.
- Always show *all* the data.
- Always label axes clearly, with the correct units.
- Always show the *name* of the person who produced the data, and the graphic.

Some Logical Fallacies, and How to Avoid Them

How to Think Like a Scientist

One of the best things that any university can do for its students is to give them practice in thinking rationally. A *fallacy* is a false idea. Fallacies are often called logical fallacies, because many of them involve errors in logic. Often, people who are otherwise intelligent think things that are demonstrably wrong, because their thinking goes wrong.

The following is a list of common fallacies. It is by no means a complete list, of all possible false ideas. Sadly, there seems to be no end of nonsense designed to fool us all.

Many of the following are used in advertising. Buyers should beware of advertisers who stoop to using them. Many fallacies have Latin names, because they've been in use since ancient times. They are still in use today, and may continue to exist until human beings become smart enough to recognize how wrong they are. Don't let them harm you!

- *An argument from authority* has little weight in science. This is because nature is the ultimate authority in science: really, nature is the only authority in science. An example is, "I must be right, because I am the professor." This is wrong because, despite their often lengthy educations, professors can indeed be wrong. Nobody's perfect.

Arguments from authority can take many forms:

- An *ad baculum* argument is an appeal to a threat. Examples include: "Vote for me, or you will be killed by terrorists," or, "Join our religion, or your soul will be damned," or, "Fund my research, or you will be killed by an asteroid impact."
- An *ad crumenam* argument is the false belief that money alone determines what is right. An example is "That book must be true, because it was a best-seller." Another example is to ask, "If you're so smart, why aren't you rich?"
- An *ad lazarum* argument is the opposite of an *ad crumenam* argument. It claims that poor people have special knowledge or wisdom or virtue, only because they are poor. Whenever a politician talks about "common sense," it is often an *ad lazarum* fallacy.

- An *ad misericordiam* argument is an appeal to sympathy. It is therefore the opposite of an *ad baculum* argument. An example of an appeal to sympathy would be a student who is late for class often, doesn't do the homework, doesn't do the reading assignments and so can't do well on exams, and then comes to the instructor and cries. This behavior is very childish: when students try this with me, I am *not* sympathetic, because they are taking up resources that other students could use better.
- An *ad hominem* argument attacks the speaker, instead of the speaker's argument.

  Scientists should refrain from calling other scientists incompetent. Even the best scientists, including Newton and Einstein, were sometimes wrong. Even if a scientist is incompetent, the scientist may still be right.
- *Poisoning the well* is a type of ad hominem argument. Suppose that a scientist, Doctor X, goes to a conference. Before a rival of his, Doctor Y, can say anything, Doctor X tells the other scientists, "Doctor Y is totally incompetent. Don't believe any results that he presents." This isn't fair, is it? It's called poisoning the well. Doctor X has tried to cast doubt on Doctor Y's results, before Doctor Y even had a chance to present them.
- *A tu quoque argument* is also called "two wrongs don't make a right," or a "you-too-ism." It is a type of ad hominem argument.

  An example is the objection against using methanol as an automobile fuel, because it is poisonous if swallowed. This is a fallacy, because gasoline is not a health drink.

  Another example is saying that global warming must be wrong because warnings of the consequences of it have been advocated by Al Gore, only because you don't like Al Gore. This doesn't disprove global warming. (That said, Al Gore damaged his own case, by overstating fears of global warming. It still doesn't make them go away.)

  Another example is that the worst-ever nuclear accident, at the Chernobyl reactor in 1986, is estimated to have caused about 10,000 deaths, mainly from cancer over the long term. Burning coal to generate energy causes about the same number of deaths every year, from the air pollution from burning the coal. This is not an argument that we should prefer nuclear power over coal: it argues that both have problems.
- An *ad ignorantiam* argument is also called an "appeal to ignorance." This fallacy has two forms:

  1. It must be true, because it has not been proved false.
  2. It must be false, because it has not been proved true.

  Neither form is true. Absence of evidence is not evidence of absence.

  An example of the first form is a fallacy sometimes used by teachers: "It can't be wrong, because no one ever complained about it before."

  An example of the second form is the fallacy that says that just because scientists can't prove that UFOs don't exist means that they must be spacecraft from other worlds. The fallacy is that, even if UFOs exist, it doesn't necessarily mean they are spacecraft from other worlds.

  Another example is a favorite of UFO enthusiasts: "Can you prove it didn't happen?" This is a fallacy because this isn't the point: the burden of proof lies with the person making the extraordinary claim, not with the skeptic.

  Another example is the fallacy that something must be true, because it is difficult to understand, or complicated. This fallacy is surprisingly common, particularly in science. Not all science is difficult to understand, nor does it need to be. In principle, all science should be understandable by a rational mind, although admittedly this may take time.

  An example of this is saying, "That must be wonderful, I can't understand a word of it." If you can't understand a word of it, it should raise suspicion, not admiration.

- *Proving a negative* is closely associated with this. It can be difficult, or impossible, to prove that anything does not exist. I can't prove that unicorns do not exist: there may be an invisible one behind you now. I can't prove that the fabled "lost" continents of Atlantis or Lemuria do not exist, but there is no geological, oceanographic, or archaeological evidence that they ever did exist.

  (Occam's razor may provide an explanation here. The legend of Atlantis may have its origin in poorly recorded accounts of the explosion in about 1600 BC of the volcano on the island of Thera, now called Santorini. The tsunami from this destroyed the Minoan civilization on the island of Crete.)

- An *ad antiquitatem* argument is the fallacy that if it's old, it must be valid. One example is the worst reason to do anything: "We've always done it that way."

- An *ad novitatem* argument is the fallacy that if it's new, it must be valid. Newer isn't necessarily better: better is better.

- *A post-hoc argument* confuses correlation with causation. Just because some things sometimes go together doesn't necessarily prove that one causes the other. Here's an example:

  Psychiatrist: "Why do you jump up and down that way?"

  Patient: "It is to keep the vampire bats away."

  Psychiatrist: "But there aren't any vampire bats here."

  Patient: "Effective, isn't it?"

  "*Post hoc*" means "after this." The full saying is "*post hoc ergo propter hoc,*" which is Latin for "after this, therefore because of this." Another example is much of mythology, such as the myth of Dionysus, discussed at the beginning of the chapter on scientific method.

  Another example is the fallacy that there is no way to observe a solar eclipse safely, because every time there is a solar eclipse, people get blinded from observing it improperly. People do get blinded from observing solar eclipses improperly, but it's not true that there is no safe way to do it.

- *Weasel words* have built-in meanings, or connotations. When Italian astronomer Giovanni Schiaparelli reported in 1877 that he saw linear features, or channels, on the planet Mars, this was translated into English as "canals." Canals are artificial structures, dug in the ground. Because of this, some English speakers, including Percival Lowell, assumed that Mars must be inhabited. Modern spacecraft have shown that it isn't.

  Another example is when one is told that "people" say something. Which people? Was this researched, or was it just made up?

  Likewise, beware the words, *doctor approved.* Which doctor? Is there more than one doctor? Is this the entire medical profession, or a governing body of doctors, such as the American Medical Association? Or is it only one doctor? Is this even a doctor of medicine, with a current license to practice medicine? No idea is so ridiculous that one can't find at least one Ph.D. somewhere who agrees with it.

- *The fallacy of personal validation* recognizes that flattery can be a powerful way to fool people. Have you ever noticed that newspaper astrology columns almost always say only good things?

  An example is graphology, the pseudoscience of telling personality traits from handwriting. It's notoriously inaccurate. When I was 13, I was fooled by a handwriting analysis computer at a state fair. It told me that I was "dominant, outspoken, and attractive to the opposite sex." *Of course* I believed that, and I was bitterly disappointed later to learn that it was a sham: that machine had told me exactly what I wanted to hear.

- *False analogy*: Analogies can be useful as ways of explaining difficult ideas. One must remember, though, that sooner or later all analogies break down. This is because no two situations are exactly the same.

  Two fallacies often made when using analogies are a sweeping generalization and its opposite, a hasty generalization.

  - A *sweeping generalization* is also called *dicto simpliciter*. It happens when a general rule is applied to a particular example where it doesn't apply. An example would be to say that everyone would benefit from a college education. Not everyone wants to go to college, though: these people may not benefit from a college education.
  - A *hasty generalization* is generalizing from too few examples to too many. If you see a brown cow, it is incorrect to conclude that all cows are brown. The larger the sample size, the more representative it may be, if it's a fair sample.

- *Confusing quantity with quality:* It is a sweeping generalization to suppose that "more must be better." Too much of a good thing can be bad. For example, doctors sometimes get child patients who are being poisoned from taking too much Vitamin A. This happens because these children have well-meaning parents who are giving their children too many vitamins. These parents' rationale for doing this is that if vitamins are good, then more vitamins must be better. This is false: too many vitamins can be toxic. Quantity isn't the same as quality.

  Another example is being unable to distinguish weather and climate. It's not possible to use a single local, low temperature to disprove global warming. Measurements taken worldwide over the past century indicate a steady trend to higher temperatures, even when one accounts for the growth of cities and their heat-island effect.

- A *hypothesis contrary to fact* tries to draw a conclusion from a premise that is false or unproved in the first place. An example would be to say that if Napoleon had a B-52 "bomber aircraft" at Waterloo, we'd be speaking French today. But Napoleon didn't have a B-52 at Waterloo, or even an observation balloon. Historians avoid supposing what might have happened in history, because it didn't happen that way. Many other things might have prevented it anyway: Napoleon might have been defeated, even if he did have a B-52. (Suppose that the Duke of Wellington had a surface-to-air missile?)

- *Special pleading*, also called an *ad hoc* fallacy, introduces new, custom-made excuses for every logical refutation one can make. These can become elaborate, and bizarre.

- An *excluded middle* is a fallacy that considers only the extremes in a range of possibilities. This isn't necessarily true: moderate ideas can be true, even if their extremes are false. Didn't anyone ever tell you that too much of anything isn't good?

- A *slippery slope* is a type of excluded middle. A slippery slope claims that an idea must be false (or true) if its extreme is also false (or true). An example is the old saying, "give them an inch and they'll take a mile." It doesn't follow that moderate positions must lead down the slippery slope to the extreme.

- *The gambler's fallacy* is common because many people don't understand the nature of random numbers. Random numbers have no connection to each other, by definition.

  Every time a slot machine is played, the chance of it paying out is the same. There is nothing keeping track of how many times you've played it. Therefore, the chance that it will pay out is the same, whether you've put one quarter into it, or whether you've put a thousand quarters in it. (Admittedly, the jackpot could be larger if you'd put a thousand quarters into it. Alternatively, you could keep your quarters in the first place.)

  An example of the gambler's fallacy would be if a police officer were to get shot in the line of duty, to suppose fallaciously that this officer can't get shot again, because although the chances of getting shot once are pretty high, the chances of getting shot twice are tiny. This is a fallacy: the chances of getting shot the second time are the same as for getting shot the first time, all else being equal. (This author hastens to add that he hopes that no police officers come to harm in the line of duty.)

- *False choices* are a favorite of politicians, and are related to exclude middle and slippery slope fallacies. An example of a false choice is to say, "If you're not with us, you're against us!" This is a fallacy, because being for them or against them aren't the only two choices. One might be neither for them nor against them.

  Another example of a false choice is: "If they can put a man on the Moon, why can't they...[do something else]." This is a fallacy because putting a man on the Moon was a specific, well-defined, technically plausible goal (although admittedly one so difficult it required $25 billion). Often, when someone says, "If they can put a man on the Moon, why can't they...[do something else]," the something else is not specific or well defined. Do you really think that if we can put a man on the Moon, we can achieve world peace? I can tell you how to get to the Moon: use a rocket with enough energy to overcome Earth's gravity, and computers to steer the rocket. I have no idea how to achieve world peace.

  Another example of a false choice is: "What does science know? Science can't even cure the common cold." This is a fallacy because the seriousness of a disease is *not* related to how easy it is to cure it. Some deadly diseases, such as bubonic plague, can be cured with antibiotics; some mild diseases, such as the common cold, are complicated.

  The deeper issue here is that *just because we don't know* ***everything*** *doesn't mean that we don't know* ***anything***. Also, just because we don't know the answer today doesn't mean that we will never know the answer.

- *Begging the question,* also called *a leading question,* is a type of false choice. Examples of leading questions include when a police officer says, "Do you know why I stopped you?" or, "Do you know how fast you were going?" Other examples are: "Do you still pick your nose?" or "Does Gary still stick ferrets in his pants?"

  TV interviewers do this disturbingly often. An example of this is:

  "I'm JUST asking a question..."

  What they mean when they say this is: "I'm JUST asking a question, even though it's entirely based on false premises that convey a suspicion of guilt where no real offense exists."

  A closely related leading comment too often used on TV is:

  "I'm sorry but that's JUST my opinion..."

  What they mean when they say this is:

  "I'm sorry but that's JUST my opinion, also based on flawed assumptions about cause and effect."

  Yet another example too often used on TV is: "I'm just saying..."

  This is used to appear to justify any position, no matter how ridiculous.

- A *straw-man argument* is another type of false choice. In a straw-man argument, one creates the illusion of refuting an opponent's argument by mischaracterizing the argument, and then refuting the mischaracterization.

  A classic example of a straw-man argument is the myth that a bumblebee can fly because it doesn't know that it can't fly. This was faulty logic from the beginning. When this myth was made up in the 1930s, *people* didn't understand how bumblebees fly. We do now: the bumblebee has been vindicated.

Another example of a straw-man argument is to ask: How can global warming be true? Scientists were wrong in the 1970s when they predicted that a new ice age would begin soon, because of global cooling.

This argument is false, because scientists in the 1970s *didn't* predict that a new ice age would begin soon. During the 1970s, scientists predicted that a new ice age would begin *10,000 years in the future.* It's still expected to happen, because of the same changes in Earth's orbit that caused previous ice ages—but problems with global warming will not negate this, and are becoming significant *now*, or at the latest, *within the next century.*

▶ *Cherry picking*, also called *cooking the data*, is the fallacy of including information that agrees with one's argument, but ignoring or omitting data that disagree.

In 2010, some people who wanted to disprove global warming showed a list of cities that experienced unusually low temperatures. During this time, Seattle and Portland were having their highest temperatures ever. Of course, this list didn't mention them. What matters is the average temperature for all cities, compared to the average in previous decades.

▶ *A circular argument* uses a term in its own definition. This can be noticeable in badly written dictionaries.

An example of a circular argument is, when I was 6 or 7 years old, I asked my Dad why the ocean was blue. He said, "Because it's reflecting off the sky." I said, "OK, why is the sky blue?" He said, "Because it's reflecting off the ocean." This doesn't explain where the blue came from in the first place.

(To be fair to my dear old Dad, he then told me the answer, and we had a good laugh about it. If he hadn't known the answer, he wouldn't have pretended that he did, but he wasn't above having a little fun, if he did know the answer.)

Another example of a circular argument is to claim that, "Science is what scientists agree is science." This is an oversimplification, because it doesn't answer the questions of who are scientists, and why do they agree.

▶ *A tautology* is a type of circular argument. An example of a tautology is the Anthropic Principle, which in its strong form says that because the Universe is capable of supporting life, it must have developed for this express purpose. As Alfred Russell Wallace noted, "Such a vast and complex universe as that which we know exists around us, may have been absolutely required...in order to produce a world that should be precisely adapted in every detail for the orderly development of life culminating in man." Another way to say this is, "We exist, because we exist."

This is a fallacy, because if the Universe were not capable of supporting life, no one would be around to ask why not. Carl Sagan made fun of the Anthropic Principle by calling it, "The Anthropocentric Principle."

▶ *Confusing a necessary condition with a sufficient condition:* A necessary condition is something required for a given outcome. A sufficient condition is something that will make the outcome happen. Don't confuse them.

An example of this is that going to class is necessary to get a good grade in a college course, but it isn't sufficient. A student also needs to learn the material, and to be able to demonstrate it, on exams and other class exercises.

Another example is that, if one asks many people these days whether humans are alone in the Universe, they will point out how enormous the Universe is, and remark that it isn't plausible that we could be alone in any place so vast. This may be appealing, but it's a fallacy. We do know of things that are so rare, for example, black holes, that there aren't more than a few hundred of them per galaxy. (If there were more, we would know it, from the X-rays they give off when things that fall into them become very hot.) The fact is that we don't know how common intelligent life is, throughout the Universe: it may be common, or it may not.

- *Plausibility and evidence*: SETI is the Search for Extra-Terrestrial Intelligence. It is the science of searching the sky for radio, laser, or other signals from intelligent life forms in space. So far, it still hasn't found anything, but the search continues. Why is this considered science, even in the absence of evidence? It is because SETI is plausible: life developed on Earth, we developed radio and lasers, and the Universe is very large, so it is plausible that intelligent life occurred more than once in the Universe.

  On the other hand, claims that Project Apollo was a hoax are not plausible. For Apollo to have been a hoax, it had to have escaped the notice of the Soviet Union, which tracked every spacecraft with their own radars. The Soviets would have loved the opportunity to humiliate the United States, since they were constantly threatening us with nuclear war.

  Apollo was also carried out under the watchful eye of every television network and nearly every journalist in the world. NASA was run as an open program, with extensive public and press access to its every activity. If Apollo was a hoax, it's remarkable that the 400,000 people who worked it have kept so quiet about it for all these years.

  To this day, no authentic part of the supposed sound set on which the hoax was recorded has ever turned up. Neither has any incriminating memo from the President of the United States to the Administrator of NASA. There are also 842½ pounds of intelligently sampled Moon rocks returned by the astronauts, many of which have been on public display for many years. As Neil Armstrong observed, "It would have been harder to have faked it than to just gone ahead and done it."

- *Confusing facts, inferences, and judgments*: Facts are items of information that are objectively true. Facts can be checked, and will be the same for everyone. An example of a fact is that Earth is round. ("Round" here means within 1% of being a sphere.)

  An inference is a conclusion derived from observation of facts. An example of an inference is that Columbus's idea that it might be possible to reach the east by sailing west, because Earth is round.

  A judgment comes with an opinion, which may or may not be based on facts and inferences. An example of a judgment is that sailing west to reach the east might open up a lucrative trade route, and that we would enjoy the profits we'd make from it.

  You're entitled to your own opinions. You are not entitled to make up your own facts. You need to get your facts by observing the reality around you, honestly.

- *A non-sequitur* means "it doesn't follow." A non-sequitur isn't a logical argument: it's just two phrases that appear to be connected logically, only because they're next to each other.

  Non-sequiturs are often used in advertising. An example is "Buy Brand X toothpaste, because it's got hexachlorophene!" Hexachlorophene may sound scientific and impressive, which is why non-sequiturs work so often in advertising. This is a fallacy, though: hexachlorophene is just a big word that doesn't mean anything to most people.

  A big, funny word isn't an explanation. People have been making this mistake since ancient times. Aristotle, the Greek philosopher in the fourth century BC, did this more than once in his writings. For example, he wrote that opium makes people sleep because it has "somatic properties." Try looking up "somatic" in a dictionary: it means "inducing sleep."

  Another example we've all heard is that grass is green because of chlorophyll. But why is chlorophyll green? Chlorophyll does reflect the wavelengths of green light more than at other wavelengths, but why should the substance that converts solar energy into plant food be green? No one really knows: it may be telling us about something that happened early in the history of life on Earth. Don't be surprised if, on other planets, the vegetation will be black (to absorb as much energy as possible from the local star), or blue (to reflect back as much energy as possible, from a local star that is too hot).

Another example of a non-sequitur is not being able to identify something as X is taken to mean it must be Y. This is a fallacy, because it might be Z. For example, just because one can't identify a UFO doesn't mean that it must be an alien spacecraft, since it might instead be a natural, but unusual, sky phenomenon.

- *"All points of view are equally valid."* This fallacy is the opposite of an argument from authority. It can be particularly dangerous in TV news. Disturbingly often, TV news will interview a scientist, and then they'll interview a crank. Then, the TV news service will claim that their coverage is "fair and balanced," by saying, "we report, you decide."

  Do you think that the "opinions" that 2 + 2 = 4 and that 2 + 2 = 5 are equally valid?

  If you do think this, do you have change for five bucks? How about change for 50 bucks?

- *Magical thinking*: Magical thinking is the belief that things are connected, which cannot possibly have a connection. (Magical thinking is therefore a type of *post hoc* fallacy.) Related to this is the fallacy that good things can come true only by "believing in" them, or that bad things will disappear only by ignoring them or wishing them away.

  It's easy to assume that only children and insane people believe magical thinking, but it's surprisingly common among otherwise sane adults. Ancient people worldwide believed that spirits inhabited every tree or stream, and old beliefs die hard, even today.

  Examples of magical thinking include any kind of belief in good luck or bad luck, or lucky charms, or jinxes. An example is the sports fan who observes that the last time he walked into the stadium backward, his team won. He therefore falsely thinks that his walking in backwards caused his team to win the game, and that if he walks in backwards again, his team will win again. His walking in backward had nothing to do with his team winning.

  Another example of magical thinking is a student who thinks he needs a big, fancy graphics calculator, when all he needs it for is simple arithmetic. He can't use the graphics functions, he doesn't even need them, and he certainly can't program the calculator. This student seems to think the calculator itself has magical power. Of course, a fancy calculator is useless, unless it's in the hands of someone who understands how to use it.

- *Dismissing an indirect argument only because it is indirect.* Much of science is indirect. This is sometimes held against it, and it shouldn't be. Seismology is an example: Earth is not transparent to light, so unaided human eyes can't see what's inside Earth. Geologists still know quite a lot about what's inside Earth because their instruments can detect sound, also called seismic waves, which travel through Earth. What confirms this is that geologists can use this to tell where in the ground one should drill to find oil, and the oil is real enough.

  Another example is quarks. Even in principle, a free quark, by itself outside of a proton or neutron, cannot exist, because to isolate one would require so much energy, other quarks would be made, in combination with the first quark. Nevertheless, the quark model does explain the properties of protons, neutrons, and other particles well. It also predicts that other particles exist, and these particles have been observed.

- *Outright lies*: These are often non-sequiturs, and are particularly nasty. Outright lies don't need to follow any logic at all. They just need to sound snappy. An example of an outright lie is the story about how NASA invented pressurized pens, also known as "the astronaut pen."

  The story goes that, when faced with the problem of astronauts having to write while weightless in the cabin of an orbiting spacecraft, NASA invented a pen in which the ink inside the pen was under pressure. Because of this, the ink would still flow out of the pen when the astronaut wrote: it didn't need gravity for the ink to flow. NASA spent about $1 million developing the pen. The pen worked, and has since enjoyed some success as a novelty item. The Russians, when faced with the same problem, used a pencil.

This story is a lie. NASA did *not* invent the pressurized pen: they'd been around for some years. Also, the Russians *didn't* use a pencil: the Russians used pressurized pens too, since shavings from sharpening pencils are a choking hazard in weightlessness.

Another example of an outright lie is to claim that "anything is possible," or that "nothing is impossible." Do you think you can get drunk while drinking ginger ale? How about leaping over a tall building in a single bound? How about bending the steel bars of a padded cell?

Another example of an outright lie is the claim that it is possible to prove that 1 = 2. You might have seen this done in high-school math classes, but it's not true: it depends on making a division-by-zero error. Many people miss this error, because they need to be fluent in algebra in order to spot it. This is related to an *ad ignorantiam* argument.

- A *delusion* is a fallacy that a person continues to hold, even after it is rationally proved wrong by compelling evidence, or shown to have no evidence supporting it. Delusions are a form of insanity: beware of them.
- *Mass hysteria* is when many people become convinced of a delusion, and start to act like a panicky herd of animals. The most dangerous thing on Earth is large numbers of frightened human beings: an angry mob is a bad place to be.

> "Men, it has been well said, think in herds; it will be seen that they go mad in herds, while they only recover their senses slowly, and one by one."
>
> –Charles Mackay, in *Extraordinary Popular Delusions and the Madness of Crowds*

The Phantom Gasser of Mattoon was an example of mass hysteria. In the 1940s, numerous reports were made in the town of Mattoon, Illinois, of a maniac who was attacking sleeping citizens in their homes by pouring under their doors a gas such as chloroform, used by doctors to put patients to sleep before surgery.

Police investigation found no physical evidence for any of the claimed attacks. Indeed, the police complained that there was nothing more than "vivid imaginations." When groups of people come to believe a made-up story, even if it's ridiculously false, the story can seem to take on a life of its own.

Another example of mass hysteria is how, when they were children, your present author and his friends would get very excited about UFOs. The things we fooled ourselves into seeing were nothing more than eye floaters, which are the shadows of dust particles on the surface of the eye.

Mass hysteria, often called "groupthink," isn't something that always happens to other people. Chances are good that it has happened to you.

This is by no means a complete list. Murphy's Law states that, "Anything that can go wrong will go wrong." This applies to thinking, too. Don't let it happen to you!

CHAPTER

8

The Motions of the Planets and the Beginning of Science

"To be ignorant of what occurred before you were born is to remain always a child."

– Cicero

"The past is never dead. It's not even past."

– William Faulkner

The **Ancient Chinese** had the most advanced astronomers in the world, until the invention of the telescope. Their earliest observations may have preceded 6000 BC, during the New Stone Age. They certainly had sophisticated observatories by 2300 BC. They had a complex lunisolar calendar, and they kept records of their observations for many centuries.

The **Babylonians** lived in Mesopotamia, now in Iraq. They began observing the Sun, Moon, and planets after 3500 BC, and kept records of this after 1800 BC. They originated the seven-day week, as well as dividing circles into 360°, hours into 60 minutes, and minutes into 60 seconds.

The **Ancient Egyptians** had an astronomical calendar with 365 days in a year in 2400 BC. They aligned their pyramids to the stars, and originated dividing days into 24 hours. The Library at Alexandria in Egypt was the center of learning of the classical world, beginning in about 300 BC.

The **Ancient Greeks** (585 BC–529 AD) began asking scientific questions about nature, such as "What is the world made of?" Scientific questions seek answers that can be understood rationally, from observations of the physical, natural world—not from the supernatural, intuition, or revelation. Science also follows ideas wherever they may lead. One's preferences, or preconceived notions, must not interfere with any investigation, for it to be considered scientific.

The Greeks didn't develop true science, though. They thought that any work with their hands was beneath them, so they couldn't test their ideas by experiment. Science is fundamentally something one does with one's *hands*. This is why lab should be an essential part of any science course.

Thales of Miletus is widely considered to have been the first Greek philosopher. He was famous for predicting the solar eclipse of 585 BC. From this, he concluded that *every observable effect has a physical cause.* He also argued that the world is made of water. Thales thought this because he knew how mud can gather in a river. He also had heard the Babylonian myth in which their god Marduk made land appear from the waters. Thales tried to find a *natural* way for this to happen, without involving Marduk. We now know the world is *not* made of water, but Thales deserved credit for asking a scientific question, "What is the world made of?" *Thales expected a rational answer, which can be observed and tested.*

► **Figure 8-1** Mercury and Venus (shown here) move through Earth's sky, over several months, in a *"yo-yo"* pattern. (Image courtesy of Tunc Tezel)

► **Figure 8-2** Mars (shown here), Jupiter, and Saturn move through Earth's sky, over several months, in a *retrograde loop* pattern. (Source: NASA/JPL)

▶ **Figure 8-3** Greek natural philosophers thought that any work with their hands was beneath them. Since this meant that they couldn't do experiments to test their ideas, they never invented science, even though they did come close. (ScienceCartoonsPlus.com, Used by permission)

Aristotle was the most influential of the Greek philosophers. He wrote widely, on ethics, politics, logic, and nature: one might read his works in many college courses. He reasoned (correctly) that Earth was round, but (incorrectly) that it didn't move. He thought that reason alone was enough to understand nature, so he missed the importance of experimenting to test one's reasoning. He did stress *empirical observation*, which is watching nature to see what it does. He died in 324 BC.

Aristotle had four proofs for why Earth is round, all of which are correct:

1. During a lunar eclipse, the shadow the Earth casts on the Moon is *round*. (This is different from the Moon's monthly cycle of phases: lunar eclipses are rare.)
2. Ships sailing out to sea disappear over the horizon hull first and then masts second, as they sail beyond the horizon. When sailing home, they re-appear masts first and hull second. (This is often attributed to Columbus, but Aristotle knew it.)
3. As one travels north, the circumpolar stars become higher in the sky. Another ancient Greek, Eratosthenes, would use this to measure the Earth's circumference, surprisingly accurately.
4. Why we have time zones: the Sun rises earlier in England than it does in the eastern United States, which rises earlier than in the western United States, which rises earlier than in Hawai'i. How could Aristotle have known about this, without modern telecommunications? He noticed that solar eclipses were recorded as occurring earlier in the day, the farther west they were.

Aristotle also had many arguments for why he thought Earth did not move. All of these later turned out to be incorrect: for example, Aristotle thought that if Earth rotated, there would be a wind opposite the direction of rotation. He also thought that falling objects would be deflected sideways, for the same reason. Both these arguments are now known to be wrong, because Aristotle didn't know about inertia. Aristotle also never saw the connection between physics and mathematics: he never understood that the laws of nature are mathematical.

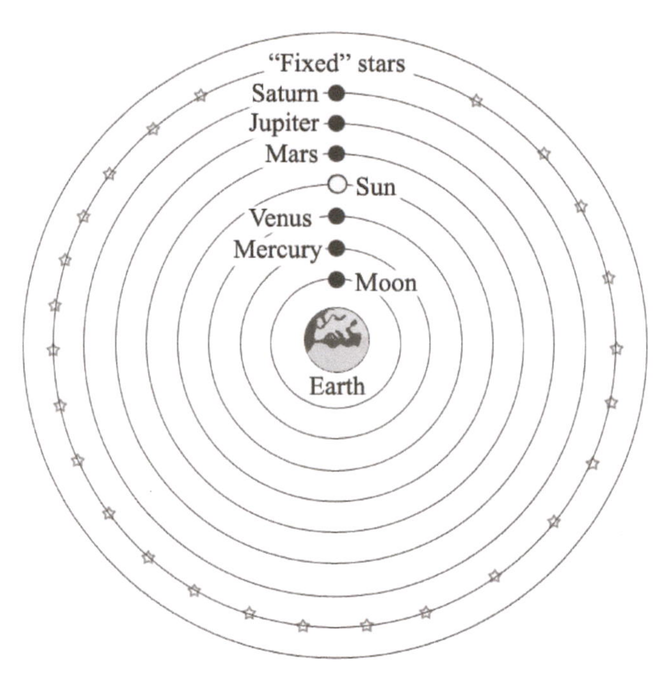

► **Figure 8-4** *Left:* **Aristotle was the most influential of the Greek philosophers because he was perceived during ancient and medieval times to be the most authoritative.** (Image © Panos Karas, 2013. Used under license from Shutterstock, Inc.) ***Right:*** **Aristotle had an Earth-centered (geocentric) model of the Solar System.** (Image courtesy of Frederick Ringwald)

► **Figure 8-5** Aristotle had four arguments that Earth is round. All four arguments are now known to be correct. The first argument is that, during a lunar eclipse, Earth always casts a round shadow on the Moon, as shown here. (Image courtesy of Frederick Ringwald)

► **Figure 8-6** Ships disappear on the horizon hull (bottom) first, because Earth is round. This has been attributed to Columbus, but Aristotle knew it much earlier. (Images courtesy of Frederick Ringwald)

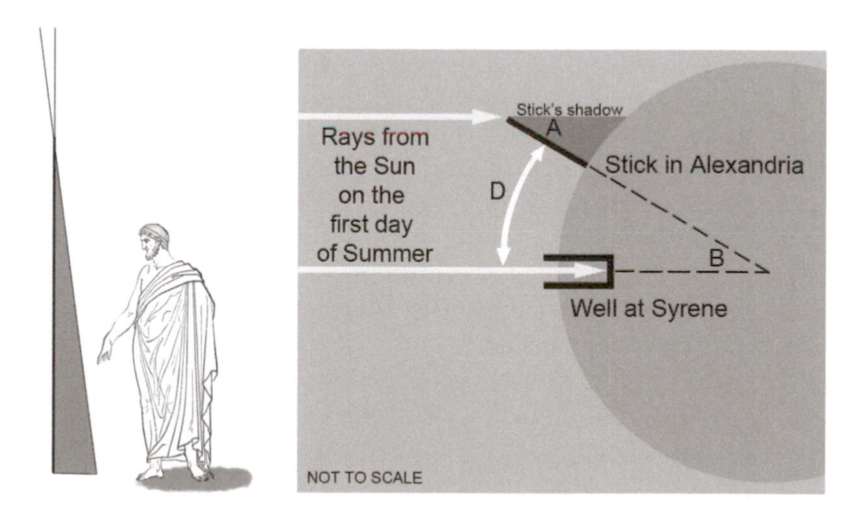

► **Figure 8-7** *Left:* Eratosthenes had heard that a well in Syrene, Egypt, directly south of him, had sunlight reach its bottom only at noon at the Summer Solstice. Eratosthenes realized that it must be on the Tropic of Cancer. He then measured the shadow of a stick at Alexandria to have an angle of 7 1/5°. This is 1/50 of a circle, so Eratosthenes reasoned that the distance from Alexandria to Syrene is 1/50 of the circumference of Earth. *Right:* Eratosthenes measured Earth's circumference to within 10% of its modern value by measuring angle A, and noticing that angle A = angle B. He then measured distance D, and calculated: Earth's circumference / distance D = 360° / angle B. (Image courtesy of Frederick Ringwald)

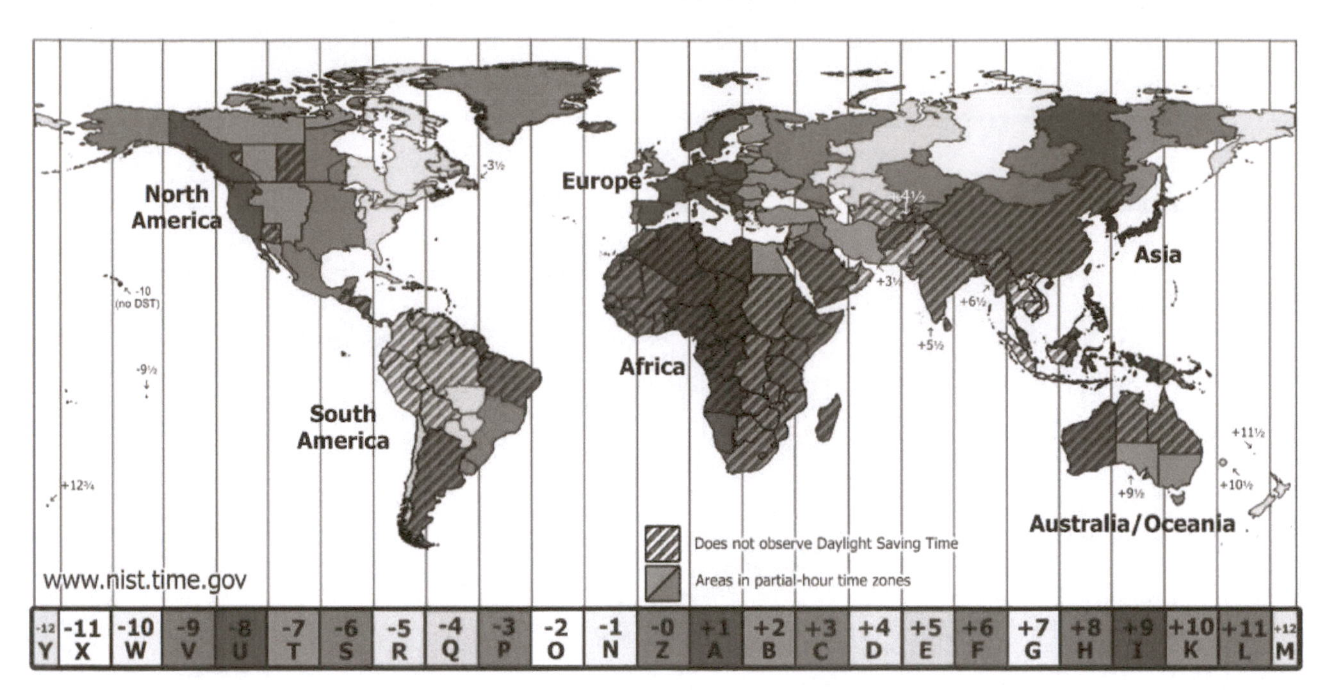

► **Figure 8-8** Time zones show that Earth is round. (Source: National Institute of Standards and Technology)

Aristarchus of Samos was a Greek astronomer who discovered that the Earth moves around the Sun, and not the other way around, as everyone had previously assumed. This result was suppressed and forgotten, however, until it was rediscovered by Copernicus, over 1700 years later. Aristarchus died in 210 BC, around the time the Roman Empire began to take over Greece.

Claudius Ptolemy was a Greek astronomer who worked in Egypt. He assumed that the Sun moved around Earth, in his geocentric (Earth-centered) model of the Solar System. He also thought he could explain the observed motions of the planets if they moved around Earth in circular paths, called epicycles, which rode on other circular paths, called deferents. His book on this came to be called *The Almagest* (*The Greatest*) because it summarized most of Greek astronomy when it was published in AD 140.

The Arabs kept classical astronomy alive after the Greek and Roman civilizations ended.

In 900–1000 AD, they invented algebra and gave many stars names still used today, such as Betelgeuse, which is Arabic for "giant's shoulder."

In the following centuries, continuing observations of planetary motion increasingly showed that Ptolemy's Earth-centered model was inaccurate. More complex versions of Ptolemy's model were proposed, involving "epicycles on epicycles."

Nicolaus Copernicus (in his native Polish, Nikolai Kopernik) simplified models of the Solar System with his heliocentric (Sun-centered) model, by applying Occam's razor. ("The simplest solution is usually the most likely.") He found he could explain planetary motion if the Sun was at the center of the Solar System, with Earth and the other planets orbiting it. He was therefore among the first to realize that Earth is a planet, too. He published this in his book *De Revolutionibus* (*On the Revolutions*) in 1543. Many people were upset by the idea that Earth wasn't the center of the Universe, and that his ideas contradicted Aristotle and Ptolemy. Legend has it that Copernicus was shown a final copy of his book only on his deathbed.

Giordano Bruno was an Italian monk. He was burned alive at the stake in 1600 for defending the Sun-centered theory of Copernicus, among other things. Bruno taught that the Sun is a star, and that "innumerable stars" have planets, some of which may be inhabited.

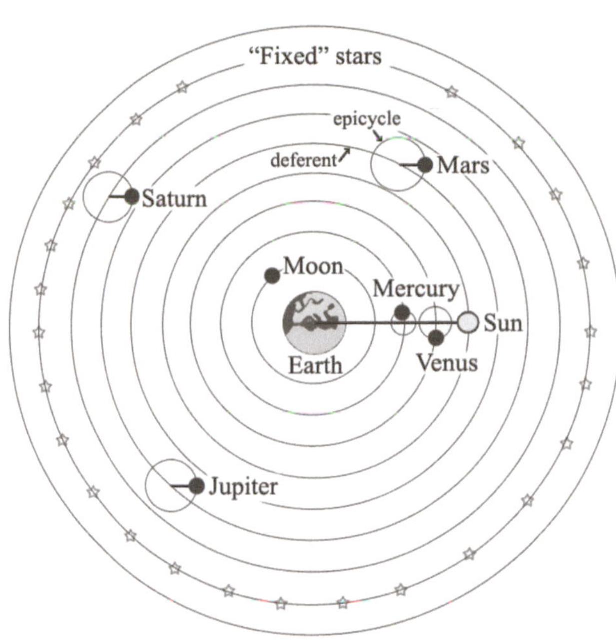

▶ **Figure 8-9** *Left:* **Claudius Ptolemy had a geocentric model, which assumed that Earth was at the center of the Solar System. It was published in 140 AD.** (Public domain image by an unknown artist in the Baroque period. No authentic images of Ptolemy exist: he may not have looked like this.) ***Right:*** **In Ptolemy's geocentric model, the Moon and Sun moved around Earth. So did the planets, which moved on circles called epicycles that moved on other circles, called deferents.** (Image courtesy of Frederick Ringwald)

▶ **Figure 8-10** **Throughout medieval times, astronomers continued to observe the sky. They found Ptolemy's model to be increasingly inaccurate.** (Image © Brian Maudsley, 2013. Used under license from Shutterstock, Inc.)

▶ **Figure 8-11** **Nicolaus Copernicus published a Sun-centered (heliocentric) model of the Solar System in 1543.** (Image © Iryna1, 2013. Used under license from Shutterstock, Inc.)

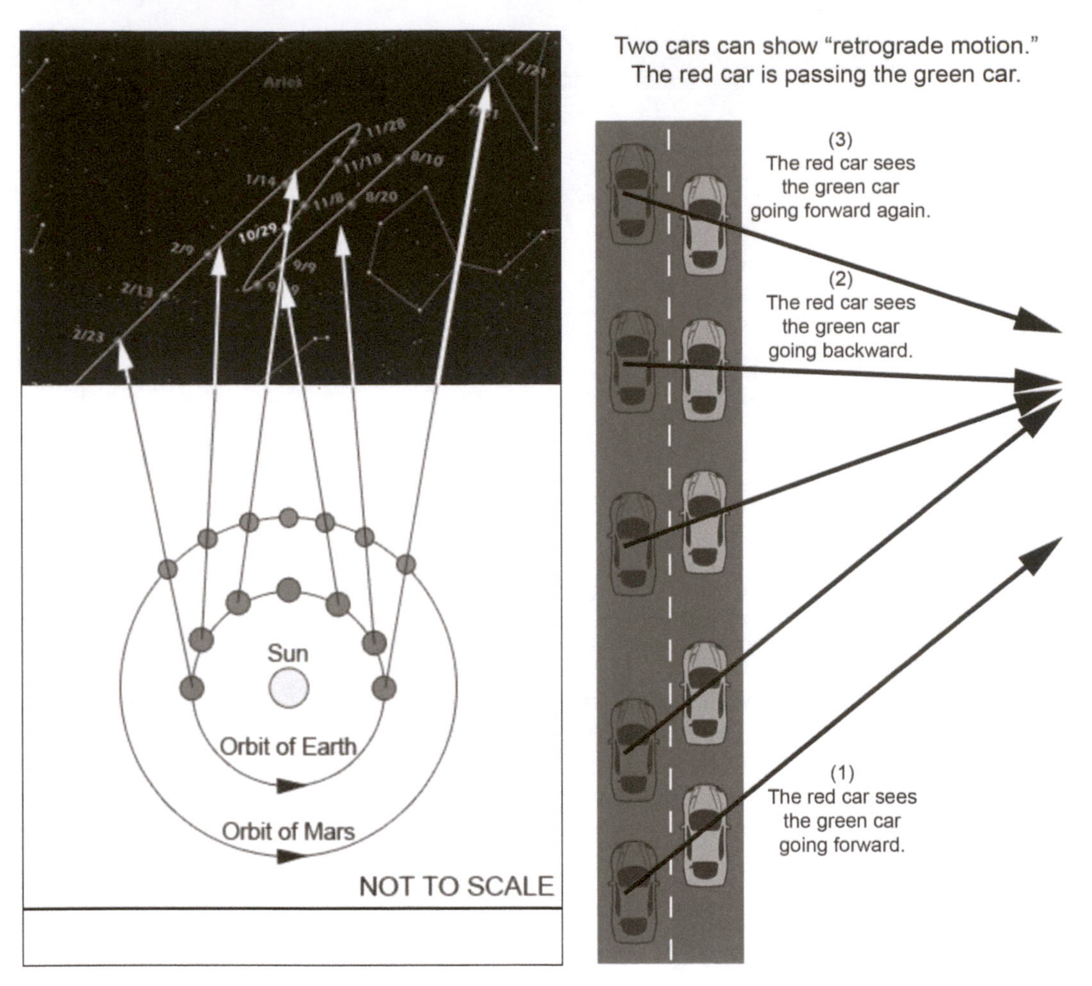

► **Figure 8-12** *Left:* The heliocentric model of Copernicus provided a natural explanation for the apparent retrograde motion of the planets. As shown here, from Earth we see Mars appear to go forward, then appear to reverse its motion as Earth passes it, and then go forward again after Earth has passed. (Image courtesy of Frederick Ringwald) *Right:* As a red car passes a green car, it looks to people in the red car like the green car briefly goes backward, as the red car passes it. That's all the retrograde motion is. (Image courtesy of Frederick Ringwald)

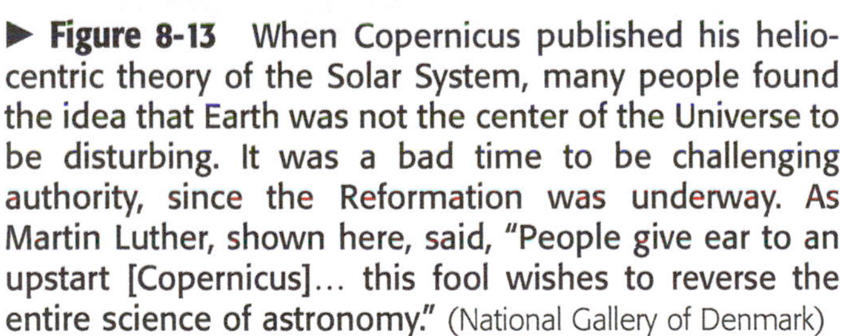

► **Figure 8-13** When Copernicus published his heliocentric theory of the Solar System, many people found the idea that Earth was not the center of the Universe to be disturbing. It was a bad time to be challenging authority, since the Reformation was underway. As Martin Luther, shown here, said, "People give ear to an upstart [Copernicus]… this fool wishes to reverse the entire science of astronomy." (National Gallery of Denmark)

► **Figure 8-14** Giordano Bruno, this statue of whom stands in the Campo De Fiori in Rome is shown, was burned alive at the stake in 1600. One reason for this was that he advocated the ideas of Copernicus, stating that:

> "There are countless Suns and countless Earths all rotating around their Suns in exactly the same way as the…planets of our system. We see only the Suns because they are the largest bodies and are luminous, but their planets remain invisible to us because they are smaller and non-luminous. The countless worlds in the Universe are no worse and no less inhabited than our Earth." Modern science now knows all of this, except the last sentence, to be true, and we're working on this last part. (Image © Oleg Milyutin, 2013. Used under license from Shutterstock, Inc.)

Tycho Brahe was a Danish nobleman. For much of his life, he wore a gold and silver cover over his nose, to hide a scar from a swordfight during his youth, over who was the better mathematician. He was the greatest astronomical observer in history, before the invention of the telescope.

Tycho made thousands of observations of the positions of the planets that in his time were unprecedentedly precise, at the unaided-eye limit of one arcminute (1′). He died in 1601: his assistant Johannes Kepler, a theorist, got his notebooks.

Johannes Kepler was a German mathematician who analyzed Tycho's notebooks. This was among the first clear divisions of labor in science between a scientist who was primarily an experimenter (Tycho, who was a mediocre mathematician) and another scientist who was primarily a theorist (Kepler, who had poor eyesight and so wasn't a good observer). Kepler discovered three laws of planetary motion:

1. The planets travel in ellipses, not circles, as everyone had previously assumed (because Aristotle liked circles, for mystical reasons).
2. The planets sweep out equal areas, in their orbital planes, in equal times.
3. The square of a planet's orbital period is proportional to the cube of its semi-major axis. In other words, P^2 / a^3 is the same for all planets, where P = orbital period, and a = semi-major axis.

In other words, how long it takes the planet to go around the Sun (the orbital period, P) depends on how big the orbit is (the semi-major axis, a), to the 3/2 power.

In other words, if one plots P^2 against a^3 for all the planets, all the points will fall along a straight line, which clearly shows a connection! Any time a scientist discovers a simple mathematical relation in nature, like this, nature is trying to tell the scientist something, because the Universe follows orderly, mathematical laws. Still, Kepler's laws were strictly **empirical:** Kepler did not know why the laws worked, just that they did work. It would be up to Isaac Newton, a generation later, to explain why.

▶ **Figure 8-15** *Left:* Tycho Brahe was the greatest astronomical observer in history, before the invention of the telescope. (Public domain image by an unknown artist) *Right:* Tycho Brahe's observatory in the 1590s was called Uraniborg. It was here that he measured the motions of the planets to unprecedented precision. (Public domain image by an unknown artist)

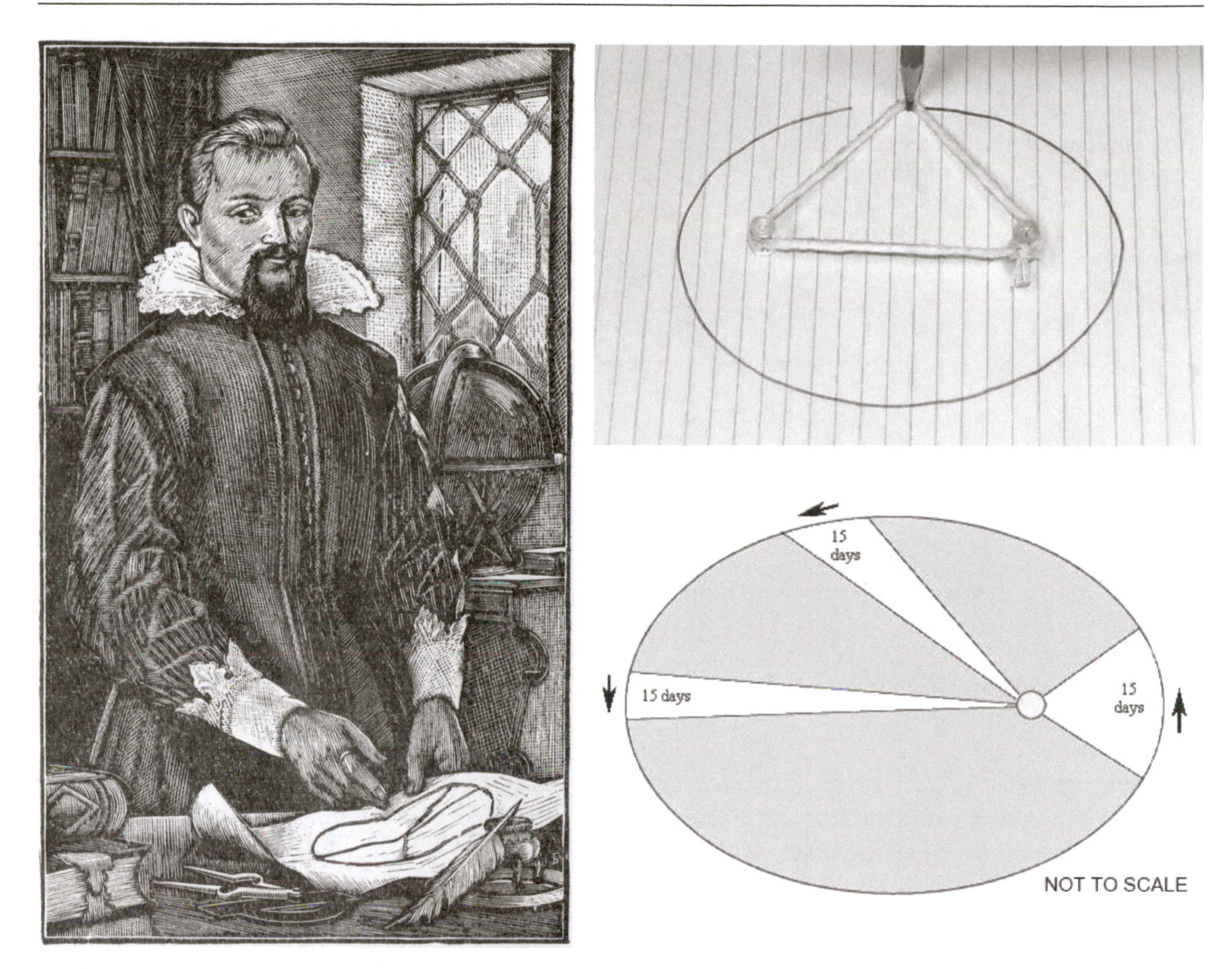

► **Figure 8-16** *Left:* Johannes Kepler was a German mathematician who worked for Tycho Brahe. Using Tycho's observations, Kepler discovered three laws that govern the motions of the planets. (Image © Iryna1, 2013. Used under license from Shutterstock, Inc.) *Top right:* Kepler's First Law is that planet orbits are *ellipses,* not circles, with the Sun at one focus. *Bottom right:* Kepler's Second Law is that a line joining a planet and the Sun will sweep out equal areas in equal times. In other words, a planet goes faster when it's closer to the Sun, and in a precisely predictable way. (Images courtesy of Frederick Ringwald)

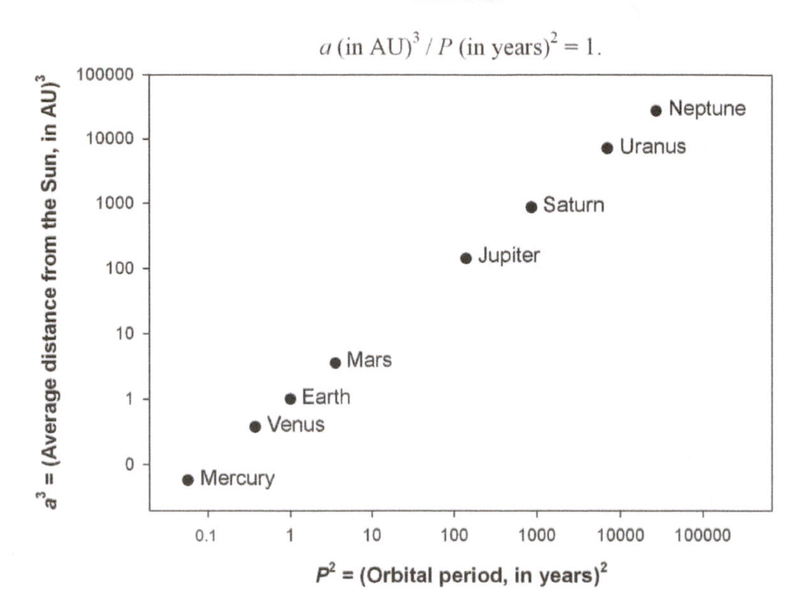

▶ **Figure 8-17** Kepler's Third Law also shows that the closer a planet is to the Sun, the faster it moves, and in a predictable way. Anytime scientists discover a mathematical relationship like this, they have discovered a new law of nature, because nature follows orderly, mathematical laws. (Image courtesy of Frederick Ringwald)

Galileo Galilei was an Italian mathematics professor. He was a contemporary of Kepler, Francis Bacon, and Shakespeare. One reason Galileo made such an impact was that he was a clear writer. His books are full of wide-eyed wonder, not unlike the works of Richard Feynman. He studied motion, although the story that he studied the acceleration of gravity by dropping weights from the Leaning Tower of Pisa is false. (One of Galileo's rivals, Giorgio Coressio, did this when attempting to disprove Galileo.) Galileo also didn't invent the telescope: Hans Lippershey, a Dutch spectacle maker, was first to claim a license to manufacture telescopes, in 1608. Galileo did read about telescopes, and started making them in 1609. Most binoculars today have optics superior to any of Galileo's telescopes. Nevertheless, Galileo discovered "wonderful things":

- The Milky Way is made of "innumerable stars," too faint to see individually with the unaided eye;
- Saturn is not round. Although he couldn't quite resolve the rings, this still contradicted Aristotle's idea that anything outside of Earth had to be a perfect circle;
- The craters of the Moon, which contradicted Aristotle's idea that the Moon had to be a smooth and "perfect" circle;
- Sunspots, which contradicted Aristotle's idea that the Sun was "perfect";
- The four largest moons of Jupiter, which show there were centers of motion other than Earth, which contradicted Aristotle, who thought that things fall toward Earth because it was the only possible center of motion;
- The phases of Venus, which could only be the way they were if Venus orbited the Sun, as predicted by Copernicus.

Galileo became convinced that Copernicus was right and that Aristotle and Ptolemy were wrong. Because of this, the Inquisition tried and convicted him of "vehement suspicion of heresy." They forced him to recant his findings, and ordered him to stop holding the opinion that Earth moved. Legend has it that he muttered, "Even so, it does move," as he rose from his knees. This may not be true. What's certain is that, to make sure that Galileo would obey, the Inquisition showed him the instruments of torture, although Galileo was not tortured because of his advanced age.

► **Figure 8-18** Galileo Galilei didn't invent the telescope, but he began making them in 1609. (Image © Iryna1, 2013. Used under license from Shutterstock, Inc.)

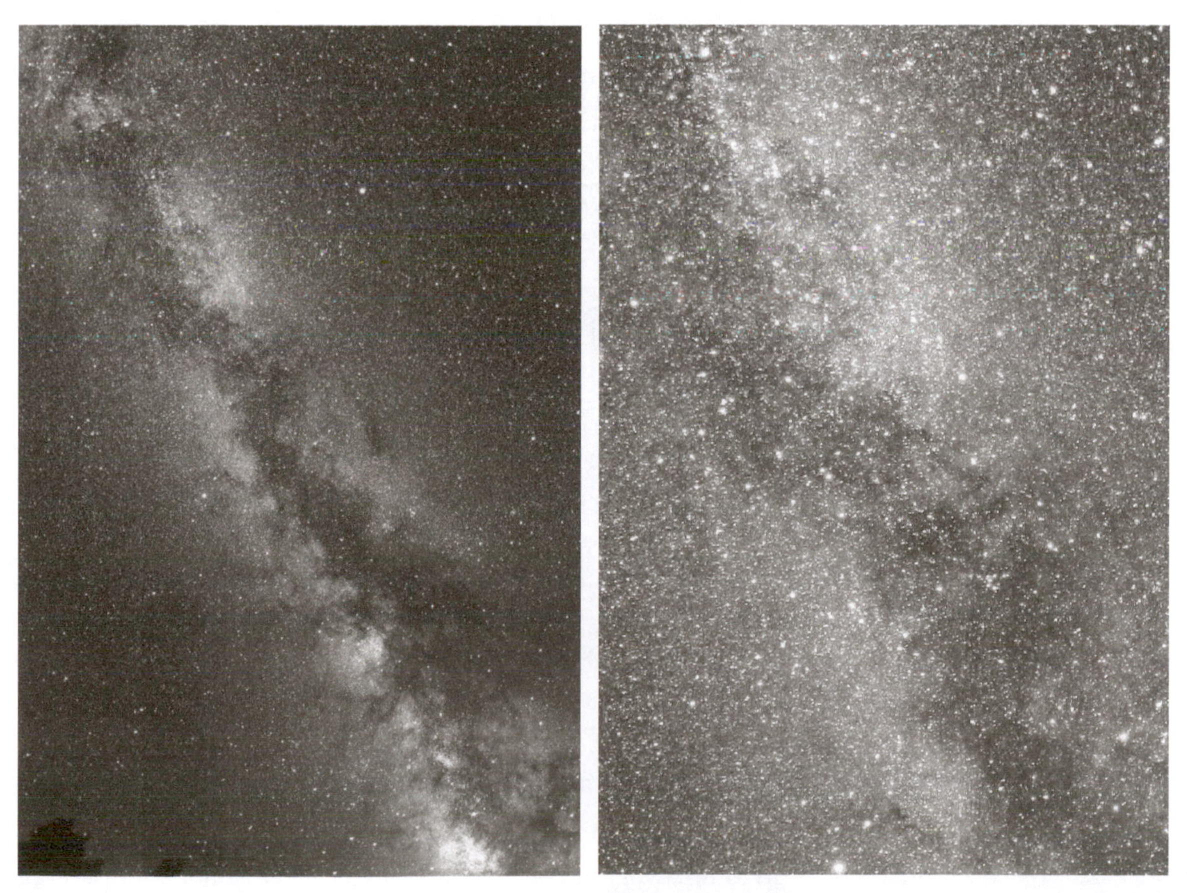

► **Figure 8-19** *Left:* The Milky Way is a faint band of light that stretches across a dark, country sky. (Image courtesy of Frederick Ringwald) ***Right:*** Galileo used a telescope to discover that the Milky Way is made of "innumerable stars." (Image courtesy of Frederick Ringwald)

► **Figure 8-20** Galileo discovered that Saturn is not round, or a circle, as Aristotle would have insisted it must be. This is because Saturn has rings. (Source: NASA/STScl)

► **Figure 8-21** Galileo used a telescope to discover the craters of the Moon. This also contradicted Aristotle, who thought that the Moon and anything in the sky must be smooth and "perfect." (Image by the Frederick Ringwald at Fresno State's station at Sierra Remote Observatories)

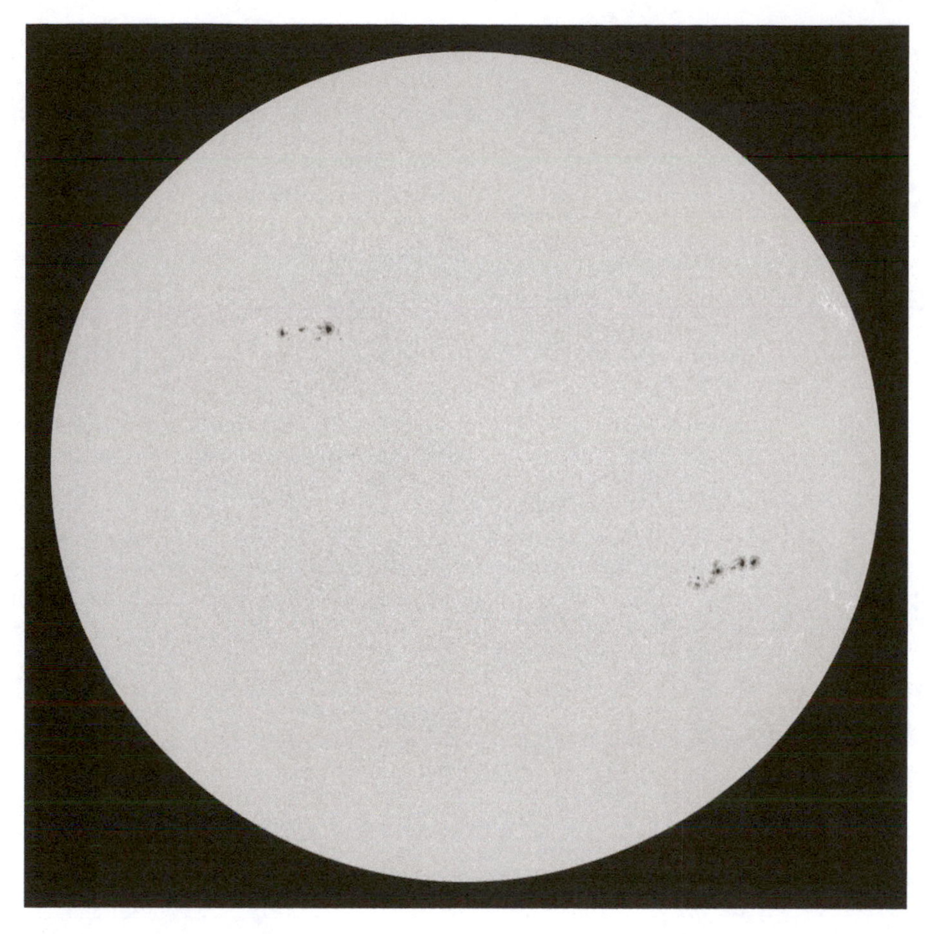

► **Figure 8-22** Galileo also discovered sunspots. Never look at the Sun through a telescope: Galileo knew he would be blinded if he did. Instead, he focused the image onto a surface, and looked at the reflection. That the Sun could have spots, or blemishes, also contradicted Aristotle's idea that anything in the sky must be "perfect." (Source: NASA/Solar Dynamics Observatory)

► **Figure 8-23** In 1610, Galileo discovered the four largest moons of Jupiter—"the Galilean satellites." (Image courtesy of Frederick Ringwald with the 40-inch Montgomery Ward telescope at Lindheimer Astronomical Research Center)

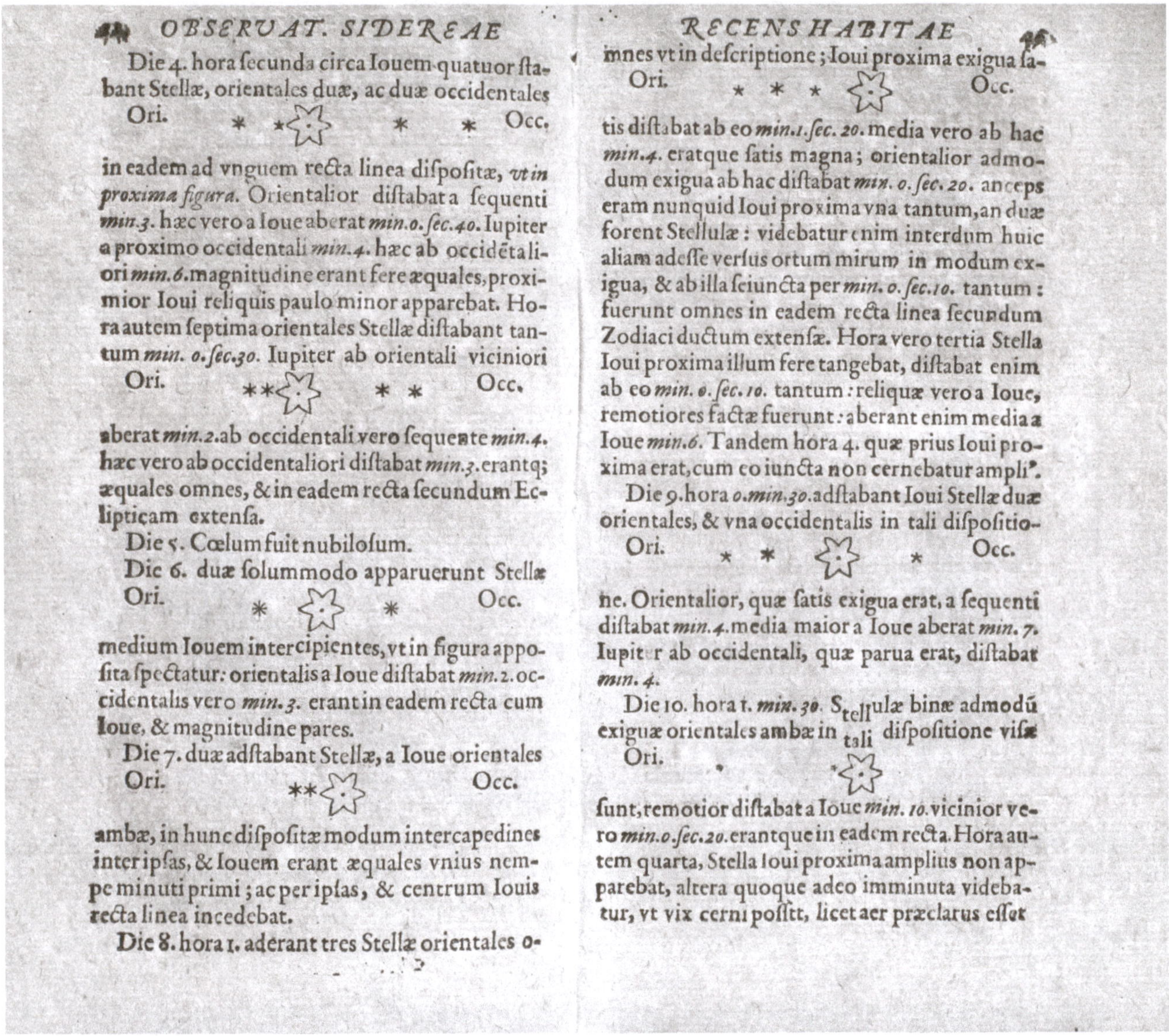

OBSERVAT. SIDEREAE

Die 4. hora ſecunda circa Iouem quatuor ſtabant Stellæ, orientales duæ, ac duæ occidentales

Ori. * * * * Occ.

in eadem ad vnguem recta linea diſpoſitæ, *vt in proxima figura*. Orientalior diſtabat a ſequenti *min. 3*. hæc vero a Ioue aberat *min. 0. ſec. 40*. Iupiter a proximo occidentali *min. 4*. hæc ab occidentaliori *min. 6*. magnitudine erant fere æquales, proximior Ioui reliquis paulo minor apparebat. Hora autem ſeptima orientales Stellæ diſtabant tantum *min. 0. ſec. 30*. Iupiter ab orientali viciniori

Ori. ** * * Occ.

aberat *min. 2*. ab occidentali vero ſequente *min. 4*. hæc vero ab occidentaliori diſtabat *min. 3*. erantq; æquales omnes, & in eadem recta ſecundum Eclipticam extenſa.

Die 5. Cœlum fuit nubiloſum.

Die 6. duæ ſolummodo apparuerunt Stellæ

Ori. * * Occ.

medium Iouem intercipientes, vt in figura appoſita ſpectatur: orientalis a Ioue diſtabat *min. 2*. occidentalis vero *min. 3*. erant in eadem recta cum Ioue, & magnitudine pares.

Die 7. duæ adſtabant Stellæ, a Ioue orientales

Ori. ** Occ.

ambæ, in hunc diſpoſitæ modum intercapedines inter ipſas, & Iouem erant æquales vnius nempe minuti primi; ac per ipſas, & centrum Iouis recta linea incedebat.

Die 8. hora 1. aderant tres Stellæ orientales o-

RECENS HABITAE

mnes vt in deſcriptione; Ioui proxima exigua ſa-

Ori. * * * Occ.

tis diſtabat ab eo *min. 1. ſec. 20*. media vero ab hac *min. 4*. eratque ſatis magna; orientalior admodum exigua ab hac diſtabat *min. 0. ſec. 20*. anceps eram nunquid Ioui proxima vna tantum, an duæ forent Stellulæ: videbatur enim interdum huic aliam adeſſe verſus ortum mirum in modum exigua, & ab illa ſeiuncta per *min. 0. ſec. 10*. tantum: fuerunt omnes in eadem recta linea ſecundum Zodiaci ductum extenſæ. Hora vero tertia Stella Ioui proxima illum fere tangebat, diſtabat enim ab eo *min. 0. ſec. 10*. tantum: reliquæ vero a Ioue, remotiores factæ fuerunt: aberant enim media a Ioue *min. 6*. Tandem hora 4. quæ prius Ioui proxima erat, cum eo iuncta non cernebatur ampli*.

Die 9. hora *0. min. 30*. adſtabant Ioui Stellæ duæ orientales, & vna occidentalis in tali diſpoſitio-

Ori. * * * Occ.

ne. Orientalior, quæ ſatis exigua erat, a ſequenti diſtabat *min. 4*. media maior a Ioue aberat *min. 7*. Iupiter ab occidentali, quæ parua erat, diſtabat *min. 4*.

Die 10. hora 1. *min. 30*. Stellulæ binæ admodũ exiguæ orientales ambæ in tali diſpoſitione viſæ

Ori.

ſunt, remotior diſtabat a Ioue *min. 10*. vicinior vero *min. 0. ſec. 20*. erantque in eadem recta. Hora autem quarta, Stella Ioui proxima amplius non apparebat, altera quoque adeo imminuta videbatur, vt vix cerni poſſet, licet aer præclarus eſſet

► **Figure 8-24** This is a page from Galileo's notebook. The moons of Jupiter proved that Earth was not the only center of motion, as Aristotle had written. (SSPL via Getty Images)

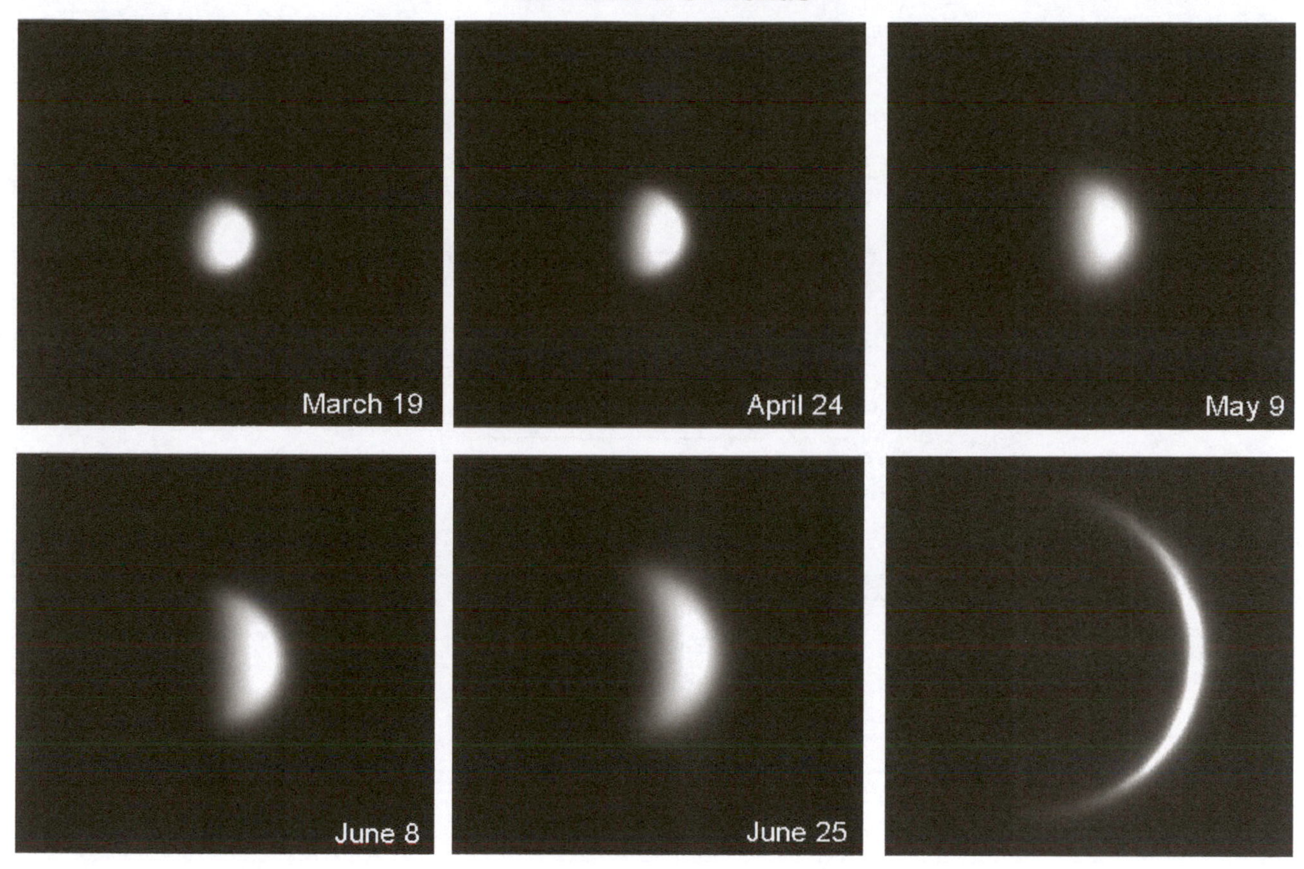

► **Figure 8-25** As Venus orbits the Sun, it goes through a cycle of phases, similar to those the Moon takes as it orbits Earth. (Images courtesy of Frederick Ringwald at Fresno State's Campus Observatory and courtesy of Greg Morgan)

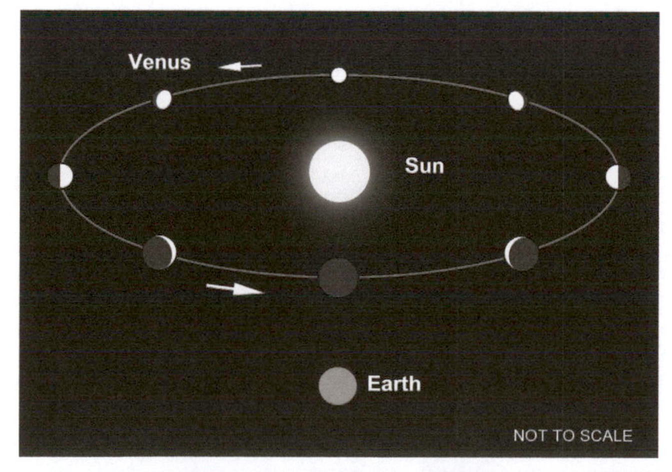

► **Figure 8-26** Why Venus has phases is shown here. Since Galileo was a mathematics professor, he understood this geometry. This convinced Galileo that Copernicus was right. (Image courtesy of Frederick Ringwald)

► **Figure 8-27** Galileo was put on trial by the Inquisition and convicted of "vehement suspicion of heresy" for teaching that the Earth moved. He was forced to recant, or take back his discoveries. This image takes some liberties with the historical record, since he wouldn't have looked defiant, as shown here—he would have been in a penitent's white, and on his knees. Rumor has it that when he was finally freed, he muttered "Even so, it does move." (@Bettmann/CORBIS)

It isn't honest to say, "You didn't see what you saw, because I disagree with it." Humans do *not* thrive when they look away from evidence.

Beware of confusing the following:

Precision	is not	accuracy.
Information	is not	understanding.
Uncertainty	is not	*significant* uncertainty.
Computer power	is not	human reason.
Quantity	is not	quality.
Big	is not	important.
Perception	is not	reality.
Opinion	is not	the same as fact, no matter how many people hold the opinion.
Complex	is not	profound.
Newer	is not	better. (Better is better.)

Above all:

Confidence	***is not***	**certainty.**

> "Ignorance more frequently begets confidence than does knowledge."
>
> –Charles Darwin

Beware this especially when scientists themselves do this, because they really ought to know better:

> "When a distinguished but elderly scientist states that something is possible, he is almost certainly right. When he states that something is impossible, he is very probably wrong."
>
> –Arthur C. Clarke, from *Profiles of the Future*

Elderly scientists who become too convinced of the rightness of their views have gotten in the way of younger scientists too many times. Lord Kelvin, who died in 1907, was among the worst offenders:

> "There is nothing new to be discovered in physics now. All that remains is more and more precise measurement."
>
> –Lord Kelvin

> "X-rays will prove to be a hoax."
>
> –Lord Kelvin, president, Royal Society, 1895

> "Radio has no future."
>
> –Lord Kelvin

> "The disintegration of the radium atom is wantonly nonsensical."
>
> –Lord Kelvin

> "Heavier than air flying machines are impossible."
>
> –Lord Kelvin

► **Figure 8-28** Lord Kelvin (Image © Panos Karas, 2013. Used under license from Shutterstock, Inc.)

▶ Figure 8-29 Isaac Newton letting in the light. This shows him doing an experiment with light in 1665–67, his "miracle year." This was the greatest burst of scientific creativity by a single person in history: only Einstein rivaled it, in his "miracle year" in 1905. As Newton said, "If I have seen farther, it is because I have stood on the shoulders of giants." This referred to Galileo, Kepler, Tycho Brahe, and others. (© National Geographic Society/Corbis)

Isaac Newton was an English mathematician. He was born in 1642, the same year Galileo died. In 1665–67, an outbreak of plague forced him to go into hiding at his family's cottage in the country. To pass the time, he experimented with optics, invented the Newtonian reflecting telescope, discovered the laws of motion and gravity, and invented calculus. (Notice the different words used: Newton discovered the laws of motion, but he invented the Newtonian reflector.) In 1687, at the urging of his friend, Edmund Halley, he published his book *The Principia,* or *Mathematical Principles of Natural Philosophy.* Newton showed that the same natural laws govern both the heavens and the Earth—the ***universality of physical law.*** This was in direct contradiction to Aristotle, who assumed that Earth and the heavens had to obey different physical laws.

Newton's Laws of Motion:

1. The Principle of Inertia (originally Galileo's idea): A body at rest will remain at rest, and a body in motion will remain in uniform motion (in a straight line, at constant speed), unless it is acted upon by an external force.
2. Force = mass × acceleration
3. For every action, there is an equal and opposite reaction (for example, a rocket).

Newton's Law of Gravity:

All objects exert a force attracting all other objects, such that:

The Force of Gravity is: $$F(\text{gravity}) = \frac{GmM}{r^2}$$

where:

m = mass of first object (e.g., you)
M = mass of second object (e.g., Earth)
r = distance between the two objects

G = the gravitational constant (just a consequence of the units we use: it shows how strong gravity is).

The story about Newton having the idea for the law of gravity when he saw an apple fall out of a tree is probably true. (His niece said she saw it happen.) There is an orchard at his family's cottage, Woolsthorpe Manor, which is now a museum.

The question in Newton's mind was: Why does an apple fall to Earth, but the Moon *doesn't* fall? Aristotle would say the apple and the Moon obey different physical laws, since the heavens were perfect and the Earth was not. Newton found this isn't true—the apple and the Moon obey the same law, the law of gravity. The only difference is that the Moon also moves sideways, in its orbit. Although the Moon does fall toward Earth, it never hits Earth, because of its sideways motion.

The force of gravity is truly universal. It connects everything in the Universe with everything else. The "weightlessness," or "zero-g," that astronauts experience isn't because "they are outside of Earth's gravity." If that were true, the astronauts and their spacecraft wouldn't orbit Earth in the first place. Astronauts float around in a spacecraft because they are orbiting Earth in the same way the spacecraft does: there is no relative motion between the astronauts and the spacecraft. One can also do this in an airplane that is falling out of the sky, but only for about 30 seconds (or else the plane will hit the ground).

Nature follows orderly, mathematical laws. The laws of nature are different from the laws that govern humans: there is no penalty for breaking the laws of nature, because one simply *can't* break the laws of nature. Nature simply will not work that way.

> "Nature, to be commanded, must be obeyed."
>
> –Francis Bacon

The laws of nature therefore perhaps shouldn't be called "laws." They should be called "observations," or better, "generalizations," about how nature works. It is astonishing foolishness to think that one is exempt from the laws of nature!

> "...When a [physical] law is right, it can be used to find another one."
>
> –Richard Feynman, in *The Character of Physical Law*

► **Figure 8-30 Newton discovered the laws of motion and gravity, invented calculus, experimented with optics, and invented the Newtonian reflecting telescope in this house, Woolsthorpe Manor. It still exists: it is a good day trip from London. Notice the apple trees on the premises.** (Image © Panos Karas, 2013. Used under license from Shutterstock, Inc.)

Newton's laws have great predictive power. In 1846, French mathematician Urbain Le Verrier used them to predict the existence of the planet Neptune, because its gravity was perturbing the motion of the planet Uranus. Today, we use Newton's laws to navigate spacecraft.

An example of Newton's First Law is the trick of pulling a tablecloth out quickly from under some dishes, without breaking the dishes. Ask your instructor to demonstrate this, it's fun!

- Example 1 of Newton's Second Law is how heavy cars tend to get poor fuel efficiency. Since Force = mass × acceleration, to achieve a given acceleration in a car with a large mass, the engine will need to burn lots of fuel, since it needs to exert a large force to move the large mass.
- Example 2 of Newton's Second Law is how the faster one throws a ball, the farther it goes.
- Example 3 of Newton's Second Law is "Newton's cannon." Newton reasoned that if one fires a cannon ball fast enough, it will travel all the way around Earth. The cannon ball will therefore be in orbit: orbital speed is about 17,500 miles/hour (7 km/s) for low-Earth orbit. If a satellite goes slower than orbital speed, it will fall back to Earth. Escape speed is 25,000 mph (11 km/s): if a spacecraft goes faster than this, it will break free of Earth's gravity, and fly into deep space.
- Example 4 of Newton's Second Law is the acceleration of gravity. Two dropped objects of different mass will hit the ground at the same time. This is because, although the force of gravity is more for the heavier object, its mass is also more, so the acceleration is the same. (Remember, Force = mass × acceleration.)

- Example 1 of Newton's Third Law is a rocket, which moves forward by expelling hot gas out its rear—or a toy balloon, inflated and then let loose.
- Example 2 of Newton's Third Law is the recoil of a rifle against one's shoulder.
- Example 3 of Newton's Third Law: Every time you take a step, Earth moves in the opposite direction—but only a little, since Earth is much more massive than you are.
- Example 4 of Newton's Third Law is a "Newton's cradle." Ask your instructor to demonstrate one.

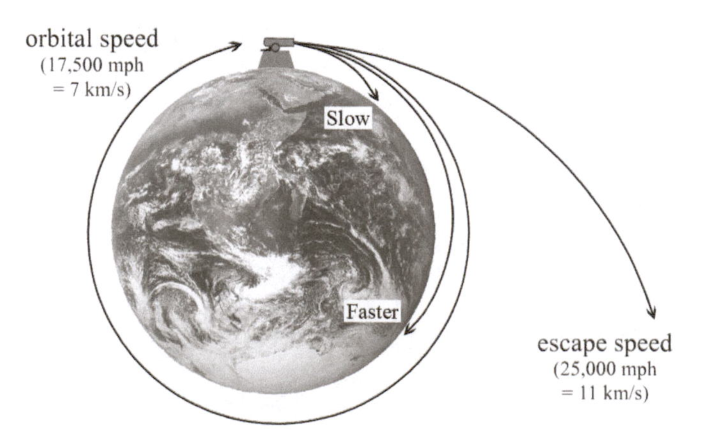

▶ Figure 8-31 **Newton's cannon: if shots were fired out of a cannon faster and faster, eventually a shot would fly all the way around the world. It will do this at orbital speed, which is 17,500 miles per hour, or 7 km/s. Orbiting is like throwing yourself at the ground and *missing,* because you're moving sideways fast enough so that you never hit Earth. Escape speed, needed to escape Earth's gravity, is 42% even faster.** (Image courtesy of Frederick Ringwald with an Earth image by NASA)

► **Figure 8-32** Weightless during the Skylab 4 mission. Commander Jerry Carr is pretending to be a muscle man by holding up Ed Gibson with one finger, in fact, he weighs nothing. (NASA/Skylab 4 crew/William Pogue)

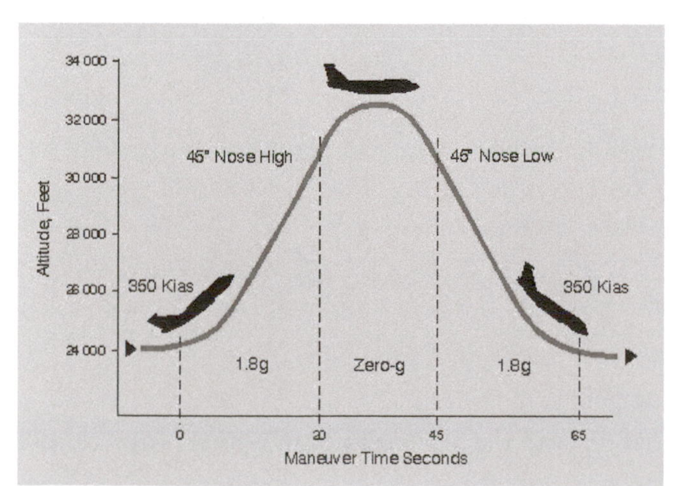

► **Figure 8-33** It's not necessary to be in Outer Space to be weightless. This plane, known as the "Vomit Comet," was used to train astronauts. By falling out of the sky like a stone *(left)*, inside the aircraft the astronauts floated around weightless *(right)*. (Source: NASA)

Implicit in Newton's laws of motion are:

1. The law of conservation of energy:

   Energy cannot be created or destroyed. It always has to come from somewhere, and go somewhere.

   This was amended in the 20th century by Albert Einstein, who showed that $E = mc^2$, or that matter could be converted into energy, and that energy could be converted into matter. This law is therefore now called the law of conservation of mass-energy.

2. The law of conservation of linear momentum:

   The total amount of momentum ($p = m\,v$) of a system is constant.

   This is just a slightly different way of stating Newton's Second Law. It's why it really hurts to be hit by a car: the car has greater mass (m) and velocity (v) than you do, so you go flying.

3. The law of conservation of angular momentum:

   Any spinning body will have angular momentum, $l = m\,v\,r$ = (mass) · (velocity) · (radius).

   Angular momentum tends to be constant, unless a force acts on the system.

   Example: an ice skater spins slowly with arms out. With arms in, the same skater will spin fast.

This explains Kepler's laws, particularly Kepler's second law (equal areas in equal times) and Kepler's Third Law (P^2 / a^3 = constant). Whenever any orbiting body is closer to its parent body (for example, Earth around the Sun, or the Moon around Earth, or one of Jupiter's moons around Jupiter), it goes faster when it's closer, because angular momentum is conserved.

(In other words, $v1 \cdot r1 = v2 \cdot r2$, since l (angular momentum) and m (mass) don't change.)

Kepler's first law naturally follows, since a circular orbit has the slowest speed an orbit can have. Orbits that are faster will be ellipses.

Remember this. We'll see it again several times in this course, when we cover stars, black holes, and dark matter.

This theory has *wonderful* predictive power. When the *Voyager 2* spacecraft arrived at Neptune, after having flown 6 billion miles in over 12 years, it was within 20 miles on course and within 1 second on schedule.

Newton's laws of motion and gravity are now used for everything from building bridges, to making cars, to flying airplanes, to navigating spacecraft. After Newton, it would take almost 200 years for anyone to improve substantially on Newton's work—and it was by Albert Einstein, whose General Theory of Relativity we will cover later in the course.

> "Perhaps no discovery or opinion ever produced a greater effect on the human spirit than did the teaching of Copernicus. No sooner was the Earth recognized as being round and self-contained, than it was obliged to relinquish the colossal privilege of being the center of the Universe."
>
> –Johann Wolfgang von Goethe

No doubt, but why are we so sure Copernicus was right?

Well-educated people often know that Earth goes around the Sun and not the other way around, mainly because their professor said so.

What *evidence* makes us so sure?

1. Copernicus's system is much simpler than the epicycles-on-epicycles nightmare into which Ptolemy's system had evolved. Still, this is only an aesthetic argument: it doesn't *prove* the Sun definitely *is* in the center.
2. Galileo's discovery of the moons of Jupiter showed there were centers of motion other than Earth. This was in direct contradiction to Aristotle, who argued that the Earth couldn't move, because it was the center of all motion. (Aristotle thought that things fell toward Earth because Earth was the center.)

   However, this only showed that Earth was not necessarily stationary.

   It didn't *prove* that Earth *is* moving.
3. Galileo's discovery of the phases of Venus showed that Venus orbited the Sun.

   However, it didn't necessarily prove that *Earth* was moving. Again, it was an indirect argument. During Galileo's trial in 1633, Galileo tried to argue that the tides were the result of the Earth's motion, but we now know that this argument was quite wrong. (Nobody's perfect.)
4. Tycho Brahe reasoned that if Earth moved around the Sun, then the stars should appear to move back and forth in the sky. This effect is called **parallax:** you can demonstrate it with your two eyes. Try it: open one eye and close the other eye. Then switch eyes, and notice how objects near you appear to shift. Of course, they didn't move at all: they only appeared to, since there is some distance between your two eyes.

   Surveyors use parallax to measure how tall mountains are, without having to climb them. (The mathematics involved is called *trigonometry*.) Tycho tried to measure the distance to the stars by observing how they shift back and forth throughout the year, but he found no shifts. He of course didn't know how far away the stars are: aside from the Sun, all known stars have parallaxes of less than 1 arcsecond (1″). With his unaided eyes, the smallest angles Tycho could measure were about 1 arcminute (1′), or 60 times larger. Tycho couldn't have seen the stellar parallax that we now know is there.
5. **The aberration of starlight** was discovered by James Bradley, who published the discovery in 1728, the year after Newton died. It was *the first direct proof that Earth moves*, and that it travels around the Sun. The aberration of starlight is the same reason that raindrops appear to move sideways when one runs in the rain. As one runs, one's motion and the straight-downward motion of the raindrops add together. This makes the raindrops appear to deflect sideways as one runs.

   Bradley was using a telescope to search for stellar parallax, but he accidentally found that all stars move back and forth across the sky, by as much as 20.49 arcseconds near the ecliptic, and near zero near the ecliptic poles. Bradley correctly reasoned that this implies that Earth must move, making the stars' light appear to deflect.
6. **The Coriolis effect,** explained by Gaspard Coriolis in 1835, shows that Earth spins.

   Hurricanes have spiral shapes because they are rotating on the surface of Earth, which also rotates. The Coriolis effect is why it's hard to walk in a straight line on a Merry-Go-Round. As you take a step, the floor moves under your feet. It is a myth that one can see the Coriolis effect in water going down a toilet or a sink: one needs something larger.
7. In 1838, Friedrich Bessel finally became the first astronomer to **measure stellar parallaxes.** This was partly because he used a telescope of unprecedented precision, which was built by Joseph von Fraunhofer. This happened just three years after the works of Galileo and Copernicus were taken off the Index of Forbidden Books.

8. In 1851, over 200 years after Galileo and over 120 years after Newton, **the Foucault pendulum** was invented by French physicist, Jean Bernard Foucault. It's essentially a long pendulum that hangs from a pivot, so it's free to move in any direction. Although a Foucault pendulum seems to swing around the pivot throughout the day, it is really the floor beneath it that rotates, as Earth spins.
9. Since the 1960s, astronomers have been able to track planets and spacecraft with Doppler radar, using instruments such as the Arecibo radio telescope. We can therefore directly observe the motion of Earth and other planets about the Sun. We also now have spacecraft that have left the Solar System and can look back at Earth. It's not easy to think of evidence more direct than this.

CHAPTER

9

Matter and Energy

We need to know as much about light as we can, because our ability to travel into space and take samples is still quite limited. Most of what we know about the Universe is from information carried by starlight. We also need to know how objects in space emit light. For this, we need to understand something of the properties of matter, energy, and atoms.

> "If you look larger or smaller than the skinny realm of life, all you see is physics."
>
> –Stewart Brand

Atoms, Isotopes, and Radioactivity

The smallest piece of any substance that still is that substance is a **molecule**. In 1776, Antoine Lavoisier founded the science of chemistry by deducing this, from how chemical reactions work. Lavoisier also discovered about 30 chemical elements, which he found are substances that can't be broken down into other substances. He also found that fire is a chemical reaction, in which molecules of fuel release energy when they combine rapidly with molecules of oxygen.

Molecules are made of **atoms**. John Dalton figured this out in 1800, although the idea of an atom was anticipated by Greek philosophers (the "atomists") in about 430 BC.

Dalton and Lavoisier inferred that atoms and molecules exist because of how chemical reactions work. One mole of oxygen and two moles of hydrogen always make two moles of water, implying that chemical reactions involve particles. In 1827, Robert Brown noticed that pollen grains move around randomly in water, for no apparent reason. In 1905, Albert Einstein explained this "Brownian motion" by realizing that the pollen grains moved because they were being bumped around on all sides by molecules. This proved that molecules exist. Even better evidence that molecules and atoms really do exist is that today *we can see them*, by using scanning tunneling microscopes that were invented in the 1950s. Scientists today routinely work with single atoms: this field is called **nanotechnology**, because individual atoms are only about a tenth of a nanometer (10^{-9} m) in size.

Organic molecules are molecules that contain **carbon**. Carbon bonds to itself uniquely well. Because of this, molecules containing carbon can form long chains that can store energy, and can also carry information. Living things are made of organic molecules.

An atom is the smallest piece of a **chemical element**. Over 100 elements are now known. They make up the periodic table, which reflects regularities in the structure of atoms.

Nearly all the mass in an atom is in its nucleus. The nucleus is made of two types of particles: positively charged **protons** and neutral (uncharged) **neutrons**. Ernest Rutherford discovered the nucleus of the atom in 1911; he and Niels Bohr devised the modern theory of the atom, in which electrons orbit the nucleus.

Electrons have negative charge. How they interact among atoms determines how molecules are made, and is the subject of the science of **chemistry**.

How electrons move through crystals and other solids is the basis for modern **electronics**.

Protons and neutrons are made of smaller particles, called quarks. We suspect that quarks are made of smaller particles, called strings, but we need better experiments to prove it.

Electrons appear to be fundamental, with no smaller particles inside them. They belong to a family of particles called leptons. Leptons may be made of strings, but we're not sure.

The best answer today to a child's question "What is everything made of?" is therefore "Quarks and leptons." We still don't understand quarks very well, though. There's still much to be learned about the science of the very small, also called high-energy physics.

All atoms of any particular chemical element have the same number of protons.

This is the **atomic number** of the element.

The **atomic weight** is the atom's number of protons and neutrons.

Isotopes are atoms of the same chemical element (same number of protons), but with different numbers of neutrons.

Ordinary hydrogen, deuterium, and tritium are all isotopes of the same chemical element, hydrogen, since they all have the same number of protons. *Helium is a different chemical element*, since it has two protons in its nucleus.

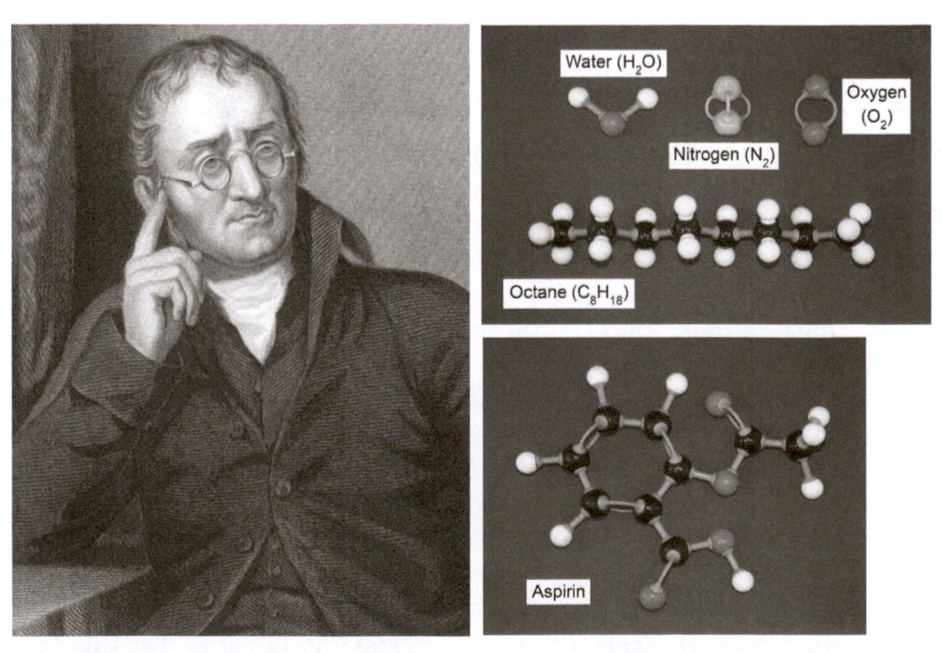

► **Figure 9-1** Molecules are made of atoms. *Left:* In 1800, John Dalton showed that molecules are made of atoms by a series of experiments that counted and weighed all components before and after chemical reactions. Atoms are the smallest pieces of chemical elements; Lavoisier had identified about 30 of them. (Image © Georgios Kollidas, 2013. Used under license from Shutterstock, Inc.) *Top right:* Some molecules in everyday life. Molecular nitrogen (N_2) and molecular oxygen (O_2) make up most of the air we breathe. Octane is a component of gasoline. (Image courtesy of Frederick Ringwald) *Bottom right:* Aspirin is an organic molecule, since it contains carbon. (Image courtesy of Frederick Ringwald)

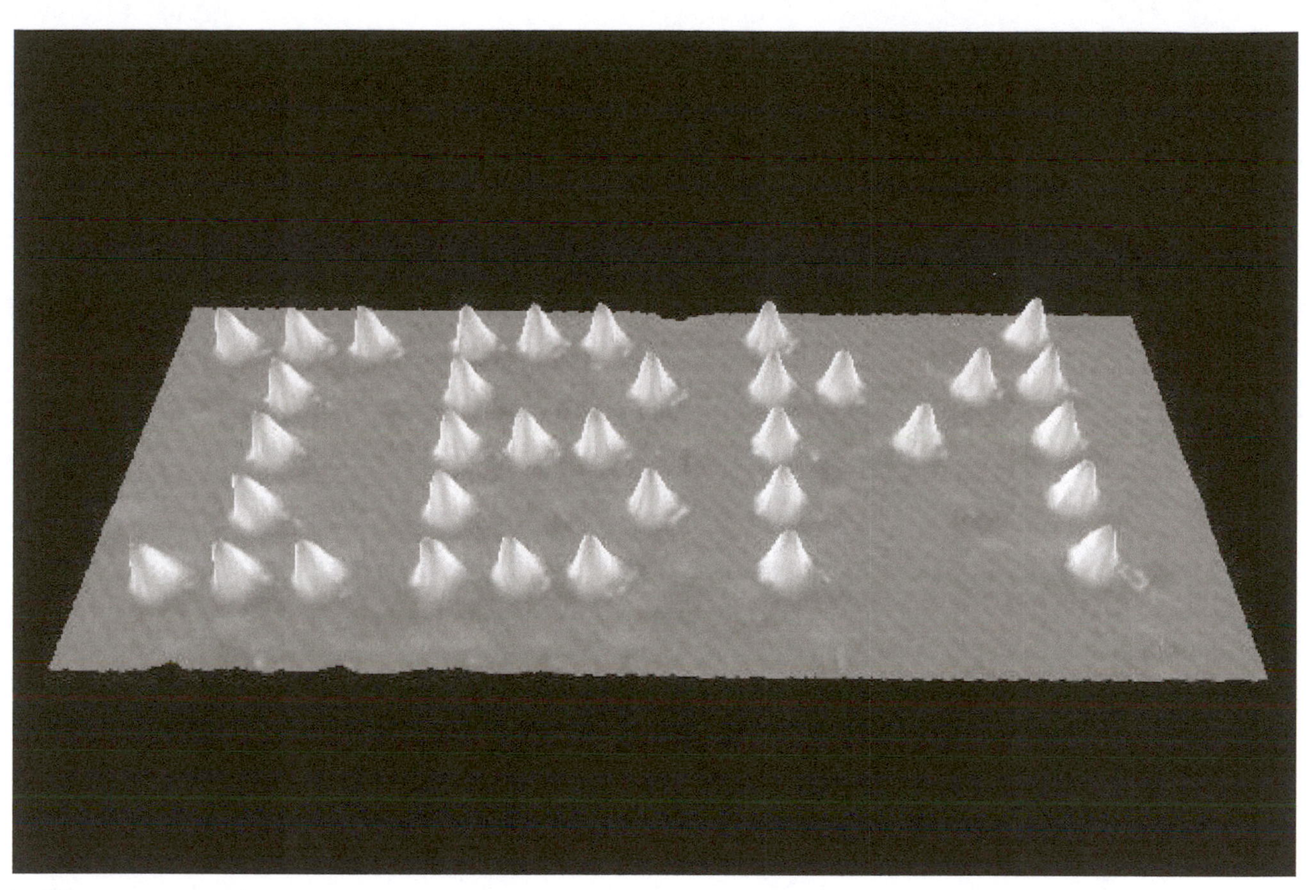

► **Figure 9-2** We know that atoms exist because *we can see them.* This image was made with a scanning tunneling microscope. Here, "IBM" is spelled out in Xenon atoms, which are stuck to a surface of other atoms that are smaller, and so aren't visible. (Courtesy IBM Archives)

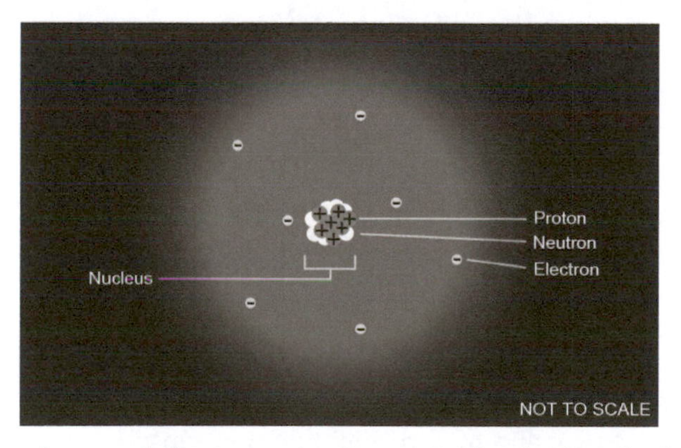

► **Figure 9-3** In the modern theory of the atom, by Ernest Rutherford and Niels Bohr, electrons orbit the nucleus, which has most of the mass. (Image courtesy of Frederick Ringwald)

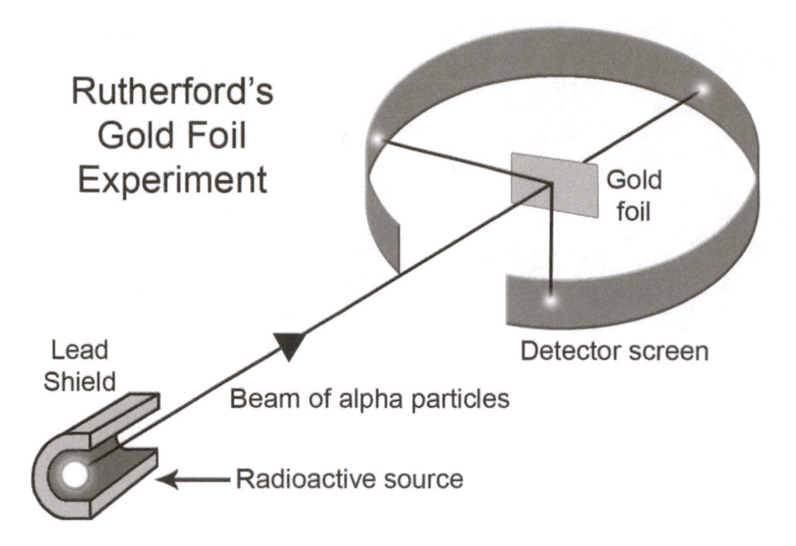

▶ **Figure 9-4** Ernest Rutherford discovered the nucleus of the atom in 1911. Rutherford also mentored many of the scientists who would figure out the atom. He said more than once to a student, "We haven't got the money, so we've got to think!" He also said: "If you can't explain to a bar[tender] in five minutes what you're doing, you don't really know what you're doing."

Rutherford was measuring the size of gold atoms by shooting radioactive alpha particles through a thin sheet of gold. He accidentally discovered that some of the particles bounced back in the direction from which he shot them. This showed that there is something small and hard inside the atom—the nucleus. (Image courtesy of Frederick Ringwald)

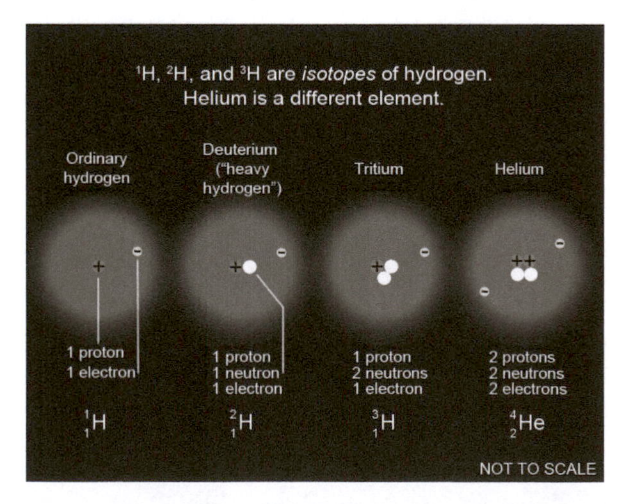

▶ **Figure 9-5** Isotopes are atoms of the same chemical element that have different numbers of neutrons. Hydrogen comes in three forms, or isotopes: ordinary hydrogen (with no neutrons), deuterium (also called "heavy hydrogen," with one neutron), and tritium, with two neutrons. Helium is a different chemical element, since a helium atom has two protons in its nucleus. (Image courtesy of Frederick Ringwald)

Ordinary hydrogen and deuterium are **stable**: their atomic nuclei don't change.

Tritium is **unstable**: tritium nuclei do change over time, or **decay**.

This process of spontaneous nuclear decay, or in other words, when nuclei break down by themselves, is called **radioactivity**. This makes atoms change from one element to another. Henri Becquerel discovered radioactivity in 1896 by accident when he found high-energy rays coming from a rock containing uranium. We now know that radioactive isotopes emit three kinds of energetic particles: alphas, betas, and gamma rays.

Marie and Pierre Curie shared a Nobel Prize with Becquerel in 1903. The Curies discovered radium, an element a million times more radioactive than uranium. They also coined the word *radioactivity*, from the Latin word *radius*, which means "ray." (Radioactivity never had anything to do with radio.)

We now understand the nature of radioactivity. The nuclei of certain isotopes spontaneously disintegrate (break down), and emit three kinds of energetic particles: alphas, betas, and gamma rays.

Pierre died in a carriage accident in 1906. He never got to use the fine laboratory built with the Nobel Prize money. Marie would later die of leukemia from all the radiation.

► **Figure 9-6** *Left:* High-energy physics experiments like this one (at Fermi National Accelerator Laboratory near Chicago) are how we know that quarks exist. They use magnets to shoot tiny particles (protons) at high speed into other particles. This makes even littler particles, including jets of quarks. There is now an even larger experiment like this, the Large Hadron Collider near Geneva, Switzerland, but it's too big to get a good picture of it. (Source: US Department of Energy) *Right:* You are made of molecules. Molecules are made of atoms. Atoms are made of protons, neutrons, and electrons. Protons and neutrons are made of quarks. Quarks and leptons (such as electrons) may be made of even smaller things, called strings. Scientists need a new generation to prove that strings exist, however: this is why they are shown here with question marks. (Image courtesy of Frederick Ringwald)

► **Figure 9-7** *Left:* Henri Becquerel discovered radioactivity in 1896. Mysterious high-energy rays were coming from a rock containing uranium. (Source: Department of Energy) *Top right:* Becquerel had a piece of pitchblende, which is uranium ore, which was next to some photographic plates in a desk drawer for months. (Source: U.S. Geological Survey) *Bottom right:* Because the photos were fogged in a shape and size similar to that of the piece of pitchblende, Becquerel deduced that energy was coming from the uranium in the pitchblende. (Public Domain)

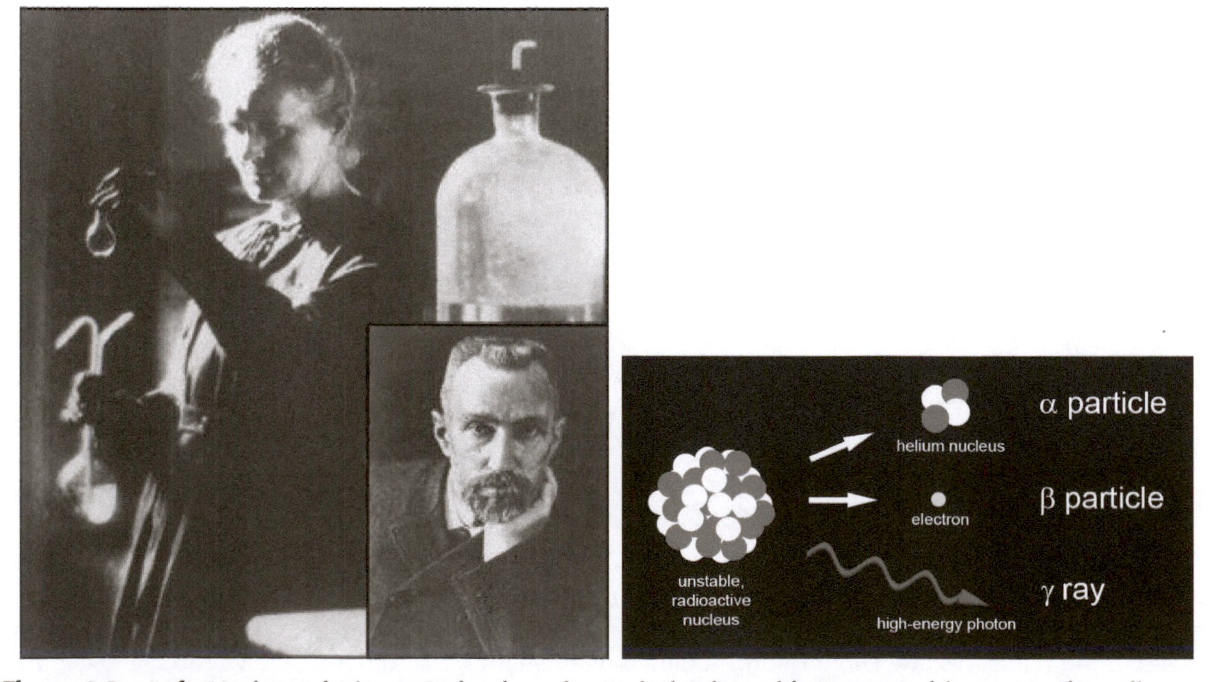

► **Figure 9-8** *Left:* Marie and Pierre Curie shared a Nobel Prize with Becquerel in 1903. They discovered radium, an element a million times more radioactive than uranium. (Source: Radium Institute) *Right:* We now understand the nature of radioactivity. The nuclei of certain isotopes spontaneously disintegrate (break down), and emit three kinds of energetic particles: alphas, betas, and gamma rays. (Image courtesy of Frederick Ringwald)

Chemical reactions don't turn elements into other elements. The alchemists of the middle ages tried to turn lead into gold, but this was doomed to failure because they didn't understand how chemistry works. Chemical reactions rearrange the electrons that bind atoms together into molecules.

The "dance of the electrons" in chemical reactions can have profound effects. A favorite example is how atoms of the highly reactive and harmful element sodium can combine with atoms of the poisonous element chlorine to make molecules of ordinary table salt, which is safe enough to eat. The properties of salt are set by how the electrons in the sodium and chlorine atoms combine to make salt molecules: the nuclei don't change, so they're still sodium and chlorine atoms.

Nuclear reactions, including radioactive decays, do change atoms into other atoms, because they change the nuclei.

Tritium has a **half-life** of 12.3 years. This means that if one starts with one gram of pure tritium, 12.3 years later only half a gram of tritium will remain. The other half a gram will have decayed into other substances. Some radioactive substances have long half-lives: for example, plutonium-239 has a half-life of 26,100 years.

Radiometric Dating

Radioactive decays provide accurate clocks for dating ancient events. Carbon-14 has a half-life of 5730 ± 40 years and is commonly used in archaeology and biology. Potassium-40 has a half-life of 1.3 billion years; uranium-238 has a half-life of 4.5 billion years. Both potassium-40 and uranium-238 are used in geology to find the ages of very old rocks. What's usually measured isn't just the concentration of the radioactive isotope—it's the ratio of this isotope and its **daughter product**, or what it becomes after it decays.

Most, but not all, of the dates in the Cosmic Calendar (covered in Chapter 5) come from radiometric dating. A variety of other methods provide independent checks that agree with radiometric dating, including observed rates of sedimentation and erosion.

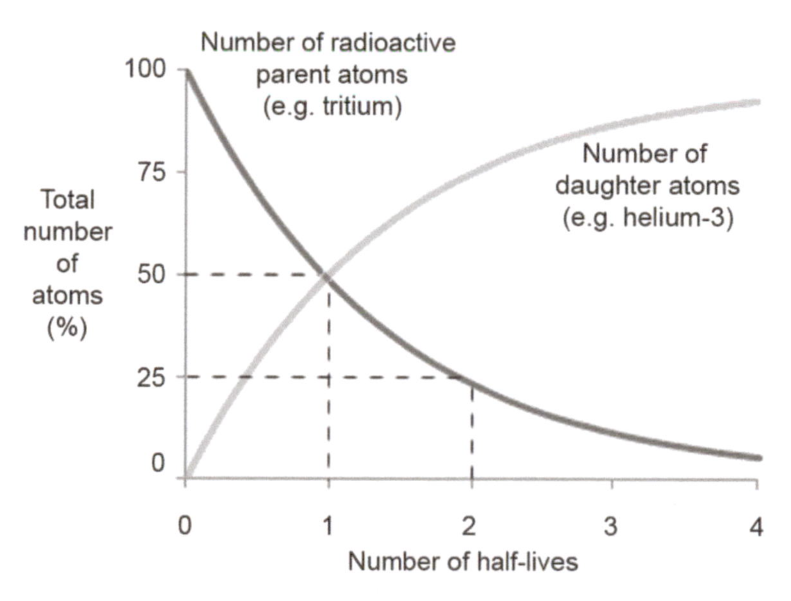

► **Figure 9-9** Radioactive decay can show how old things are. For tritium (which is hydrogen-3) to become helium-3, the half-life is 12.3 years. For carbon-14 to become nitrogen-14, the half-life is 5,730 years. For uranium-238 to become thorium-234, the half-life is 4.5 billion years. (Image courtesy of Frederick Ringwald)

Temperature and Matter

Atoms move, and the faster they move, the higher their *temperature.*

Some Important Temperatures

	Fahrenheit	Celsius	Kelvin	
	T(°F)	T(°C)	T(K)	
Water boils	212	100 (defined)	373	T(K) = T(°C) + 273.15
Body temperature	98.6	37	310	
Room temperature	70	21	295	T(°F) = (9/5) T(°C) + 32
Water freezes	32	0 (defined)	273	(But avoid using
Liquid nitrogen	–321	–196	77	Fahrenheit!)
Liquid helium	–452	–269	4	
Absolute zero	–459.67	–273.150	0 (defined)	

Absolute zero is the temperature at which all molecular motion would stop.
This is the same as having no thermal energy—when all heat has been removed.

Kelvin temperatures make more sense for low and for high temperatures:

The surface of the Sun has: T = 5800 K, about as hot as a lightning bolt (≈ 10,000°F).

In gas this hot, the atoms can't hold onto their electrons. They break free, making a fluid of electrons and ions. (Ions are atoms with electrons missing.)

Ionized gas is called **plasma**, which is Greek for "jelly."

"Plasma" is the same word used for the clear component of blood, which also resembles jelly.

Plasma is sometimes called "the fourth state of matter":

Low T						High T
Solid	→	Liquid	→	Gas	→	Plasma

Most of the Universe that we can see easily, including stars and nebulae, is mostly primarily plasma.

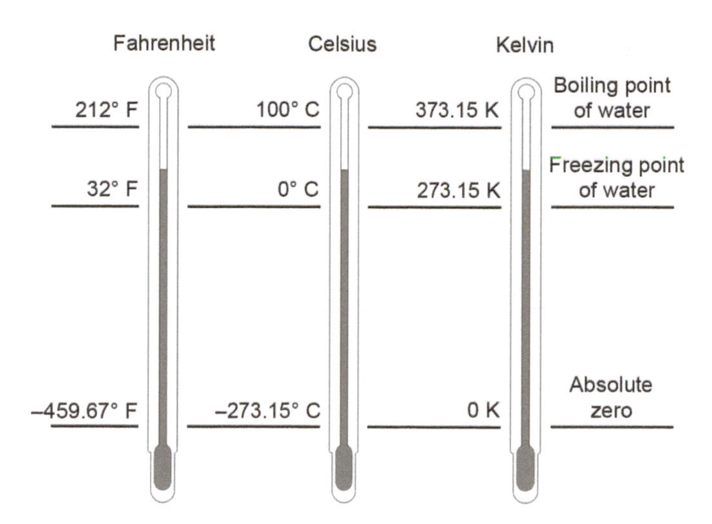

▶ **Figure 9-10** Temperature depends on how fast the molecules move. They would stop moving at absolute zero. Comparing temperature scales, T(°F) = (9/5) T(°C) + 32, T(K) = T(°C) + 273.15. (Image courtesy of Frederick Ringwald)

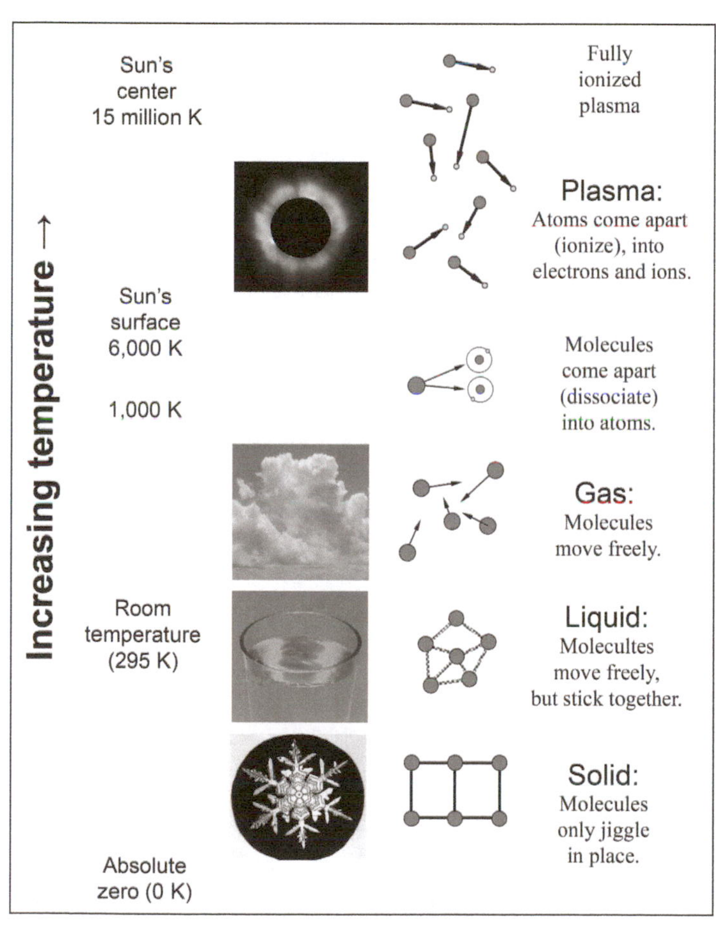

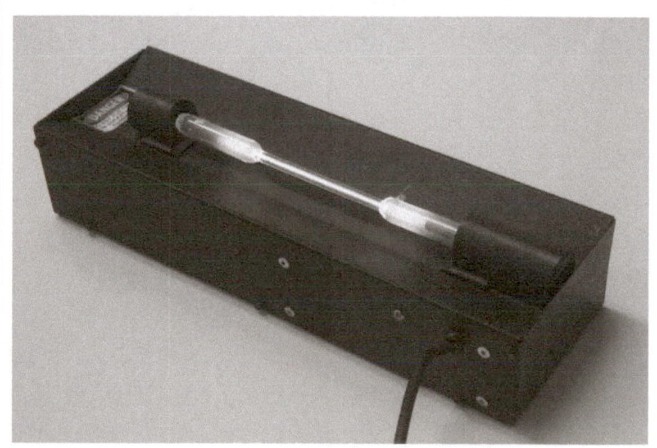

▶ **Figure 9-11** *Left:* Phase changes happen as temperature increases. Solids turn into liquids, which turn into gases. At temperatures over 1000 K, molecules come apart, or *dissociate*. At even hotter temperatures, electrons break free of atoms, making *plasmas*. (Image courtesy of Frederick Ringwald) *Right:* This tube of plasma is heated by electrons, as in a fluorescent lamp. Plasma is gas that is so hot, it's *ionized:* the electrons break free of the atoms. Most stars are plasma. Most of the visible light in the Universe comes from plasma. (Image courtesy of Frederick Ringwald)

The Essentials of Thermodynamics

"thermo" = energy; "dynamics" = motion

Thermodynamics is the study of energy. It is a basic branch of physics. It was developed in the 1800s, for improving the efficiency of steam engines—yet it is amazingly general, because it applies to *all* processes that involve energy. These include life, stars, black holes, the cosmic background radiation, and even the direction that time goes, called "the arrow of time."

The Laws of Thermodynamics

The Zeroth Law is **the definition of temperature.** Temperature is essentially how fast the molecules are moving. The temperature at which all molecular motion stops is absolute zero, or 0 Kelvins, which is –273.15°C, or –459°F. This is one reason why it is called "the zeroth law." Another reason is that it was thought of after the 1st and 2nd laws.

Heat is different from temperature. Heat is the energy flowing from an object at high temperature to another object at a lower temperature.

It's possible to have high temperature with not much heat, for example, in a thin gas that is very hot. Here, the molecules are moving fast, but there aren't many molecules, so the gas doesn't carry much heat. It's also possible to have tremendous heat at low temperature, for example, a big tank of lukewarm water, which is losing heat into a room-temperature room.

Believe it or not, there isn't really any such thing as *cold*: cold is just the absence of heat. Touching something cold gives one a distinctive sensation, because it's the heat flowing *out* of one's fingers.

First Law: In any system with no energy inputs or losses, **energy is conserved.** It cannot be created from nothing, and it cannot be destroyed. It can only be converted into other forms of energy. Einstein amended this to conservation of mass-energy, since $E = mc^2$ shows that mass can be turned into energy, and that energy can be turned into mass.

Second Law: This law is so general there are several equally useful ways to say it. Essentially, though, it says that **energy has a preferred direction.**

You can't warm yourself up by making contact with an object that has a temperature lower than you. Another way to say this is that, by itself, heat moves from an object at high temperature to an object at low temperature: never from a cool object to a hotter object. Of course it's possible to make a refrigerator, the inside of which is cooler than its surroundings, but notice: one must plug in the refrigerator, for it to work.

Another way to say this is that no process involving heat can be 100% efficient. There will always be losses, and waste heat. This is because of the random motion of the molecules.

***Never* believe *anyone* who claims they can power a car with water!** Hydrogen can be used as fuel, and hydrogen is present in water. It doesn't follow that *water* can be used as a fuel—for the same reason that a fire can't be fueled with a pile of ashes.

To use hydrogen as a fuel, one needs to extract the hydrogen from water. The problem with this is that it takes energy to extract the hydrogen from water. Where does this energy come from? Please don't say burning hydrogen! Frauds or fools might point to something vague, such as "a catalyst," or "a bubbler," or "a membrane," but none of these can work, ***without even more energy put into them***.

Another way to say this is that, **in any system *with no energy inputs or losses*, the amount of disorder, or entropy, always increases.** Of course, it's possible to make a disorderly system more orderly—for example, one can always tidy up a room—but notice: one *must apply energy* to do it.

Life, and the evolution of life, are sometimes claimed to be violations of the second law, but they aren't: plants grow when they take in energy, by absorbing sunlight. Animals also take in energy, by eating food. Notice that the second law states that entropy always increases, *in a closed system*. Living things are *not* closed, since they take in energy.

Einstein once commented that the First and Second Laws of Thermodynamics are probably the only two physical laws that will never be overturned, because they're just good accounting. The First Law counts up **energy**. Energy doesn't appear from nothing, and it doesn't just disappear.

The Second Law counts up **states**. For example, there are many ways to smash a china cup into 1000 pieces (with entropy, or disorder, increasing), but there's only one way to assemble 1000 pieces into a cup. A thousand pieces turning into a cup by themselves (with no energy input) is so unlikely, it's practically equivalent to saying it never happens.

Third Law: By no finite series of processes is absolute zero possible.

This is due to the wave nature of molecules, discovered by quantum mechanics in 1900–1930. Molecules aren't just particles: they also have wave properties. Because of this, they can never hold still completely. This property of molecules and anything smaller is *Heisenberg's Uncertainty Principle.*

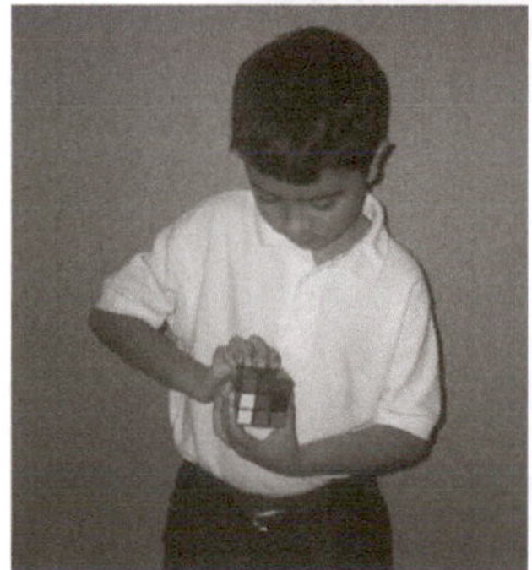

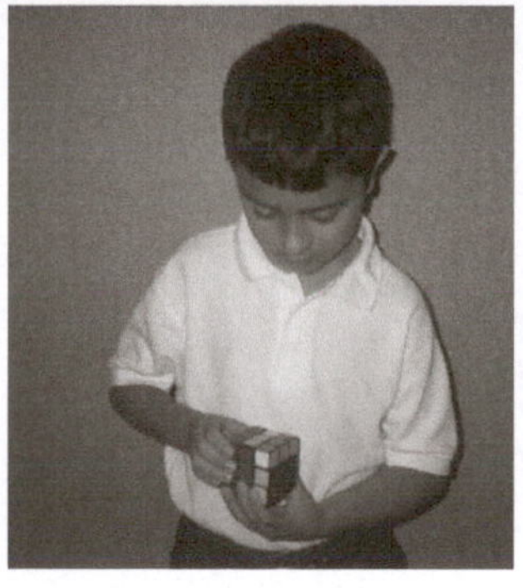

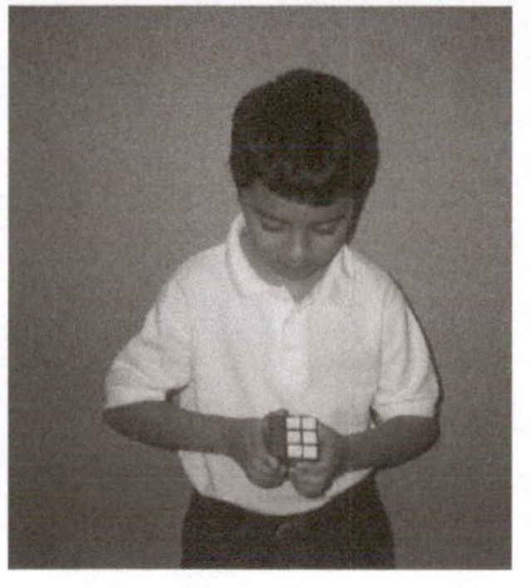

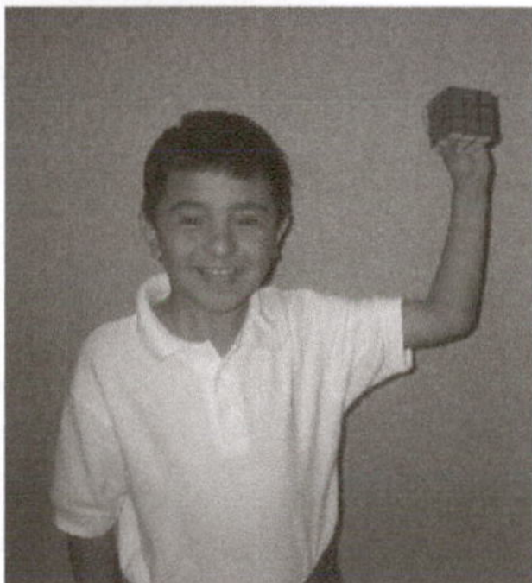

► **Figure 9-12** *Top:* Thermodynamics means the motion ("dynamics") of energy ("thermo"). The First Law of Thermodynamics is easy to understand: it says that energy isn't free. Energy never "just appears," nor does it ever "just go away." Because of this, nuclear plants like the one shown here are often built on rivers, to carry away their tremendous heat. The towers are for air cooling. (Image © hornyak, 2013. Used under license from Shutterstock, Inc.) *Bottom:* The Second Law of Thermodynamics has applications both practical, about the efficiency of steam engines, and esoteric, such as the direction, or "arrow," of time. The Second Law of Thermodynamics says that nature has a preferred direction. Whenever a movie runs backwards, it's obvious: it shows many things that are ridiculously improbable. (This kid really didn't solve this puzzle in four moves: the pictures are shown backwards in time, as he randomized it.) (Image courtesy of Frederick Ringwald)

▶ Figure 9-13 *Left:* One example of the irreversibility of the Second Law of Thermodynamics is the nursery rhyme:

"Humpty Dumpty sat on a wall
Humpty Dumpty had a great fall
All the King's horses, and all the King's men
Couldn't put Humpty together again."

Entropy, or the amount of disorder, is the arrow of time. The Second Law of Thermodynamics is that, in any system with no energy inputs, entropy tends to *increase* with time. (From *Through the Looking Glass* by Lewis Carroll: image by Sir John Tenniel. Image © Morphart Creation, 2013. Used under license from Shutterstock, Inc.)

Right: Entropy is the amount of disorder in a system. If you don't clean up your room, it will become more disorderly over time: its entropy will increase. (ScienceCartoonsPlus.com. Used by permission)

Don't misinterpret the Second Law. One *can* make a room more orderly (and decrease its entropy), if one puts energy into it.

CHAPTER

10

Light and Spectra

Light is electromagnetic radiation, or the combination of electric and magnetic fields, which travels through space. Radiation is any kind of microscopic particle that can move through a vacuum, or empty space, such as the space between the stars.

Electromagnetic radiation, also called e/m radiation, isn't the only kind of radiation. Particles from the decays of radioactive nuclei, such as protons or neutrons, are also radiation. This is the well-known nuclear radiation, which is feared because it can be unhealthy. Not all radiation is unhealthy: visible light isn't as long as it isn't too bright.

The speed of light, in a vacuum c = 300,000 km/s
= 186,000 miles/s = 670 million miles per hour.

The speed of light in air is slower. It's slower still in denser materials, such as glass.

James Clerk Maxwell published his Theory of Electromagnetism in 1865. This theory explained that electricity, magnetism, and light were all essentially different aspects of the same phenomenon. It also predicted that visible light was only one kind of electromagnetic radiation, and that there are many kinds of electromagnetic waves that human eyes can't see.

Light is distinguished by its wavelength (λ, or lambda):

λ = distance between the crests of the waves.

Visible light has wavelengths between:

Violet: 0.4 microns = 4×10^{-7} m = 400 nm = 4000 Ångströms
(Blue light has a wavelength of about half a micron.)

Red: 0.75 microns = 7.5×10^{-7} m = 750 nm = 7500 Ångströms.
Violet light therefore has shorter wavelengths than red light.

$c/\lambda = f$ frequency = the number of cycles (or wavelengths) that pass by each second.
1 cycle/s 1 Hertz = 1 Hz.

This was named for Heinrich Hertz, who in 1887 discovered radio waves, which had been predicted by Maxwell's electromagnetic theory.

Radio waves were the first kind of e/m radiation known that the eye can't see. Look at a radio: the dial is measured in frequency units, in kHz (kilohertz, or 10^3 Hz) and in MHz (megahertz, or 10^6 Hz).

► **Figure 10-1** *Left:* Benjamin Franklin (and his son William) found that electricity flows through wires in 1752, by flying a kite in a lightning storm. It is a miracle they weren't killed doing that. Someone was killed reproducing this experiment in 1976. (Public domain image) *Top right:* Alessandro Volta invented the electric battery in 1799. (Image © Georgios Kollidas, 2013. Used under license from Shutterstock, Inc.) *Bottom right:* In 1831, Michael Faraday invented the electric generator, by finding that a rotating magnet makes electricity. This implied that there is a connection between electricity and magnetism, even though they don't look similar, at first. (Image © YANGCHAO, 2013. Used under license from Shutterstock, Inc.)

► **Figure 10-2** In 1865, James Clerk Maxwell found four equations that provide a complete description of electricity, magnetism, and light: Maxwell's *electromagnetic theory*. It gave us radio, television, cell phones, microwave ovens, and X-rays. (Image © Nicku, 2013. Used under license from Shutterstock, Inc.)

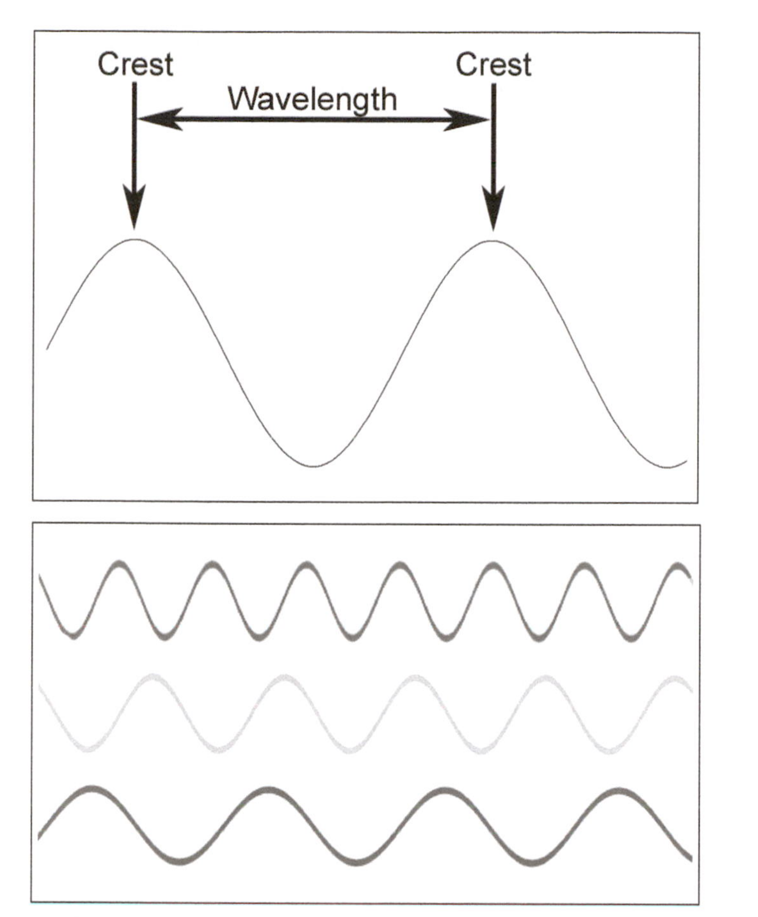

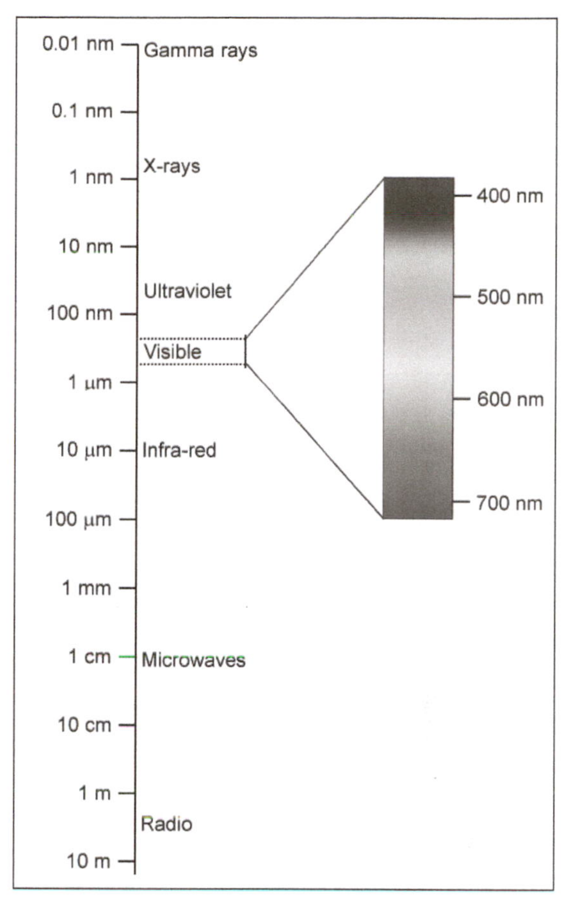

► **Figure 10-3** *Top left:* Different *colors* of light have different *wavelengths*. Wavelength is the distance from crest to crest of a wave. (Image courtesy of Frederick Ringwald) *Bottom left:* Violet light has a short wavelength. Red light has a long wavelength. The other colors are in between. (Image courtesy of Frederick Ringwald) *Right:* Maxwell's electromagnetic theory predicted the existence of radiation with wavelengths the eye *can't* see: *the electromagnetic spectrum.* (Image courtesy of Frederick Ringwald)

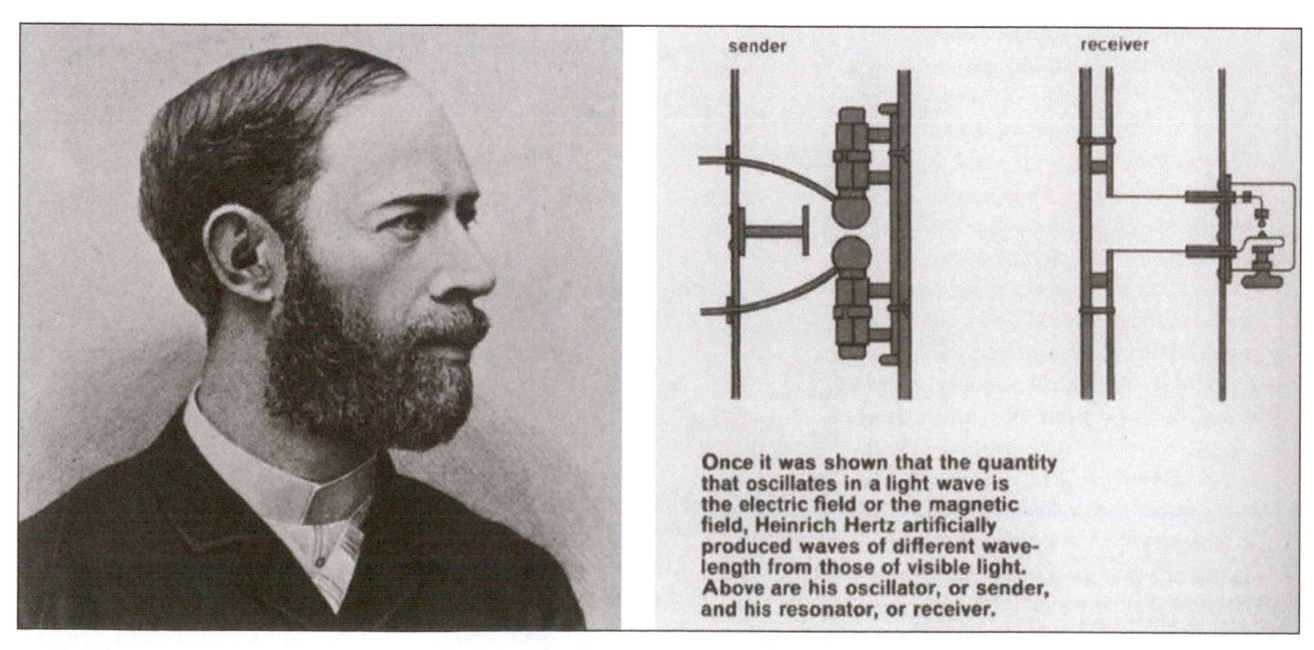

► **Figure 10-4** In 1887, Heinrich Hertz confirmed the predictions of Maxwell's electromagnetic theory, by being the first to transmit and receive *radio waves.* (Source: NRAO: National Radio Astronomy Observatory)

Light has both wave **and particle** properties. In 1905, Albert Einstein found that *photons* are particles of light. The shorter the wavelength of light, the higher the energy of the photons: each photon has an energy $E = h f = hc/\lambda$, where h = Planck's constant.

So, while violet light has shorter wavelengths than red light does, photons of violet light have higher energies than red photons have. X-rays have wavelengths too short for the eye to see. X-rays also have such high energy they can go through the human body, and so are used in medicine to take pictures of people's bones.

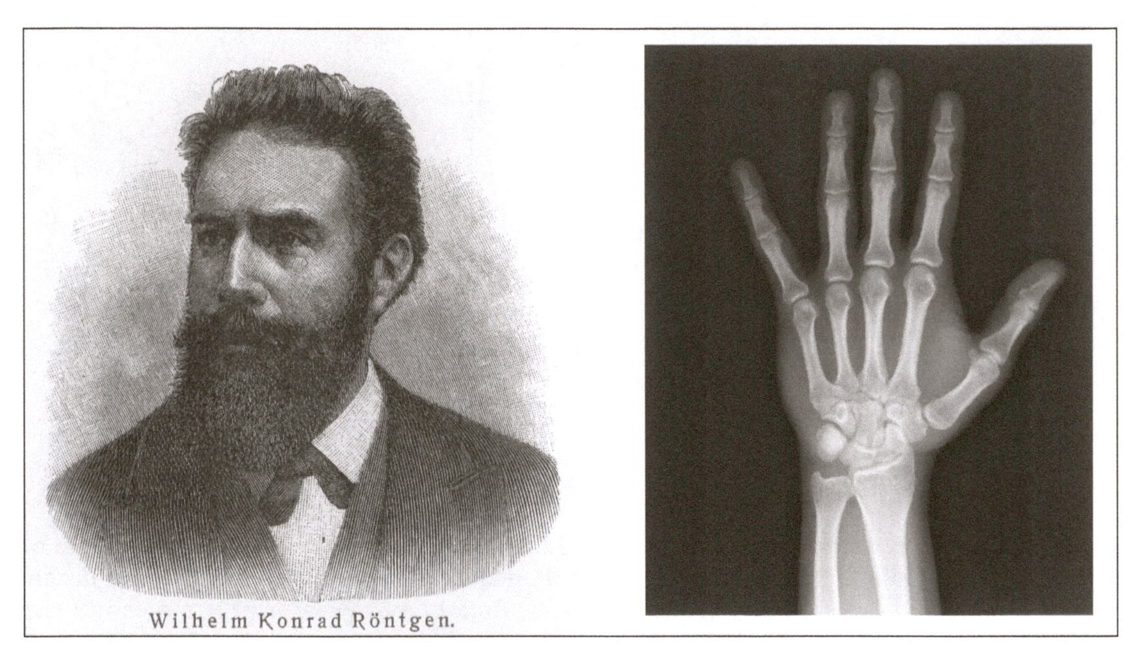

► **Figure 10-5** Wilhelm Roentgen discovered X-rays in 1895. (Left image © Nicku, 2013. Used under license from Shutterstock, Inc. Right Image © itsmejust, 2013. Used under license from Shutterstock, Inc.)

► **Figure 10-6** In 1905, Albert Einstein showed that light isn't just a wave: it's also made of particles. Particles of light are called *photons*. The shorter the wavelength of the light, the *higher* the energy of the photons. *(For example: X-rays have such high energy, they go through a human body's soft tissues.)* (Image © Pete Spiro, 2013. Used under license from Shutterstock, Inc.)

The Electromagnetic Spectrum

Visible and invisible light together make up the electromagnetic spectrum, also called the e/m spectrum. Visible light, which the unaided eye can see, makes up only a tiny portion of the possible kinds of e/m radiation, the difference between which is their wavelength (and energy):

Region of the spectrum	Wavelength (λ)	Photon Energy
Gamma rays	10^{-13} m and shorter	> 10^6
X-rays	10^{-10} m = 1 Ångström	1000
Ultraviolet radiation	0.01 – 0.4 microns	
Visible light	0.4 – 0.7 microns	1
Infrared radiation	0.7 – 100 microns	
Microwaves	millimeters	10^{-3}
Radio waves	centimeters & longer	Low

In other words, visible light, radio waves, X-rays, etc., are all essentially the same thing. The essential difference between them is their wavelengths which are colors which are their energies.

The Spectrum of Visible Light

Don't confuse the electromagnetic spectrum (which is all the wavelengths of light, whether or not the unaided human eye can see them) with the spectrum of *visible* light, which is breaking visible light into its component colors (or wavelengths):

R O Y G. B I V = red, orange, yellow, green, blue, indigo, violet
One can break visible light into its component colors with a prism, or a grating.

For most non-luminous objects, such as planets, we see *reflected* light. Another example of this is green grass, which reflects green light but absorbs (and doesn't reflect) other colors. Stars (and light bulbs) are luminous: they emit their own light.

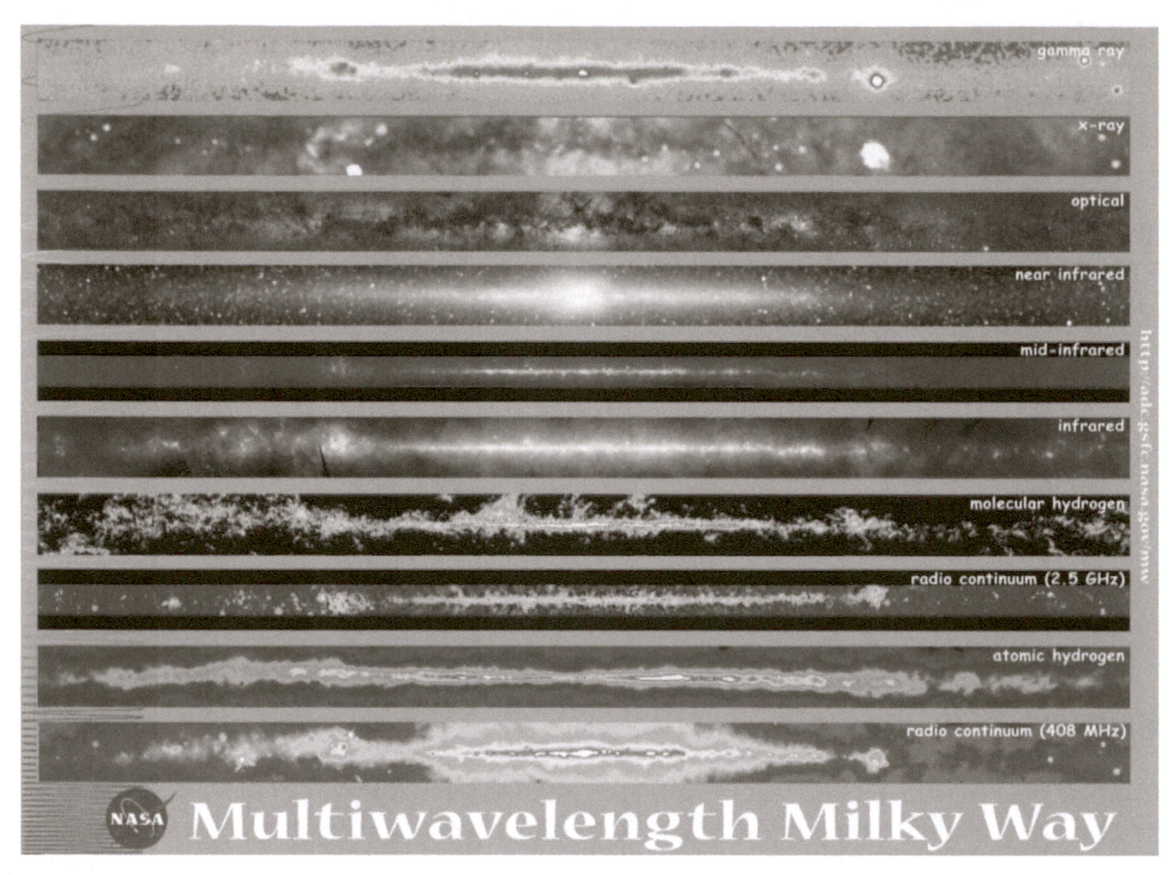

▶ **Figure 10-7** The Milky Way is shown in wavelengths across the spectrum, from high-energy, short-wavelength gamma rays at top to low-energy, long-wavelength radio waves at bottom. Visible light, labeled "optical," is in between these extremes.

Astronomers used to be able to detect only visible light, like our eyes can see. In the 1930s, they started to detect radio waves. In the 1960s, X-rays and infrared followed. Now astronomers can observe all wavelengths. This is the single biggest advance in astronomy for the past two generations. We learn about the Universe much faster now, since we can detect all wavelengths, visible and invisible, of the electromagnetic spectrum. (Source: NASA)

▶ **Figure 10–8** In 1665-67, during his "miracle year," Isaac Newton showed that white light is composed of all the colors of light. (Image © Iryna1, 2013. Used under license from Shutterstock, Inc.)

▶ **Figure 10-9** Newton showed that white light is made of light with all the colors with *two* prisms. The first prism turned the white light into its colors. The second prism joined the colors back into white light. This showed that the colors were in the white light, and that the colors weren't in the prism before the white light passed through it. (Image courtesy of Frederick Ringwald)

▶ **Figure 10-10** A rainbow is a natural example of a spectrum. In a rainbow, white light from the Sun is broken into its colors by water droplets in the air, just after a rain storm. (Image courtesy of Frederick Ringwald)

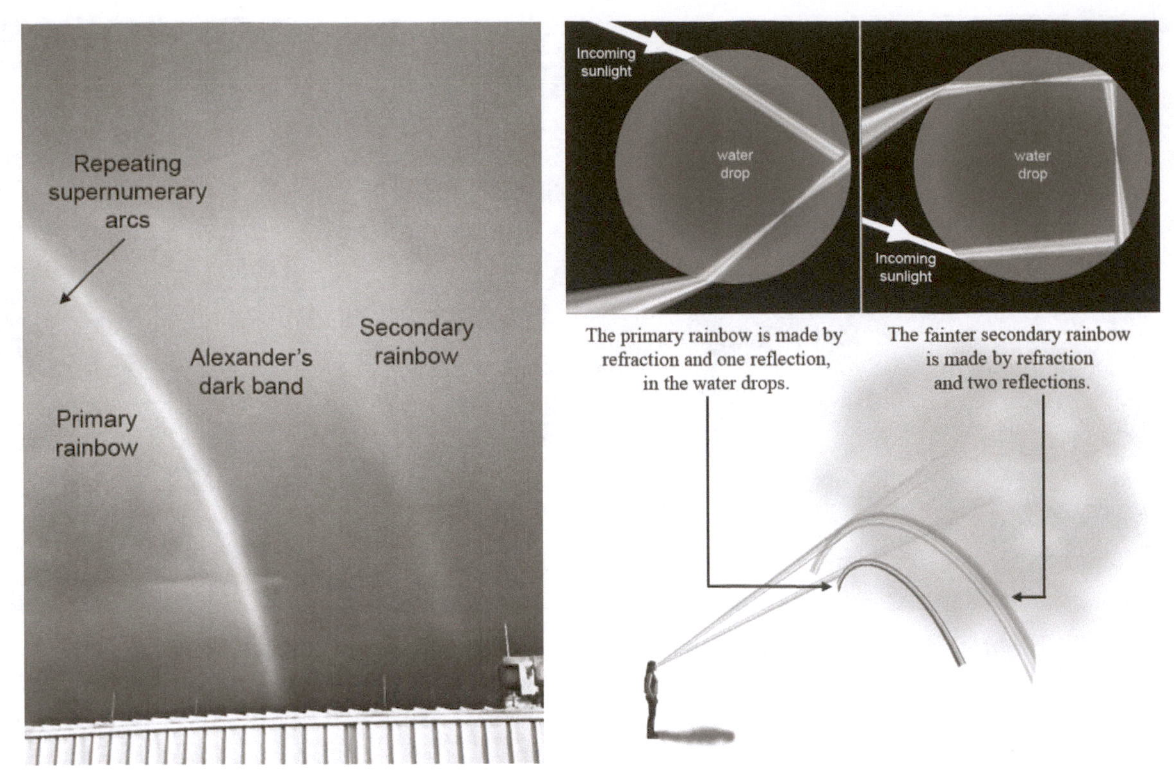

▶ **Figure 10-11** Science now understands rainbows quite well. This doesn't detract from their beauty in the slightest. Indeed, understanding rainbows may help you see subtle phenomena you might otherwise have missed, some of which will be shown in the following pages. (Images courtesy of Frederick Ringwald)

▶ **Figure 10-12** "I can appreciate the beauty of a flower. But at the same time, I see much more in the flower. I can imagine the cells inside, which also have a beauty... There are all kinds of interesting questions that come from a knowledge of science, which only adds to the excitement and mystery and awe of a flower. It only adds, I don't understand how it subtracts." –Richard Feynman

And so also with rainbows. Understanding them helps one appreciate their beauty more, not less. (Image courtesy of Frederick Ringwald)

► **Figure 10-13** This is a moonbow. Notice the stars in the background: this image was taken by the light of a Full Moon. (Image courtesy of Frederick Ringwald)

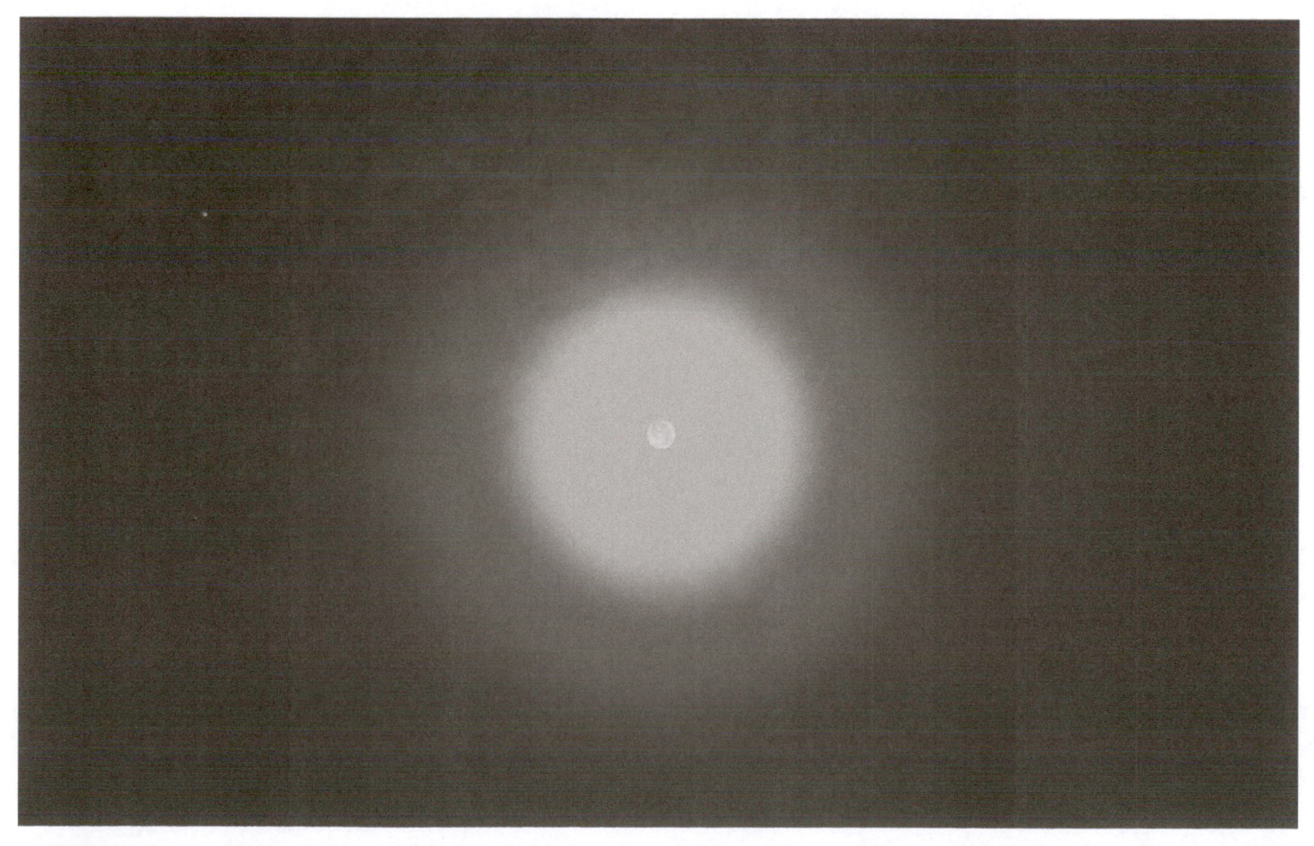

► **Figure 10-14** This is an aureole around the Moon. (Image courtesy of Frederick Ringwald)

► **Figure 10-15** This is a lunar halo. They don't always forecast rain. (Image courtesy of Frederick Ringwald)

► **Figure 10-16** This is a solar halo, caused by a trail from a jet airplane, with the author's fingers blotting out the Sun to show the halo's colors. (Image courtesy of Frederick Ringwald)

► **Figure 10-17** These are sundogs, on both sides of a solar halo. They are caused by ice crystals in the air. (Image © isarescheewin, 2013. Used under license from Shutterstock, Inc.)

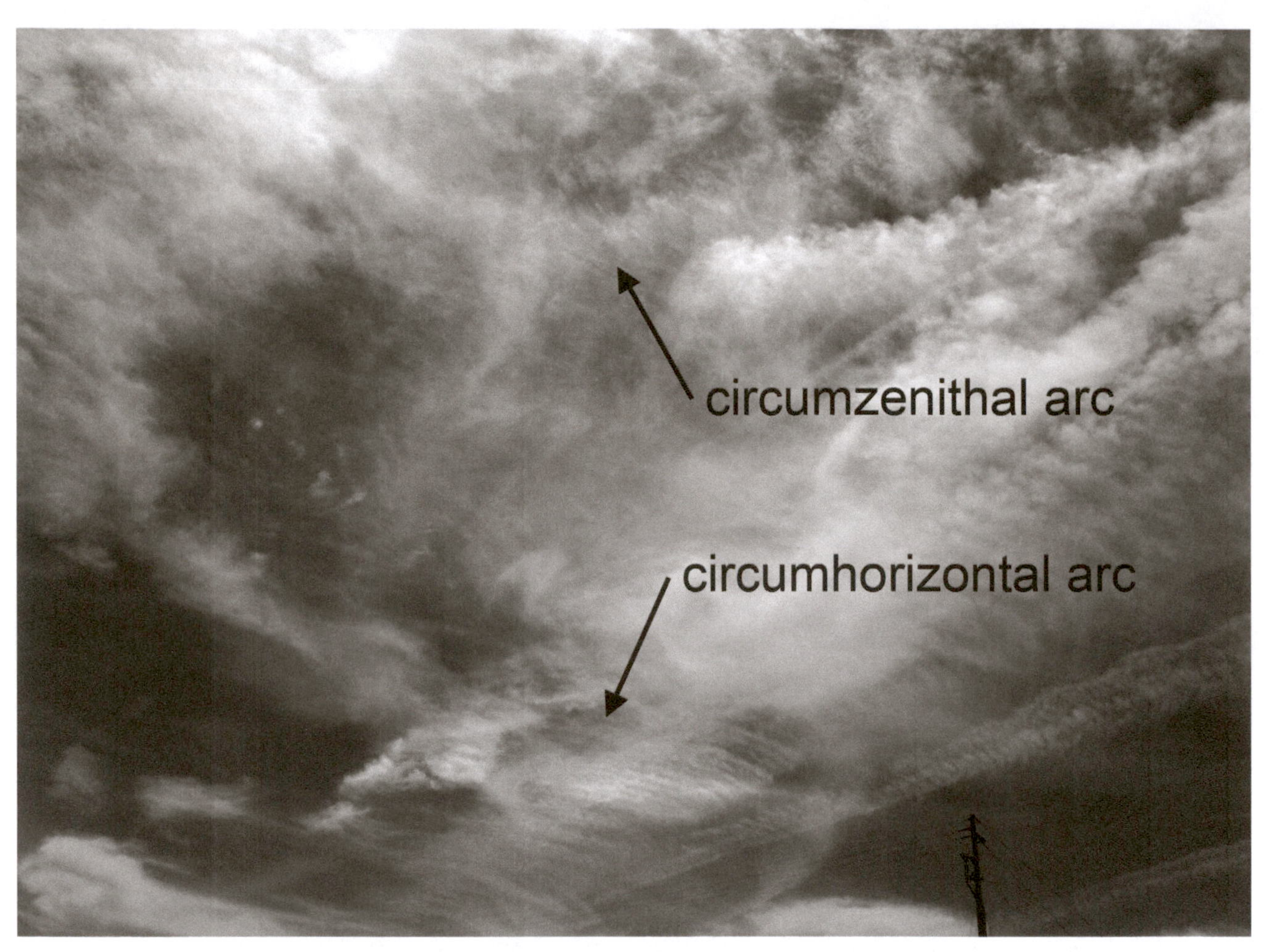

▶ **Figure 10-18** Circumzenithal and circumhorizontal arcs, shown here, are not particularly rare, but they are often mistaken for rainbows. (Image courtesy of Frederick Ringwald)

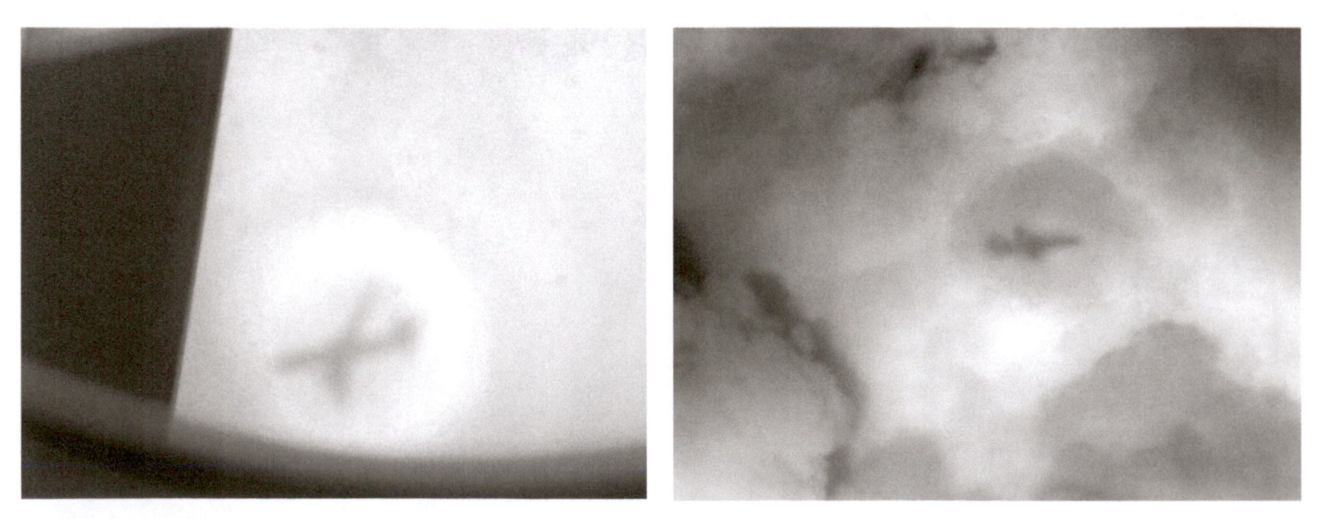

▶ **Figure 10-19** A glory is a ring of colors around an observer's shadow, when the shadow falls in clouds. Both photos were taken through airplane windows. (Images courtesy of Frederick Ringwald)

► **Figure 10-20** This is a sunset. (Image courtesy of Frederick Ringwald)

► **Figure 10-21** This is a sunrise. (Image courtesy of Frederick Ringwald)

CHAPTER

11

More Tricks of the Light

(or further ways astronomers use information in starlight)

Thermal Radiation is the light that any hot object gives off, because of its *temperature*. An example of this is a red-hot poker in a fire: it glows because it's hot.

A human body radiates about 100 Watts of power, not in visible light but in lower-energy infrared radiation. The Sun is much hotter: it emits mostly visible light. Stars that are hotter than the Sun radiate high-energy ultraviolet radiation, or even X-rays.

▶ **Figure 11-1** *Thermal radiation* is the light that any hot, opaque object radiates, because it's hot. (This is sometimes called blackbody radiation, since an object at T = 0 K would be black.) Examples include a red-hot poker in a fire, the fire itself, and the Sun and the stars. (Image courtesy of Frederick Ringwald)

Hotter bodies emit higher-energy radiation, which has shorter wavelengths—this is called **Wien's law**. Because of this, astronomers can tell the **temperatures** of stars and other objects **just by looking at them.**

Thermal radiation is sometimes called *blackbody radiation*, because if an object had a temperature of absolute zero, or zero Kelvins (T = 0 K), it would be black, because it wouldn't radiate any thermal radiation. The Sun isn't black, of course, because its temperature isn't 0 Kelvins: the Sun is quite hot. Thermal radiation has a distinctive pattern over a wide range of wavelengths, or in other words, a continuous spectrum.

Thermal Radiation, the e/m Spectrum, and Temperature

As shown above, the wavelength, color, and energy of light are connected. Temperature is also connected with all of these. This is because thermal radiation is the visible and invisible light (or in other words, e/m radiation) that is given off by hot objects, because they are hot:

Region of the Spectrum	Effect of Earth's Atmosphere	Wavelength (λ)	Photon Energy	What Radiates It? How Hot Is It?
Gamma rays	Absorbed by ozone	10^{-13} m and shorter	$> 10^6$	Nuclear reactions $T = 10^9$ K
X-rays	Absorbed by ozone	10^{-10} m = 1 Ångström	1000	Near black holes $T = 10^6 - 10^7$ K
Ultraviolet radiation	Absorbed by ozone	0.01 – 0.4 microns		The hottest stars $T = 10^4 - 10^5$ K
Visible light		0.4 – 0.7 microns	1	The Sun: T = 5800 K
Infrared Radiation	Absorbed by water vapor	0.7 – 100 microns		Human body: T = 310 K = 37°C = 98.6°F
Microwaves		millimeters	10^{-3}	Cosmic background radiation: T = 2.7 K
Radio waves		centimeters and longer	Low	From cold gas between the stars, and non-thermal processes.

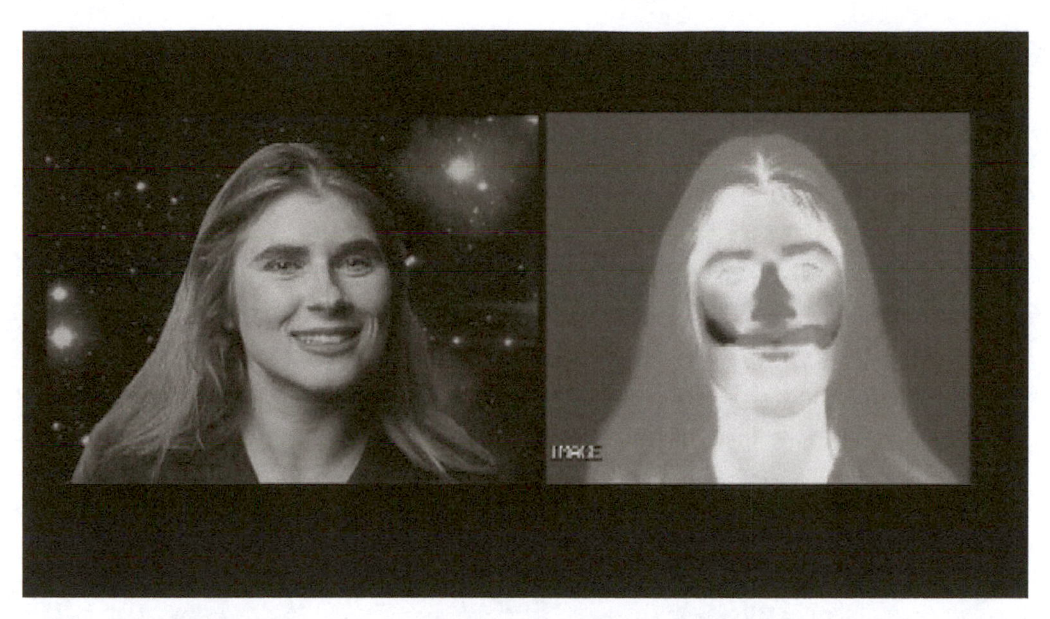

► **Figure 11-2** *Left:* **Visible light:** our unaided eyes see this. It is mostly light from objects hotter than you, such as the Sun, that reflects off you. *Right:* **Infrared radiation:** your body makes, or radiates, 100 Watts of infrared radiation, because of your body temperature. Your clothes don't make you warm—it's you. (Source: NASA/IPAC/Spitzer Space Telescope/Dr. Michelle Thaller)

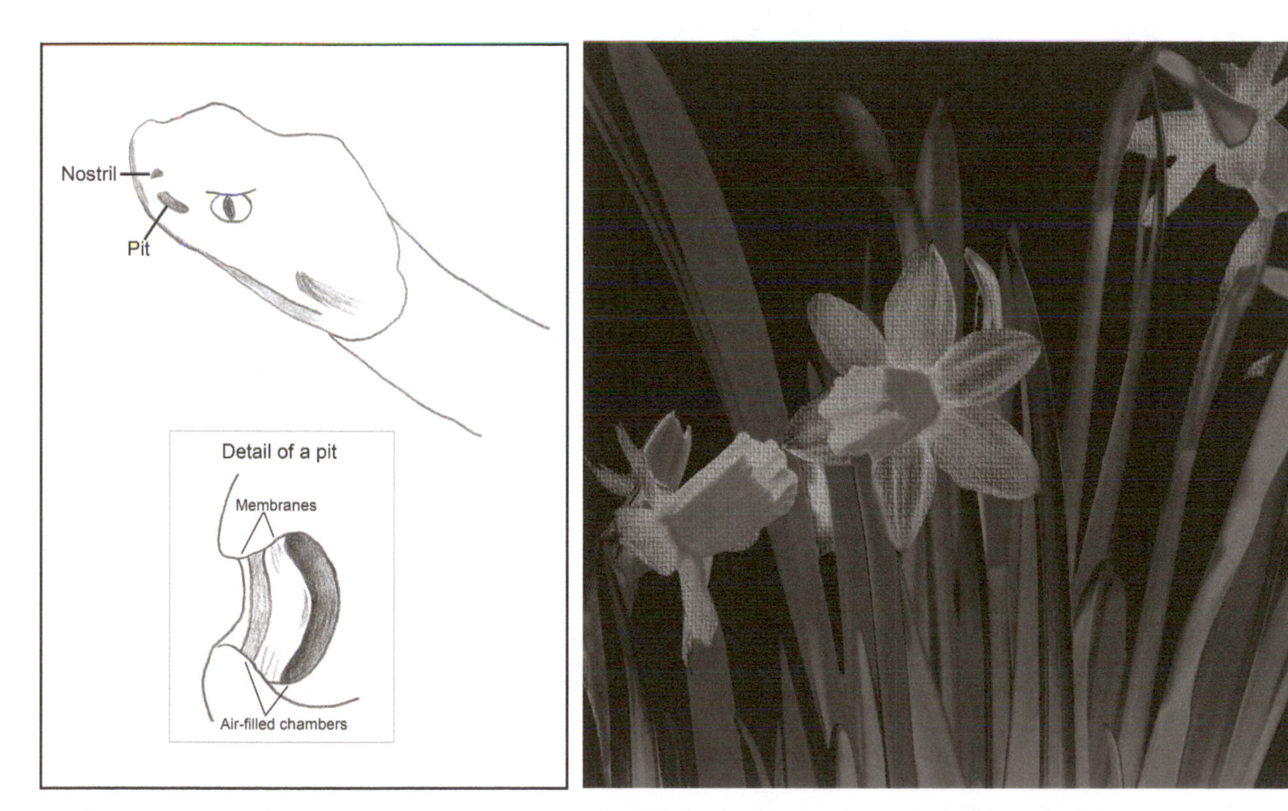

► **Figure 11-3** *Left:* Snakes can sense infrared radiation with specialized organs called pits. (Image courtesy of Frederick Ringwald) *Right:* Bees can see ultraviolet. That's right, bees can see colors of flowers that you can't. If one could "see with someone else's eyes," the world probably *would* look quite different! (Image © Katarzyna Mikolajczyk, 2013. Used under license from Shutterstock, Inc.)

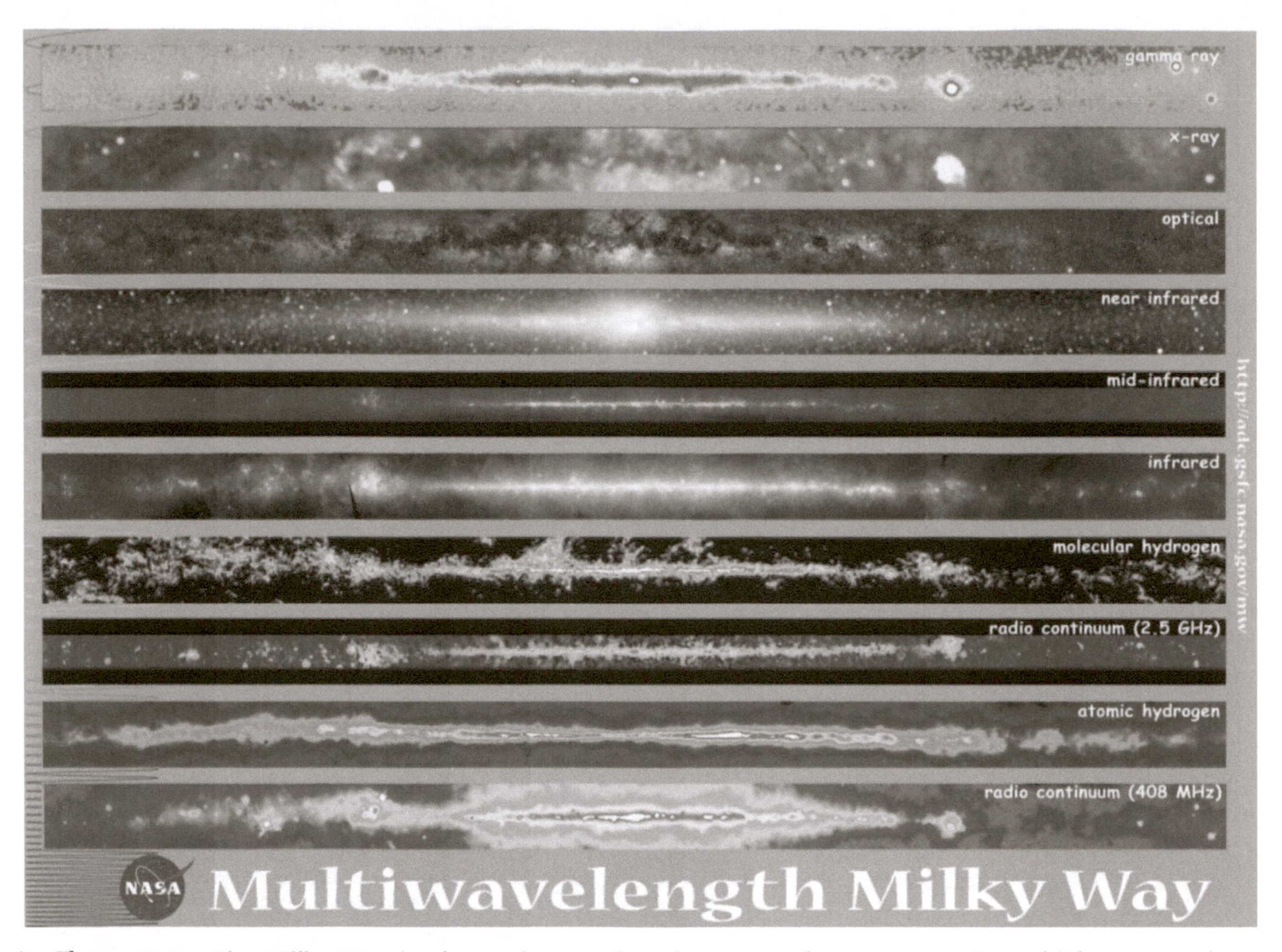

► **Figure 11-4** The Milky Way is shown in wavelengths across the spectrum, from high-energy, short-wavelength gamma rays at top to low-energy, long-wavelength radio waves at bottom. Visible light, labeled "optical," is in between these extremes.

This provides a way of telling how hot objects are just by looking at them. Visible light, labeled "optical," is from normal stars, such as the Sun. X-rays come from hotter things, such as the gas going down black holes, or shocks from exploding stars. Gamma rays *(at top)* come from even hotter things, such as matter/antimatter annihilation. Radio waves *(at bottom)* come from the cold gas between the stars. Infrared comes from room-temperature objects, such as dust and planets. Near-infrared comes from stars that are cooler than the Sun, which are the most common stars. (Source: NASA)

Light carries other information, too.

Kirchhoff's laws were discovered by Gustav Kirchhoff and Robert Bunsen in 1860:

1. Any hot object that isn't transparent, or in other words is *opaque*, radiates a rainbow of colors, because it's hot. This is called a continuous spectrum, because this light has many different wavelengths. A red-hot poker in a fire emits a continuous spectrum.

2. Hot gas that is transparent acts differently. It emits only specific wavelengths of light, which are called spectral lines. A *nebula* has such an emission spectrum.

3. Hot, opaque objects that are surrounded by cooler, transparent gas emit spectra that have dark lines on a bright continuum. A *star* has such an absorption spectrum, because it's opaque on the inside, and transparent on the outside. This is because a star is a ball of gas that is held together by gravity.

► **Figure 11-5** *Left top:* Gustav Kirchhoff (Image © Nicku, 2013. Used under license from Shutterstock, Inc.) *Left bottom:* Robert Bunsen (Image © Nicku, 2013. Used under license from Shutterstock, Inc.) ***Right:*** Kirchhoff's laws of spectra (Image courtesy of Frederick Ringwald)

► **Figure 11-6** Light can also tell astronomers what stars are made of. Philosopher Auguste Comte, shown here, wrote in 1835 that "We may determine [the stars'] forms, their distances, their sizes, and their motions—but we can never know anything of their chemical composition…" He shouldn't have written this, because shortly afterward astronomers started using spectra to do exactly this. (Public Domain)

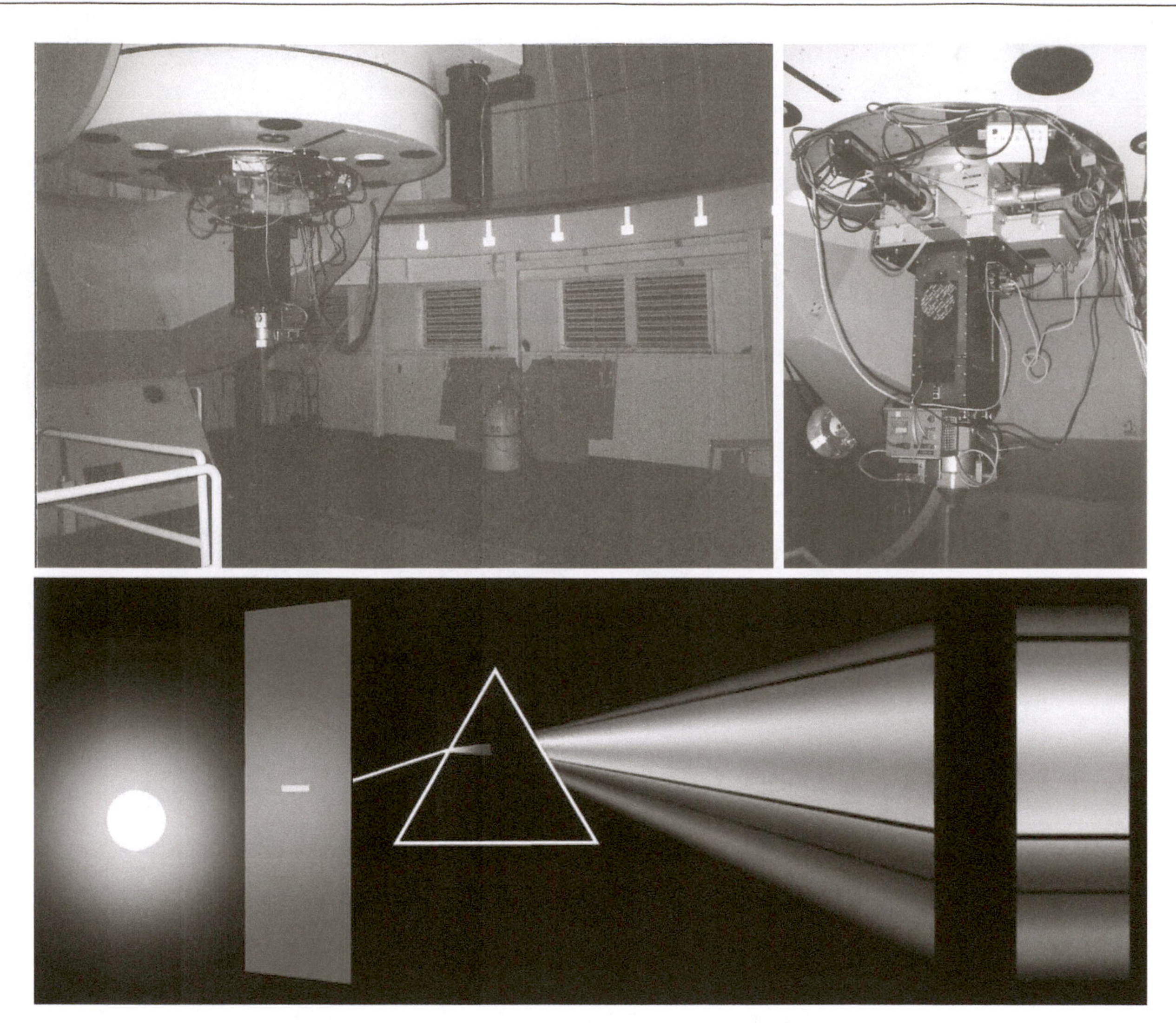

► **Figure 11-7** *Top left and right:* This is a **spectrograph** on a large telescope. Astronomers use it to record the spectra of planets, stars, nebulae, and galaxies *(bottom).* (Images courtesy of Frederick Ringwald)

Chemical Fingerprinting

Different chemical elements have different lines in their spectra, either in emission or absorption. Astronomers can tell what stars are made of by observing these lines. A star made of pure hydrogen will have only hydrogen lines in its spectrum. A star made of both hydrogen and helium will show both hydrogen and helium lines.

Different elements have different lines because the elements' atoms are different. The electrons in atoms can only have specific amounts of energy. This is because electrons have both particle and wave properties, similar to how photons have both wave and particle properties. This is called "the wave-particle duality" of nature. This is the essence of quantum mechanics, one of the most influential ideas of 20th-century physics. (Relativity is another.) Quantum mechanics was developed by Einstein, Niels Bohr, and others. The dark lines in absorption spectra (from stars) come from the electrons in atoms absorbing photons. The bright lines in emission spectra (from nebulae) come from electrons in atoms emitting photons.

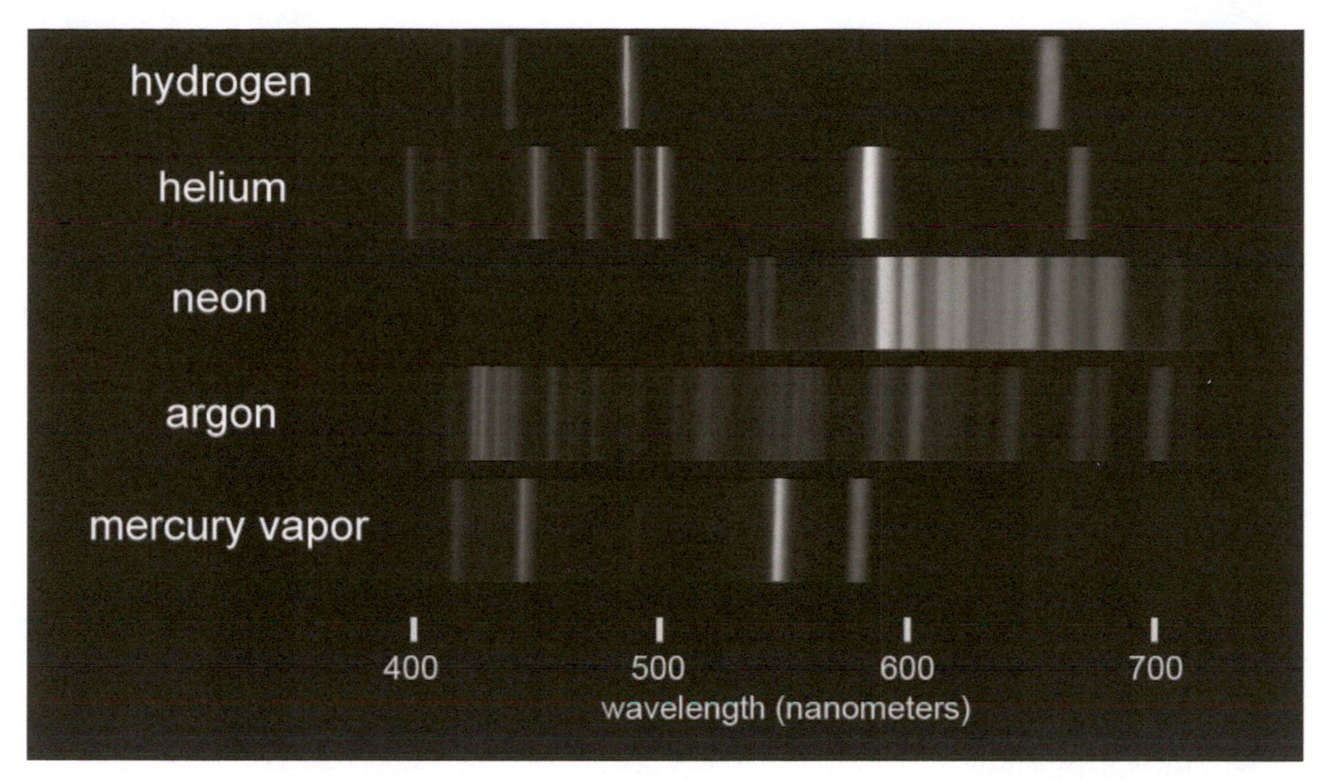

► **Figure 11-8** Chemical fingerprinting can tell what stars are made of. This works because different elements have different lines in their spectra. In other words, different hot gases give off different colors, or wavelengths, of light. (Image courtesy of Frederick Ringwald)

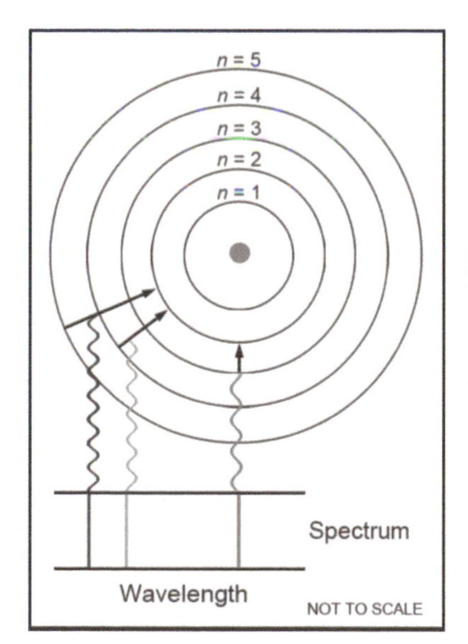

► **Figure 11-9** *Left:* In 1913, Niels Bohr devised the modern theory of the atom. It explains spectral lines. Electron orbits have specific energies. Electrons jump between orbits when atoms gain or lose energy. Atoms therefore emit or absorb only specific amounts of energy, or *wavelengths of light*. (Image courtesy of Frederick Ringwald) *Right:* Niels Bohr worked out his theory of the atom when he was working as a postdoctoral researcher for Rutherford. (Image © Antonio Abrignani, 2013. Used under license from Shutterstock, Inc.)

The Doppler Effect

Briefly: Objects coming *toward* you look *bluer* than they are.
Also: Objects moving *away* from you look *redder* than they are.

This is because of the wave nature of light. The Doppler effect works with all kinds of waves, both light and sound waves. The lonesome wail of a train whistle changes pitch, because of the Doppler effect.

The lines in an object's spectrum will be shifted in wavelength, by the object's motion. Objects moving toward you will have lines that are shifted toward shorter wavelengths, or blue-shifted (or in other words, shifted toward the blue). Objects moving away from you will have lines that are shifted toward longer wavelengths, or red-shifted (or in other words, shifted toward the red). Formally,

$$\frac{\text{The shift in wavelength}}{\text{The normal wavelength}} = \frac{\text{Speed toward or away from the observer}}{\text{The speed of light}}$$

or: $$\frac{\Delta\lambda}{\lambda} = \frac{v}{c}$$

With this single technique, we can:

- Detect extrasolar planets
- Measure masses of stars
- Detect black holes
- Observe the expansion of the Universe

This is a powerful technique. It's a good example of what scientists like: one method that can provide information about many different kinds of phenomena. Christian Doppler deserved a Nobel Prize for discovering this, but he died in 1853. The first Nobel Prizes weren't awarded until 1901, and may not be awarded posthumously.

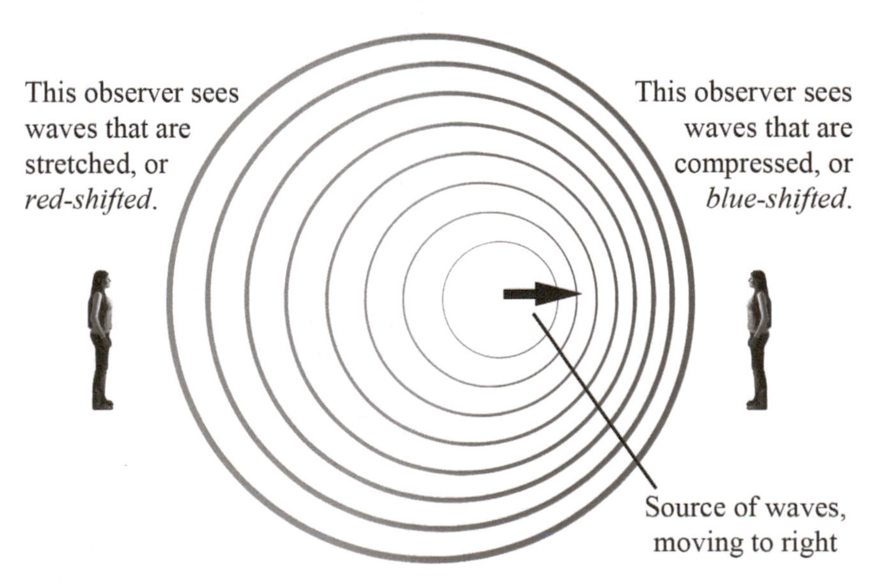

Figure 11-10 The Doppler effect works with light waves. As a star comes toward an observer, it looks bluer than it would if it were stationary, because the light waves in front of it are squashed, or blue-shifted. As a star moves away, it looks redder, because the light waves in back of it are stretched, or red-shifted. (Image courtesy of Frederick Ringwald)

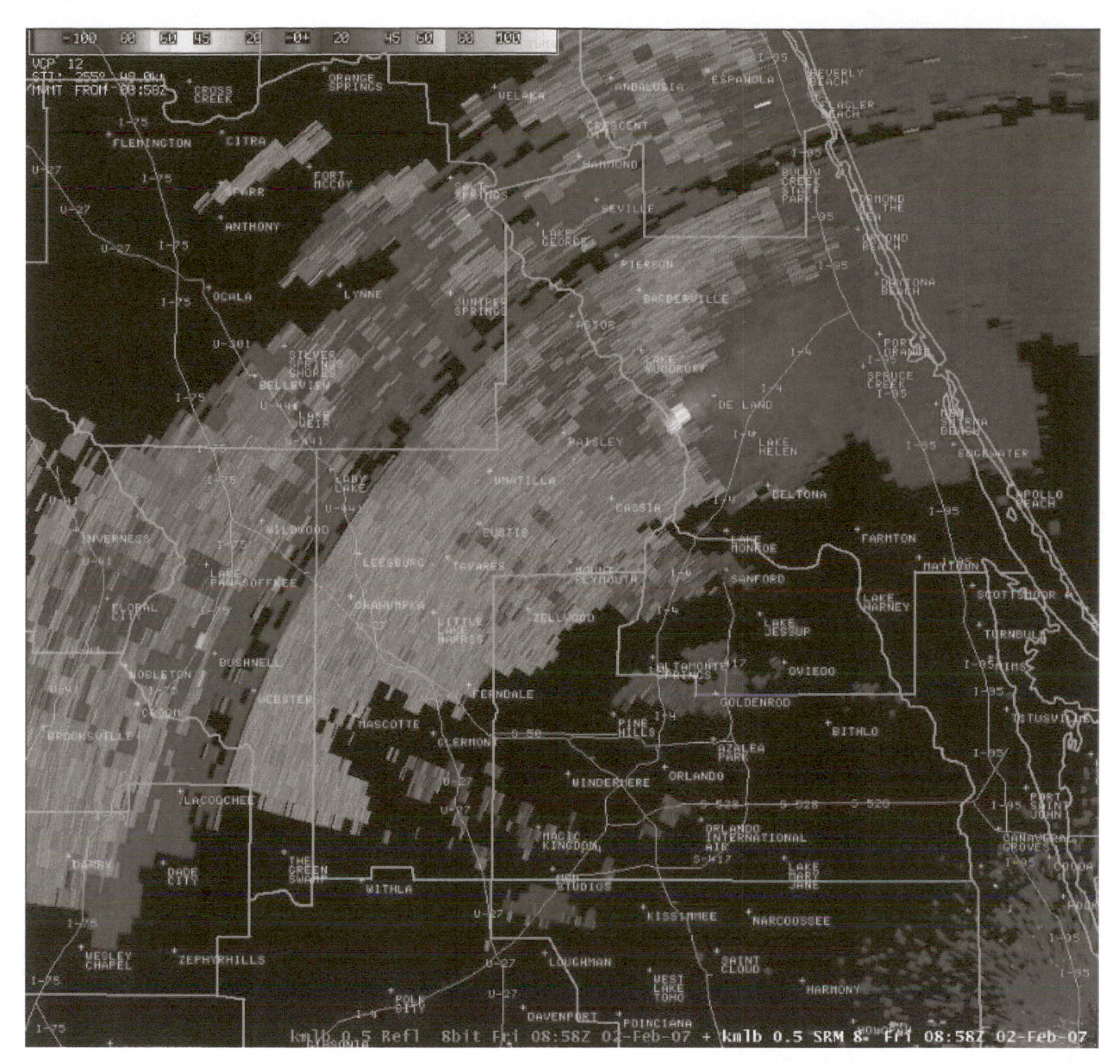

▶ **Figure 11-11** The Doppler Effect: A Doppler radar can detect *motion*. This one has found a tornado. (Source: NOAA)

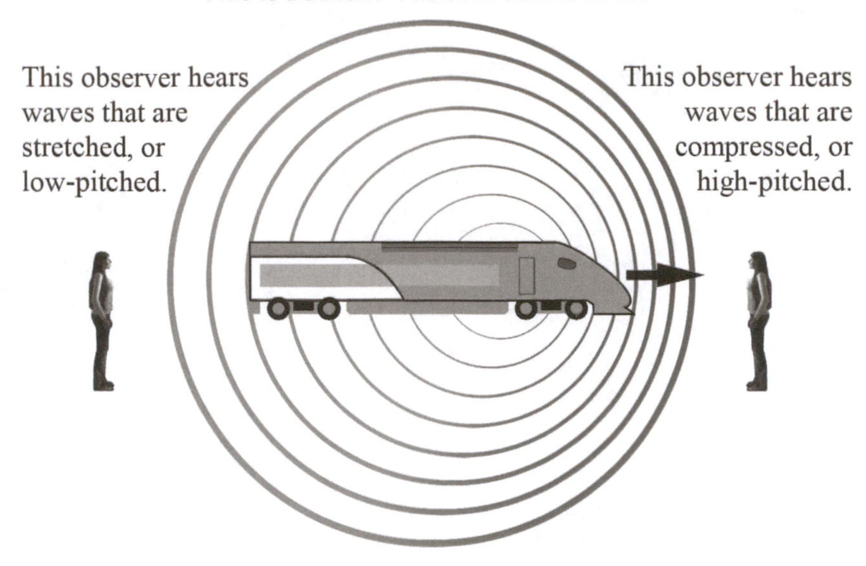

► **Figure 11-12** The Doppler effect works with sound waves, too. (Image courtesy of Frederick Ringwald)

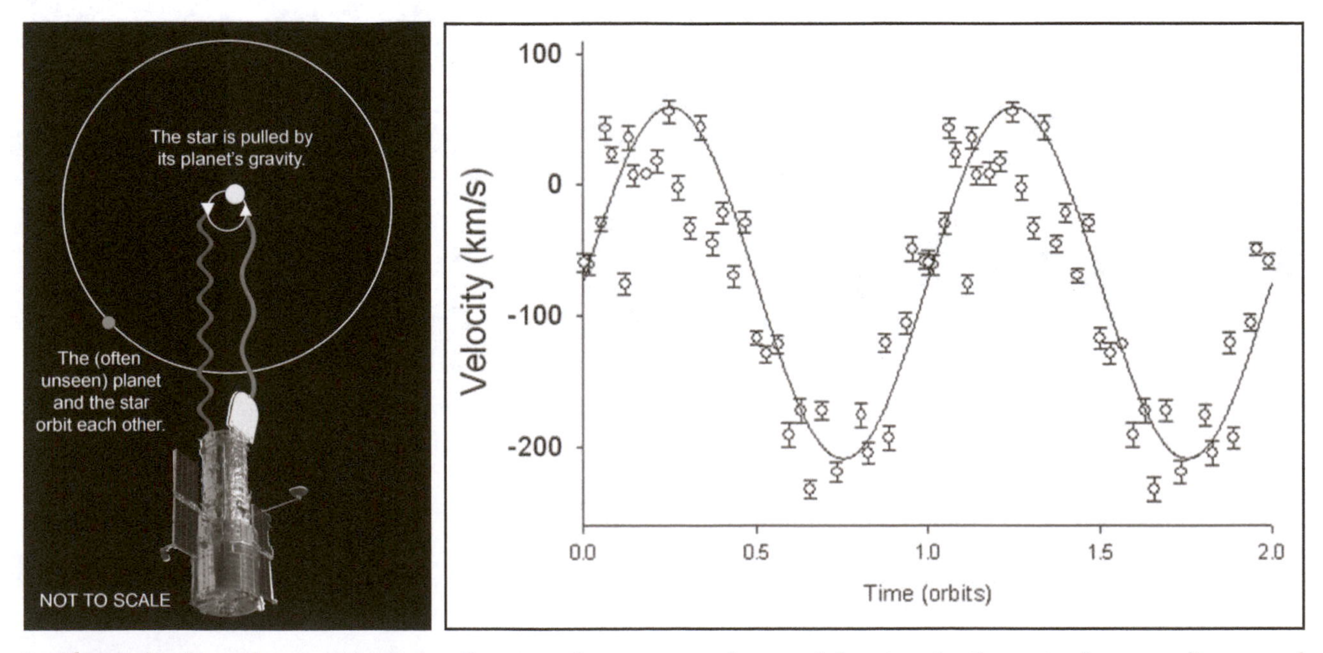

► **Figure 11-13** About 400 of the first 500 known exoplanets (planets of other stars) were discovered indirectly, by detecting motion in their spectra by using the Doppler effect. A telescope on or near Earth detects blue-shifted starlight when the star moves toward Earth, and red-shifted starlight when the star moves away from Earth. This is also how astronomers detect black holes. This is also how Edwin Hubble discovered that the Universe is expanding. (Images courtesy of Frederick Ringwald and NASA)

CHAPTER

12

Telescopes

Telescopes come in two basic kinds: refractors and reflectors. Refracting telescopes use lenses. Reflecting telescopes use mirrors.

Why Refracting Telescopes Reached Their Maximum Practical Size with the Yerkes 40-inch Refractor in 1897

Refractors use lenses, which refract, or bend, light. They also split light into its component colors, much as prisms do. This is called chromatic aberration.

To reduce chromatic aberration, many refractors have a long focal length. This means that refractor tubes are very long, with no way to fold them up, as with reflectors.

This means that refractors often have:

- Long tubes,
- Tall piers,
- Domes like great cathedrals, and
- HUGE COSTS.

Refractors also often have bad "dome-effect" seeing, from their huge domes trapping lots of air.

Other problems with refractors are that:

- Lenses bend out of shape over time, due to their own weight. Mirrors don't, since they can be supported from the back.
- Glass from which lenses are made must be perfect, with no bubbles or cracks, and perfectly transparent.
- Mirrors are more forgiving. Only one surface matters.
- Mirrors only need to be polished and aluminized (formerly silvered) on one side. Lenses need at least two sides, and so are at least twice as expensive, and typically a lot more. An 8-inch reflector costs about $500; an equivalent 8-inch refractor can cost $10,000.

► **Figure 12-1** *Left:* The 40-inch (1.0-meter) refracting telescope at Yerkes Observatory has a lens 40 inches in diameter. It is the world's largest *refractor*, and was built in 1897. (Image courtesy of Fred Lusk) *Right:* This is the objective lens of a 4-inch refracting telescope. It has a 4-inch aperture, because it is 4 inches in diameter. The Yerkes refractor's lens is 10 times larger. (Image courtesy of Frederick Ringwald)

► **Figure 12-2** Refraction is why this pencil looks like it is bent. Refraction happens because the speed of light in water is less than the speed of light in air, so the light rays bend. (Image courtesy of Frederick Ringwald)

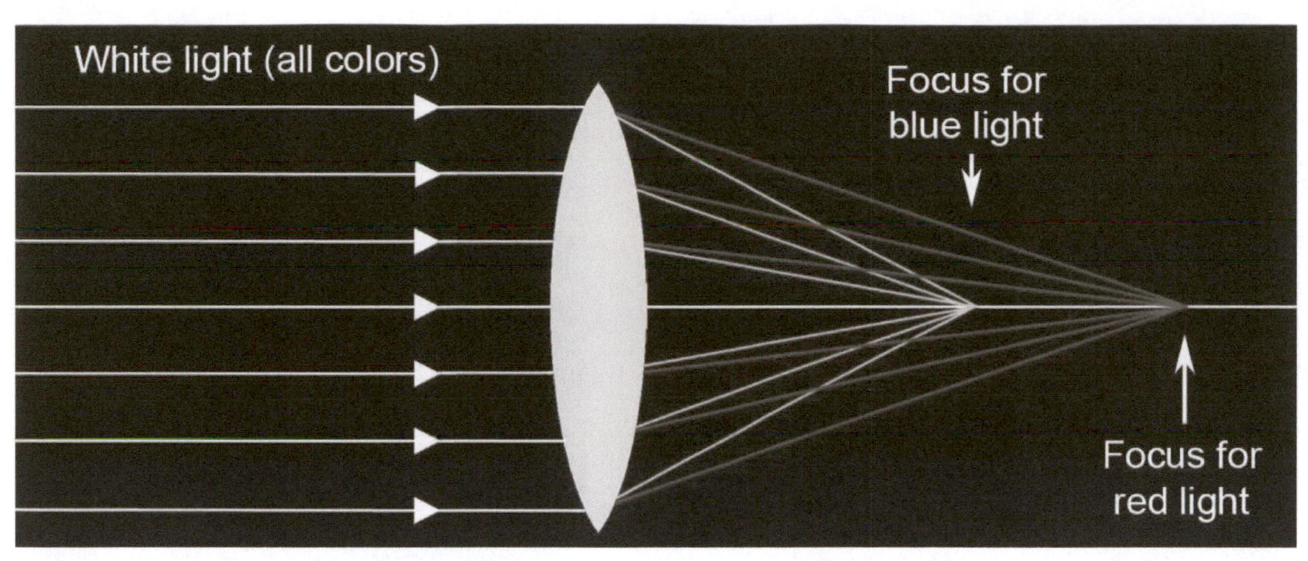

► **Figure 12-3** Chromatic aberration, or color error, happens because lenses break light down into its component colors, much as how prisms do. It makes color photography difficult, since it prevents different colors of light from being in focus at the same time. (Image courtesy of Frederick Ringwald)

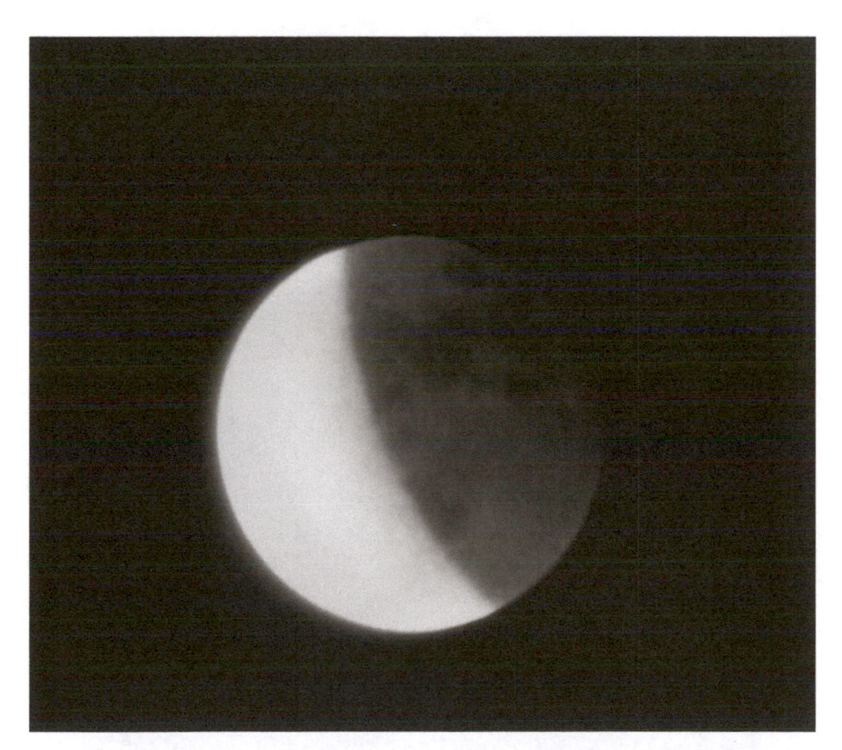

► **Figure 12-4** This shows what chromatic aberration does to an image. This photo of a lunar eclipse was taken through a cheap refracting telescope that suffers from chromatic aberration. Notice the violet halo around the left edge (or limb) of the Moon. The Moon doesn't really have a violet halo—this violet light is focused wrongly by the telescope. (Image courtesy of Frederick Ringwald)

► **Figure 12-5** *Left:* Isaac Newton invented the Newtonian *reflector*. A reflecting telescope uses mirrors, not lenses. It therefore does not suffer from chromatic aberration. (Image © Iryna1, 2013. Used under license from Shutterstock, Inc.) *Right:* This photo was taken looking down the tube of a Newtonian reflector. Notice the mirror at the bottom of the tube. Light is reflected by this mirror to another mirror on the the black disk in the top of the tube, which reflects it out the eyepiece, at top right. (Image courtesy of Frederick Ringwald)

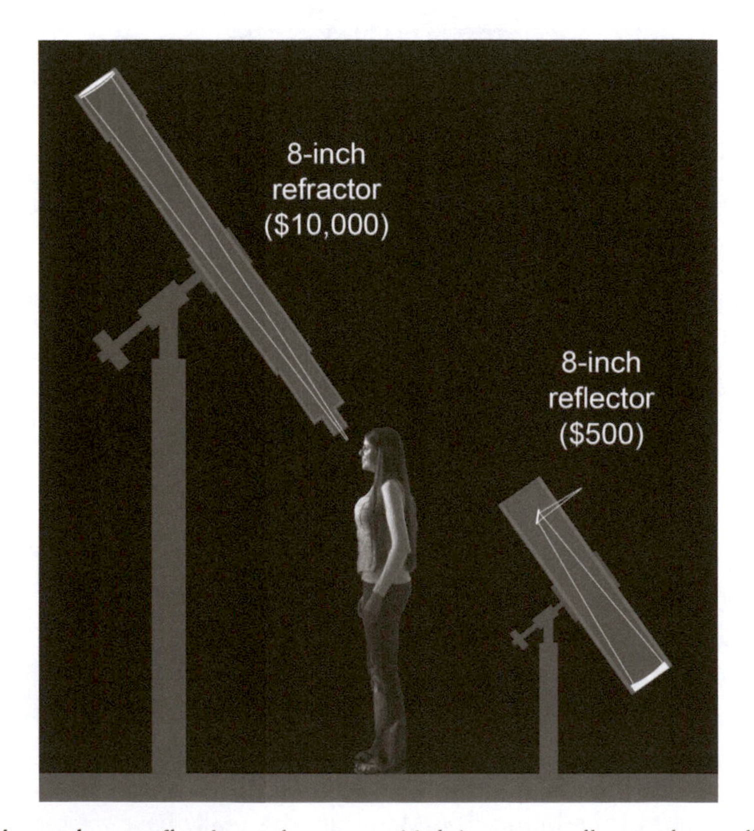

► **Figure 12-6** This shows how reflecting telescopes *(right)* are usually much smaller and more portable than refracting telescopes *(left)*, since both telescopes shown have 8-inch apertures. Reflectors are also usually cheaper; an 8-inch reflector like the one shown on the right costs about $500. An 8-inch refractor like the one at left costs about $10,000.

Catadioptric telescopes are combination refracting and reflecting telescopes, with both lenses and mirrors. They are popular with amateur astronomers since they can be small and portable. They can also be fairly expensive; an 8-inch catadioptric telescope like this costs about $1,200. (Image courtesy of Frederick Ringwald)

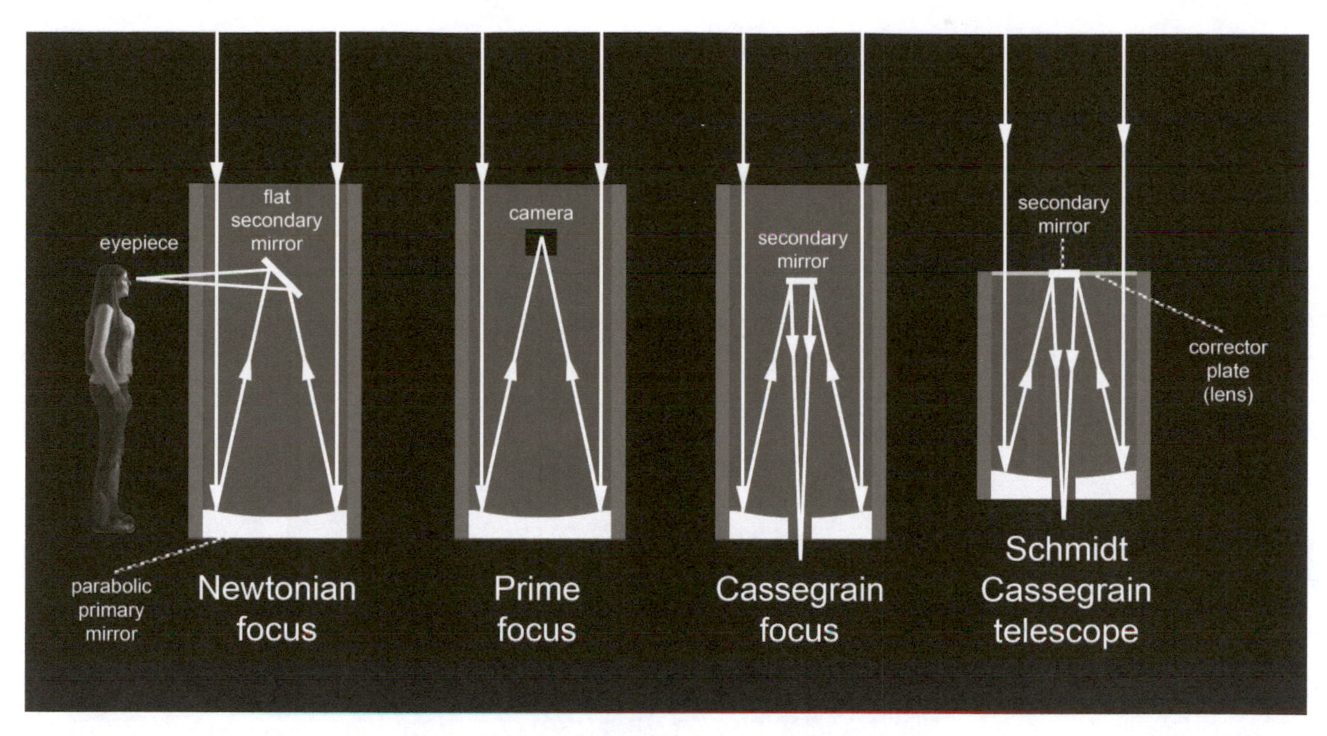

► **Figure 12-7** Another advantage of reflecting telescopes is that, since they are folded up because of the mirrors, they can be built in many ways. (Image courtesy of Frederick Ringwald)

► **Figure 12-8** Nearly all large observatory telescopes built since the Yerkes 40-inch refractor in 1897 have been reflectors. This is a 0.9-meter (36-inch) telescope for visible light. It has nearly the same aperture (diameter) as the Yerkes refractor but cost about 1/25th as much. (Image courtesy of Frederick Ringwald)

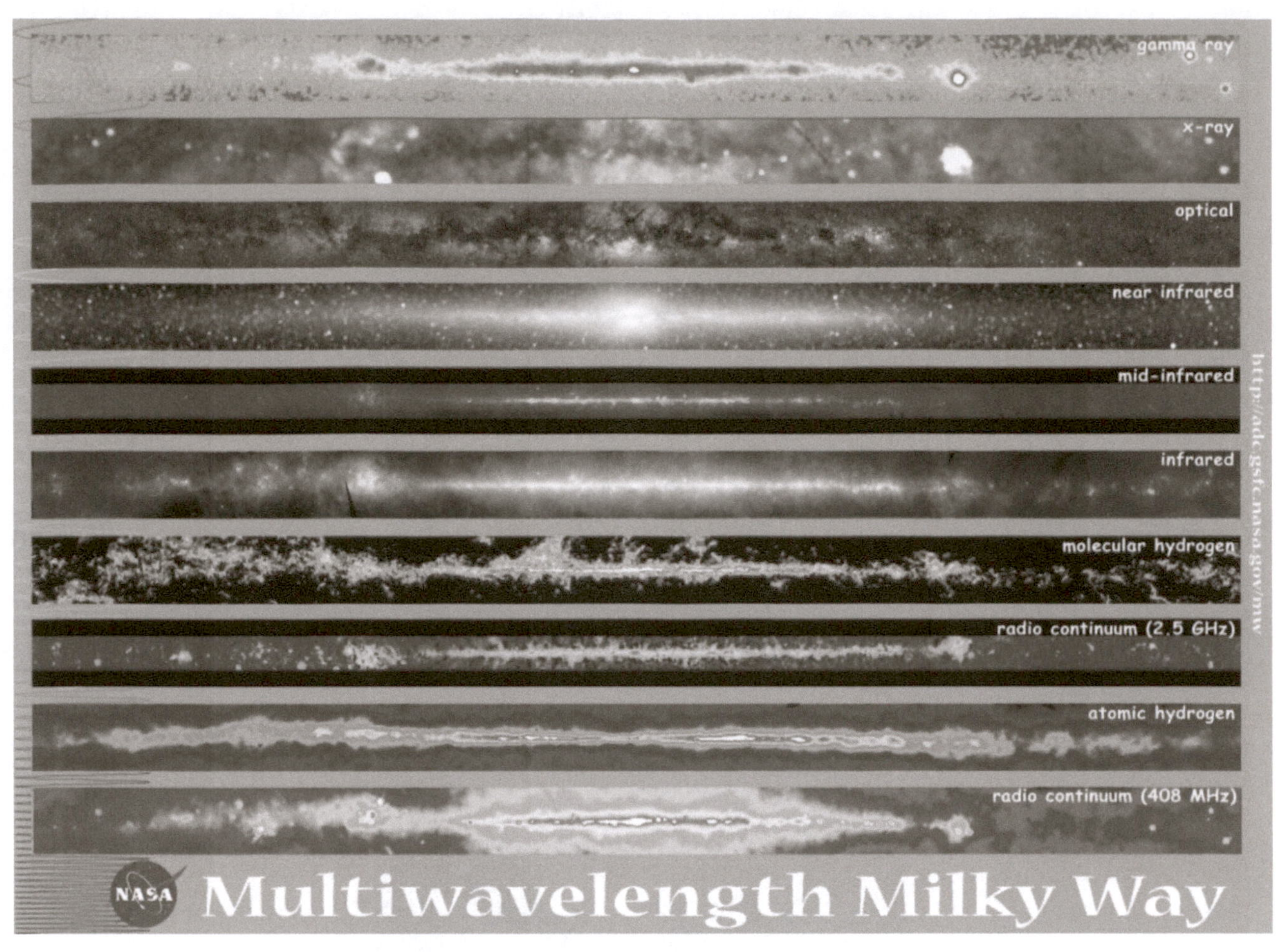

► **Figure 12-9** The most influential advance in astronomy in the past two generations has been the opening of the electromagnetic spectrum. Astronomers can now detect light at both visible and invisible wavelengths. This means that there are radio telescopes, microwave telescopes, infrared telescopes, ultraviolet telescopes, X-ray telescopes, and gamma-ray observatories. (Source: NASA)

► **Figure 12-10** *Left:* Radio telescopes detect radio waves from space. Since radio waves have such long wavelengths, their surfaces do not have to be as smooth and precise as the mirrors of visible-light telescopes. Radio telescopes can therefore be bolted together out of sheet metal, and so can be made much larger than visible-light telescopes. Because of their large aperture, radio telescopes are now the most sensitive telescopes of any kind in existence. The radio telescope shown here is the 250-foot radio telescope at Joddrell Bank, in England. Notice the clouds. Clouds are transparent to radio waves, unlike visible light. Radio telescopes can also observe during daylight. (Image courtesy of Frederick Ringwald) *Right:* The reflective dishes of radio telescopes are similar to, although much larger than, the dish antennae commonly used for satellite TV and Internet. (Image courtesy of Frederick Ringwald)

► **Figure 12-11** This is a microwave telescope. It has a 12-meter aperture. (Source: NRAO)

► **Figure 12-12** Many visible-light (optical) and infrared telescopes are at Mauna Kea Observatory in Hawai'i. (Image courtesy of Frederick Ringwald)

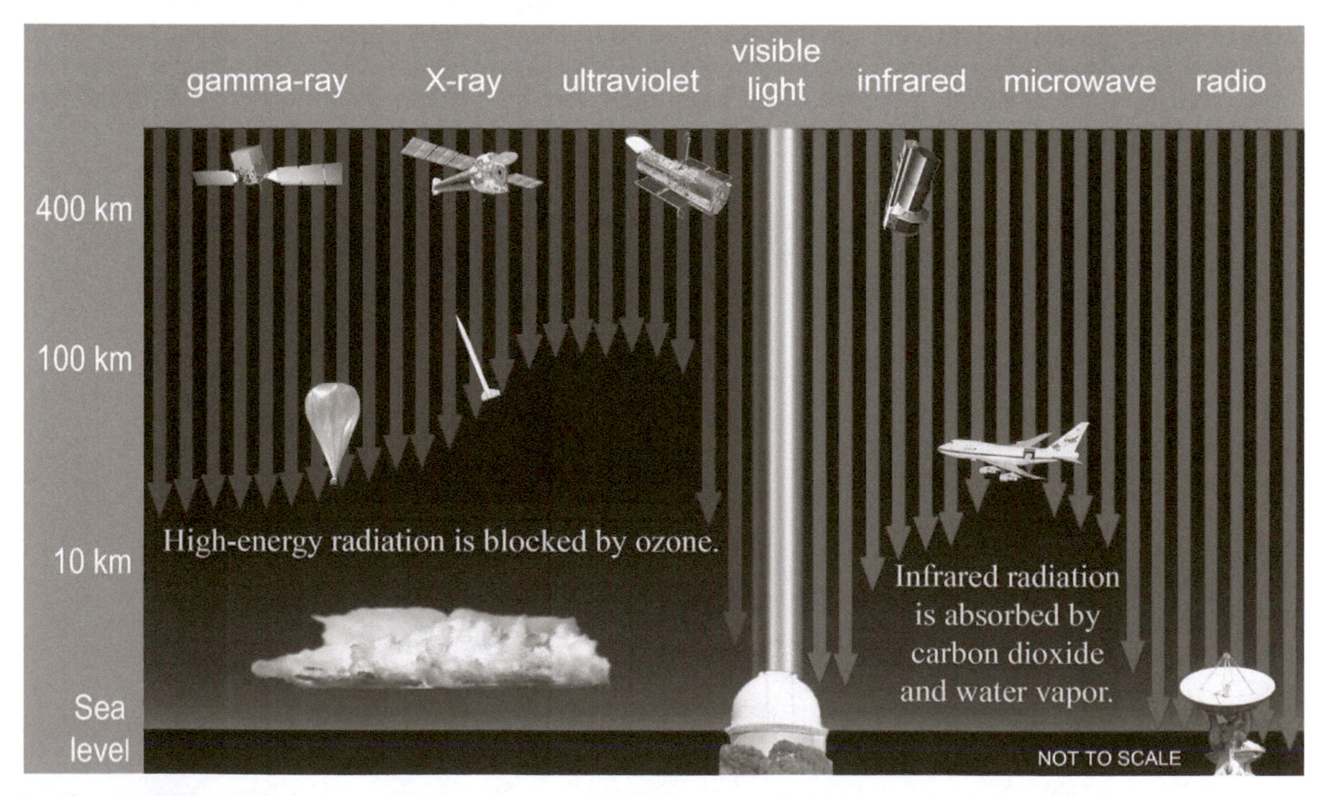

► **Figure 12-13** Earth's atmosphere is transparent only to visible light and to radio waves. (Image courtesy of Frederick Ringwald)

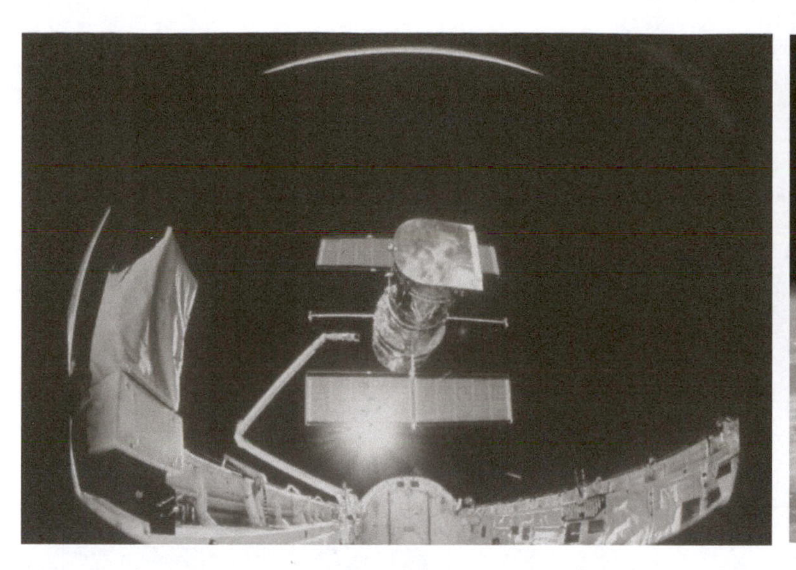

► **Figure 12-14** *Hubble Space Telescope* is above Earth's atmosphere. It detects wavelengths of light not observable from the ground, including *ultraviolet* and *infrared* radiation. *Hubble* was launched into Earth orbit by a Space Shuttle in 1990. (Source: NASA)

The properties of astronomical telescopes, in order of importance:

Most important:

1. **Collecting area,** which sets how much light a telescope gathers:

   This depends on **aperture**: a telescope with larger aperture (or in other words, diameter) can gather more light.

   More precisely, collecting area is the area of the objective:

   Area = π (Aperture/2)^2

   Example: a 5-m telescope (with an aperture of 5 meters) gathers $5^2 = 25$ times more light than a 1-m telescope.

2. **Resolution,** also called resolving power, or image resolution:

   Resolution is how **clear** the images are.

   How small is the smallest detail visible? It depends on the quality of the telescope optics and the observing site. Telescopes on tall mountains are above much of Earth's atmosphere, and are less affected by its obscuring effects.

3. For amateurs: **Portability**!

   Large telescopes have more aperture, but if they're too heavy to pick up, they don't get used much. How steady the telescope's mount is, and how easy it is to point, are also important. An equatorial mount with a drive can compensate for Earth's rotation and keep objects in the field of view.

Least important:

4. **Magnification**: One can change the magnification easily, just by changing eyepieces. High magnification can be a hindrance, not a help. It's hard to find things at high magnification.

► **Figure 12-15** The radio telescope at Arecibo, Puerto Rico has an aperture of 1000 feet (305 meters). It is therefore *very* sensitive. (Source: NAIC)

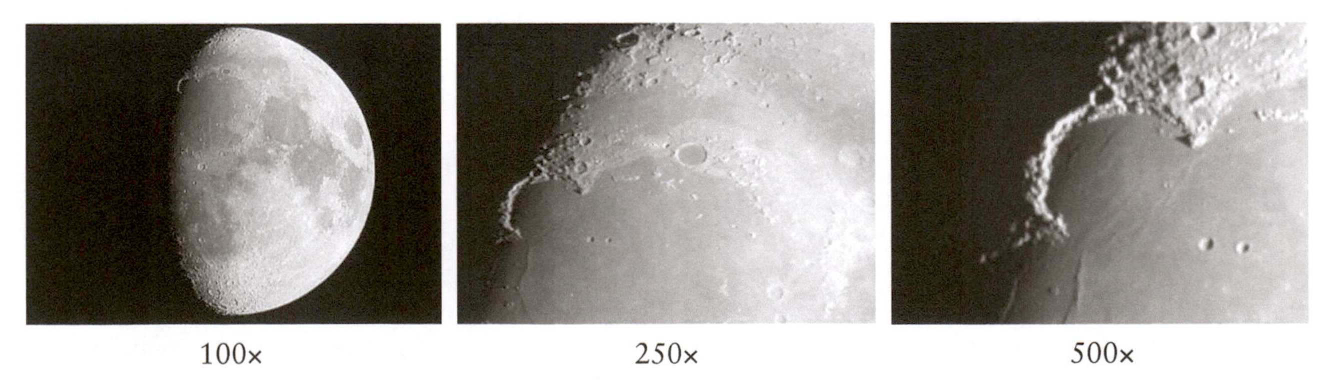

► **Figure 12-16** **Magnification** is the size of the image. Increasing magnification decreases the field of view. It also makes the image fainter. (Images courtesy of Frederick Ringwald at Fresno State's Campus Observatory)

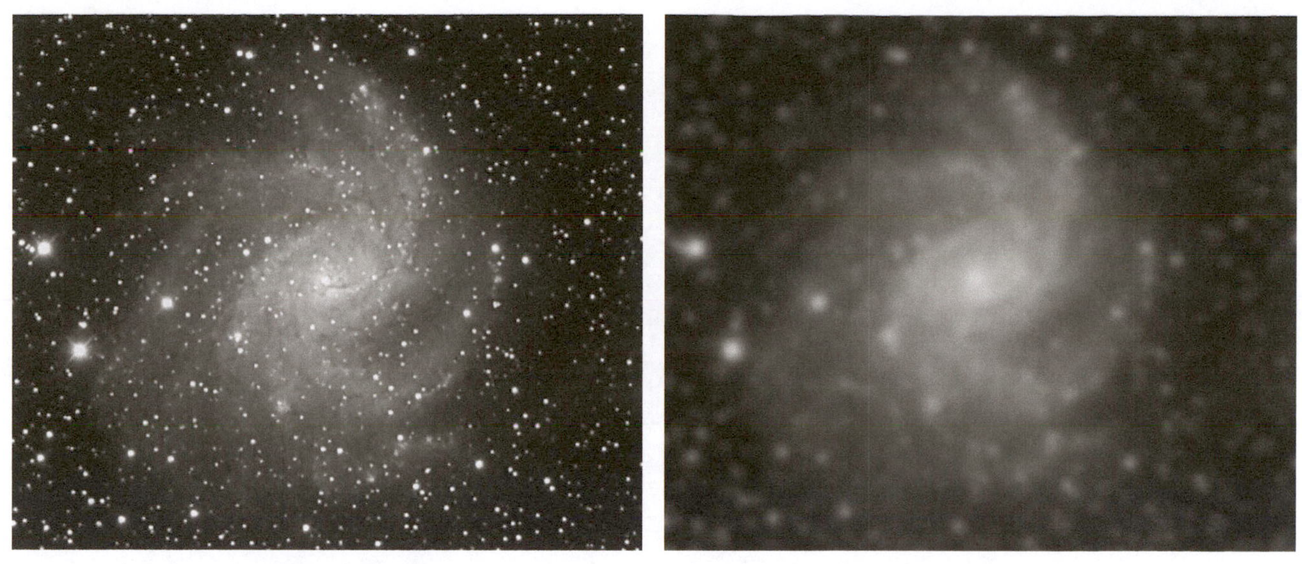

▶ **Figure 12-17** **Resolution** is how clear the image is, or in other words, the size of the smallest discernible detail. Notice also how fainter stars are visible at high resolution because their light isn't spread out so much. (All images courtesy of Frederick Ringwald, made at Mount Laguna Observatory)

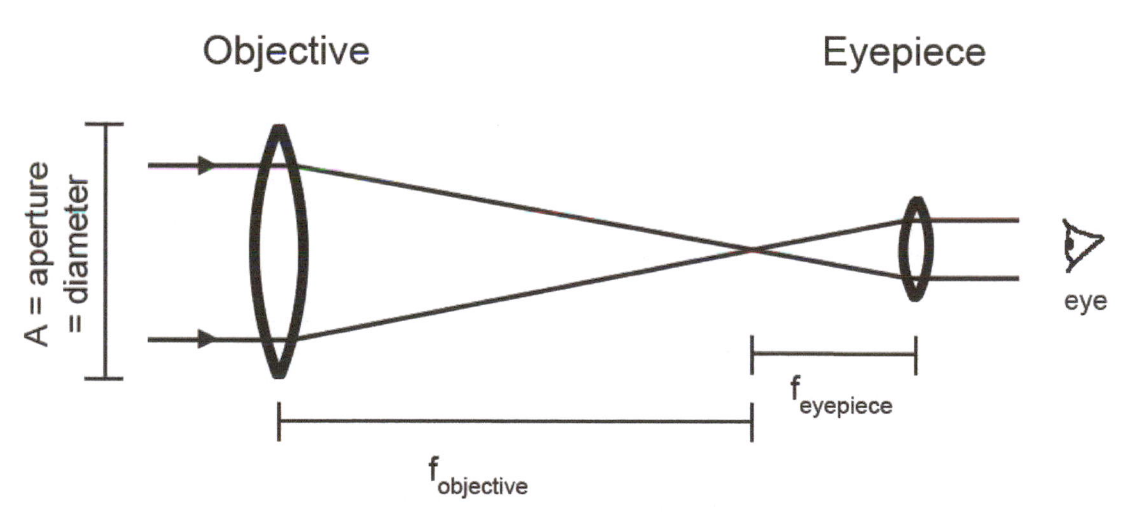

▶ **Figure 12-18** Telescopes have essentially two parts: (1) the objective, which in a refracting telescope is the big lens in the front, and (2) the eyepiece. In a reflecting telescope, the objective can be a mirror in the back. The diameter of the objective is the *aperture*. This figure shows the light rays going through a refracting telescope from left to right. The rays come in through the objective, cross at the focus, and go out the eyepiece, which makes them parallel again to give the eye an undistorted image. The distance from the objective to the focus is the focal length of the objective ($f_{objective}$). The distance from the focus to the eyepiece is the focal length of the eyepiece ($f_{eyepiece}$). (Image courtesy of Frederick Ringwald)

Magnification = $f_{objective} / f_{eyepiece}$ = focal length of objective / focal length of eyepiece.

For example, for a telescope with a focal length of 120mm with an eyepiece of 25mm,

Magnification = $f_{objective} / f_{eyepiece}$ = 1200 mm / 25 mm = 48 x, which is also called 48 power.

If one puts in a higher-power eyepiece, with $f_{eyepiece}$ = 10 mm,

Magnification = 1200/10 = 120 x.

So the images are 2.5 times larger, **but** the field-of-view (how much of the sky you can see) is 2.5 times narrower.

So, higher magnification **isn't** always better. As you increase the magnification, you **decrease the field of view**, which is how much of the sky you can see.

Therefore, use **low** power to find things, since you can see a wide area of the sky. Only after you find what you're looking for can you switch to high power, for a close-up view.

Magnification is not the same as resolution.
Magnification is how **big** the image is.
Resolution is how **clear** the image is

It is perfectly possible to have a large, unclear image. (This can be demonstrated easily with an overhead projector.) Therefore, don't buy a telescope because of the magnification the sellers say that it can attain. All one needs to do to change the magnification is just change the eyepieces!

► **Figure 12-19 Portability** can be a problem, for telescopes that are not housed permanently in observatories. This 16-inch reflector delivers excellent images, but it weighs over 200 pounds. This requires its owner to be so dedicated, he has a mini-van to transport this telescope. (Image courtesy of Frederick Ringwald)

► **Figure 12-20** A smaller, more portable telescope *can* show you more, if you use it more often—and if it's well-made. Binoculars can be excellent telescopes for beginners. (Image courtesy of Frederick Ringwald)

More on Resolution

Any telescope on the surface of Earth is called a "ground-based telescope." The best ground-based telescopes are on high mountaintops, to get above the obscuring effects of Earth's atmosphere. A "space telescope" is onboard a spacecraft, such as *Hubble Space Telescope*, which orbits above Earth's atmosphere.

The resolution of a ground-based telescope is limited by the **seeing. Seeing is turbulence in Earth's atmosphere.** Seeing blurs the images. This is because the images move and twinkle in turbulent air. Many students notice this motion when looking through a telescope for the first time.

Seeing:	3–4"	Typical backyard ("poor seeing")
	1"	Typical mountaintop observatory
	0.2"	Superb for a ground-based telescope
	0.0455"	*Hubble Space Telescope,* which is above Earth's atmosphere, and so isn't affected by it.

Don't confuse **seeing** with **atmospheric transparency**. Poor seeing is caused by too much turbulent motion in Earth's atmosphere, which blurs images.

Poor atmospheric transparency is different. It's caused by clouds, haze, fog, dust, smoke, or smog, all of which prevent light from passing through the air. Sometimes, the best seeing happens on nights that are a little hazy.

Adaptive optics can allow ground-based telescopes to see almost as clearly as if Earth's atmosphere wasn't there. Adaptive optics uses computer-controlled moving mirrors in the telescope to take the twinkle out of starlight, literally.

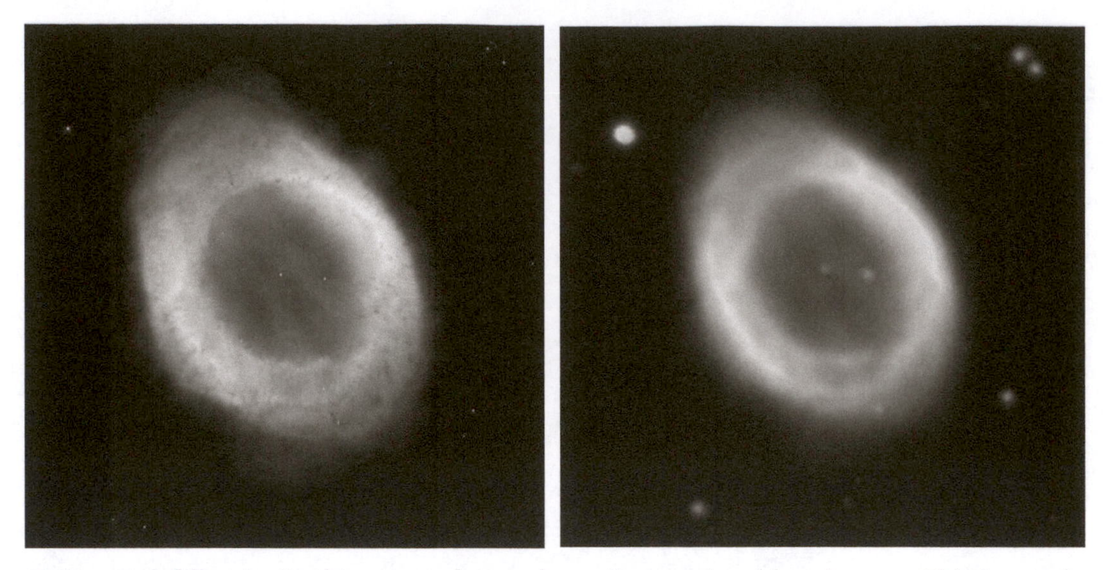

► **Figure 12-21** **Twinkling** in Earth's atmosphere, also called *seeing,* blurs images. This is another reason that astronomers put telescopes on mountains. *Left:* This is perfect seeing, taken with *Hubble Space Telescope*, which is onboard a satellite above Earth's atmosphere. (Source: NASA/STScI) *Right:* This is poor seeing, taken at a Campus Observatory, near sea level. (Image courtesy of Frederick Ringwald at Fresno State's Campus Observatory)

The diffraction limit is the absolute physical limit of the resolution that any telescope can deliver, whether or not it's looking through Earth's atmosphere, or has adaptive optics. This is because light waves spread out when passing through a telescope's tube. The diffraction limit depends on the telescope's aperture and the wavelength of the light.

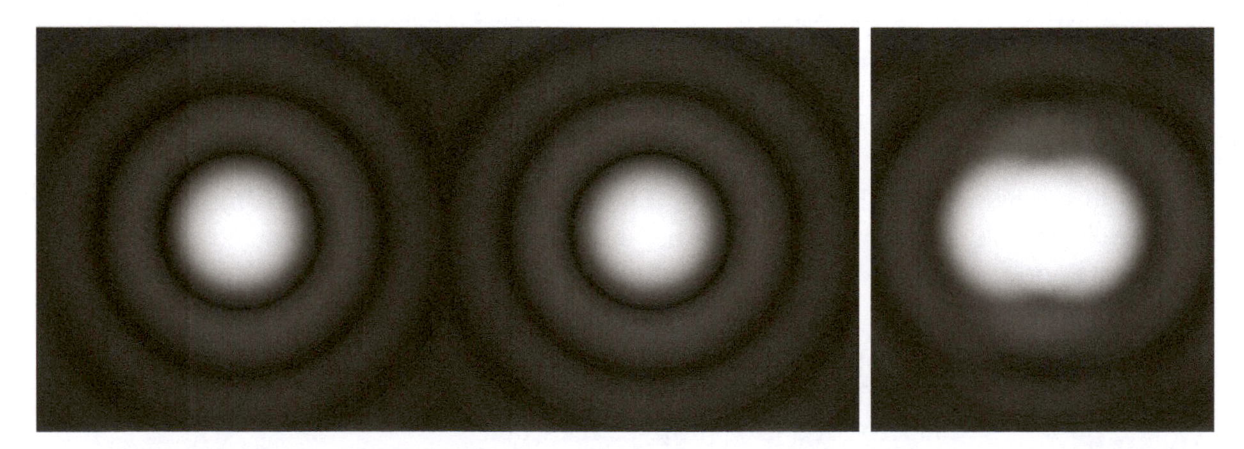

► **Figure 12-22** The wave nature of light limits the resolution of any telescope. This is the *diffraction* limit. *Left:* The diffraction patterns from two point sources that are well separated, or resolved. *Right:* Diffraction patterns from two point sources that are so close they are unresolved. Notice how, if they were any closer, it would become difficult to tell whether they are one or two stars. (Images courtesy of Frederick Ringwald)

Interferometers can take images more detailed than even the diffraction limit. They do this by connecting multiple, widely separated, telescopes together.

An example is the Very Large Array (VLA) radio telescope in New Mexico.

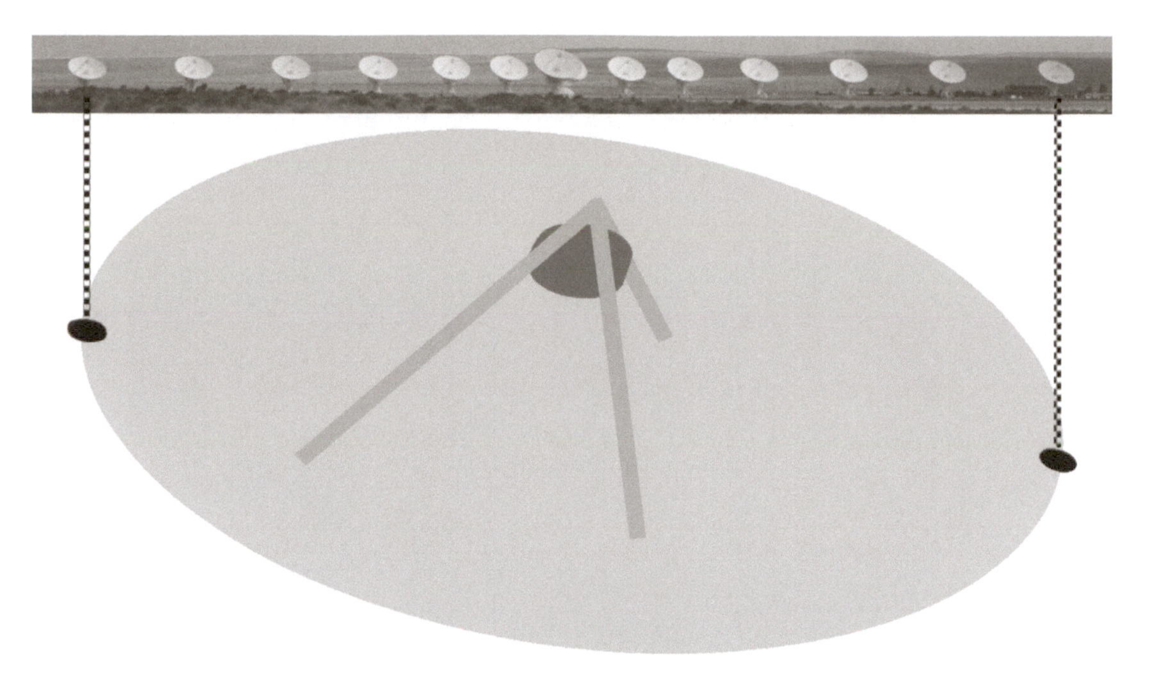

► **Figure 12-23** Many telescopes connected together can have the same resolution as a much larger one. This is called an *interferometer*, or an *array*. (Image courtesy of Frederick Ringwald)

► **Figure 12-24** The Very Large Array (VLA) is a radio interferometer. (Image courtesy of Frederick Ringwald)

CHAPTER

13

Eyes, Small Telescopes, and Photography

The Human Eye is like a camera, with a lens in front to focus the light, and a detector in back to detect the light.

In the eye, the detector is the retina. In an old-fashioned camera, film was the detector. In a modern digital camera or camcorder, it's a CCD detector.

The retina has two types of cells:

1. The **cones** are in the center of the eye's field-of-view. We use these for seeing high resolution, such as when reading.
2. The **rods** are around the edge of the vision field. The rods are larger than the cones, and so have less resolution. They are more sensitive, though. This is useful for seeing dangerous things at night out of the corner of one's eye, such as predators or cars.

One can use this to see fainter objects, when looking through a telescope. Since there are more rods around the edge of the retina, try looking a little to the side of where the target is, and one can often see it better. This is called **averted vision.** It takes practice to learn to look in one direction while concentrating one's mind in a slightly different direction, but it's worth it.

Human eyes also become **color-blind at night**. The brightest stars can activate human color vision, in a dark sky.

The human eye still **isn't** particularly sensitive to **red light**, in any case. Looking through binoculars or a telescope, most nebulae and galaxies won't show much color, even if they're spectacularly red or blue in pictures. The eye will see them as ghostly blue-white, unless one looks through a *very* large telescope.

Optical Illusions

Many people assume that the human eye and brain are like a video camera, which sees the world exactly the way it is. Many people also assume that human memory is like a video recorder, which makes records that are always reliable and complete, and that don't change. Both assumptions are *quite wrong*.

Since it can take about 0.1 seconds for nerve signals to travel from the eye to the brain, the eye often "guesses" at what it is seeing. Sometimes it guesses wrong. Many optical illusions demonstrate this; for example, closely spaced lines that appear to shimmer or move when they clearly cannot, because they are stationary patterns that are printed on paper. *Eye and Brain*, by Richard L. Gregory, has many examples of this.

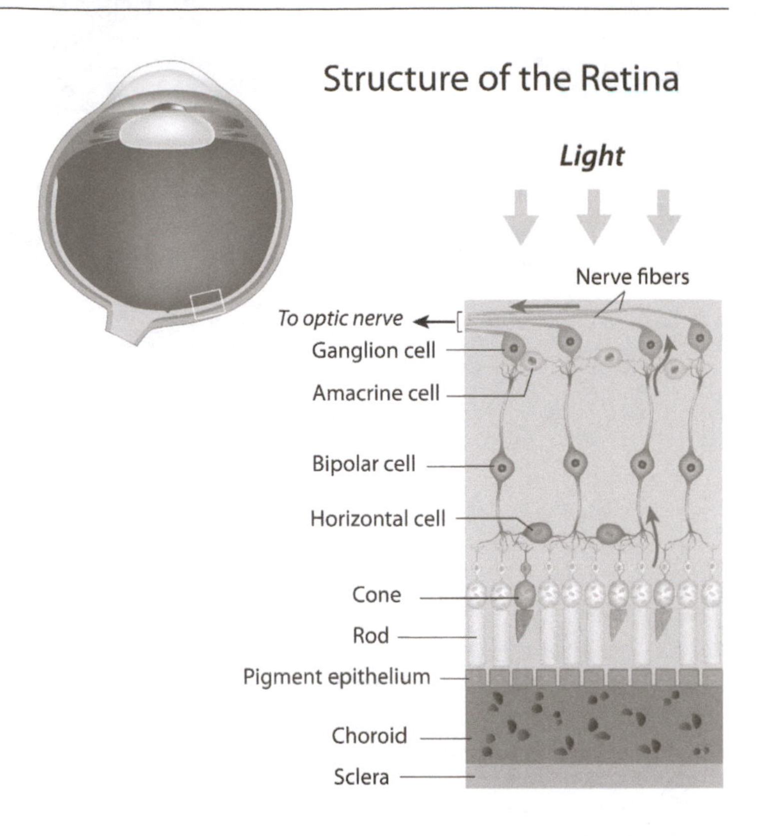

▶ **Figure 13-1** The human eye. The retina, in the back of the eye, has *rods* and *cones.* (Image © Alila Medical Images, 2013. Used under license from Shutterstock, Inc.)

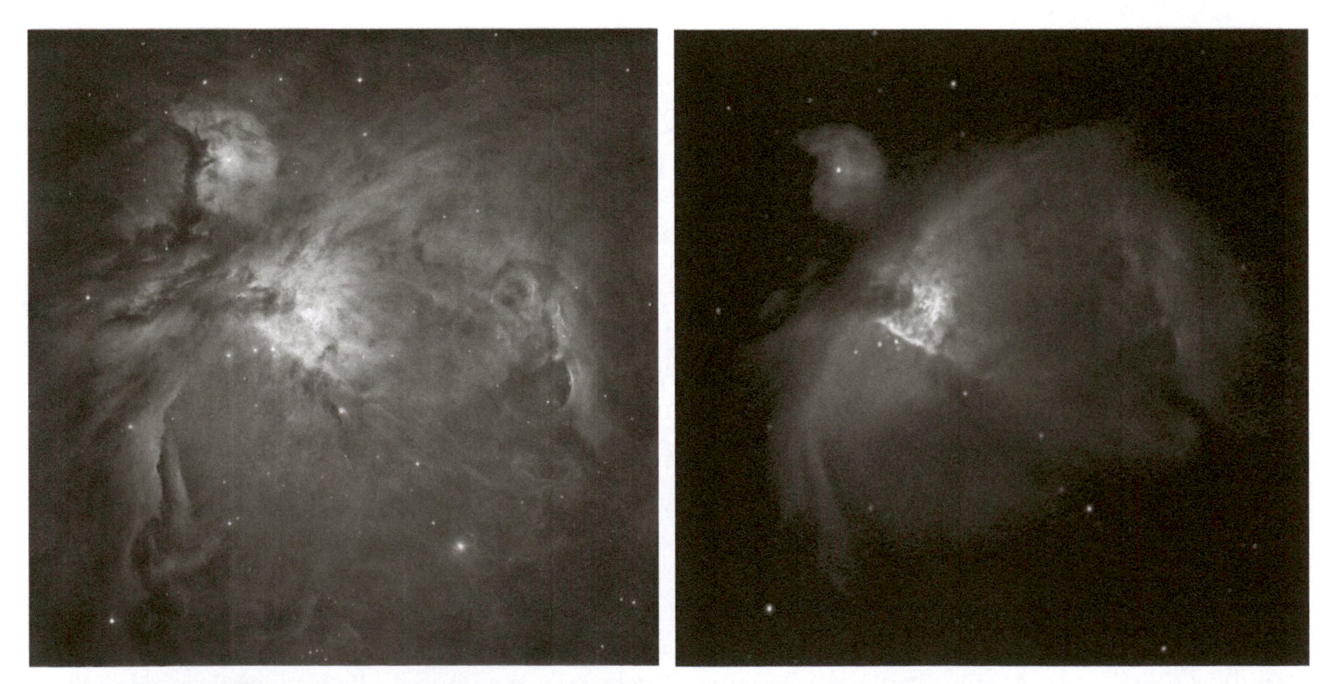

▶ **Figure 13-2** The human eye is not well suited for night-time use, one reason being it becomes color-blind at low light levels. Often, nebulae (clouds in space where stars form) look very different in photos than they do when seen in a telescope's eyepiece. *Left:* What a camera sees. (Source: NASA/STScI/Robert Gendler) *Right:* What the eye sees. (Image courtesy of Frederick Ringwald) This isn't a case of the camera telling lies. The camera is showing what is really there. The eye simply can't see it.

▶ **Figure 13-3** *Left:* The motion you see in this optical illusion isn't real. *(If you can't see the motion, move your eyes around.) Right:* Most **optical illusions** happen because the eye and brain are "guessing" incorrectly at what they are seeing and perceiving. (All images courtesy of Frederick Ringwald)

▶ **Figure 13-4** The Moon illusion: the eye and brain perceive the Moon to be larger when it's near the horizon. (Image courtesy of Frederick Ringwald)

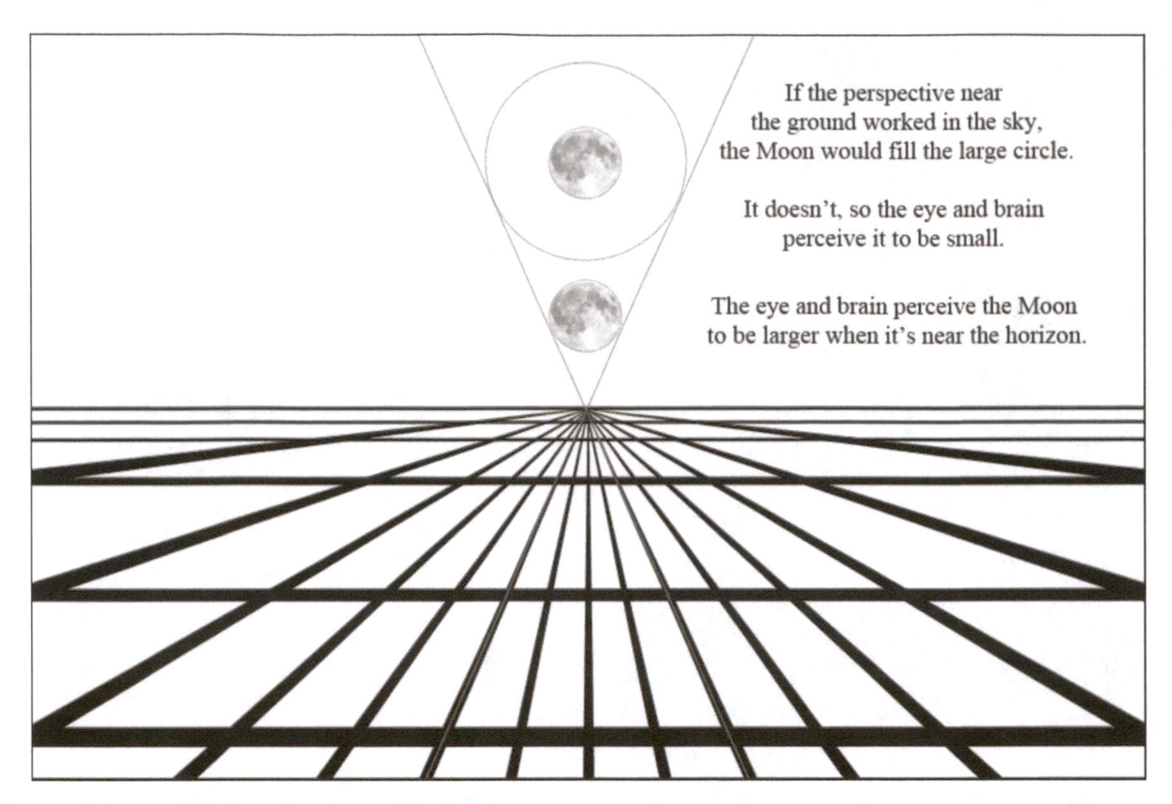

► **Figure 13-5** Explaining the Moon illusion. (Image courtesy of Frederick Ringwald)

► **Figure 13-6** The yellow in this Harvest Moon (in September) is from dust in Earth's atmosphere. *(The yellow isn't an illusion, it's real.)* (Image courtesy of Frederick Ringwald)

► **Figure 13-7** The Moon can appear to be blue, when smoke is in the air. *(This isn't an illusion, it's real.)* "Blue Moon" also means the second Full Moon in one month. On average, this happens once every six years. This gives us the saying, "once in a blue Moon." (Image courtesy of Frederick Ringwald)

Never buy a telescope marked "600 magnification" (or "600 power" or "600 x") on the box.

This is because:

- One can change the magnification to any desired value, by changing eyepieces.
- The size of the image is less important than how sharp and detailed the image is. Cheap plastic lenses don't deliver sharp image resolution!
- Because of the turbulence in Earth's atmosphere (called "seeing"), the maximum usable magnification of most telescopes is only about 150–200×.
- That any telescope manufacturer shows off such an irrelevant benchmark shows that either they're clueless, or worse, they think you're clueless.
- Cheap, semi-toy, "department store" telescopes often have flimsy plastic mounts and tripods. These are hard to point and hold still on anything, and shake and vibrate at the slightest touch. This can easily cause a headache!

(A department store can be a poor place to buy a telescope, because the staff may not know much about optics. The best camera store in town might be better, but be careful, since camera stores can be expensive.)

It's useful for a telescope to **track** the stars, or move so that it compensates for the rotation of Earth. This makes stars rise in the east, and set in the west.

To track, one needs an **equatorial mount**, which moves east-to-west. An equatorial mount needs to be **polar aligned**, or have its polar axis pointed at the North Celestial Pole. An equatorial mount should also have a **drive**, to rotate the telescope once per 24 hours. Since this is half as fast as the hour hand on a clock, a drive is sometimes called a **clock drive**.

Be aware that many telescopes, including many low-cost ones for beginners, **don't** track. They have **alt-azimuth mounts**, which are the simplest mounts possible: the telescope moves up-and-down and left-and-right, the way a cannon moves.

Alt-azimuth mounts are therefore simpler and about $150 cheaper than equatorial mounts. Alt-azimuth mounts may not be easier to use, though, since it's easier to lose the object one is looking at, in the eyepiece. Equatorial mounts set up by knowledgeable instructors are probably best for total novices.

For more on "How to Choose a First Telescope," see several online sources, including the web pages by *Sky & Telescope* and *Astronomy* magazine.

Choosing a First Telescope

When I was 11 years old, my parents noticed my interest in astronomy. They did the right thing and asked about buying a telescope at the local planetarium.

Alas, we got some poor advice. We were told, "Don't buy a telescope, unless you're willing to spend at least $1800 for it." (They said $300, but this was in the early 1970s.) My poor mother made a sound like she'd been kicked in the stomach, something like, "Ooof!"

The next day, she gave me her 7 × 35 birdwatching binoculars. As it turned out, that was exactly the right thing to do. I learned to observe with those binoculars.

Today, there do exist good telescopes for children that cost only about $200. There are also plenty of crummy, "department store" telescopes—the buyer must beware. A good way to see many different telescopes in use at the same time is to go to a star party hosted by the local amateur astronomy club.

Binoculars have many advantages as a first telescope:

- They're inexpensive. A good pair of 7 × 50 binoculars can cost only $50–$100.
- Binoculars can be used for other things: nature watching, sports, etc. If the kid's interests change, therefore, they can still be useful.
- Portability is excellent, except for the largest giant binoculars.
- The magnification is LOW, only about 7× or 10×. It's therefore easier to find things.
- The image is right-side-up and not left/right reversed, as in a telescope. Again, this makes it easier to find things.
- Using both eyes, the human eye and brain can just plain see better.
  Advanced amateur astronomers often prefer binoculars.

For binoculars, 7 × 50 means 7× magnification and a 50-mm aperture.

7 × 35:	Birdwatching binoculars are small and light.
7 × 50 or 10 × 50:	These are good for astronomy, since they have more aperture.
11 × 80 or 20 × 70:	Giant binoculars require a tripod, being too heavy to hold steadily.

Image-stabilized binoculars use SteadyCam technology to eliminate the effect of one's hands shaking. They're more expensive, but well-made ones are worth it.

Photography, with film or plates, is now largely obsolete, thanks to CCDs. It was the mainstay of astronomy from the 1890s to the 1980s.

Charge-Coupled Devices (CCDs) **are nearly perfect detectors of light.**

CCDs are the electronic chips that take the picture in many digital cameras, camcorders, and webcams. Their use was pioneered by astronomers in the 1980s because of their high *sensitivity* and *linearity*.

1. CCDs are very *sensitive*. More precisely, CCDs have ***high quantum efficiency*** (QE). Over 90% of the starlight falling on a CCD is recorded.

   Compare this with older TV sensors: (10% QE), the human eye (2% QE), or photographic film or plates (< 1% QE).

2. CCDs are highly *linear*. If one leaves the camera shutter open (or in other words exposes) for twice as long, one gets twice as much signal.

   Photographic film is notorious for having problems with this. This is called reciprocity failure: if one exposes twice as long with film, one rarely detects stars that are twice as faint. CCDs are therefore *photometrically accurate*: brightness measurements of stars are reliable.

Also because of their linearity, CCDs have a high *dynamic range*. This means they can see both bright and faint stars in the same image.

An example of where film photography fails are photos taken by astronauts in space, which almost never show stars in the background. This is because most pictures are taken during daylight, and the bright Sun overpowers the faint stars.

CCDs are also all digital, with the images going directly into a computer. There's no need to scan, as with photographic film.

▶ **Figure 13-8** Eyepiece observing is fun and educationally valuable. (Image courtesy of Frederick Ringwald)

► **Figure 13-9** **This stereotype for an astronomer, peering through the eyepiece of a telescope, has been out of date since the invention of photography.** (Engraving from "Selenografia sive Lunae Descriptio" by Johannes Hevelius, 1647)

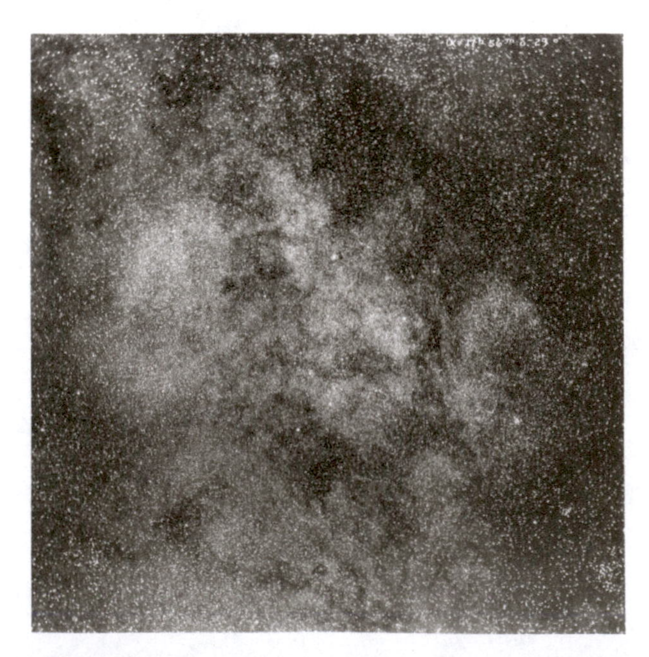

► **Figure 13-10** **Star clouds in the Milky Way photographed by Barnard. The old-fashioned chemical print is starting to turn yellow with age.** (Georgia Institute of Technology Library and Information Center "(Great Star Clouds in Sagittarius-2)", Edward Emerson Barnard's *Photographic Atlas of Selected Regions of the Milky Way,* June 2004, Georgia Institute of Technology, (June 25, 2013). http://www.library.gatech.edu/Barnard_Project_W/plate/Bar-pt1-pl027_sm.jpg)

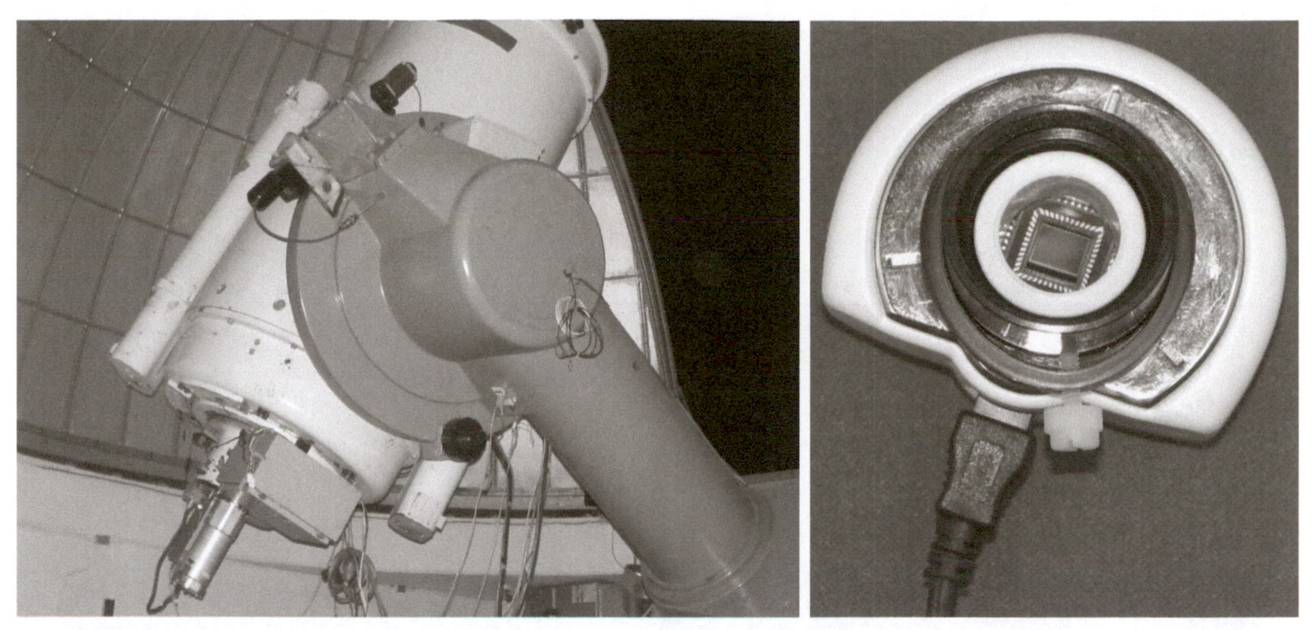

► **Figure 13-11** *Left:* Since 1980, astronomers have used digital CCD cameras. The gold-colored instrument on the tail of this 40-inch telescope is the CCD camera. (Image courtesy of Frederick Ringwald at Mount Laguna Observatory) *Right:* Digital CCDs are silicon chips that are *very* sensitive to light, often over 90% quantum efficient. The camera in your cell phone is a descendent of CCD cameras. (Image courtesy of Frederick Ringwald)

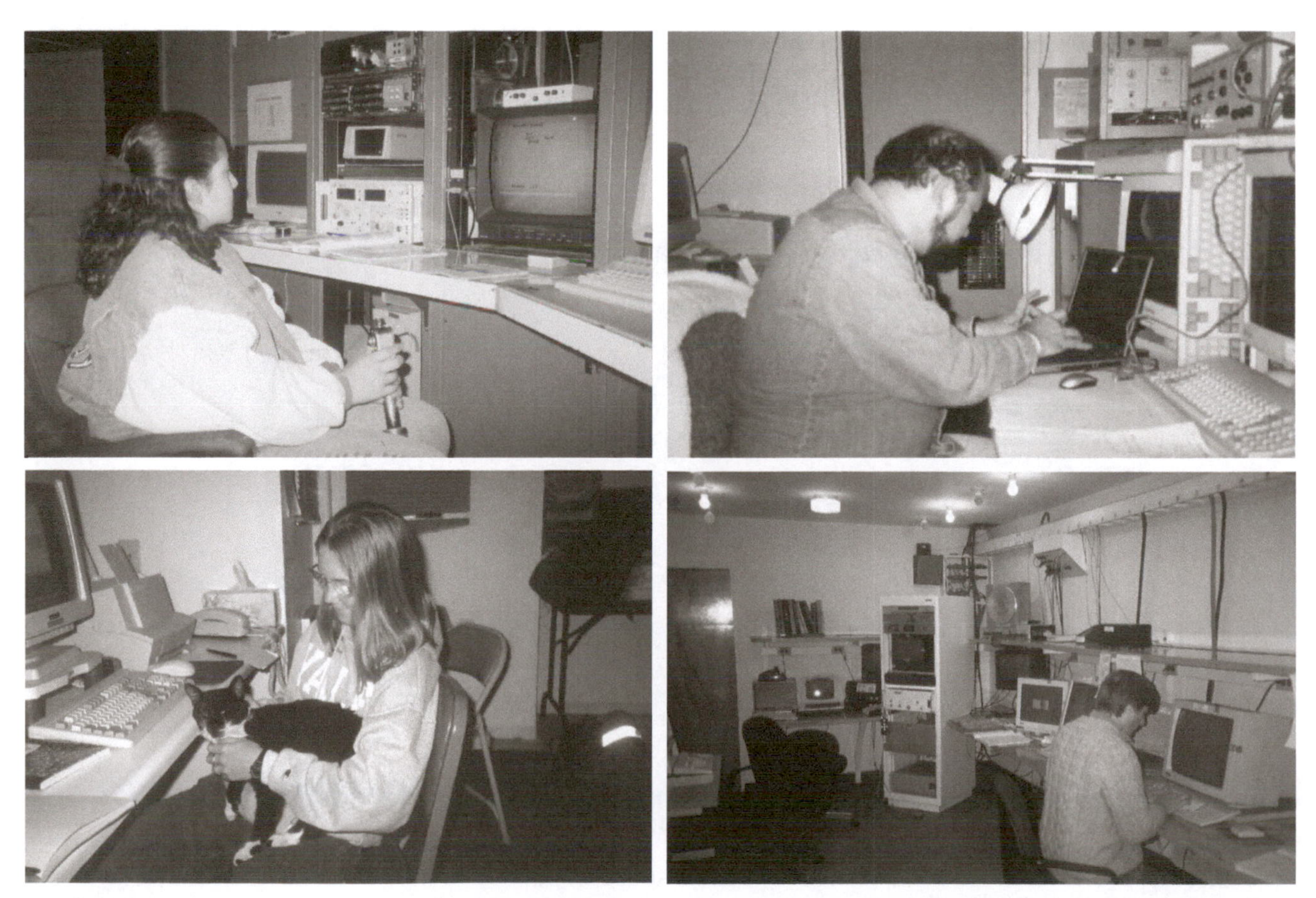

► **Figure 13-12** Because of digital CCD cameras, astronomers today rarely peer through telescopes. They mainly look at computer screens, preferably with a cat on the lap. (Images courtesy of Frederick Ringwald)

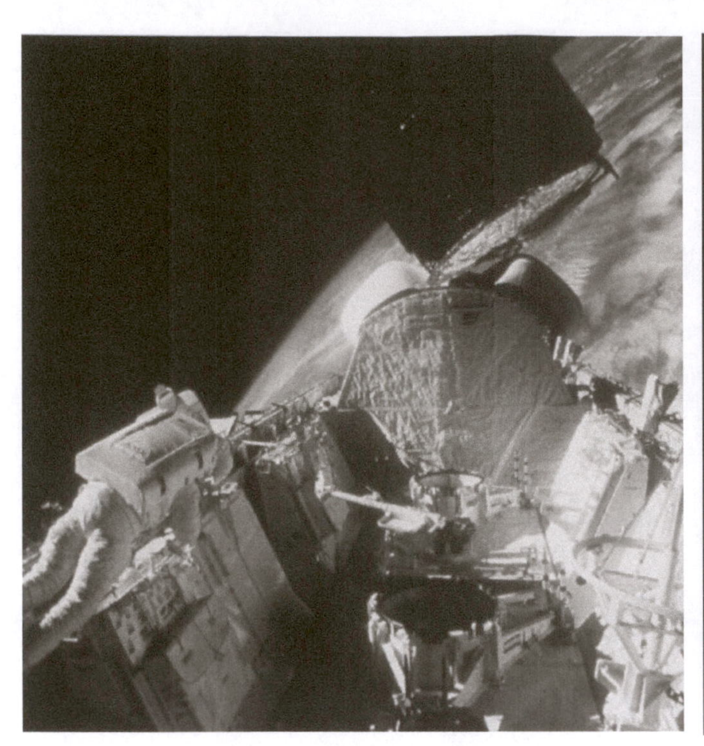

► **Figure 13-13** *Left:* Photos taken in space often don't show stars in the background. Despite what some conspiracy enthusiasts would say, this isn't because they're faked. It's because most space pictures are taken in daylight, with the camera set for daylight conditions. The blackness of space in the background is from there being no air in space to scatter sunlight. (Source: NASA) *Right:* You can do this yourself easily. Go outside at night, stand under a bright light, and take a picture of the sky with an automatic camera. Chances are good that the camera will set itself by the light from the lamp, which is nearly as bright as daylight. No stars are visible in the sky here. The Moon is visible, but it's as bright as an asphalt parking lot in daylight. (Image courtesy of Frederick Ringwald)

► **Figure 13-14** It is possible for an astronaut in space to take a picture that does show stars in the background, if the astronaut is on the night side of Earth, turns off the lights in the spacecraft, and holds the camera steadily for a long-time exposure. This image of the constellation Orion was taken by an astronaut on board a Space Shuttle. (Source: NASA/STS-35 crew/Dr. Samuel T. Durrance)

For Photographers

Focal Length, f-ratio, and Time Exposures

f = focal length = distance between objective and focus

f-ratio = focal length / Aperture = f/A

It may seem surprising, but any telescope (or camera lens) with the same focal length will form images that are the same size, regardless of aperture.

So why do we like larger aperture? Because the images are *brighter*: there's more light.

To make images brighter, so we can detect fainter objects, we often take a long time to take a picture. Leaving the camera shutter open for long periods of time is called making a **time exposure.**

To do this without blurring the picture, we need to hold the camera steady. Since Earth moves, the sky appears to move, with objects rising in the east, and setting in the west.

The telescope therefore should **track** the stars, or move so that it compensates for the rotation of Earth. One therefore needs an **equatorial mount** (which moves east-to-west) and a **clock drive** to track.

One also needs to be able to **guide.** Guiding is moving the telescope precisely by *small amounts* in any direction. This compensates for drive error, gusts of wind, vibrations from people moving near the telescope, etc. Guiding can be done manually (and is very tedious), or automatically with an autoguider, which requires an equatorial mount that can autoguide.

Astrophotography is a rewarding but challenging art. Don't pick a first telescope to do astrophotography. That should wait for a third or fourth telescope, since astrophotography requires a high skill level.

Nevertheless, a good, imaginative astrophotographer can take good pictures with even simple equipment, such as a camera on a stationary tripod, in much the way that a good, imaginative photographer of any kind can take good pictures with almost any camera.

CHAPTER

14

The Solar System

Recall Kepler's Third Law (that P^2 is proportional to a^3):

		a (average distance from the Sun, in AU)	P (orbital period, around the Sun)	Discovered:
My	Mercury	0.4	88 days	
Very	Venus	0.7	243 days	
Educated	Earth	1 AU (by definition)	1 year (by definition)	
Mother	Mars	1.52	687 days	
Just	Jupiter	5.2	12 years	
Showed	Saturn	9.6	29 years	
Us	Uranus	19	84 years	1781 William Herschel
Nine	Neptune	30	165 years	1846 Johann Galle
Planets	Pluto	40	248 years (?)	1930 Clyde Tombaugh

Mercury–Saturn can all be seen with the unaided eye, and so were known to the ancients.
Nearly all (99.7%) of the mass in the Solar System is in the Sun.
Jupiter has 3 times more mass than all the other planets put together.

The terrestrial planets are the Earth-like planets (Terra = "Earth," in Latin).
They're relatively small (with diameters less than 6400 km), and rocky.

The Jovian, or giant planets, include Jupiter, Saturn, Uranus (you-RAN-us), and Neptune.
Jupiter and Saturn are called "gas giants"; Uranus and Neptune are "ice giants."
All are large and gaseous. All have rings and extensive systems of satellites.
They're more like small stars than terrestrial planets: none have solid surfaces.

Small bodies, with diameters less than 1000 km, include:

- **Asteroids** are from the inner Solar System, and are mostly rock.
- **Comets** are from the outer Solar System, and are mostly ice.
- **Centaurs** are objects between Saturn and Uranus that have characteristics of both asteroids and comets: they have asteroid-like orbits, but they're icy, like comets.

Between Mars and Jupiter is **the asteroid belt,** where hundreds of thousands of asteroids orbit the Sun. Not all asteroids are in the Main Belt: hundreds cross Earth's orbit. (!)

Pluto is smaller than Earth's Moon. It was discovered in 1930. Since 1992, over 1000 other, small, icy bodies have been found beyond the orbit of Neptune. These are called **Kuiper Belt Objects** (KBOs), also known as Trans-Neptunian Objects or Ice Dwarfs.

Pluto is now known to be comparable in radius to another Kuiper Belt Object, named Eris. There may be other, larger KBOs. Pluto was therefore demoted from being called a planet in 2006.

Beyond this is **the Oort Cloud,** a spherical shell of icy **comets.** The comets sometimes wander into the inner Solar System and dazzle us with their tails, which are their ice evaporating (actually sublimating) into space, in the light of the Sun.

▶ **Figure 14-1** The Solar System. This does show the relative sizes of the planets correctly: 11 Earths could stretch across the face of Jupiter, and 10 Jupiters could stretch across the face of the Sun. Still, the planets are never this close to each other. They also are never in a straight line. (Source: NASA/JPL/Annotated by Frederick Ringwald)

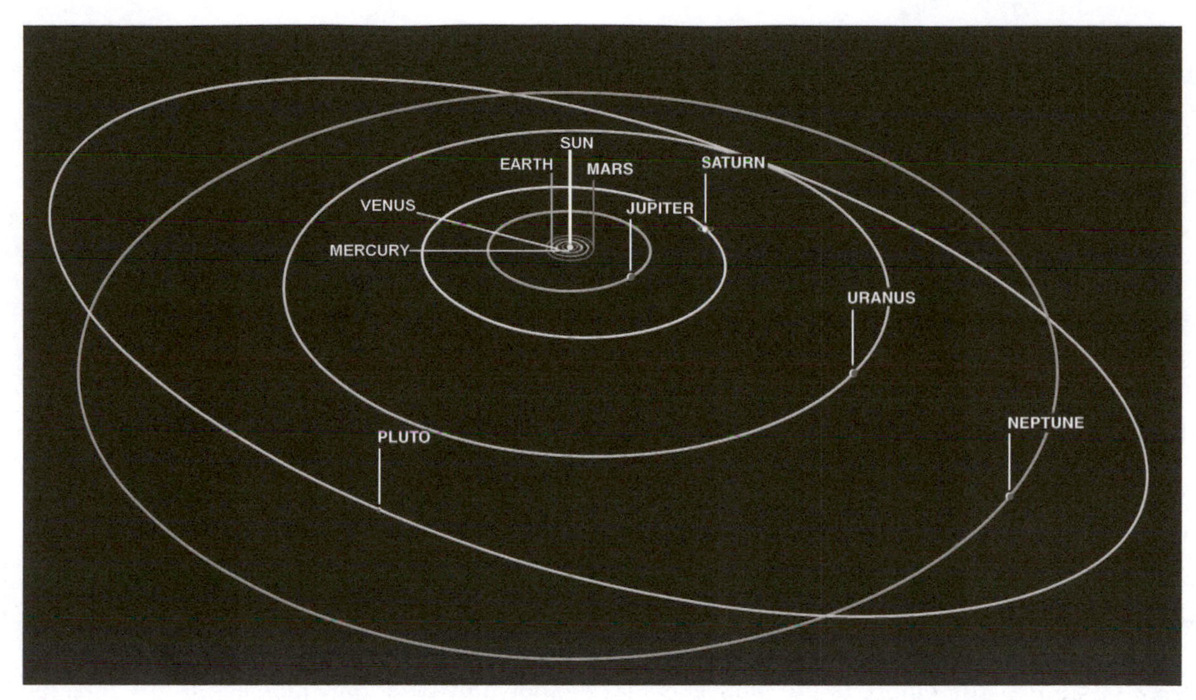

► **Figure 14-2** This is a more realistic view of the Solar System, because the planets are *very* spread out. (Source: NASA/LPI)

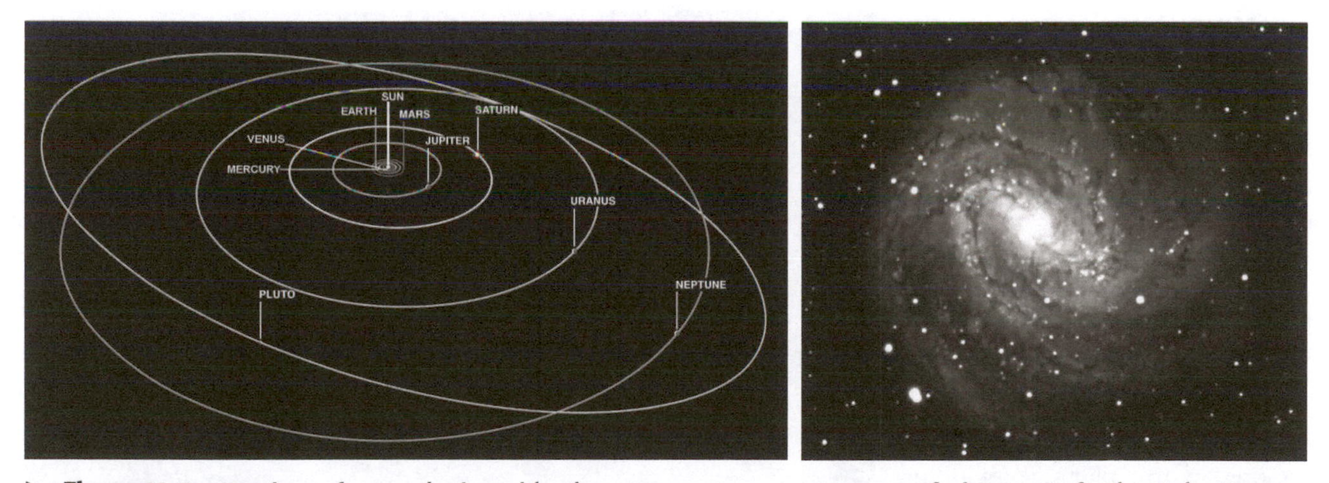

► **Figure 14-3** Don't confuse *galaxies* with *planetary systems*, or *systems of planets*. *Left:* The Solar System is our planetary system. It has one star (the Sun) and its system of planets. (Source: NASA/LPI) *Right:* This is a whole galaxy, with hundreds of billions of stars. (Image courtesy of Frederick Ringwald at Mount Laguna Observatory)

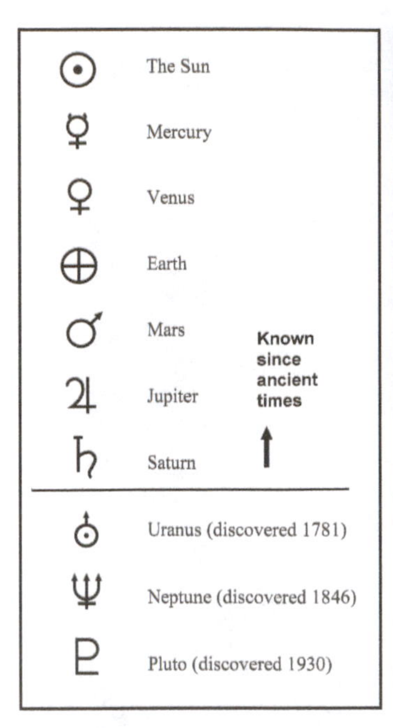

▶ **Figure 14-4** The true nature of the planets is further shown by what they look like to the unaided eye. Mercury, Venus, Mars, Jupiter, and Saturn are all bright enough to be seen by the unaided eye, and so have been known since ancient times. *Left:* These are the medieval astrological symbols for the Sun and the known planets. (Image courtesy of Frederick Ringwald) ***Top right:*** To the unaided eye, Mercury, Venus, Mars, Jupiter, and Saturn look like bright stars. They move, though, relative to the stars over weeks and months—"Planet" is ancient Greek for "wanderer." In a telescope, one can see the planets are round worlds, but to just the eye, they look like stars. (Image courtesy of Frederick Ringwald)

▶ **Figure 14-5** "Terra" means Earth, so "terrestrial" means Earth-like. The terrestrial planets, including Mercury, Venus, and Mars, are Earth-like in that they are small and rocky. It is otherwise debatable whether they are similar to Earth. Venus is hotter than a blast furnace for steel-making, for example, even though it is roughly the same size and mass as Earth. Here the terrestrial planets are shown with the larger moons of the planets, including Earth's Moon, the four largest moons of Jupiter (Io, Europa, Ganymede, and Callisto), Titan, the largest moon of Saturn, and Triton, the largest moon of Neptune. (Image courtesy of Frederick Ringwald with images by NASA)

► **Figure 14-6** The giant planets are large and gaseous. Jupiter and Saturn are often called *gas giants*, since they are largely gas. Uranus and Neptune are called *ice giants*, since they have a larger fraction of ice. Again, they are never this close to each other, or this close to Earth: this rendering is just for size comparison. (Image courtesy of Frederick Ringwald with images by NASA)

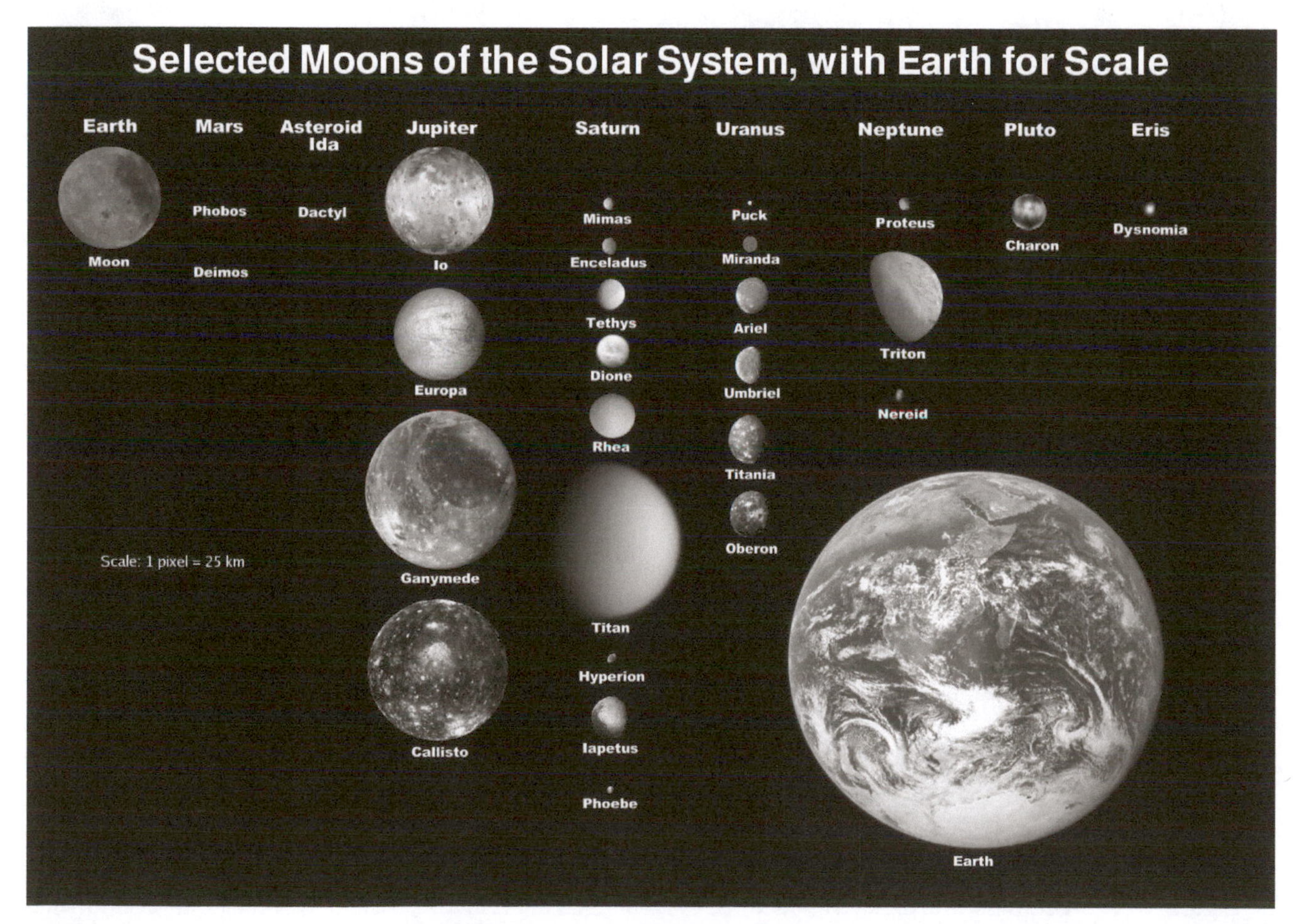

► **Figure 14-7** Moons used to be called planetary satellites, but now the word *satellite* has come to be used for "artificial satellite," so any natural object that orbits a planet is often now called "a moon." Moons come in a wide variety: there is no such thing as a boring planetary satellite, if you look at them carefully. (Source: NASA)

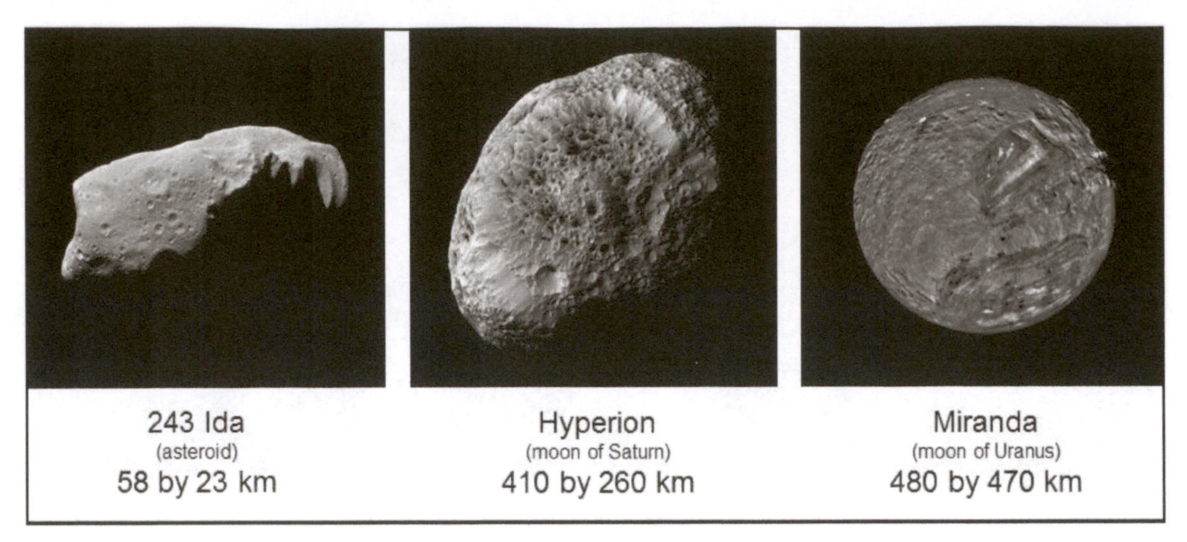

► **Figure 14-8** More massive bodies have more surface gravity, which pulls them into a round shape. (Source: NASA/JPL)

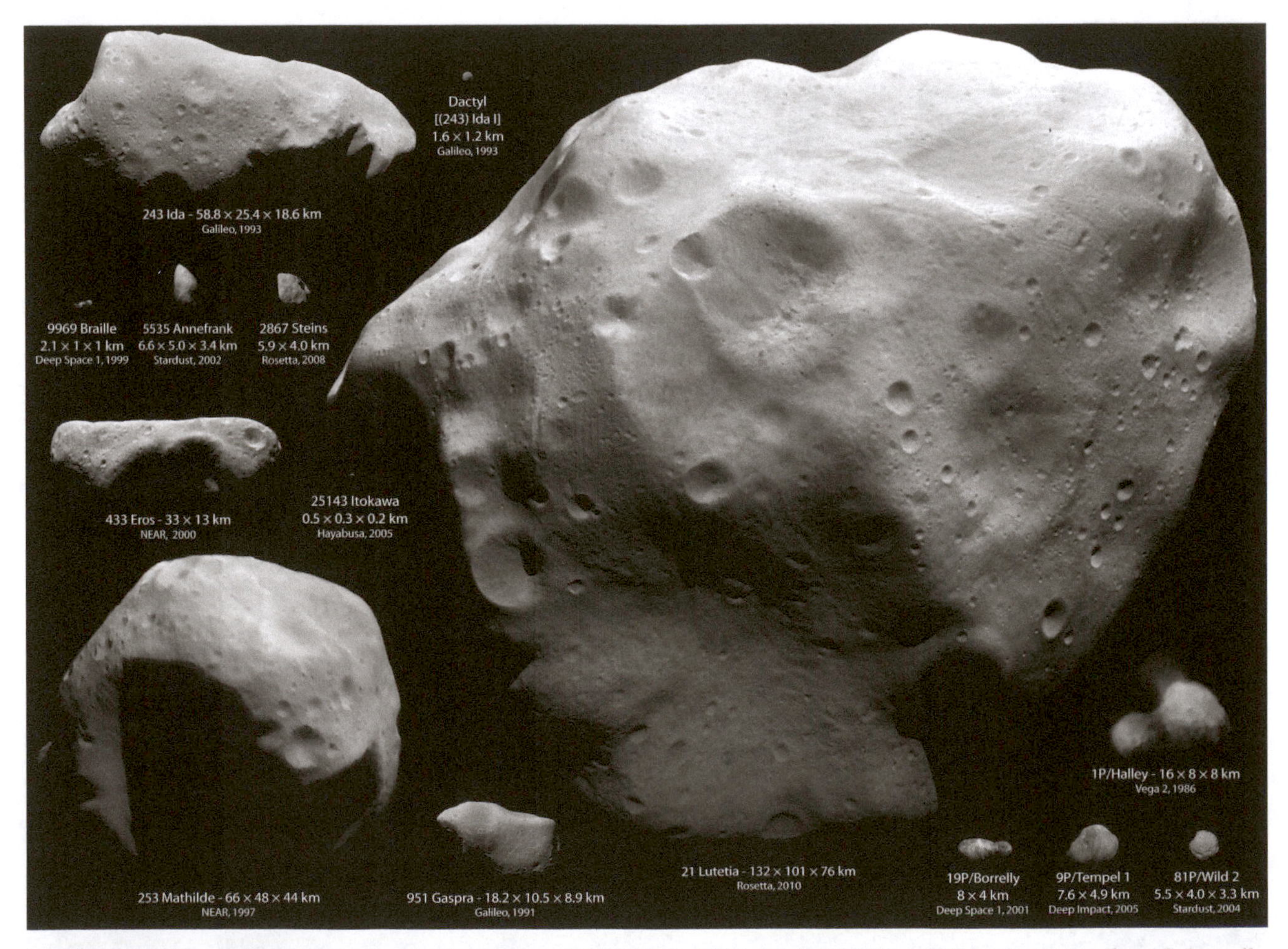

► **Figure 14-9** Asteroids are rocks in space. **(ESA-NASA-JAXA-RAS-JHUAPL-UMD-OSIRIS Emily Lakdawalla and Ted Stark)** (Montage by Emily Lakdawalla. Ida, Dactyl, Braille, Annefrank, Gaspra, Borrelly: NASA / JPL / Ted Stryk. Steins: ESA / OSIRIS team. Eros: NASA / JHUAPL. Itokawa: ISAS / JAXA / Emily Lakdawalla. Mathilde: NASA / JHUAPL / Ted Stryk. Lutetia: ESA / OSIRIS team / Emily Lakdawalla. Halley: Russian Academy of Sciences / Ted Stryk. Tempel 1, Hartley 2: NASA / JPL / UMD. Wild 2: NASA / JPL.)

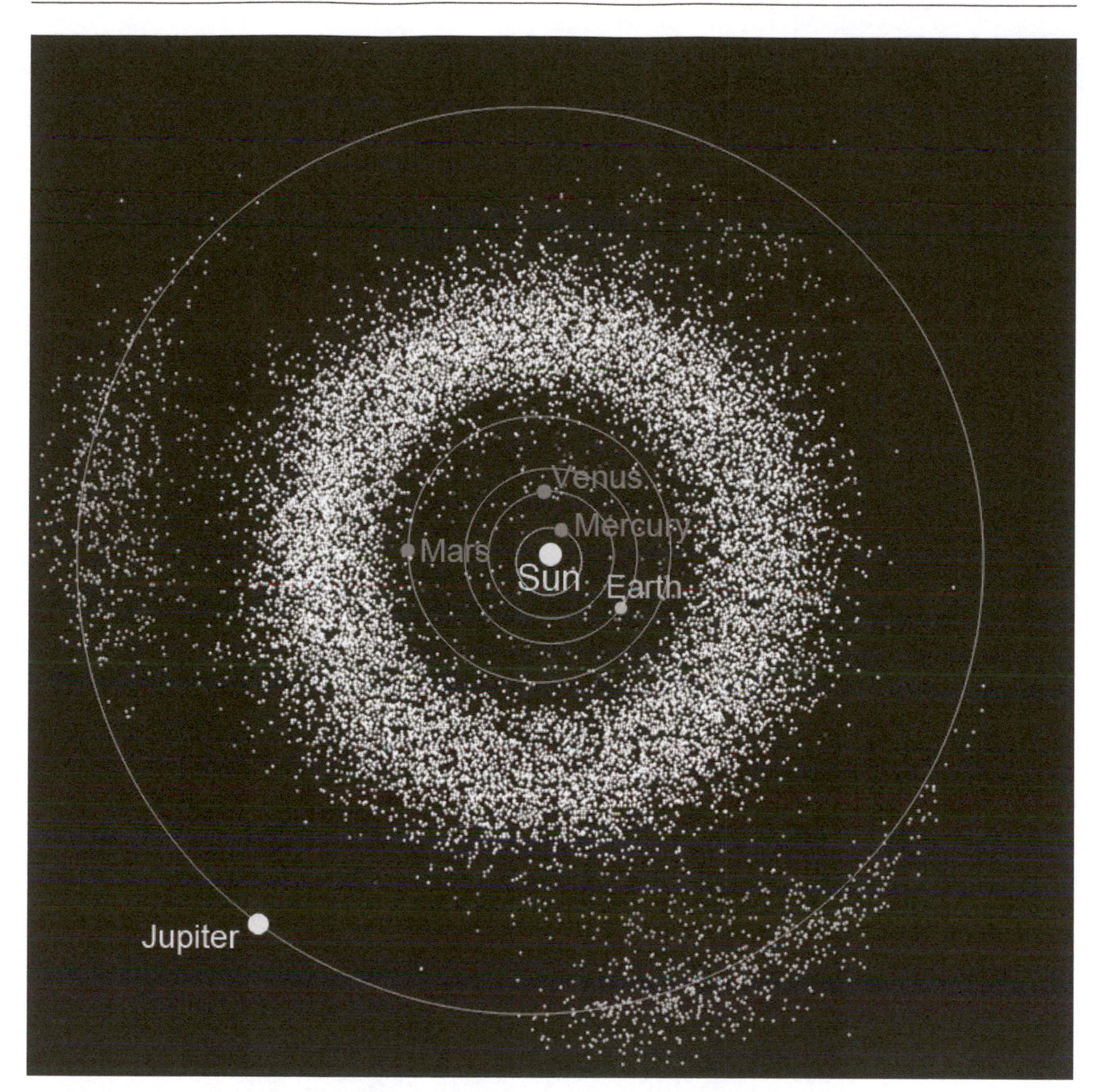

▶ **Figure 14-10** Over 300,000 asteroids are known in the main asteroid belt, between the orbits of Mars and Jupiter. Not all asteroids are in the main asteroid belt: hundreds of asteroids are known to cross the orbit of Earth, and so could possibly hit Earth. (NASA-JPL DE-405 ephemeris; annotated by author)

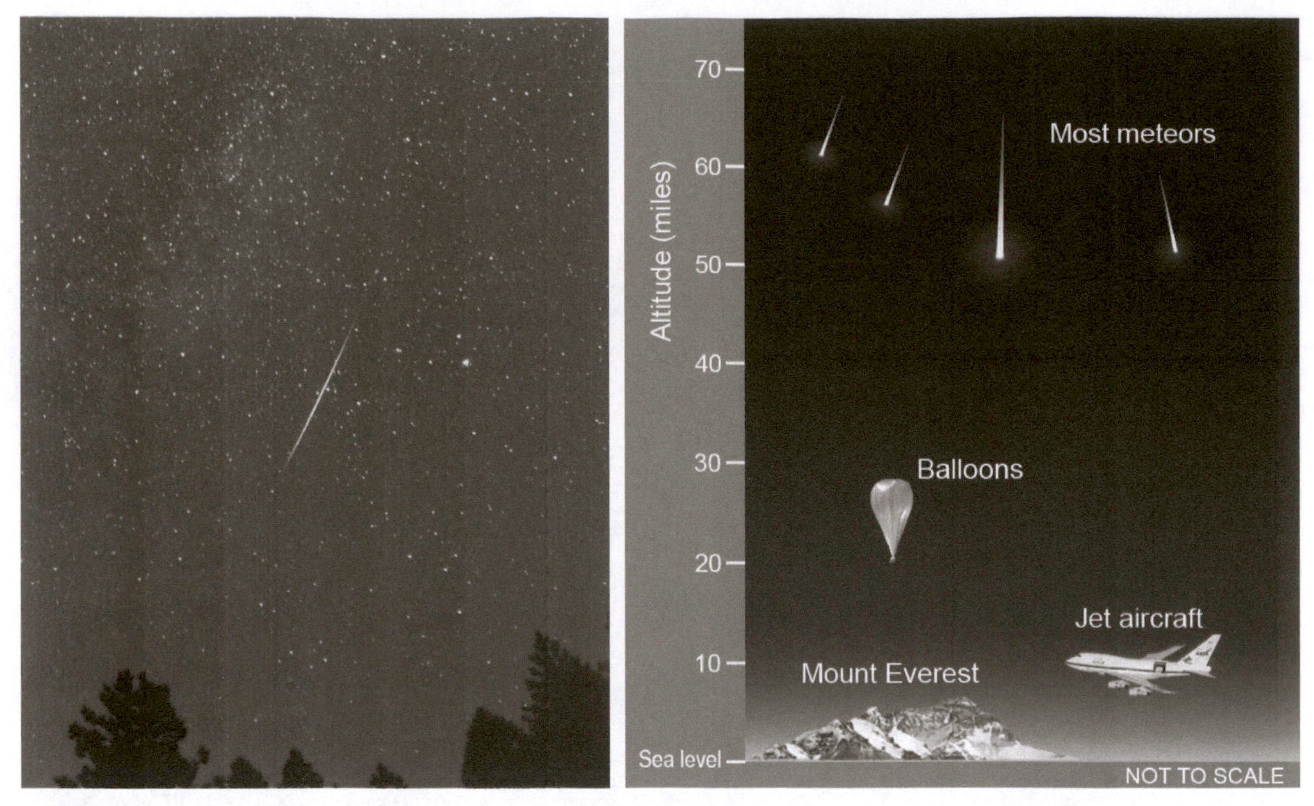

► **Figure 14-11** Meteors are often called "shooting stars" or "falling stars," although they're not stars. They're mostly dust particles or rocks from space that burn up as they hit Earth's atmosphere, because of their great speed. They last only a few seconds. *Left:* A meteor flashes across the sky, in less time than it takes for you to say, "Hey, look at that!" *Right:* Most meteors burn up over 30 km (50 miles) above Earth's surface. This is why most meteors make no sound. Some meteors are large enough to hit Earth—these are called meteorites. (Images courtesy of Frederick Ringwald)

► **Figure 14-12** Comets are icy, with gaseous tails. They can be visible for weeks. This is Comet Hale-Bopp, which was visible in 1997. It had a blue ion tail (left) blown by the solar wind. Its yellow dust tail (right) traced the comet's motion. The two tails noticeably changed from night to night. (Image © MarcelClemens, 2013. Used under license from Shutterstock, Inc.)

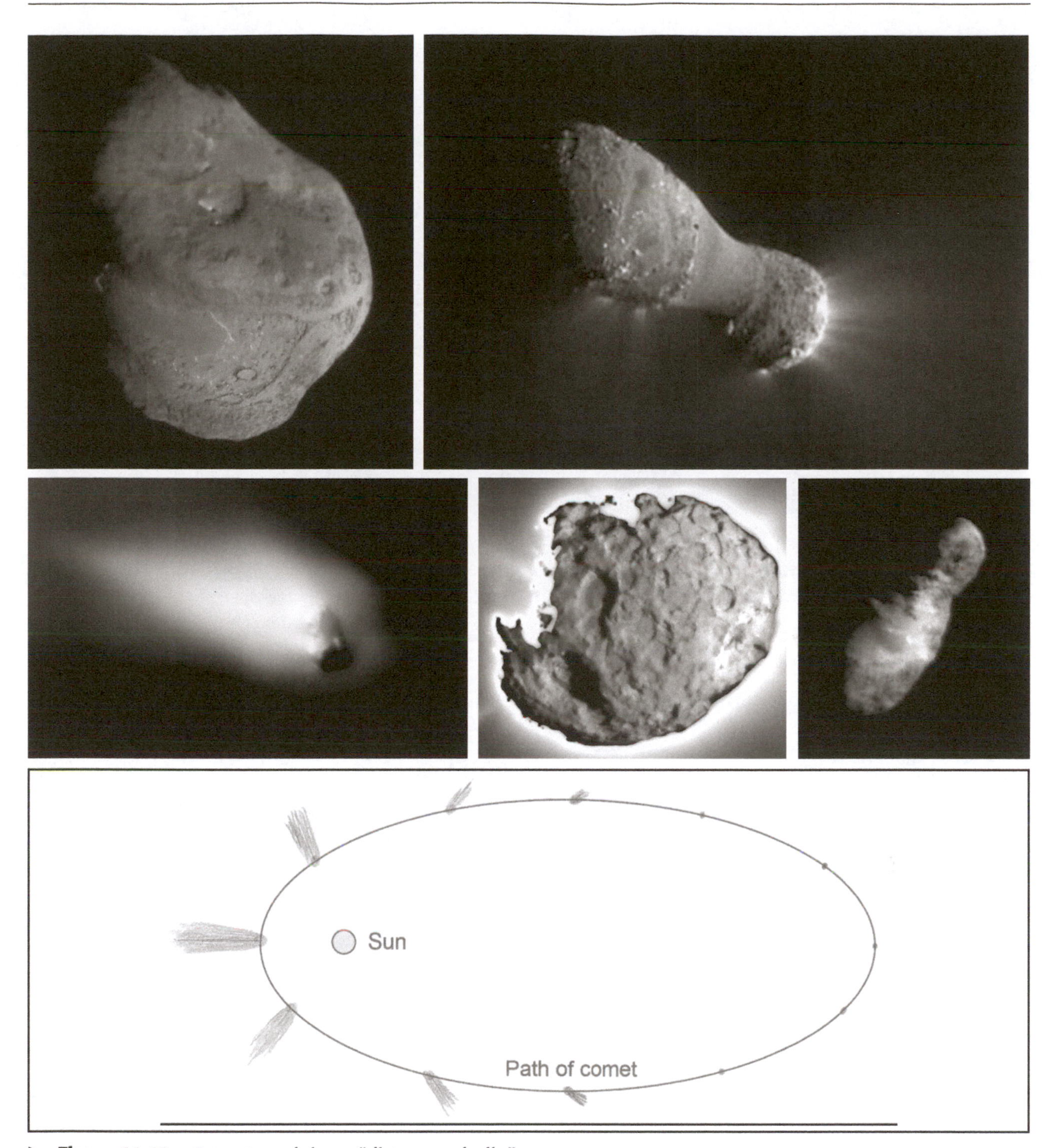

▶ **Figure 14-13** Comet nuclei are "dirty snowballs."
Top row, left: Comet Tempel 1 (Source: NASA/JPL)
Top row, right: Comet Harley 2 (Source: NASA)
Middle row, left: Comet Halley (Source: ESA)
Middle row, middle: Comet Wild 2 (Source: NASA/JPL)
Middle row, right: Comet Borelly (Source: NASA/JPL)
Bottom: As a comet comes close to the Sun, its snow and ice melts (actually sublimates), making the gaseous tail. A comet tail points away from the Sun. (Image courtesy of Frederick Ringwald)

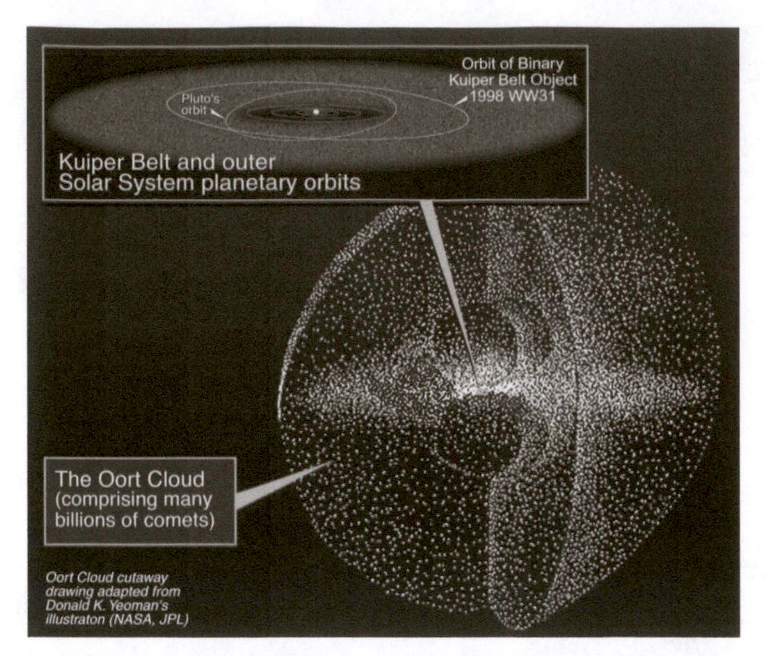

► **Figure 14-14** This is an artist's depiction of the Solar System, ringed by the Kuiper Belt, all of which is inside the (roughly spherical) Oort Cloud. Gerard Kuiper inferred that the Kuiper Belt is there because short-period comets often are in the plane of the ecliptic. (Source: NASA/JPL/Donald Yeomans)

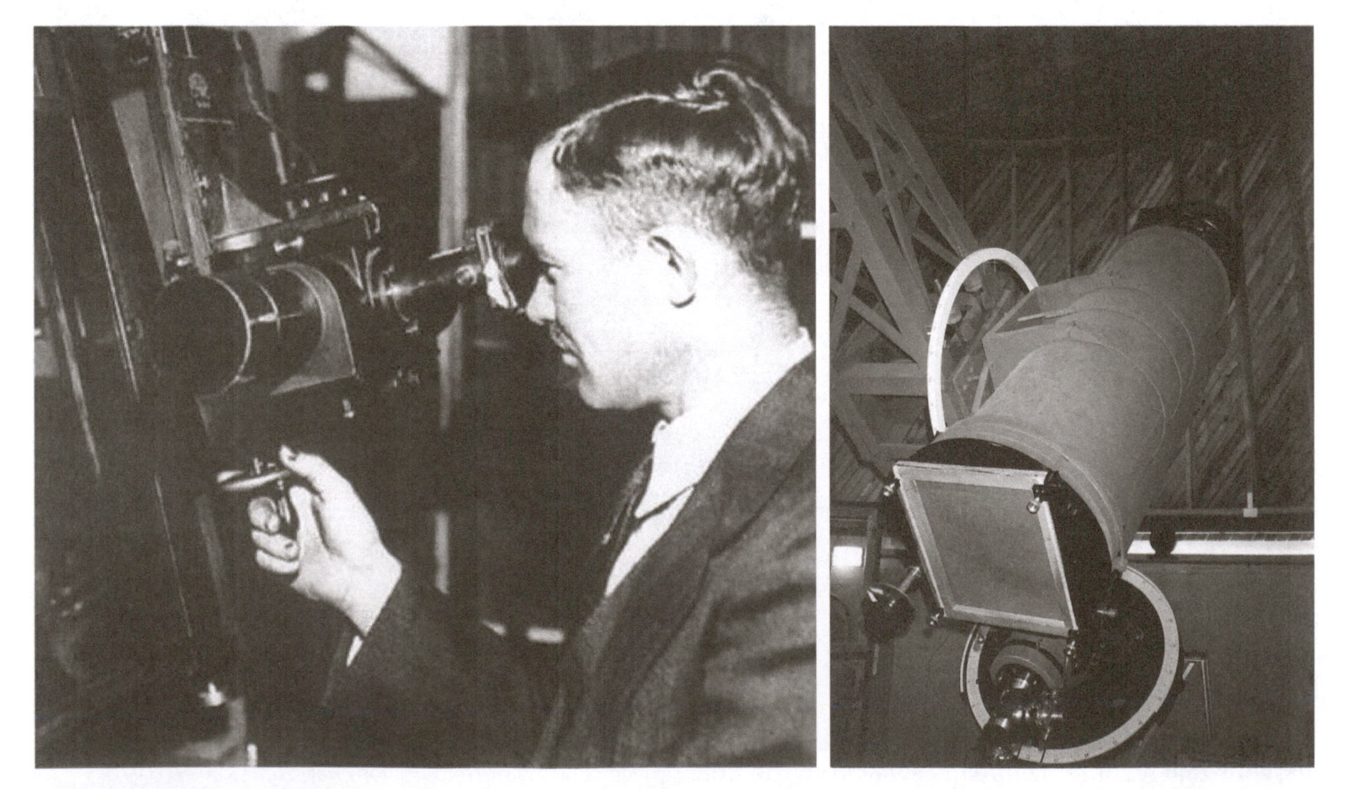

► **Figure 14-15** *Left:* Clyde Tombaugh discovered Pluto in 1930. It was announced as "the ninth planet," even though he'd been searching for a *giant* planet. (American Institute of Physics) ***Right:*** The 13-inch astrograph on Mars Hill used to discover Pluto (Image courtesy of Frederick Ringwald)

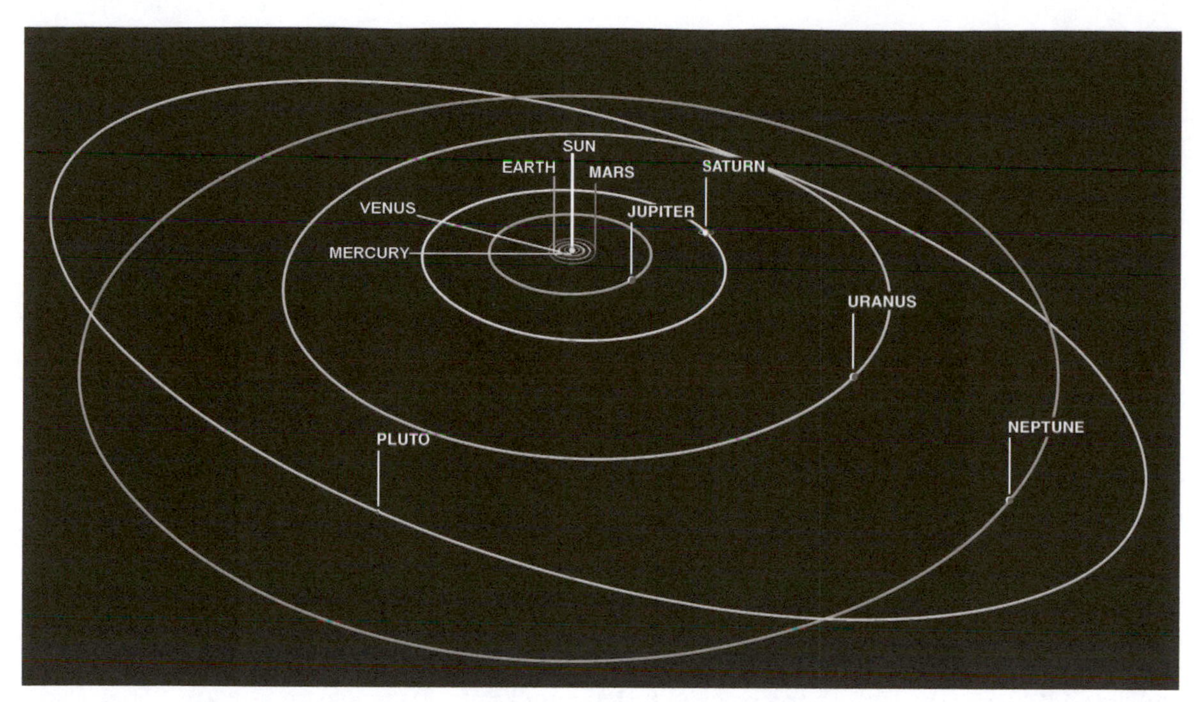

▶ **Figure 14-16** Notice Pluto's weird orbit. It is much more inclined (17°) to the ecliptic and much more elongated (or eccentric) that any of the planets' orbits. The reason for this is now known: Pluto isn't a planet. It is one of the swarm of over 1,000 icy Kuiper-Belt Objects, many of which have orbits similar to Pluto's. (Source: NASA/LPI)

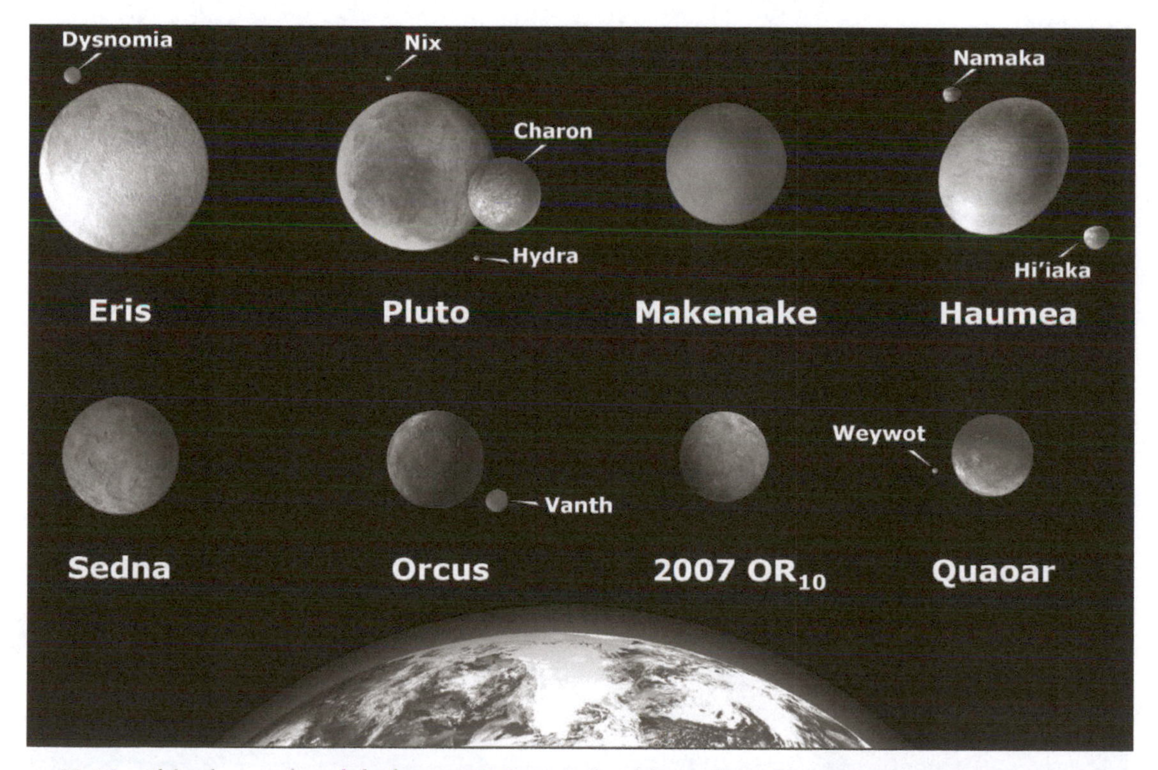

▶ **Figure 14-17** This shows the eight largest Kuiper-Belt Objects, known as of 2012, along with their moons. Notice how Pluto is number 2. Eris is now the largest known KBO. It was discovered in 2003, and caused all the fuss about Pluto no longer being considered a planet. Eris was the spirit of strife in Greek mythology. According to legend, she would ride into battle with Ares, and cause wars. It is fitting that this name was given to a Kuiper-Belt Object that caused so much trouble. (Source: NASA)

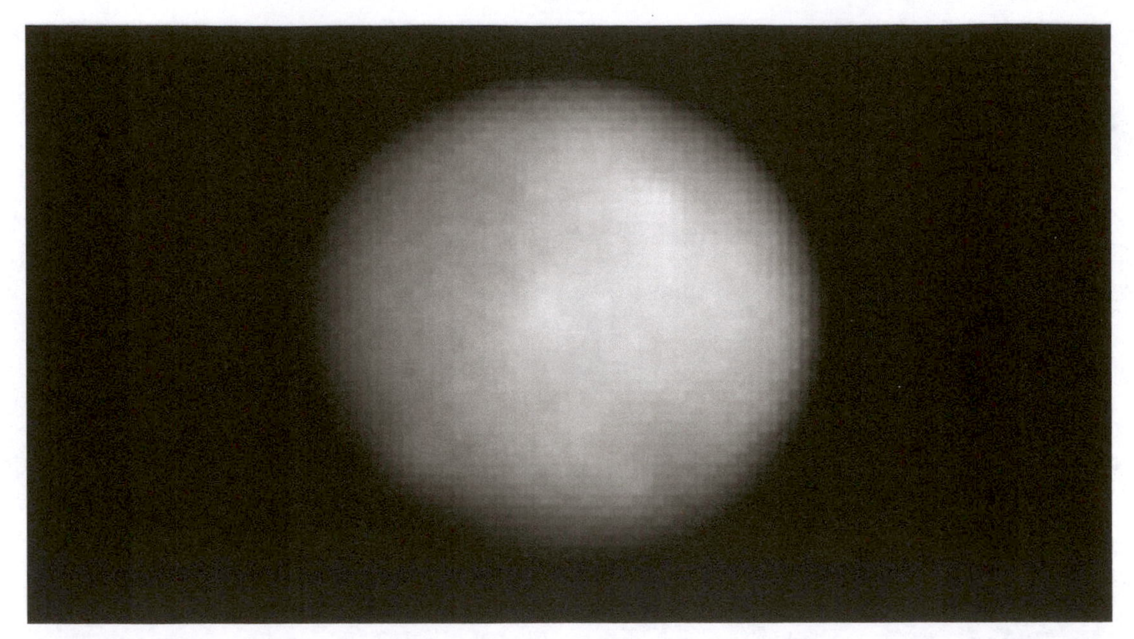

► **Figure 14-18** Realizing that something that has been called a planet really is not a planet is not unprecedented. Ceres is the largest and was the first-known asteroid, discovered in 1801. It was also called a planet, for about 50 years after it was discovered. This is the best image of Ceres that currently exists: better images may be taken when NASA's *Dawn* spacecraft arrives there in 2015. (Source: NASA/STScI)

► **Figure 14-19** NASA's New Horizons spacecraft will arrive at Pluto in 2015 (whether or not it's a planet). (Source: NASA)

CHAPTER

15

Exoplanets

Planets of Other Stars

Exoplanets = Extrasolar Planets = Planets of Other Stars, Not the Sun

> "The curtain is going up on countless new worlds with stories to tell."
>
> –Timothy Ferris

In 1991, astronomers became able to detect planets around other stars. Thousands are now known, with more being rapidly discovered.

So far, this has been mostly indirect detection. Only a few exoplanets have been seen in direct images. There are several indirect detection methods. Most of the first 300 known exoplanets were discovered with the Doppler effect. As an exoplanet orbits its parent star, its gravity pulls the star around. The Doppler effect makes lines in the star's spectrum move, and spectrographs looking through telescopes can detect this. This method may be surpassed by observing transits, when exoplanets pass between their parent star and Earth.

What have astronomers found?

1. **Pulsar planets:** These orbit dead stars, called pulsars. This was a surprise, since pulsars form in supernovae in which the stars that made them explode. The last place in the Universe one might have expected to find a planet is around a pulsar.
2. **Hot Jupiters:** These are about as massive as Jupiter but closer to their parent stars than Mercury is to the Sun. This was unexpected. We used to think we knew that Jupiter, Saturn, Uranus, and Neptune are so much more massive than the terrestrial planets in the Solar System because they are farther from the Sun, where it's cold, so the gases they're made of could condense into planets. Do giant planets form far from their parent stars, and then migrate inward?
3. **Eccentric Jupiters:** These are also about as massive as Jupiter, but with orbits that are eccentric (in other words, elongated) ellipses. These orbits are unlike those of any planets in the Solar System, but are similar to the orbits of comets. These are hot Jupiters caught in the act of migrating: these orbits cannot possibly be stable, since the gravity of other planets would throw them out of orbit, or into their parent stars.

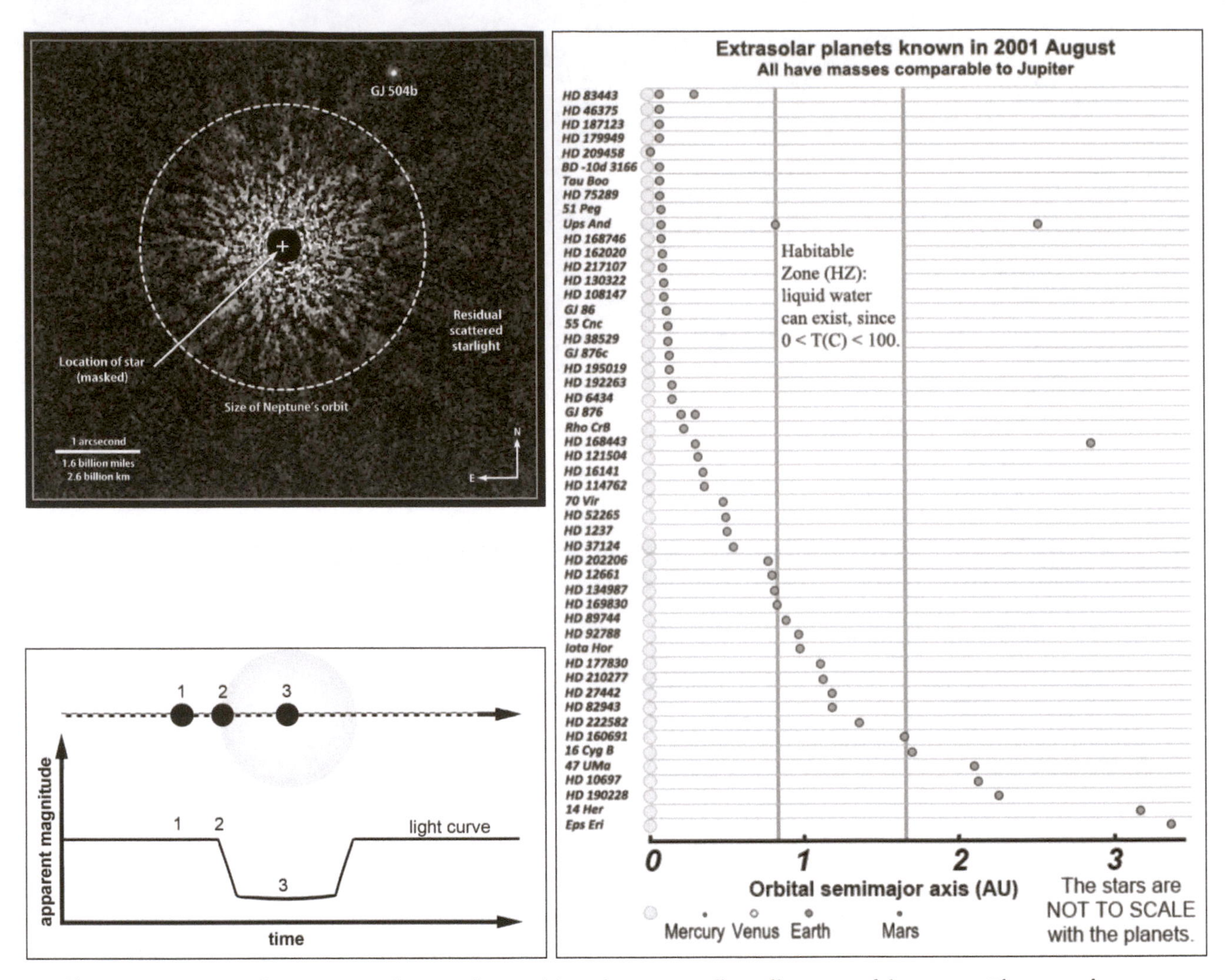

▶ **Figure 15-1** Exoplanets are planets that orbit other stars, first discovered in 1991. Thousands are now known. Many were discovered indirectly, from the Doppler effect in spectra. Increasingly many are being discovered from *transits.*

Top left: A few exoplanets have been found directly, in images. This is difficult because the parent stars are so much brighter than the exoplanets. The Sun, for example, is 200 million times brighter than Jupiter. It's like trying to see a firefly next to a searchlight. (Source: NASA/GSFC)

Bottom left: When an exoplanet transits its parent star, it causes the parent star's brightness to fade by as much as a few percent. Modern digital cameras are accurate enough to detect these fadings in brightness. (Image courtesy of Frederick Ringwald)

Right: This is a plot of the exoplanets known in 2001 August. Why use a plot that old? It is because so many exoplanets have been found since then, that an up-to-date plot would be so crowded it would be impossible to read, if jammed onto one page. (Image courtesy of Frederick Ringwald)

Notice that exoplanets have been found inside, in, and outside the Habitable Zones of stars both like the Sun (shown here) and not like the Sun. The Habitable Zones are the regions around a star where temperatures are between 0 and 100°C, where liquid water can exist.

4. **Classical Jupiters:** These are planets with similar mass and distance to their parent stars as Jupiter.
5. **Hot Neptunes:** Planets with masses > 3–5 times that of Earth are common, orbiting about 30% of stars like the Sun. (Jupiter has a mass of 300 Earths, and hot Jupiters orbit about 10% of stars like the Sun.)
6. **Super Earths:** At least one planet with a mass of 5 Earths is known to be rocky, because transits show its radius.
7. **"Rogue," or "orphan," planets:** Planets free in space may be as common as planets that orbit stars.
8. **Terrestrial planets:** Terra means "Earth" in Latin. NASA's *Kepler* spacecraft has found at least five Earth-sized planets in the habitable zones of Sun-like stars. We still don't know whether these have water or life. New observations, particularly spectra, are needed for this.

Astronomers have also found:

- Many cases in which multiple planets orbit their parent star, like our own Solar System.
- Planets around binary, or double stars. Since most stars like the Sun have companion stars, this may imply that planets are common throughout the Universe.

Figure 15-2 What everyone wants to find: a truly *Earth-like* exoplanet. It may have already been found: at least five planets the size of Earth have turned up in the habitable zones of Sun-like stars. Many more will be found in the next few years. (Image courtesy of Frederick Ringwald, using a NASA image taken on Apollo 11)

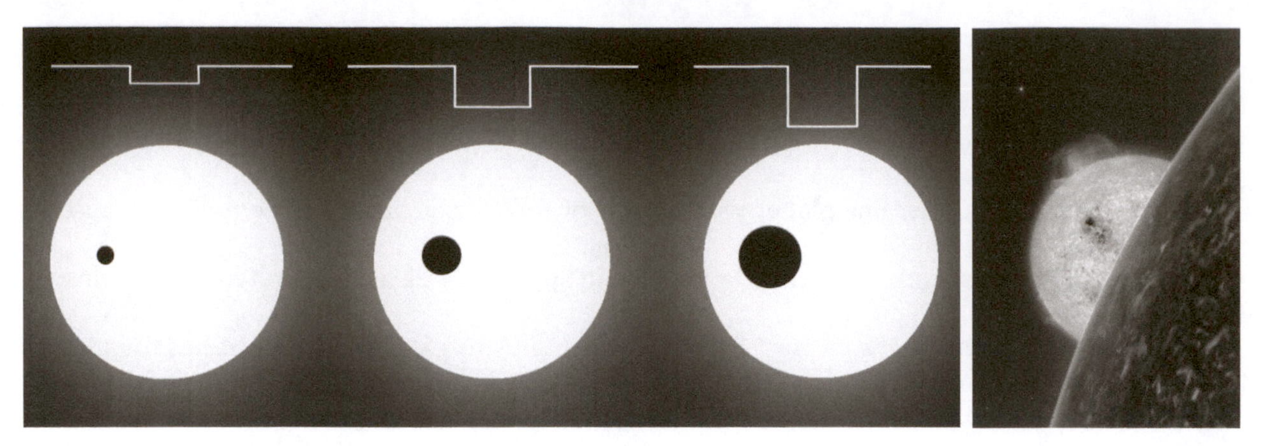

► **Figure 15-3** *Left:* **Transits show an exoplanet's size.** (Image courtesy of Frederick Ringwald) ***Right:*** **A milestone made in 2009 was the discovery of CoRoT-7b, the first confirmed Earth-sized exoplanet. It was also the *first confirmed rocky exoplanet*, a lava planet with 5 Earth masses, orbiting a normal star, closer than Mercury is to the Sun. *(It's not very Earth-like, is it?)*** (ESO/L. Calcada. Courtesy the European Southern Observatory)

► **Figure 15-4** **NASA launched the *Kepler* spacecraft in 2009. It is searching for transits from *Earth-size planets in the habitable zones* of 145,000 Sun-like stars. It is designed to find 50–500—or none. (As of January 2013, it has found 2,740 candidates. Of these, 262 are in their stars' habitable zones. Of these, 5–23 and probably many more are Earth-sized.)** (ESO/L. Calçada/Nick Risinger (skysurvey.org). Courtesy the European Southern Observatory)

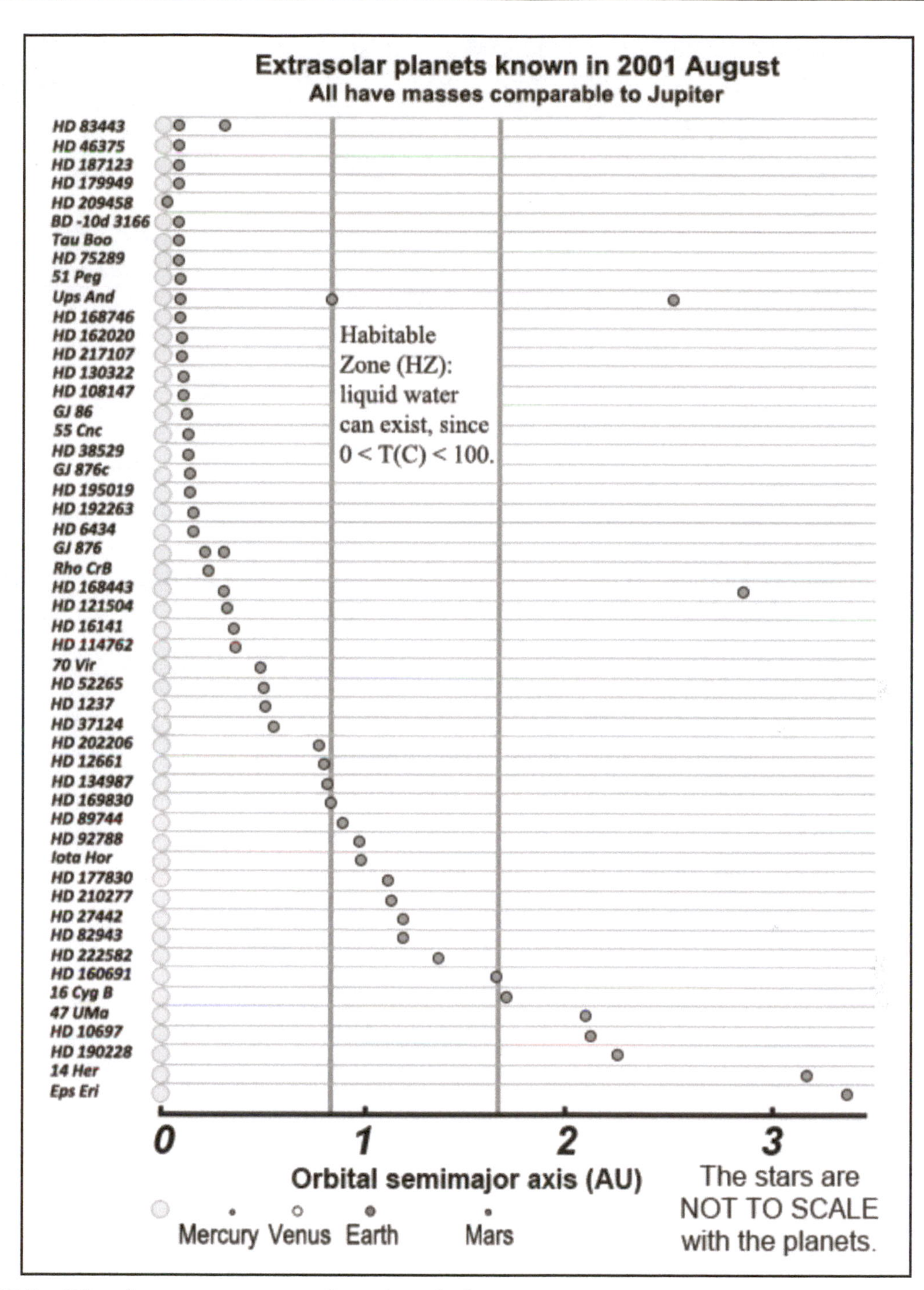

▶ **Figure 15-5** What have astronomers found, so far?

1. Pulsar planets, around dead stars
2. Hot Jupiters, around normal stars
3. Eccentric Jupiters
4. Classical Jupiters, much like Jupiter
5. Hot Neptunes, gaseous planets with masses as low as 3–5 Earths
6. Super-Earths, rocky planets with masses as low as 5 Earths
7. Terrestrial planets: "Terra" means "Earth" in Latin. NASA's *Kepler* spacecraft has found at least five Earth-size planets in the habitable zones of normal stars, like the Sun. We still don't know whether they have water or life—we need spectra. We might identify another Earth at any time now.

(Image courtesy of Frederick Ringwald)

► **Figure 15-6** Exoplanets have been found orbiting binary, or double, stars, in which two stars orbit each other. This is an artist's conception of a double sunset. (Source: NASA)

CHAPTER

16

Planet Earth

The four primary geological processes that shape landforms are:

1. **Impact cratering**
2. **Tectonism**
3. **Volcanism**
4. **Gradation**, including water and wind erosion.

Earth is highly unusual among planets, in that its surface has so few impact craters. This is because Earth is the most geologically active of all the terrestrial planets.

Over 70% of Earth's surface is covered by water, at an average depth of about two miles. Only 2% of this is fresh water. The only other world in the Solar System known to have liquid on its solid surface is Titan, a moon of Saturn, and the liquid isn't water, it's hydrocarbons such as methane and ethane.

The ocean floor, under sediment, is mainly black basalts. Most of these are less than 90 million years old. Nowhere is the ocean floor over 200 million years old, because it's resurfaced so actively.

The oldest known rocks on Earth are on the continents, and are 4.3 billion years old. Still, over 90% of Earth's rocks are less than 600 million years old.

Rocks are mostly silicates, made of silicon, oxygen, and metals. They come in three general types:

1. **Igneous** rocks form in lava. Examples include basalt, granite, and quartz.
2. **Sedimentary** rocks form in water. Examples include limestone, sandstone, and shale. Fossils are normally found in sedimentary rocks, because life needs water.
3. **Metamorphic** rocks were once igneous or sedimentary rocks, which were changed by heat and pressure in Earth's interior, and then brought back to the surface by erosion or tectonism. Examples include chalk, slate, and marble.

Tectonism

Earth's surface is divided into about 15 plates that move about over millions of years, riding on the convective currents in Earth's mantle. Plate tectonics is the unifying principle of geology, like Newton's laws of motion for physics, or evolution by natural selection for biology.

Tectonism and volcanism are driven by heat flowing out of Earth. Earth's internal heat is mostly from the decay of radioactive isotopes inside Earth, including uranium, thorium, and potassium-40.

► **Figure 16-1** Planet Earth (Source: NASA/Apollo 17 crew/Dr. Harrison H. Schmitt)

► **Figure 16-2** Typical Earth surface. (Image courtesy of Frederick Ringwald)

► **Figure 16-3** Impact craters dominate the surface of the Moon. (Image courtesy of Frederick Ringwald)

► **Figure 16-4** Impact craters are made by *impacts*. Moon rocks show that ancient, cratered terrain on the Moon formed 4.0–4.5 billion years ago, shortly after its birth. This is an artist's concept of the formation of Earth and Moon, during this time. (Image courtesy of Frederick Ringwald)

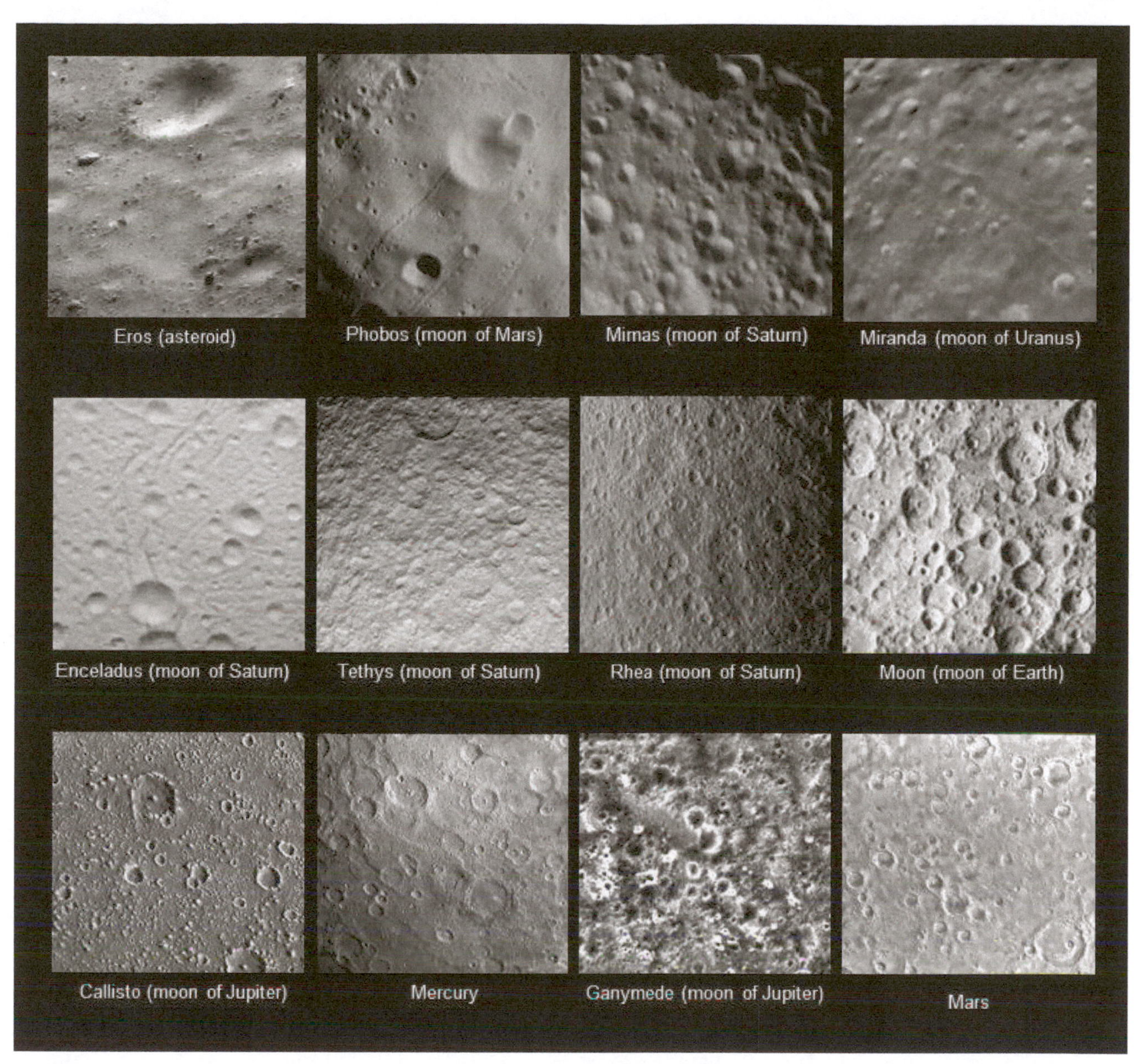

► **Figure 16-5** *Nowhere* on Earth has ancient, heavily cratered terrain like this, the way that moons and planets all over the Solar System do. (Source: NASA-JPL-JHU-APL-LPI)

► **Figure 16-6** Meteor Crater, on I-40 (formerly Route 66) near Winslow, Arizona, is the best-preserved impact crater on Earth. (Image courtesy of Frederick Ringwald)

► **Figure 16-7** Impact craters are rare on Earth. Fewer than 200 are known. There are thousands of volcanoes, and tens of thousands of folding mountains. (Source: NASA/Lunar and Planetary Institute)

► **Figure 16-8** The main reason impact craters are so rare on Earth is because Earth's surface is so active, with erosion from water, wind, tectonism, and volcanism. This is Odessa Meteor Crater, on I-20 near Midland-Odessa, Texas. A circular rim of rock is visible about 3 meters (10 feet) above the crater floor. (Image courtesy of Frederick Ringwald)

► **Figure 16-9** As this sign shows, the crater floor has been filled by water erosion. The crater was discovered because it is in a Texas oil field, which was carefully surveyed geologically. (Image courtesy of Frederick Ringwald)

► **Figure 16-10** Folding mountains are the most common mountains on Earth. They are caused by tectonic plates pushing together. These are the Himalayas, the tallest mountains in the world. They were caused by India colliding with Asia, starting about 40 million years ago. (Source: NASA)

► **Figure 16-11** Another example of folding mountains are the Andes, in South America. (Image courtesy of Frederick Ringwald)

► **Figure 16-12** Another example of folding mountains are the Appalachians. They look worn because they have been eroded more than the Andes or Himalayas. This is because they are older, 300–500 million years old. (Image courtesy of Frederick Ringwald)

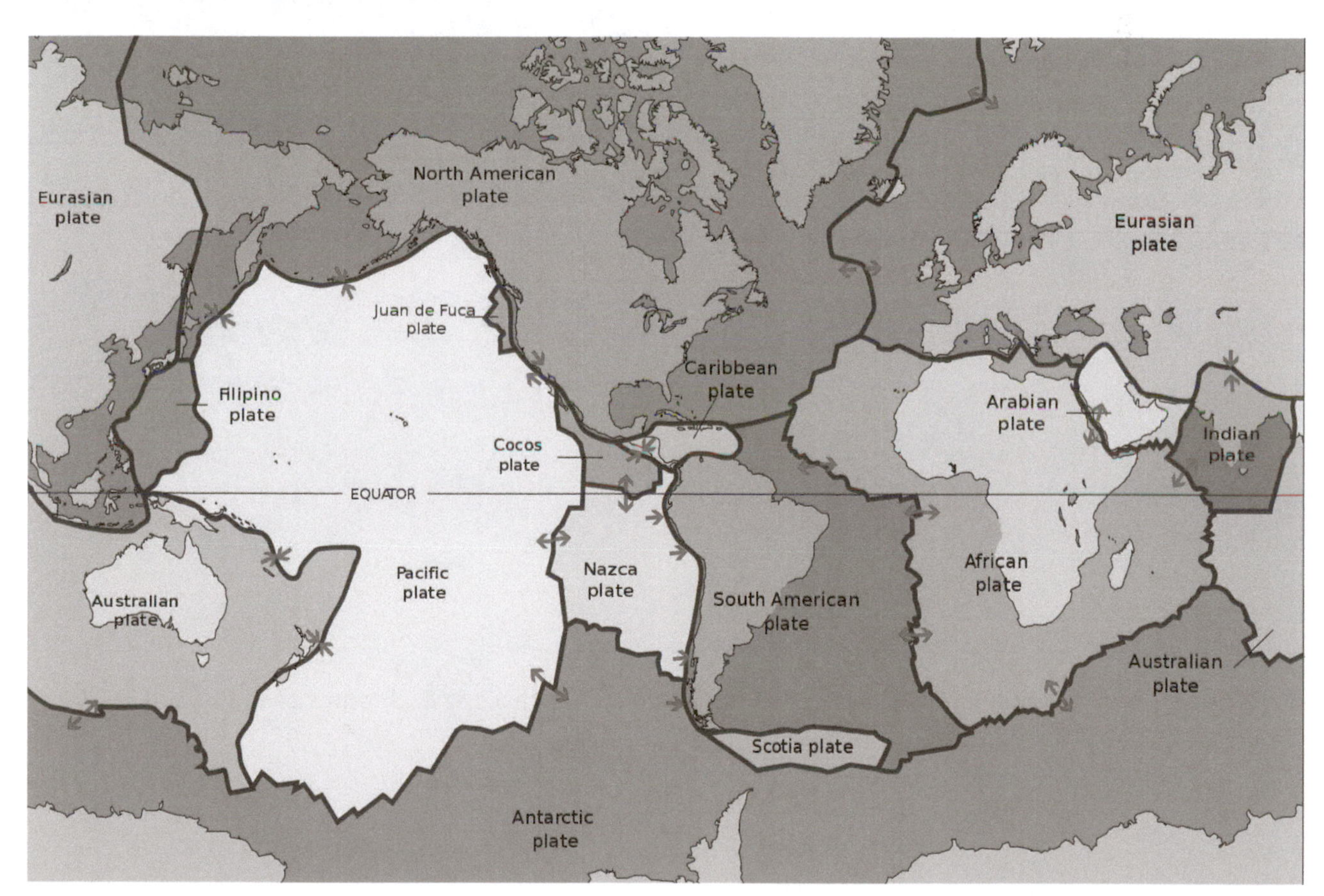

► **Figure 16-13** Earth's surface is broken into about 15 rigid plates, which move because of heat escaping from the interior of Earth. This idea is called plate tectonics. Plate tectonics is the central and unifying idea of the science of geology because it explains so many observations that by themselves and out of context would be difficult to explain. Folding mountains happen where plates push together, earthquakes happen where plates slide along each other, and seafloor spreading and continental drift happen where plates pull apart from each other. (Source: USGS)

► **Figure 16-14** Earthquakes occur when two tectonic plates slide along each other, as in California. (Source: USGS, bottom left USGS-J-K-Nakata)

▶ **Figure 16-15** Tectonism caused uplift in the Guadalupe Mountains in Texas. In the rock layers near the top of the slope shown here is a fossil reef, showing obvious signs of life that lived in shallow seawater. (Image courtesy of Frederick Ringwald)

▶ **Figure 16-16** This fossil reef in the Guadalupe mountains is most of the way up a mountain and hundreds of miles from the sea. (Image courtesy of Frederick Ringwald)

► **Figure 16-17** Volcanism occurs when heat inside Earth causes lava (which is molten rock), hot gas, cinders, and ash to flow through Earth's crust. Mount Etna, shown here, has been erupting intermittently for over 2000 years. (Source: NASA)

► **Figure 16-18** Don't confuse a volcanic *caldera* with an impact *crater*. This is Crater Lake, in Oregon. It should be called *Caldera* Lake since it was a volcano, not an impact crater. (Source: USGS/Charles R. Bacon)

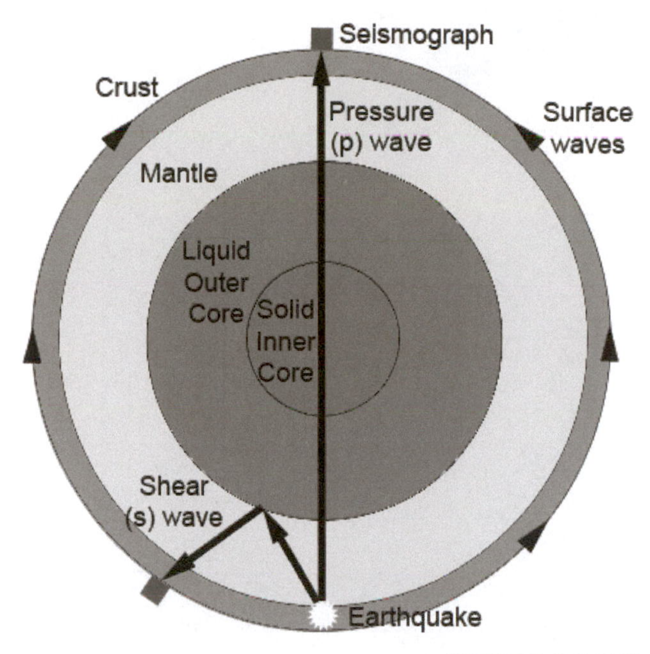

► **Figure 16-19 Deep inside Earth:** green olivine (which is olive in color) is mantle material. It's heavier (denser) than both black basalt, which lines the ocean floors, and the gray granite, which makes up the continents. *Earth's interior heat* is mostly from *radioactive decay* of heavy elements. *Seismology* is the study of earthquake waves. How they travel through Earth tells us about its interior. (Image courtesy of Frederick Ringwald)

► **Figure 16-20** Black basalt is *oceanic* crust. This is a lava plain in Hawai'i. (Image courtesy of Frederick Ringwald)

► **Figure 16-21** Gray granite is *continental* crust. These granite cliffs are in Yosemite National Park. Gray granite is lighter (less dense) than black basalt, which is *oceanic* crust. (Image courtesy of Frederick Ringwald)

► **Figure 16-22** *Left, top and bottom:* **Igneous** rocks form in lava. *Middle, top and bottom:* **Sedimentary** rocks form in water (or mud). *Right, top and bottom:* **Metamorphic** rocks were changed (metamorphosed) after being buried. (Image courtesy of Frederick Ringwald)

► **Figure 16-23 Rocks of organic origin**: Fossils are most often found in sedimentary rocks, since they form in water. Coal, petroleum, shale oil, and coquina are other examples of rocks that are by-products of life. (Image courtesy of Frederick Ringwald)

Earth completely melted soon after it formed. Heavier material sank to the center, making its iron core. This iron core generates Earth's magnetic field, which is what deflects a compass needle. The deep interior of Earth between the core and the crust is called the mantle. We know about Earth's interior because of seismology, the study of how sound waves travel through different materials inside Earth.

Plate tectonics causes continental drift, seafloor spreading, mid-oceanic ridges, and subduction, or one plate moving under another, as in the Andes. Earth is the only terrestrial planet with active plate tectonics, oceans, and abundant life on its surface.

Surprisingly, plate tectonics only became accepted as recently as the 1960s, even though Alfred Wegener first proposed it in 1912. One reason Wegener had difficulty convincing geologists was that he was not a geologist, but a meteorologist. Groups of people are often less ready to accept the views of outsiders. Louis Pasteur, the chemist who devised the germ theory of disease, had similar difficulty in convincing doctors to wash their hands.

(Wegener also supported the idea that impact craters are from meteorite impacts. This seems so obvious today, but until the 1950s, many scientists incorrectly thought they were from volcanism.)

To be fair, Wegener only had circumstantial evidence for continental drift. We also now know that his estimates for how much heat is flowing out of Earth and how fast the continents move were ten times too high.

Evidence for continental drift includes:

1. Some *geographic features fit together*, such as South America and West Africa. Wegener's critics dismissed this as only a coincidence.
2. *Similar fossils are found on different continents.* Finding fossils of palm trees in Greenland doesn't necessarily prove that Greenland drifted north, however: climate change might explain them.
3. What convinced most geologists were *magnetic stripes around spreading centers in the seafloor.* These were found near Iceland in the 1960s. As lava oozes out of the seafloor and solidifies into rock, iron and other magnetic material solidifies in the direction of Earth's magnetic field. These alternating bands of rock with similar magnetic directions show that the rock came from the spreading center.
4. We now can observe plate motion by *direct observation.* Satellite laser ranging shows the continents move by a few centimeters per year, about as fast as one's fingernails grow.

Folding mountains, from tectonism, are *the dominant form of mountain building on Earth.* Examples of folding mountains include the Sierra Nevada, the Rockies, the Appalachians, the Andes, the Mid-Oceanic Ridge of the bottom of the Atlantic Ocean (the longest mountain range in the world, even though these are mountains on the ocean floor), and the Himalayas (the tallest mountains on Earth), which were pushed up when India crashed into Asia.

Faults, such as the San Andreas Fault in California, result from sudden sliding of plates along each other. This causes earthquakes.

The Richter scale for measuring earthquake power is logarithmic. The most powerful earthquake known was an earthquake in Chile in 1960 that was rated 9.0 on the Richter scale. This was 200 times more powerful than the one that destroyed San Francisco in 1906, which was a 6.7.

Vocanism

Volcanism occurs when heat vents through Earth's crust. Volcanoes occur mainly around the rim of Pacific Ocean, the "Ring of Fire," where plates meet and subduction often occurs. Most volcanoes do not continuously erupt, but are usually dormant. Many are extinct. Volcanic eruptions are less common than earthquakes, but can still be devastating.

The ocean floor largely consists of *lava plains*, in which lava gradually oozed out of Earth's crust. There are similar basaltic lava plains on the Moon, Mercury, Venus, and Mars.

Gradation

Gradation is important in shaping most landforms, and in making sedimentary rocks. Water erosion is the dominant form of gradation on Earth, although some wind erosion occurs, for example, sandstone arches in Utah, or sand dunes in the Mojave Desert.

► **Figure 16-24** The Painted Desert in Arizona was shaped by water erosion. The cliffs in the distance were the shore of an ancient sea. (Image courtesy of Frederick Ringwald)

► **Figure 16-25** Bryce Canyon in Utah was shaped by erosion by water and ice. (Image courtesy of Frederick Ringwald)

► **Figure 16-26** This sandstone arch in Utah was carved by wind erosion. (Image courtesy of Frederick Ringwald)

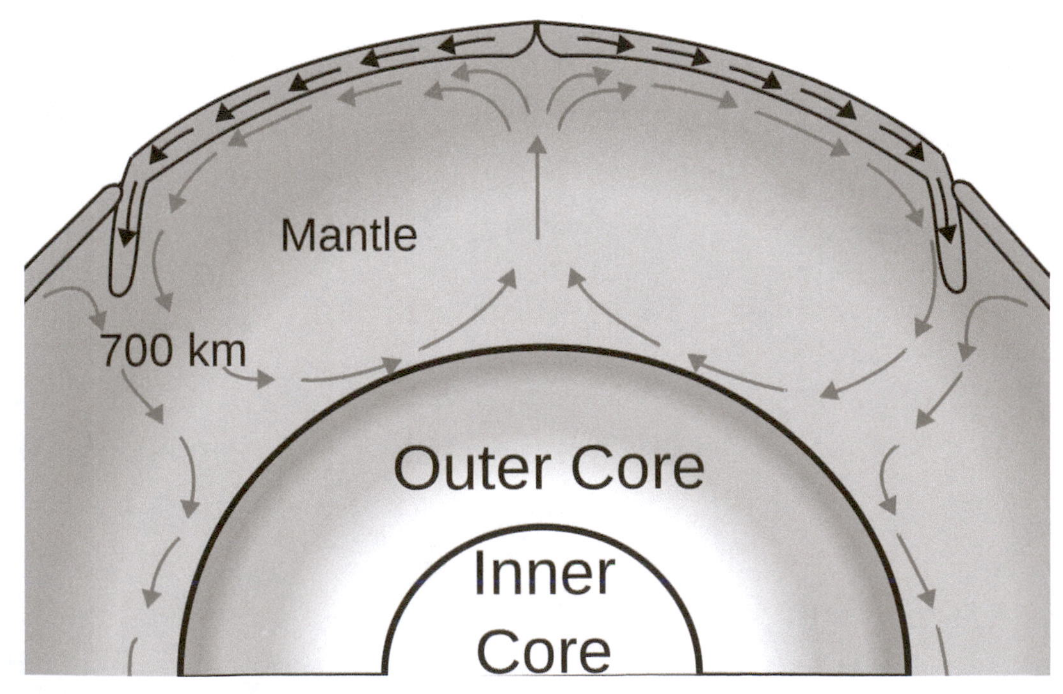

► **Figure 16-27** **The mechanism of plate tectonics** is now known to be heat flow and convective motion deep inside Earth. (Source: USGS)

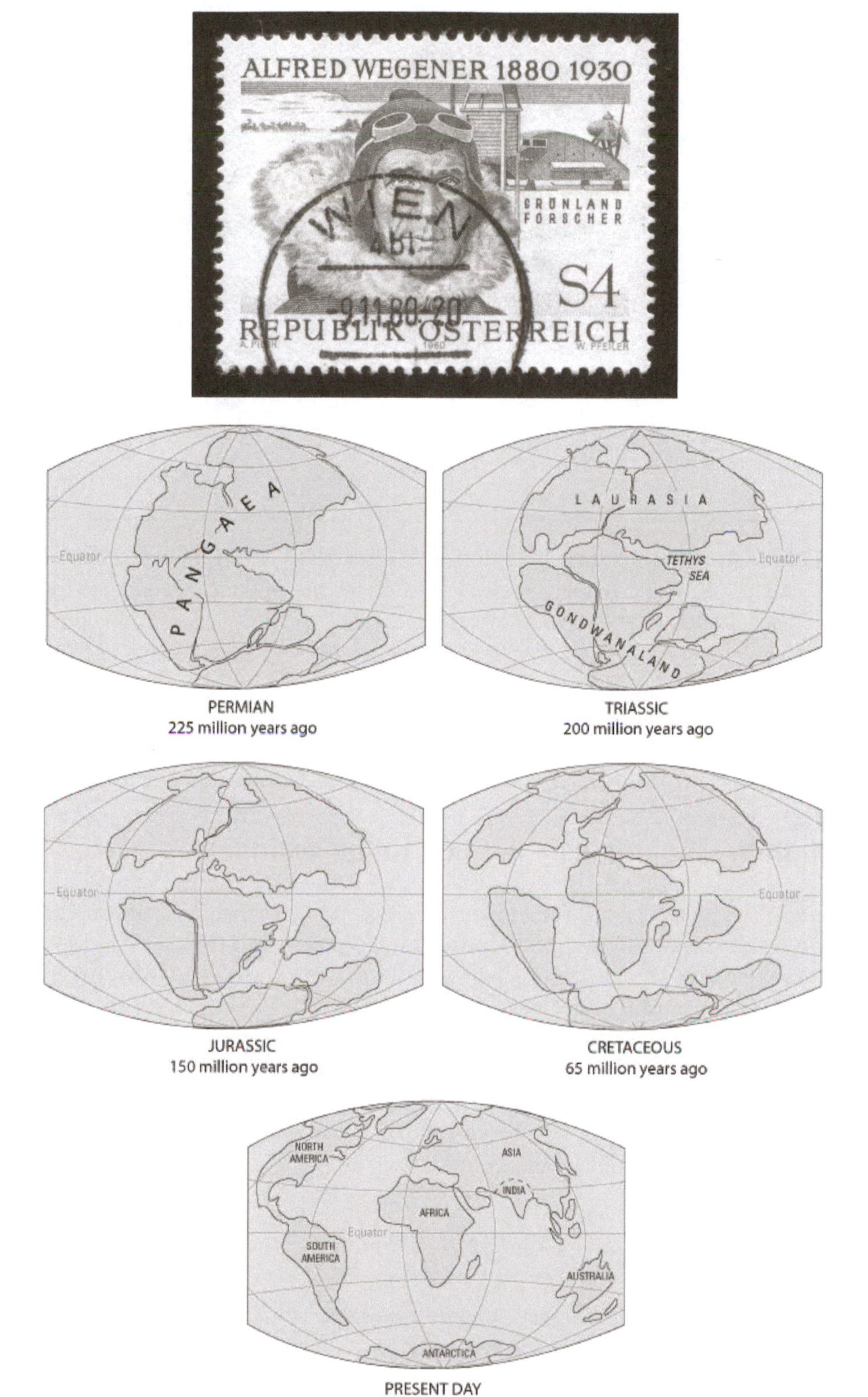

► **Figure 16-28** *Top:* Alfred Wegener had only a circumstantial case for continental drift. He didn't know how it worked, nor what caused it, nor could he measure the motion. Because of this, he had a difficult time getting geologists to accept continental drift. Wegener first suggested continental drift in 1912. He died during one of his expeditions to Greenland in 1930. Plate tectonics was not accepted until at least 1965. (Image © rook76, 2013. Used under license from Shutterstock, Inc.) ***Bottom:*** Continental drift, caused by plate tectonics, is now well established. (Source: USGS)

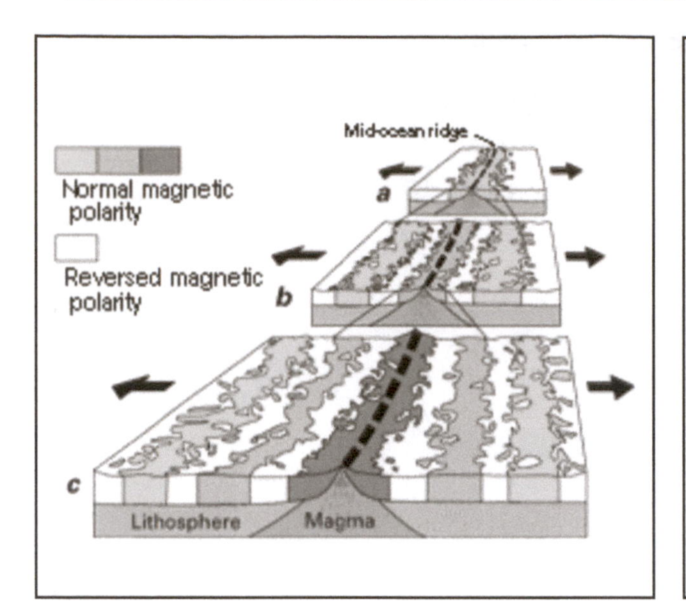

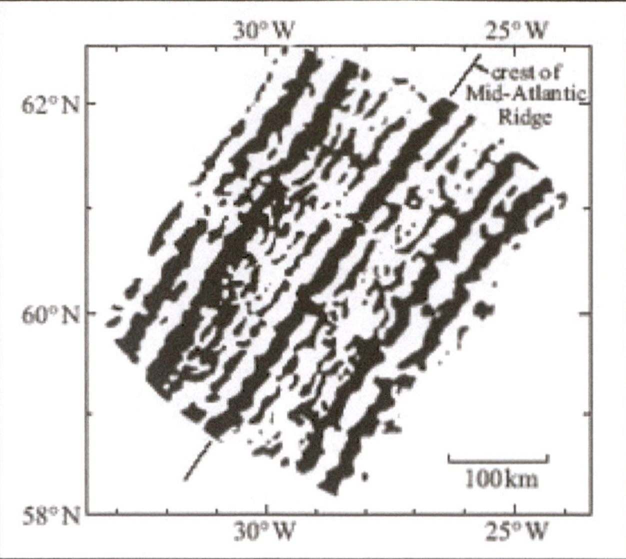

► **Figure 16-29** *Left:* Plate tectonics was accepted only after 1965, when evidence for *seafloor spreading* was discovered. (Source: USGS) *Right:* As new, melted crust oozes out of Earth, it solidifies toward the direction of Earth's magnetic field, which changes over time. (Source: USGS)

How Old is Earth? How Old is the Universe? How Do We Know?

Seashells High in the Mountains and Deep Under Sediment

Fossils were curiosities known since ancient times. Sometime in the 1480s, some peasants brought Leonardo da Vinci a bag of fossil seashells found high in the mountains. (By the way, the popular novel *The Da Vinci Code* by Dan Brown is historically inaccurate.) Leonardo inferred that an ocean had once been where the mountains are now. He ruled out that the seashells could have been carried there by a flood, since a flood would have washed seashells away from land, and not onto land.

In 1671, John Ray observed a bed of sand 30 meters deep on top of a bed of seashells. This meant that the sand had been deposited since the animals that lived in the shells were alive. Ray observed the rate at which sand was deposited, by measuring how fast rivers silt up, and so calculated that the sand must have been deposited over 10,000 years.

How Fast Earth Cooled

In 1687, Isaac Newton published the *Principia*, his book in which he presented his laws of motion and gravity. Included was an estimate of the age of Earth. Newton assumed that Earth is mainly made of iron, not a bad assumption since iron is now known to be Earth's most abundant element. Newton also considered observations of how long an iron sphere takes to cool. If an iron sphere the size of Earth started out molten, Newton calculated that it would take over 50,000 years to cool to Earth's temperature today. Therefore, Earth must be at least 50,000 years old.

In 1778, the Comte de Buffon carried out more detailed experiments to observe how fast iron spheres cool. From these, he estimated that Earth must be at least 74,000 years old. Taking into account non-uniform cooling, and that Earth is not 100% iron, led Buffon to conclude that the actual age was "millions of years."

In 1862, William Thompson, later known as Lord Kelvin, calculated how long a molten Earth must have cooled to reach its present temperature, taking into account conduction of heat through Earth's interior, and cooling by radiation into space. Kelvin calculated that Earth must be 98 million years old. In 1897 he revised this downward, to 20–40 million years.

A problem with this approach is that it gives only a lower limit to the age. Earth could be much older, because in the early 1900s, Earth was found to have an internal heat source: radioactive decay of elements such as uranium and thorium.

The Saltiness of the Oceans

In 1901, John Joly calculated the rate of delivery of salt to the oceans. River water has only a small concentration of salt, which could be measured, and rivers flow to the sea. Evaporation of seawater concentrates salt, making ocean water saltier over time. The age of the oceans is therefore equal to the total amount of salt in the oceans, divided by the rate that rivers add salt to the oceans. Joly calculated the oceans to be at least 90–100 million years old.

A problem with this is that, again, it gives only a minimum age. Earth could be much older, since Joly couldn't account for recycled salt or salt bound into mineral deposits.

The Orbit of the Moon

Starting in 1969 during the Apollo 11 mission, several reflectors were placed on the Moon's surface, by astronauts and by robotic spacecraft. Astronomers on Earth have since used lasers to bounce light off these reflectors. With precise clocks, they measure the round-trip time for the light, and can therefore measure the distance to the Moon to within a few centimeters.

The Moon is slowly moving away from Earth, at a rate of 3.82 ± 0.7 centimeters/year. Since the Moon is 385,000 kilometers (3.85×10^{10} cm) from Earth, it must have taken the Moon at least one billion years to reach its present distance from Earth.

A problem with this simple linear extrapolation is that it doesn't take into account how much energy of the Moon's motion is lost to compression of the Moon and the Earth by each other's gravity, also known as tidal heating. This will slow down the Moon's recession from Earth, which therefore gives only a minimum age estimate: the actual age will be longer.

Fossils

In the 1790s, William Smith noticed that fossils occur in layers, with the oldest layers normally at the bottom, and that the layering follows a distinctive sequence. This implies that if two rock layers contain the same fossils, they have the same age.

A problem is while the Principle of Fossil Succession can provide relative ages (or in other words tell which fossils are older than others), absolute ages required each rock layer to be measured with radioactive decay (see below). Measurements of rock layers from around the world have given ages so consistent, geologists now use fossils for dating rock layers, with enough predictive power to be able to find oil.

Stromatolites are rocks that were layered by bacteria. Stromatolites are the oldest known direct evidence of life. The oldest known stromatolites are 3.5 ± 0.1 billion years old. The oldest indirect evidence of life on Earth are carbon isotopes, estimated to be 3.8 ± 0.1 billion years old.

Erosion and Sedimentation

While engineering canals in 1514, Leonardo da Vinci observed sediment in the Po river. From the rate that silt was deposited, he calculated the deposits to be 200,000 years old, and concluded that the entire Earth was probably much older.

In 1785, James Hutton began the idea of uniformitarianism, that "the past history of our globe can be seen to be happening now." Observations on his farm convinced him that both erosion and rock formation can be understood if one assumes that the same physical and chemical processes that we can observe today have been operating over vast amounts of time. Unfortunately, he wasn't a clear writer, and it took decades for his work to get the recognition it deserved. John Playfair's writing helped: as he noted, "the mind seemed to grow giddy by looking so far into the abyss of time."

In 1830, Charles Lyell first published *Principles of Geology*, in which he presented more evidence of uniformitarianism, "the present is the key to the past." One line of evidence was in mountains that showed many cycles of erosion, subsidence, and uplift. Another line of evidence were layers of sediment over 25,000 meters thick. Since rivers are observed to deposit millimeters of sediment per year, these layers of sediment required "millions of years" (as Lyell wrote) to form. By 1917, thicker layers of sediment were found, implying an age of at least 1.25–1.7 billion years (Barrell 1917).

Problems were that this did not account for past erosion, or how ancient sedimentary rocks are metamorphosed or melted. This again gave only a minimum age. Also, while uniformitarianism does explain much of geology, scientists now recognize that some sudden processes do occur, also called catastrophism. Impact craters, for example, form only in a few seconds, and although they are rare on Earth, their effects can be important: a large impact on Earth 65 million years ago is thought to have caused the extinction of the (non-avian) dinosaurs.

Varves are uniform layers of sediment that form each year in large lakes, with light layers from summer runoff, and dark layers from winter runoff. One can determine how old varves are by counting the layers. Some varves in the Green River formation in Colorado, Utah, and Wyoming are 20 million layers thick.

Radioactive Decay

Henri Becquerel discovered radioactivity in 1896. In 1905, Ernest Rutherford and Bertram B. Boltwood used radioactive decay to measure the ages of rocks and minerals. Uranium decay produces helium, leading to an age for Earth of 500 million years. In 1907, Boltwood determined that lead was the stable end-product of the decay of uranium. Using a sample of urananite, uranium-lead dating gave an age of 1.64 billion years.

Again, a problem with this is that it gives only a minimum age. We now know that over 90% of rocks on the surface of Earth are less than 600 million years old, because of weathering and being recycled into the Earth by plate tectonics. The search was on to find the oldest rocks on Earth, which are generally found in places where they are under ice sheets, as in the Canadian Shield or Greenland, or far from plate boundaries, such as in Australia.

Many radioactive elements can be used today as geologic clocks. Each element decays at its own nearly constant rate. Once this decay rate is known, geologists can estimate the length of time over which decay has been occurring by measuring the amount of radioactive parent and the amount of stable daughter elements. There are few radioactive isotopes with half-lives less than 80 million years on Earth, because Earth is so much older than this.

Supernovae, or exploding stars, are observed to have radioactive elements because they emit gamma rays, which have energies and intensities that match those observed on Earth, even for supernovae millions of light-years away (and, because of the finite speed of light, millions of years ago). That the laws of physics, such as the strength of gravity, the speed of light, and radioactive decay rates, have not changed significantly over the age of the Universe is shown by spectra of distant supernovae and galaxies—the relative wavelengths of spectral lines are the same as they are today.

The Oldest Earth Rocks

The oldest Earth rocks now known are 4.28 ± 0.08 billion years old. These are volcanic basalts from the Nuvvuagittuq greenstone belt, on the eastern shore of Hudson Bay in northern Quebec, Canada. These rocks were dated by isotopes of neodymium and samarium, two rare-Earth elements. They may be remnants of Earth's original crust, before it had any stable crust. The previous record-holder was Acasta gneiss (pronounced "nice"), metamorphic rock from the Canadian Shield. Uranium and lead isotopes give an age of 4.03 ± 0.01 billion years old.

Zircons from the Jack Hills in Western Australia have been found by radiometric dating to be 4.404 ± 0.008 billion years old, although these are only crystals of a specific mineral, or rock substance, not intact rocks. Zircons are quite hard, making them almost indestructible, and contain uranium, thorium, and lead isotopes that show their age. Surprisingly, the isotopic composition of these zircons indicates that they must have formed in liquid water. Since liquid water can only exist at temperatures between 0° and 100°C, this shows that the Earth's temperature had to be similar to what they are today, even as early as 150 million years after the oldest meteorites formed. Observations of stars like the Sun show that young stars like the Sun are 30% less powerful than the Sun: this implies that the early Earth's atmosphere was rich in greenhouse gases. This is also implied by ancient siderite minerals, which form when carbon dioxide is washed out of the atmosphere by rain.

Moon Rocks and Meteorites

The six Apollo missions to the Moon, from 1969 to 1972, returned 842.5 pounds of Moon rocks. Several ounces of lunar dust were also recovered by robotic spacecraft from the Soviet Union. Radiometric dating with several isotopes shows that rocks from the lunar highlands are typically 4.1 to 4.4 billion years old. Basaltic rocks from the lunar maria (or lava plains) are typically 3.2 to 3.8 billion years old. The oldest known Moon rocks are 4.46 ± 0.04 billion years old. The oldest known meteorites are carbonaceous chondrites, dated at 4.57 ± 0.02 billion years old.

The oldest Earth rocks are younger than the oldest Moon rocks and meteorites because Earth is geologically active, and has a hot, molten interior. When rocks melt, their radiometric clocks are reset. Rocks on Earth's surface are acted on by erosion and weathering. Rocks on Earth's surface are not as old as the Earth, since they are "recycled" rock materials (Cole and Woolfson 2002).

The Age of the Sun

Lord Kelvin estimated the age of the Sun at 20–40 million years. This assumed that the Sun radiated at its observed luminosity, and that all of this energy came from the Sun's gravity. The problem with this estimate was that, again, it's only a lower limit to the age. The Sun's gravity isn't its only source of energy: we now know that the Sun is primarily nuclear powered.

Nuclear reactions inside the Sun turn hydrogen into helium, so the Sun's concentration of helium should increase over time. Helium is noted for its funny properties on sound, and astronomers can observe how sound waves travel through the Sun, by the ripples they make on the Sun's surface. Astronomers can therefore study the interior of the Sun in a way similar to how geologists use seismic waves to study the interior of Earth. This is called helioseismology (from "helio," which means "the Sun"). This gives an age for the Sun of 4.57 ± 0.11 billion years (Bonanno et al. 2002), which agrees with the ages of the oldest meteorites (see earlier discussion), so the Solar System is thought to have condensed under gravity out of the Solar Nebula at this time.

How do we know that the Solar System is 4.57 ± 0.02 billion years old?

1. We can see other planetary systems forming now around young stars, such as in the Orion Nebula.
2. The scars of it, *impact craters*, are on Earth's Moon and other moons and planets.
3. The age is known from radioactive dating of meteorites (see above).

How do we know that the Universe is 13.80 ± 0.04 billion years old?

1. In 1929, Edwin Hubble discovered that the Universe is expanding. If we run the movie backwards, there had to have been a time when the Universe had a dense origin. This is widely called "the Big Bang." *Hubble Space Telescope*, which was launched by Space Shuttle in 1990, was specifically designed to measure how fast this expansion is, in order to measure the age of the Universe. These observations gave an age of 13.7 ± 0.2 billion years (Freedman et al. 2001).
2. If the Universe had a dense origin, the Universe must have been hot during this time. This is because the Universe is mostly made of gas, and the denser a gas is, the hotter it is. (Some of you may remember this from chemistry classes as the ideal gas law.) If the early Universe was hot, it should have radiated light, which we should see: and we do, it's called the Cosmic Background Radiation. This is also called the Cosmic Microwave Background, since the maximum intensity of this radiation has wavelengths of a few millimeters, or in other words, microwaves.

   The reason why it's microwaves is that, as the Universe has expanded, the radiation has cooled, increasing its wavelength. This was the light the Universe emitted when it first became cool enough to become transparent to light, and detailed observations of it have been made with the European Space Agency's *Planck* spacecraft.

   Since we know from laboratory experiments at what temperature hot gas should do this, this can reveal the time at which the Universe did become transparent to light, and also the amount of time since then. The Universe first became transparent to light when it was about 380,000 years old. Since then, it has cooled for 13.798 ± 0.037 billion years (Planck Collaboration 2013). This agrees with a previous measurement made by NASA's Wilkinson Microwave Anisotropy Probe (WMAP) spacecraft of 13.73 ± 0.12 billion years (Hinshaw et al. 2009).
3. The Sloan Digital Sky Survey has made a 3-D map of the positions, distances, and speeds over 200,000 galaxies. How they cluster together shows how they condensed under gravity, which shows how long this took: 13.5 ± 0.2 billion years (Tegmark et al. 2004).
4. The speed of light is not infinite, so as we look deeply into the Universe, it's like looking onto a time machine, since we see deep into the past. The deepest images of the Universe ever made, with Hubble Space Telescope, show galaxies 13.0 ± 0.7 billion years old, the way they were during the first 5% of the age of the Universe (Illingworth and Faber 2004).
5. When a star uses up the nuclear fuel in its center, it turns into a red giant. How old a star needs to be before this happens depends only the star's mass: massive stars burn themselves out faster. Finding the least massive red giants in a star cluster can therefore give a minimum age for the cluster. Doing this with globular clusters shows that the oldest red giants are 11.5 ± 1.3 billion years old (Cheboyer 1999).
6. Two types of exploding stars, namely Type Ia supernovae and gamma-ray bursts, are so powerful they can literally be seen across the Universe. Ones far from us show explosions from over 12.7 billion years ago (Chincarini 2005).

7. Nucleocosmochronology (or nucleo-cosmo-chronology) uses measurements of the relative abundances of radioactive isotopes, in a manner quite similar to radiometric dating of rocks, described above, but using lines in the spectra of astronomical objects, taken through telescopes. The ratio of U-238/Th-232 isotopes in the spectra of halo stars, together with measurements from meteorites, gives an independent estimate for the age for the Universe of 14.5 (+2.8/–2.5) billion years (Dauphas 2005). The oldest star known in the Milky Way Galaxy, a metal-poor star in the Galactic halo known as HE 1523–0901, has several isotopic ratios (from U/Th, U/Ir, Th/Eu, and Th/Os) that all indicate an age of 13.2 ± 0.7 billion years (Frebel et al. 2007).
8. White dwarf stars, which are burned-out cinders of what used to be stars like the Sun, cool down in a characteristic manner. How cool the coolest white dwarfs are can be used to give a minimum age for the population of stars to which the white dwarfs belong. Observations of white dwarfs in a globular cluster gave an age of the globular cluster of 12.7 ± 0.7 billion years (Hansen et al. 2002).

   Notice also that 1–8 give consistent answers. This inspires confidence, since science is like a jigsaw puzzle, where the pieces fit together. Isolated incident are unusual, because everything is interconnected, even if for physically quite different reasons.

References

Badash, L. 1989. *Scientific American*, Vol. 261, No. 2 (August issue), p. 90.

Barrell, J. 1917. Rhythms and the Measurements of Geologic Time: Bulletin of the Geological Society of America, Vol. 28, p. 745.

Bonanno, A. et al. 2002. *Astronomy & Astrophysics*, Vol. 390, p. 1115.

Chaboyer B. 1999, in A. Heck, F. Caputo (eds.), Post-Hipparcos Cosmic Candles, Astrophysics and Space Science, Vol. 237, p. 111.

Chincarini, G. 2005, European Southern Observatory Press Release 22/05.

Cole, G. H. A., and Woolfson, M. M. 2002. Planetary Science, Topic B (Geochronology), p. 202.

Dalrymple, G. B. 1991. *The Age of Earth*. Stanford University Press.

Dauphas, N. 2005. *Nature*, v. 435, p. 1203.

Frebel, A. et al. 2007. The Astrophysical Journal Letters, Vol. 660, p. L117.

Freedman, W. et al. 2001. The Astrophysical Journal, Vol. 553, p. 47.

Gore, P. J. W. 1999. http://www.dc.peachnet.edu/~pgore/geology/geo102/age.htm

Hansen, B. et al. 2002. *The Astrophysical Journal*, Vol. 574, p. 155.

Hinshaw, G. et al. 2009. *The Astrophysical Journal Supplement*, 180, 225.

Illingworth, G., and Faber, S. 2004. University of California News Article 6181 (2004-03-09).

Norman, M. 2004. Planetary Science Research Discoveries: The Oldest Moon Rocks, http://www.psrd.hawaii.edu/April04/lunarAnorthosites.html

Planck Collaboration. 2013, http://adsabs.harvard.edu/abs/2013arXiv1303.5076P

Tanford, C., and Reynolds, J., 1993. The Scientific Traveler, John Wiley and Sons.

Tegmark, M. et al. 2004. Physical Review Letters D, Vol. 69, p. 103501.

Zimmer, C. 2001. *National Geographic*, Vol. 200, No. 3, p. 78 (September issue).

CHAPTER

17

Moon Phases and Eclipses

The Moon is about quarter the size of Earth, and it orbits Earth about once a month. This makes it easy to see, even in daylight.

Every month the Moon goes through its cycle of apparent shape, or *phases*, from crescent to gibbous to Full and back. This has nothing to do with the Earth's shadow: it's due to the Sun angle.

► **Figure 17-1** Earth and Moon are very much a double planet, with the Moon being about 2000 miles in diameter, about ¼ the diameter of Earth. Still, the contrast between a living world, Earth, and a dead one, the Moon, is clear in this photo taken from lunar orbit. (Source: NASA/Apollo 8 crew/William A. Anders)

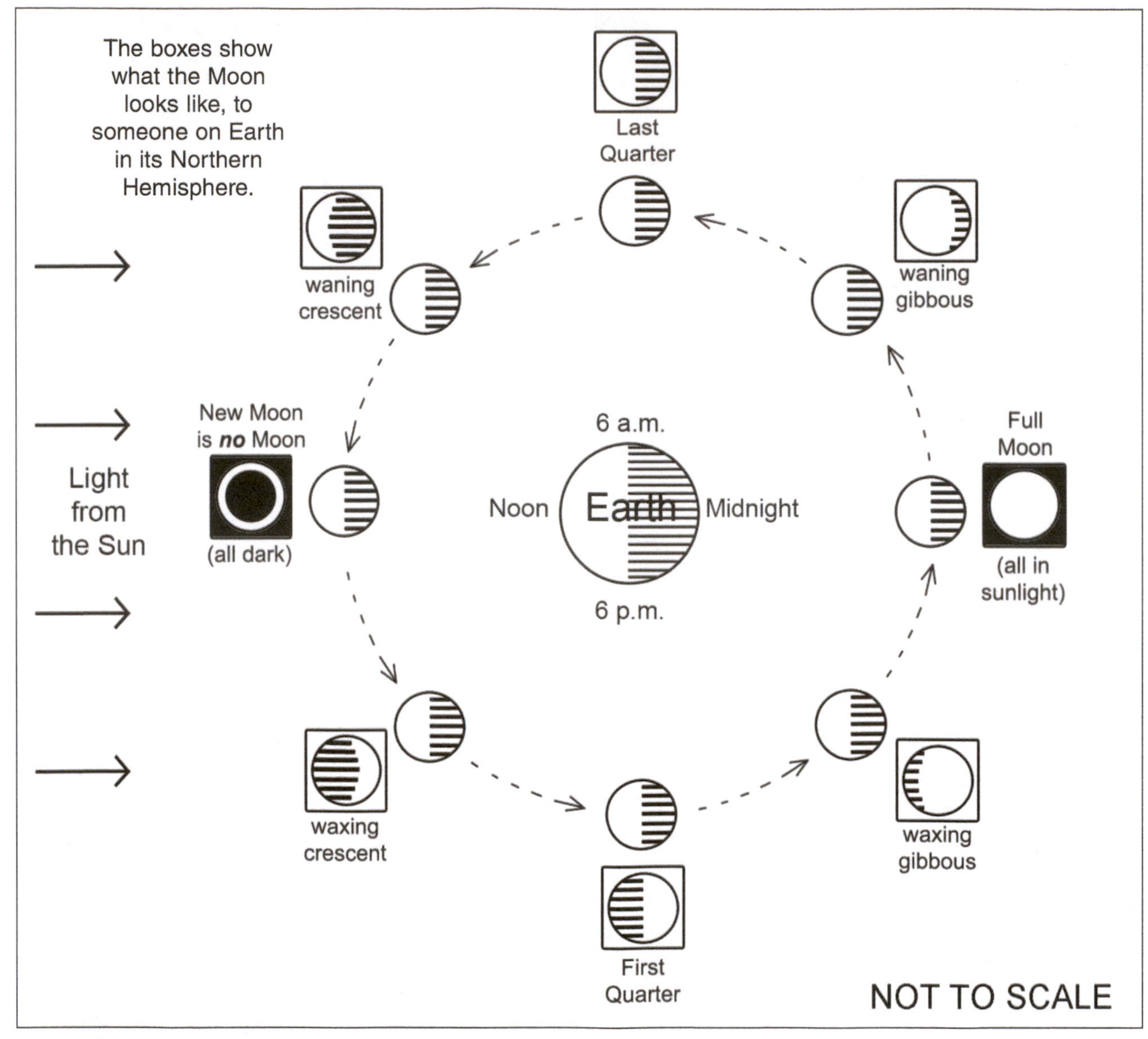

► **Figure 17-2** The Moon's monthly cycle of *phases:*
The Moon orbits the Earth once per month (about 28 days).
Each night, the Moon moves $360°/28 \approx 12$ degrees *east* in the sky.
The Moon therefore rises about 45 minutes later, each night.

For exams: Be able to reproduce this drawing, to understand it.

This drawing can be used to find what times of day or night the Moon rises and sets. For example, a Full Moon rises at sunset, and sets at sunrise. This is because a Full Moon is in the opposite part of the sky from the Sun. A waxing crescent Moon sets just after sunset, because it's just east of the Sun, in the sky. A waning crescent moon rises just before sunrise, because it's just west of the Sun, in the sky. (Image courtesy of Frederick Ringwald)

► **Figure 17-3** The Moon is about as reflective as an asphalt parking lot, which means that its side that is in the Sun's light is easily bright enough to be seen in Earth's blue, daytime sky. Indeed, about half the time the Moon is visible is during the daytime. (Image courtesy of Frederick Ringwald)

► **Figure 17-4** Every month, the Moon goes through a cycle of phases, from crescent to gibbous to Full to gibbous to crescent and back. The phases of the Moon are due to the Sun angle: they have nothing to do with Earth's shadow. (Image courtesy of Frederick Ringwald)

► **Figure 17-5** A waxing crescent Moon is visible just after sunset. Two bright planets, Venus and Jupiter, are also visible. (Image courtesy of Frederick Ringwald)

▶ **Figure 17-6** Two nights later, the Moon has moved 24° east, as it orbits Earth. Notice how now the Moon is fuller, with more of its side facing us in daylight. It will continue to get fuller each night, until it reaches Full Moon. At Full Moon, all of the side facing Earth will be in daylight. (Image courtesy of Frederick Ringwald)

▶ **Figure 17-7** A waxing crescent Moon (Image courtesy of Frederick Ringwald, at Fresno State's Campus Observatory)

▶ **Figure 17-8** A First Quarter Moon (Image courtesy of Frederick Ringwald, at Fresno State's Campus Observatory)

▶ **Figure 17-9** A waxing gibbous Moon (Image courtesy of Frederick Ringwald, at Fresno State's Campus Observatory)

► **Figure 17-10** A Full Moon (Image courtesy of Frederick Ringwald, at Fresno State's Campus Observatory)

► **Figure 17-11** A waning gibbous Moon (Image courtesy of Frederick Ringwald, at Fresno State's Campus Observatory)

► **Figure 17-12** A Last Quarter Moon (Image courtesy of Frederick Ringwald, at Fresno State's Campus Observatory)

► **Figure 17-13** A waning crescent Moon (Image courtesy of Frederick Ringwald, at Fresno State's Campus Observatory)

Eclipses

Eclipses don't happen every month because the orbit of the Moon is inclined by 5° to the ecliptic. Every now and then, on intervals of roughly six months, the Earth, Sun, and Moon do line up, so the Earth and Moon cast shadows on each other.

Lunar eclipses are when the Earth casts a shadow on the Moon. They can only occur during Full Moon. (Look at the drawing on Moon phases and convince yourself.)

During a lunar eclipse, the Moon turns black, or often red. This is because of dust in Earth's atmosphere, and is the same reason that sunsets are red. Lunar eclipses last about 3 hours, and are a great pretext for an outdoor party. Ask your instructor whether there will be any lunar eclipses this term.

Many people have seen lunar eclipses. All one needs to do is to be on the hemisphere of Earth facing the Moon when a lunar eclipse happens, and have good weather.

▶ **Figure 17-14** This is a **lunar eclipse**, shown as a series of images taken as the Moon moves through Earth's shadow. (Image © Phillip Holmes, 2013. Used under license from Shutterstock, Inc.)

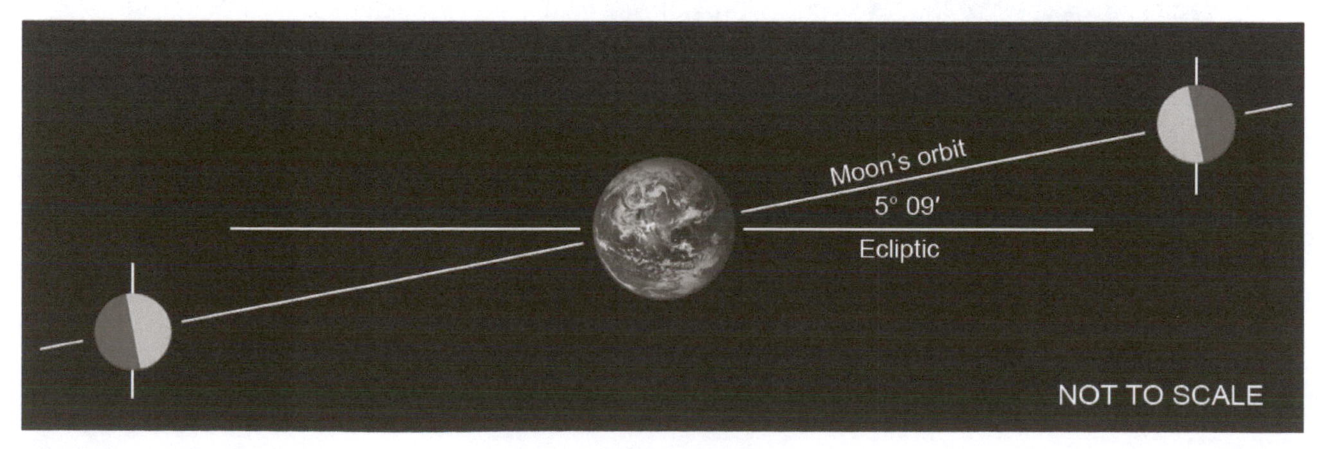

▶ **Figure 17-15** We don't have eclipses every month because the Moon's orbit is off the ecliptic plane by over 5°. (Image courtesy of Frederick Ringwald using a NASA image of Earth)

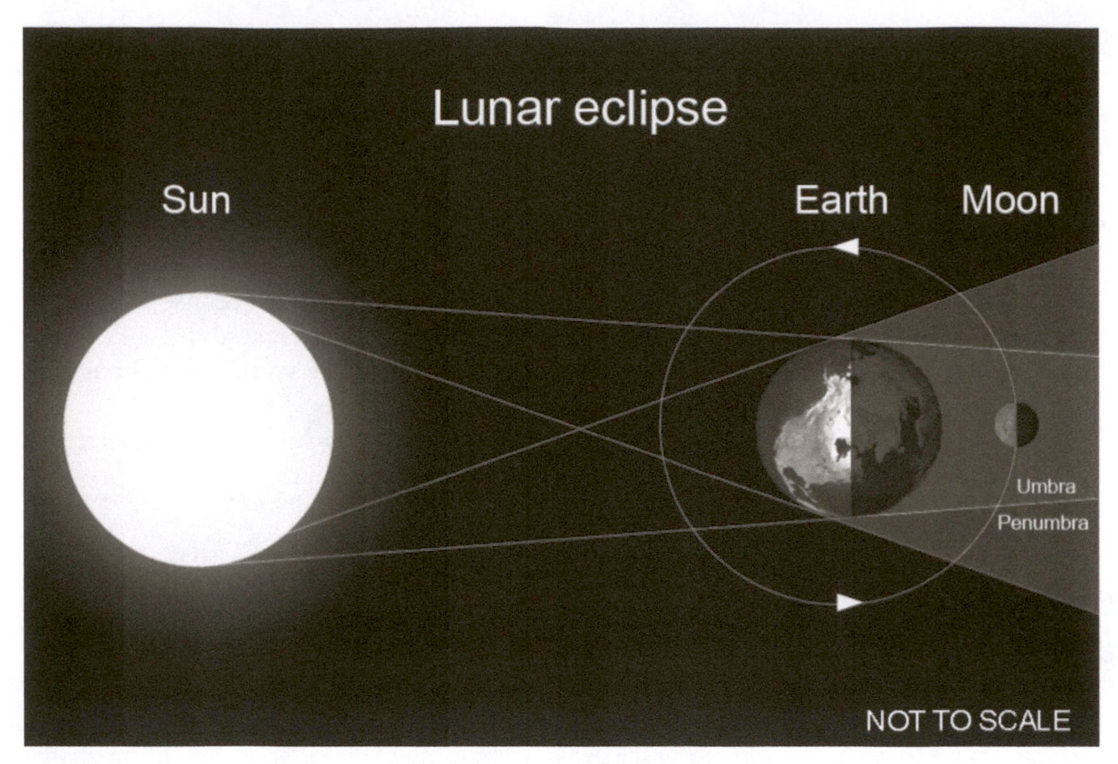

▶ **Figure 17-16** In a lunar eclipse, Earth casts a shadow on the Moon. This can only happen at Full Moon. (Image courtesy of Frederick Ringwald, using a NASA image of Earth)

▶ **Figure 17-17** A lunar eclipse can be an excellent pretext for an outdoor party, if it happens during the warmer months of the year. (Image courtesy of Frederick Ringwald, at the Downing Planetarium)

▶ **Figure 17-18** A lunar eclipse. (Image courtesy of Dr. Greg Morgan)

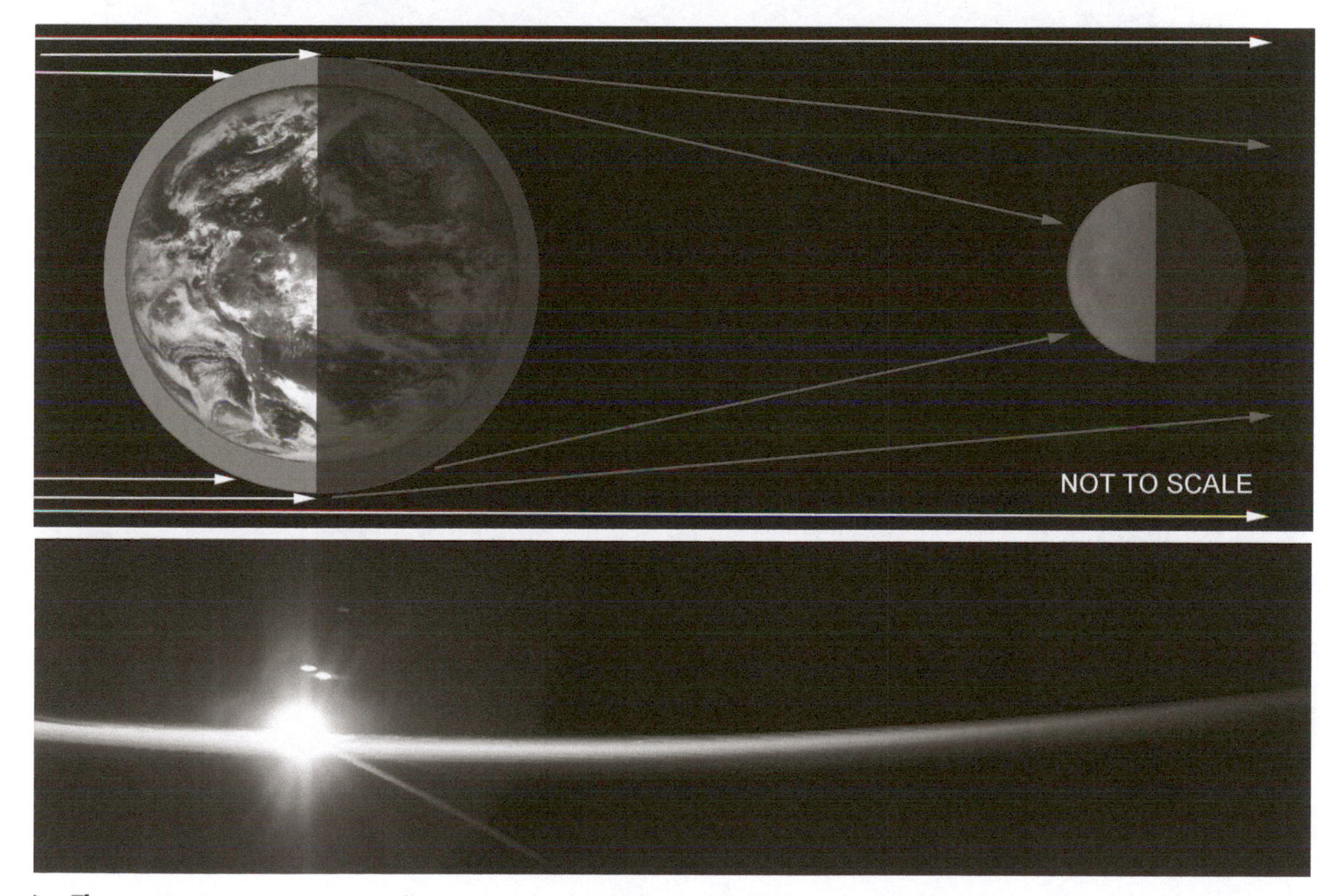

▶ **Figure 17-19** *Top:* Lunar eclipses can look red because of dust in Earth's atmosphere. This is because, during a lunar eclipse, the light from the Sun that falls on the Moon passes through Earth's atmosphere. Dust in Earth's atmosphere allows only the red light to get through, so the Moon can look red during an eclipse. This depends on the amount of dust in Earth's atmosphere, however. Eclipses that occur just after major volcanic eruptions can look black. Any red that is visible can range from fire red to blood red to brick red. (Image courtesy of Frederick Ringwald, using a NASA image of Earth) ***Bottom:*** This is a sunrise, seen from Earth orbit. Sunsets are also red because of scattering and absorption by dust in Earth's atmosphere. (Source: NASA)

Solar eclipses happen when the Moon casts a shadow on the Earth.

Partial solar eclipses happen when the Moon covers only part of the Sun's bright surface, the photosphere. Observing them can therefore harm your eyes, just as observing the Sun at any other time can harm your eyes. ***Don't*** *observe the Sun, unless you know exactly what you're doing!*

Annular solar eclipses happen when the Moon passes directly in front of the Sun, but doesn't cover the entire surface. A bright ring (annulus means "ring," in Latin) of the Sun's photosphere is still visible, making annular eclipses dangerous to look at, too. *NEVER look at the Sun's photosphere!*

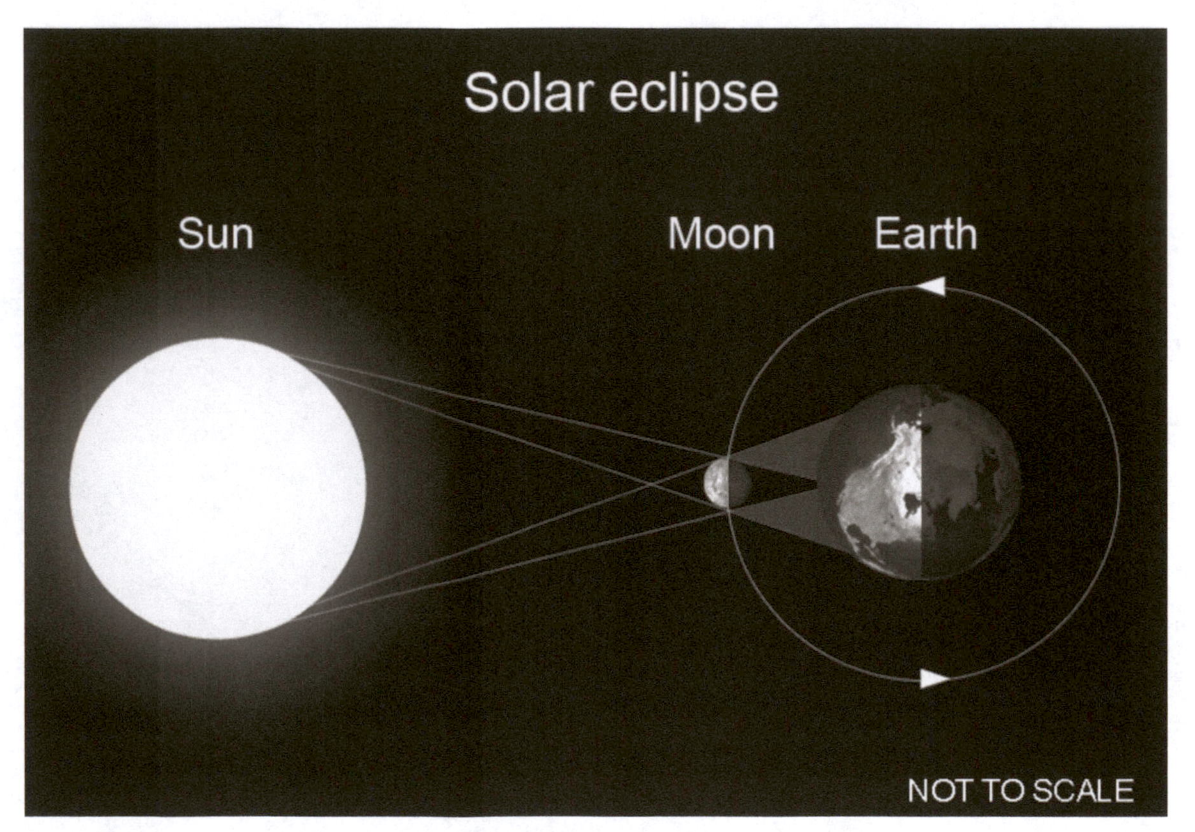

► **Figure 17-20** In a solar eclipse, the Moon casts a shadow on Earth. (Image courtesy of Frederick Ringwald

▶ **Figure 17-21** This is a partial solar eclipse, when the Moon only partly blocks the bright surface of the Sun, or solar photosphere. Careful, these are dangerous to observe! That bright part is still the surface of the Sun! (Image courtesy of Dr. Greg Morgan)

▶ **Figure 17-22** Mylar "eclipse glasses" like these are supposed to be safe, but one should be skeptical, because children can move *much* faster than adult eyes can see. (AFP/Getty Images)

► **Figure 17-23** I would prefer that you *not* observe the Sun, unless you know exactly what you're doing. Misunderstanding the directions can result in instant *permanent* blindness. *Top left:* Reflective glass filters for solar observing go in front of a telescope's aperture, so that little sunlight gets into the telescope at all. They should look like mirrors, and fit securely onto the telescope tube, so some fool cannot remove them. *Top right:* Notice the webcam in this dedicated solar telescope's eyepiece holder. It can project an image on the laptop computer that is shown. With instruments like this, there is no longer any reason whatsoever to risk anyone's eyesight while solar observing. *Bottom left:* Fresno State graduate student Dillon Trelawny observing the Sun. *Bottom right:* The screen of a laptop computer is easier to read while in bright sunlight when the laptop is enclosed in this hood, which was made by the author of cardboard and duct tape. (Images courtesy of Frederick Ringwald)

http://umbra.nascom.nasa.gov/images/latest.html

Current solar images

Click on any of the following thumbnail images for the most recent, full-resolution solar image of each type in the SDAC archive (the time and date of the image are in *square brackets* after the description).

Access to full resolution representations of EIT images in all four wavelengths is available at the EIT images page .

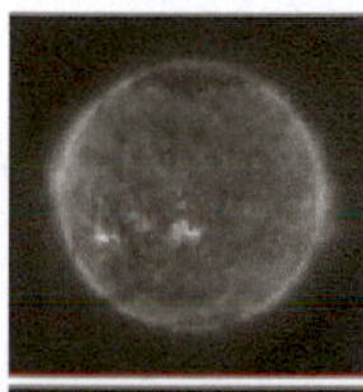

SOHO Extreme ultraviolet Imaging Telescope (EIT) full-field Fe IX, X 171 Å images from NASA Goddard Space Flight Center [2006/08/12 19:00:13]

SOHO Extreme ultraviolet Imaging Telescope (EIT) full-field Fe XII 195 Å images from NASA Goddard Space Flight Center [2006/08/12 22:24:10]

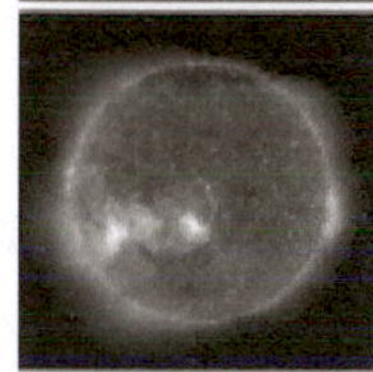

SOHO Extreme ultraviolet Imaging Telescope (EIT) full-field Fe XV 284 Å images from NASA Goddard Space Flight Center [2006/08/12 19:06:08]

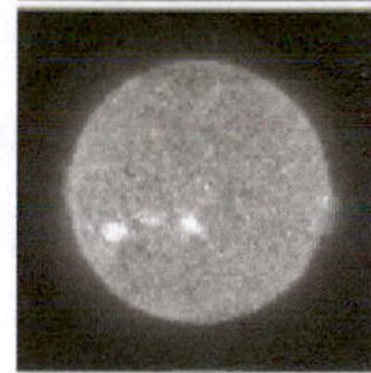

SOHO Extreme ultraviolet Imaging Telescope (EIT) full-field He II 304 Å images from NASA Goddard Space Flight Center [2006/08/12 19:19:38]

► **Figure 17-24** Another reason that it is unnecessary to risk anyone's eyesight while solar observing is that NASA maintains a web page that shows images of the Sun from various spacecraft and ground-based telescopes, updated at least daily. It is at: http://umbra.gsfc.nasa.gov/images/latest.html (Source: NASA)

► **Figure 17-25** An annular solar eclipse (*annulus* = ring). *Careful,* these are also *dangerous* to observe! (Image courtesy of Frederick Ringwald)

Eye Safety for Observing the Sun and Solar Eclipses

Disclaimer: The views expressed here are those of the author (F. Ringwald) only, who accepts no responsibility for any damage, injury, or death resulting from the following of these instructions and guidelines, which are offered as information only. Any person or persons using this information does so at their own risk.

Warning: The Sun can easily hurt you if you observe it improperly.
DO NOT observe the Sun under any circumstances, unless you know exactly what you are doing. Misunderstanding these instructions has a high price: instant permanent blindness. This can happen in less than a second.

With today's cheap webcams and digital cameras, there's no reason for anyone ever to risk their eyesight by looking at the Sun. These cameras can show an image of the Sun on the screen of a laptop computer, which is safe to look at. **IMPORTANT:** Don't wreck the camera with unfiltered sunlight, or hurt your eyes by looking at the Sun through the camera's viewfinder.

There are many ***unsafe** ways to observe the Sun*, which must never be used. Here is a partial, and by no means complete, list of **unsafe** ways:

- Absorptive filters, which go over a telescope's eyepiece, after the light has been concentrated by the telescope. These are included with many low-cost telescopes. DO NOT USE them, as they can crack or even explode, without warning.
- Welders glasses. Many sources say Type 14 and higher are safe, but this is *not* true.
- Smoked glass (despite what the Monty Python joke says).
- Sunglasses of any kind.
- Neutral density filters for cameras (which pass all the infrared radiation).
- Exposed (or unexposed) photographic negatives.
- Observing the Sun when it is near the horizon.
- Looking at a reflection in water: despite the widespread myth to the contrary, this is NOT safe.
- Mylar filters. These can show an image safely, but it's much too easy for the mylar to tear or get holes in it.
- *Anything whatsoever* that causes even the slightest discomfort to your eyes, including a reflective glass or mylar filter. If your eyes feel uncomfortable, **stop observing** at once!

Projection of the image of the Sun with a telescope onto a white surface was used by Galileo to discover sunspots, but remember, he did go blind later in life. This technique is really *too potentially dangerous* to use in most cases. All it takes is one curious child to look up the light path for a moment, and permanent blindness will be the result.

To observe the Sun safely, one will need a **safety certified solar filter** for each telescope, from a reputable dealer such as Orion Telescopes and Binoculars (www.telescope.com). The only filters that are ever safe are reflective glass filters, which look like mirrors. They go in front of a telescope's objective (which is the opening of its tube), cover the entire objective, and reflect back all but a tiny fraction of the Sun's light, so only a tiny fraction of the Sun's light gets into the telescope in the first place.

Every time one uses one of these filters, it is *essential* to check it *carefully* for holes or scratches in the reflective material. It is also *essential* that the filter fit *snugly* over the telescope's entire tube. It *must* not fall off, or be easy to remove, or have gaps that do not cover the tube. Preferably, the filter should bolt to the telescope tube: I like to duct tape the edges, to make sure. Be sure also to remove all finder scopes: don't just cover them. Find the Sun by minimizing the shadow of the telescope: don't look up the tube at the Sun!

It is now completely unnecessary to risk anyone's eyesight while solar observing, because of webcams and low-cost TV cameras that show the image of the Sun on the screen of a laptop computer. These cameras cost only $100–$150. IMPORTANTLY, a safe, reflective solar filter must *still always* be used with any camera, to avoid burning out the camera.

***Avoid* looking at the Sun through a telescope's eyepiece, even in a telescope with a "safe" filter. It takes only one mistake to lose the vision in that eye.**

A second safe way to observe the Sun is *to use a pinhole* to project an image of the Sun onto a white surface, and look at the white surface. During a partial solar eclipse, it's fun to look under trees, where the shadows can turn into thousands of little crescents.

A third safe way to observe the Sun is the Current solar images web page by the Solar Data Analysis Center at NASA Goddard Space Flight Center. It can be found here: http://umbra.gsfc.nasa.gov/images/latest.html

These images are from a variety of professional solar telescopes around the world, as well as one or more billion-dollar spacecraft. They are perfectly safe to look at on your computer screen, and are fascinating to watch, as they change from day to day. *For teachers whose students want to observe the Sun,* this author highly recommends them. After all, not only do they avoid the risk and anxiety of solar observing, but the images are marvelous, much better in quality than nearly any setup one might put together inexpensively.

Total Solar Eclipses

Total solar eclipses occur when the Moon completely covers the Sun. A total solar eclipse is without question nature's most awe-inspiring phenomenon: this is *not* an exaggeration.

> "The first time you see totality, you really understand where you are, on a piece of rock hurtling around the Sun. It's an awesome feeling–a life-changing experience."
>
> –Brian May

> "If you only *think* you saw a total eclipse, I promise you–you didn't!"
>
> –Jim Rosenstock

During totality, the Sun's photosphere is completely covered. The corona, the hot gas escaping from the Sun that is about a million times fainter than the photosphere, becomes visible. Day turns into an eerie twilight. There is a noticeable temperature drop, of as much as 8°C (15°F). Animals and people make lots of noise, because they get so excited.

Many people have seen partial solar eclipses. They shouldn't look, since they could easily hurt their eyes looking at the Sun.

Relatively few people have seen a total solar eclipse, because the Moon's shadow is only a few hundred kilometers across. They are *well* worth chasing, however!

The only time is it *ever* safe to look at the Sun is during the brief few minutes of totality during a total solar eclipse, when one is in the path of the umbra. You'll very clearly *know* it when it happens: it will look and feel like the end of the world.

▶ **Figure 17-26** **In a total solar eclipse, the Moon totally covers the bright surface of the Sun, called the photosphere. During the few minutes that totality lasts, the hot gas escaping from the Sun, called the solar corona, becomes visible.** (Image courtesy of Dr. Greg Morgan)

▶ **Figure 17-27** Total solar eclipse, 1999. **The visual impact of totality.** In 585 B.C., in what now is Turkey, two armies ended a war during a total solar eclipse. They saw it as an omen from the gods. (Image © Danshutter, 2013. Used under license from Shutterstock, Inc.)

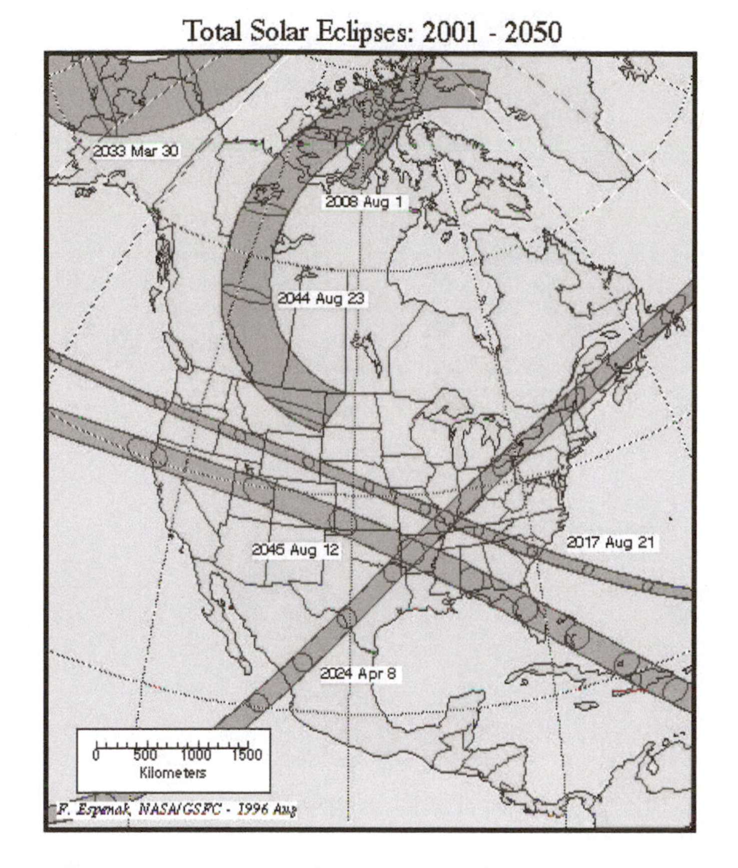

► **Figure 17-28** More people have seen a *partial* solar eclipse than a *total* solar eclipse, because the darkest part of the Moon's shadow (the *umbra*) is so narrow. Totality always lasts less than 7.67 minutes. This shows the paths of totality for North America until 2050. (Source: NASA/GSFC/Fred Espenak)

See NASA's eclipse home page, at: http://sunearth.gsfc.nasa.gov/eclipse/eclipse.html

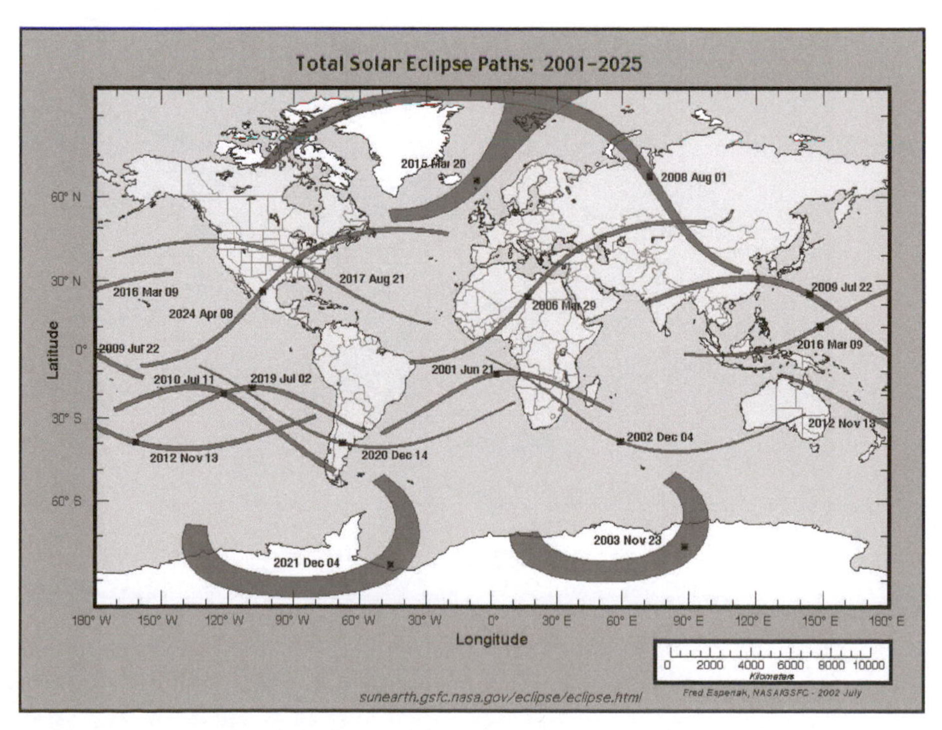

► **Figure 17-29** Eclipse tracks around the world. Updated ones are on a website maintained by NASA. (Source: NASA/GSFC/Fred Espenak)

► **Figure 17-30** Setting up to observe a total solar eclipse. The weather did not look promising: it started raining about an hour before the eclipse. (Image courtesy of Frederick Ringwald)

► **Figure 17-31** Just then, a hole in the clouds opened up. (Image courtesy of Frederick Ringwald)

▶ **Figure 17-32** ...the shadow of the Moon approaching, it started getting *really* dark... (Image courtesy of Frederick Ringwald)

▶ **Figure 17-33** When someone yells, "Diamond ring!" it means that the last tiny part of the Sun's photosphere is disappearing. (Image © Igor Kovalchuk, 2013. Used under license from Shutterstock, Inc.)

▶ **Figure 17-34** During totality, day literally turns into night. A total solar eclipse isn't called nature's most awe-inspiring phenomenon for nothing. (Image © Rachel Sanderoff, 2013. Used under license from Shutterstock, Inc.)

▶ **Figure 17-35** During totality, the Moon's shadow does not extend from horizon to horizon, so one can see sunset colors all the way around the horizon. During totality one can see the solar corona, which is hot gas escaping from the Sun. It is no wonder that the myth of a dragon eating the Sun is so common around the world, since that's what the corona looks like. As many cultures have found, if you beat on drums or smash pots and pans, the dragon will go away: it works every time. (Image courtesy of Dr. Greg Morgan)

CHAPTER

18

The Earth's Moon

The Moon

Diameter = 3476 km ≈ 1/4 Earth
Mass = 1/81.3 Earth
Surface gravity = 1/6 Earth
Distance = 384,400 km = 239,000 mi = 60 Earth radii
Earth's Moon has an area comparable to that of North America.

Why is it called "the Earth's Moon"? See the discussion in "the Basics of Astronomy." (Chapter 4). Perhaps someday it will be referred to by a proper name, such as Luna. "Lunar" means belonging to the Moon, such as "lunar craters" or "lunar dust."

Earth's Moon has almost no atmosphere, because its mass is too low for its gravity to hold onto one. With no air to even temperatures out, the surface of the Moon is essentially the same as space, with:

Temperature during daytime = 130°C = 250°F
Temperature at night, or in the shade = –200°C = –300°F

The time between Full Moons is 29.5 days. This is called a synodic *month*. Calendar months can be between 28 and 31 days long, because of the arbitrary way the Gregorian calendar that we use was made.

Tides are caused by the *difference* of forces due to gravity on opposite sides of an object. Tides from the Moon's gravity create ocean bulges on both sides of Earth. High tide therefore comes about every 12 hours. Since Earth rotates in 24 hours, and the Moon orbits Earth once a month, the tidal bulges lag behind the Moon's position. *High tide therefore comes about every 12 hours, 25 minutes.*

The Moon's rotation is tidally locked. This is why its period of rotation (day) equals its period of revolution (month).

- ▶ One side of the Moon therefore always faces Earth. It's called the Near Side.
- ▶ The other side always faces away from Earth. It's called the Far Side.

► **Figure 18-1** Earth and Moon seen together, by a spacecraft leaving them both. (Source: NASA/JPL/Northwestern University)

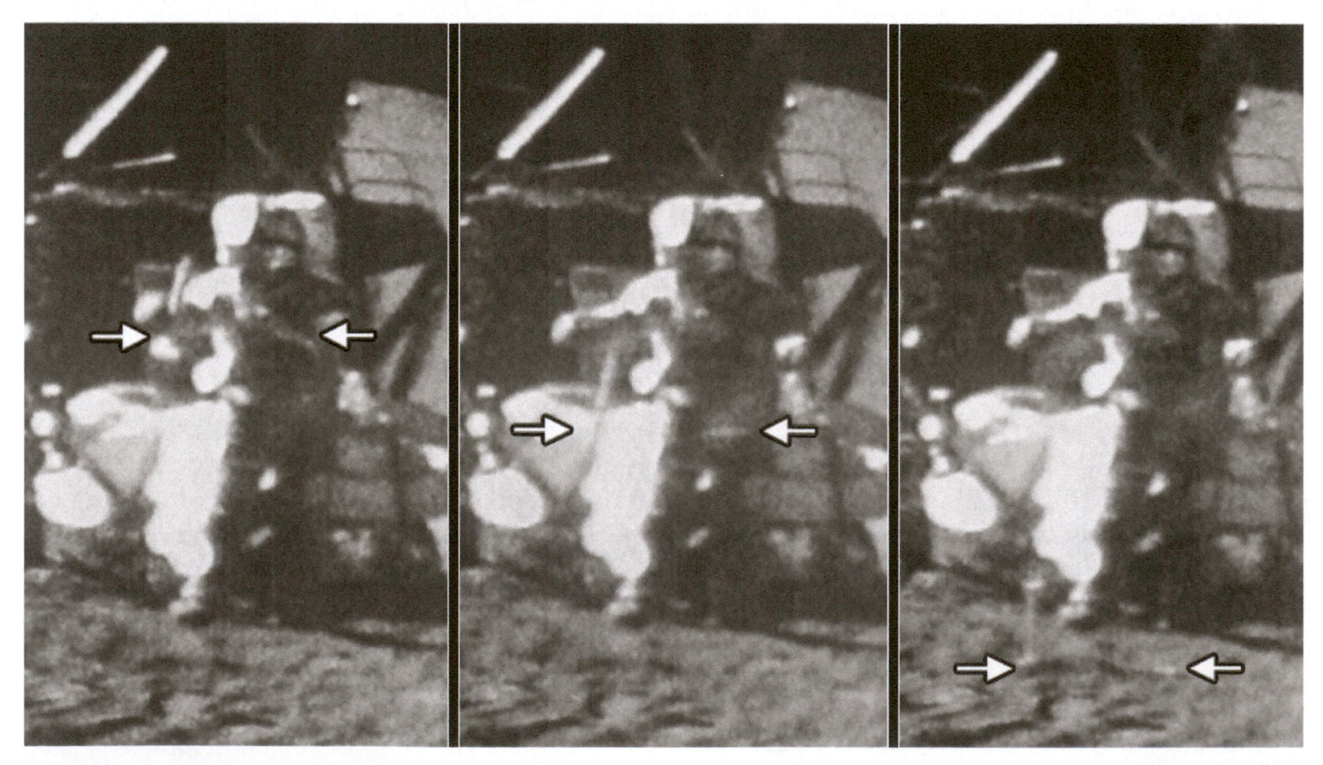

► **Figure 18-2** The Moon has no air, but it does have gravity. With no air, a dropped hammer and feather hit the Moon at the same time, the way gravity always works. They fell slowly, though, because the Moon's surface gravity is only 1/6th that of Earth. This was video-recorded during the Apollo 15 mission. Twelve astronauts, all Americans, walked on the Moon during the Apollo 11, 12, 14, 15, 16, and 17 missions, between 1969 and 1972. No humans have returned to the Moon since then, nor have humans gone anywhere else in space, except for low-Earth orbit. (Source: NASA)

► **Figure 18-3** *Top left:* High tide. *Bottom left:* Low tide was six hours later. *Center right:* Spring tides are the highest high tides, and the lowest low tides. *Far right:* Neap tides are in-between. (Images courtesy of Frederick Ringwald)

There are two basic types of lunar terrain:

1. *Highlands* are ancient cratered areas that are 4.1–4.4 billion years old.
   Many of the largest craters, or basins, are 3.8 billion years old.
   The oldest known Moon rocks are 4.46 ± 0.04 billion years old.

2. *Maria* are relatively smooth, dark, lava plains, about as dark as asphalt.
   They are younger and therefore less heavily cratered than the highlands.
   Much of the lava that formed the maria oozed out of the lunar interior over 3.0 billion years ago.

"Maria" is Latin for seas. (Mare is singular; maria is plural.) The maria never contained water: they're only called maria because they were named by astronomers in the 1600s, who wrongly thought they were seas. It isn't a total misnomer, though, since the floors of Earth's oceans are also lava plains (under the sediment). Similar lava plains are found on Mercury, Venus, and Mars.

► **Figure 18-4** This is the Near Side of the Moon, which always faces Earth. The Moon has two distinctive types of terrain: the ancient, heavily cratered highland, and the smoother, younger mare terrain (plural maria), which are lava plains. (Image courtesy of Frederick Ringwald at Fresno State's Campus Observatory)

► **Figure 18-5** The Far Side of the Moon always points away from Earth. The Far Side is unlike the Near Side, shown in the previous image, in that the Far Side has almost no maria, and so is dominated by highands. No one knows why. (Source: NASA/GSFC/Arizona State University/Lunar Reconnaissance Orbiter)

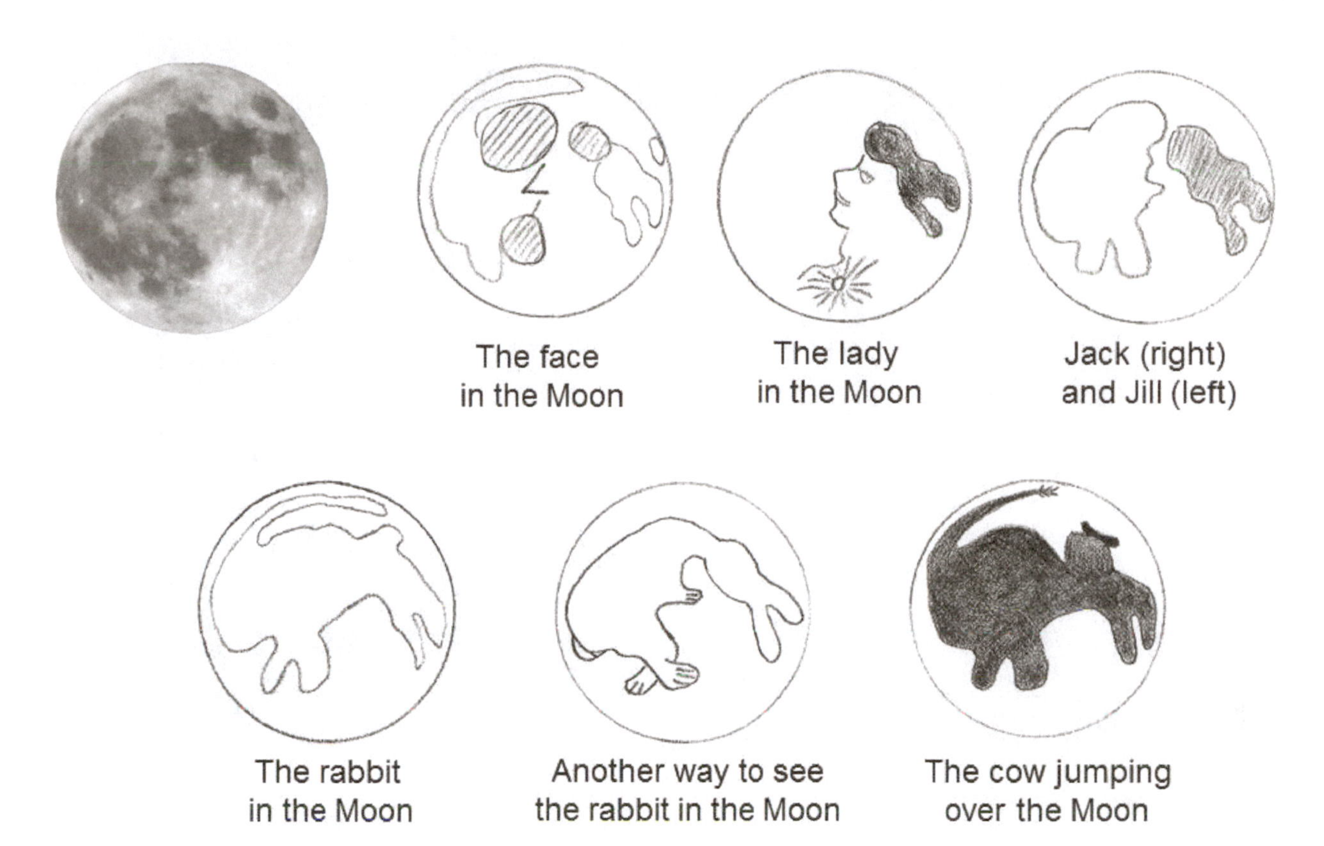

► **Figure 18-6** People all over the world see patterns in the maria, similarly to how they see patterns in constellations. Many of these are folklore. (Images courtesy of Frederick Ringwald)

► **Figure 18-7** Maria are lava plains, similar to this lava plain of black basaltic igneous rock fresh out of a volcano in Hawai'i. (Image courtesy of Frederick Ringwald)

▶ **Figure 18-8** Before the first landings on the Moon by humans, it was thought that the surface of the Moon would look like fresh lava plains, with lots of sharp, jagged peaks and rocks similar to those in Hawai'i. The reasoning behind this expectation was that since there was no air and no water, there would be no erosion.

The real Moon turned out to look very different. There are no sharp, jagged peaks: they are all rounded. There is also dust everywhere. The reason why is a form of erosion that hadn't been imagined, because it's so slow: micrometeorite impacts, over billions of years. No one had thought of this because it is so much slower than the water and wind erosion processes on Earth. It's rare to find Earth rocks that are more than 600 million years old: it's rare to find Moon rocks that are less than 3 billion years old, five times older. (Source: NASA/Apollo 17 crew/Eugene Cernan)

▶ **Figure 18-9** The Moon is a very quiet, static place. Because erosion by impacts is so slow, the footprints left behind by the astronauts are expected to last a million years. (Source: NASA/Apollo 11 crew/Buzz Aldrin)

▶ **Figure 18-10** Every one of the 12 astronauts who walked on the Moon commented about the pervasiveness of the dust. It is a very fine-grained power of sharp particles, never exposed to water or air, and got into everything. This is Gene Cernan, the Commander of the last mission. Just before this picture was taken, he wrote his daughter's name in the lunar dust, and as he told her, "It will be there for thousands of years." (Source: NASA/Apollo 17 crew/Harrison Schmitt)

Origin of the Moon

The origin of the Moon wasn't figured out until 1984, 15 years after Apollo 11.

Before the Apollo missions, there were three leading models:

1. Capture, in which supposedly the Moon formed elsewhere in Solar System, and was then captured by the Earth's gravity;
2. Fission, in which the early Earth rotated so fast, it split in two.
3. Double planet, in which the Moon formed alongside of Earth; and:

None of these models could explain all these observations:

1. The Moon rocks brought back by Apollo showed the Moon has the same ratio of oxygen isotopes as Earth, in particular, O-16/O-18. This implies that the Moon must have formed near Earth, so capture can't be correct.
2. The Moon's density is the same as Earth's mantle. Measurements of the Moon's magnetic field during Apollo showed it has little or no iron core, like Earth's. This means binary accretion is unlikely, since it predicts that Earth and Moon should both have iron cores.
3. Earth should still be rotating *much* faster than its observed 24-hour day, if it ever spun fast enough for a piece to break off it. Fission therefore has problems.
4. The Moon is rich in refractory materials, which are materials with high boiling points such as rocks and metals. The Moon is poor in volatile materials, which are materials with low boiling points, such as water and organic compounds. This implies the whole Moon was once so hot, its volatiles were baked out.

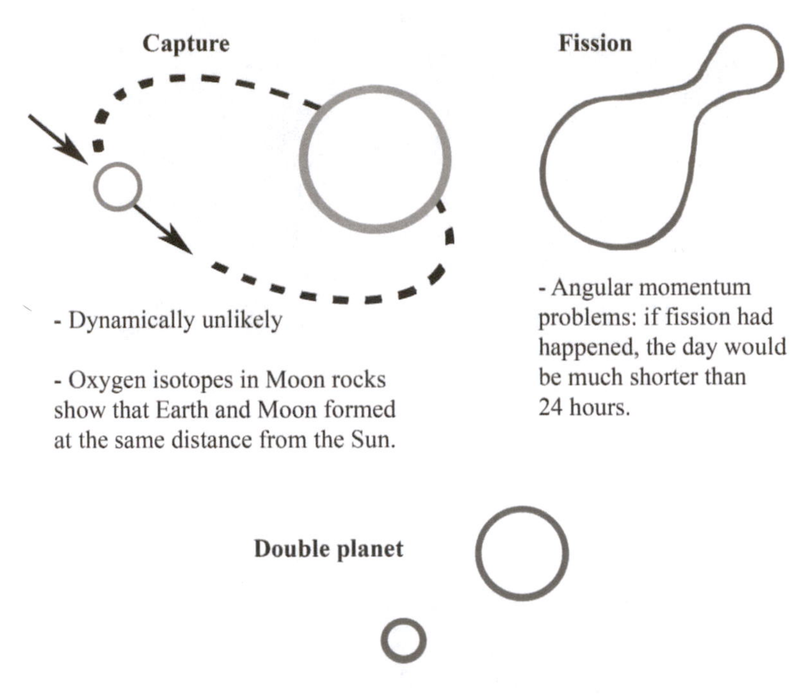

▶ **Figure 18-11** The origin of the Moon was a mystery for many years after the Moon landings. The Moon rocks brought to Earth by the Apollo missions at first seemed to confuse the issue further, since they showed that the Moon was chemically similar to Earth, but physically different. (Image courtesy of Frederick Ringwald)

The Giant Impact Hypothesis

Don Davis and Bill Hartmann thought of the Giant Impact Hypothesis in 1975, but it took until 1984 for many other planetary scientists to accept it.

The idea recognizes that at the end of planetesimal accretion, mostly big pieces were left over. This therefore resulted in big collisions.

"The Big Splat" was a collision by an object the size of Mars with an early, smaller, Earth. This giant impact is thought to have happened 4.55 ± 0.01 billion years ago, about 10–30 million (0.01–0.03 billion) years after the Sun formed. An oblique impact would eject much of Earth's mantle, which could re-condense under its own gravity to form the Moon.

This would explain the Moon's O-16/O-18 ratio, density, lack of an iron core, abundance of refractories, and lack of volatiles. Earth's rotation rate is no problem, either. Earth didn't lose its volatiles, because it is over 80 times more massive than the Moon. The giant impact hypothesis is therefore a combination of the caption and fission models. Giant impacts may also explain the backwards and slow rotation of Venus, and why Uranus is tilted on its side.

In the 1950s, Clyde Tombaugh did a photographic survey that showed the Moon is the only natural satellite of Earth. Two exceptions have since turned up:

1. Asteroid 3753 Cruithne, which is captured in an Earth-Sun horseshoe orbit,
2. Asteroid 2010 TK_7, which is captured in the L4 Earth-Sun neutral gravity point.

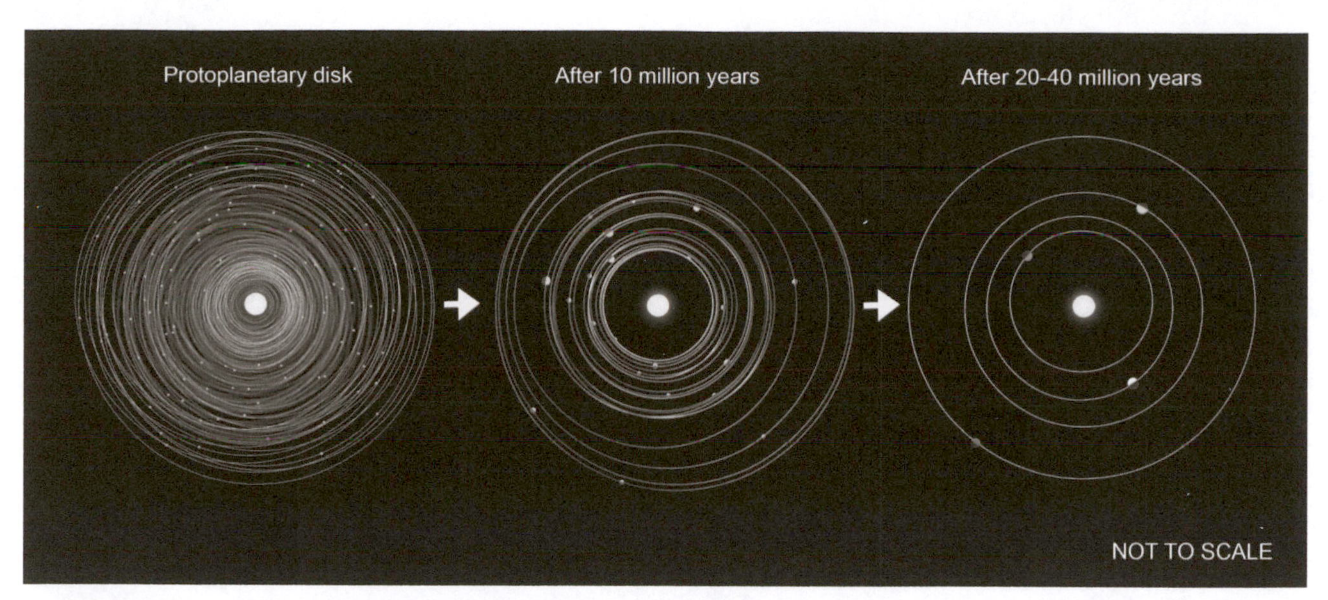

► **Figure 18-12** The Solar System formed by gravity. A clue to how the Moon formed came from considering the end of this process: the biggest pieces will be hitting the biggest pieces. (Image courtesy of Frederick Ringwald)

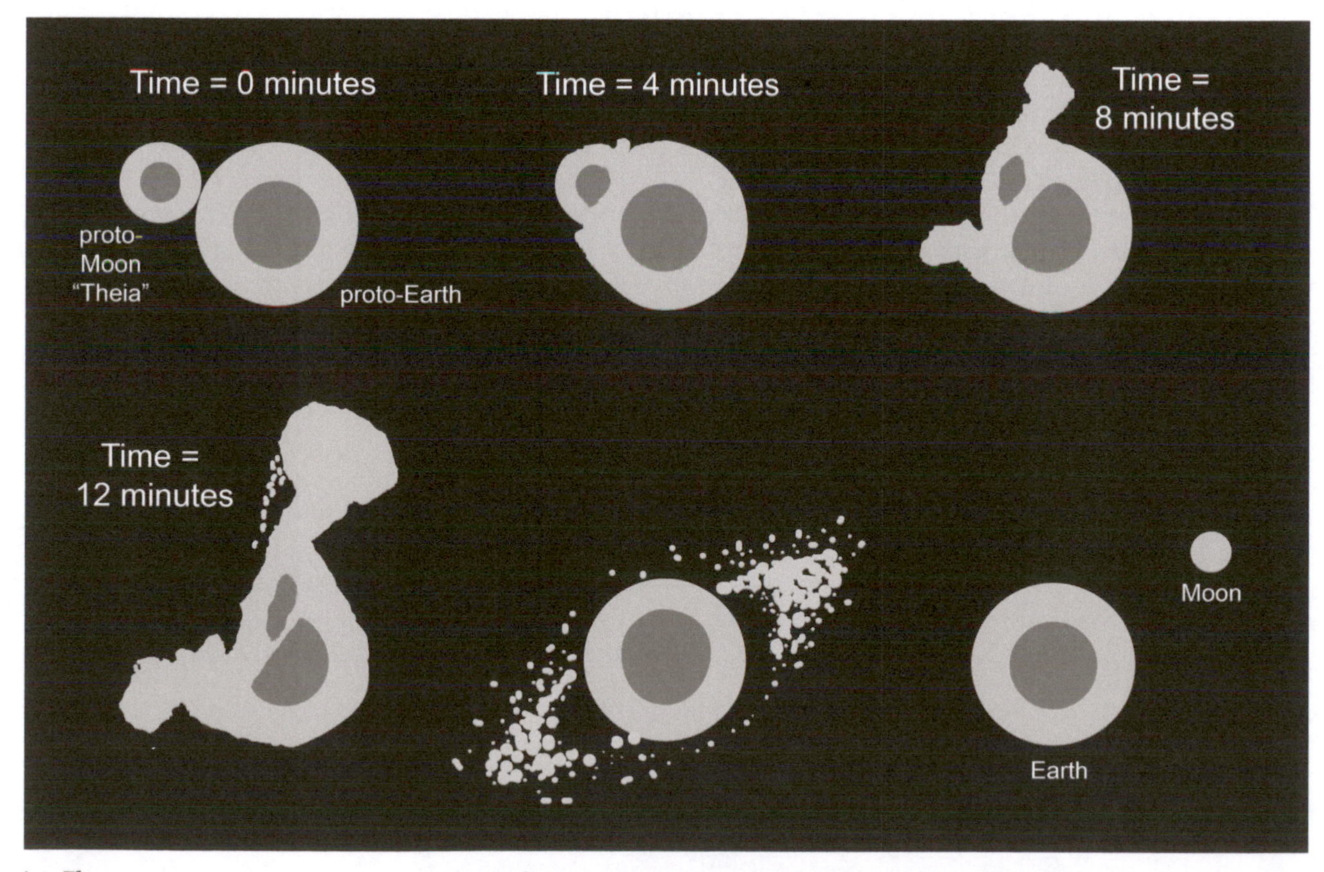

► **Figure 18-13** The giant impact hypothesis (1984) is a combination of capture and fission. The giant impact is thought to have happened 4.55 ± 0.01 billion years ago, in the Solar System's first 10–30 million years. (Image courtesy of Frederick Ringwald, based on image by M. E. Kipp and H. J. Melosh, in "Origin of the Moon," 1986.)

Most of the Moon's maria are on the Near Side, facing Earth. No one knows why. Perhaps a second giant collision occurred? Samples from the Far Side are needed to confirm this. The crust is known to be thicker on the Far Side.

Everywhere on the Moon's surface is covered with fine dust, or regolith, often wrongly called "soil," because it has no organic material at all, unlike soil on Earth. Lunar regolith is a "very, very fine-grained powder," as in the words of Neil Armstrong, made by billions of years of impacts. Pete Conrad, on Apollo 12, noted, "This is DIRT dirt."

Interestingly, nearly all of the twelve astronauts who walked on the Moon commented on its *beauty*. As Buzz Aldrin noted, "Magnificent desolation!"

Moon Rocks

Moon rocks are easy to distinguish from Earth rocks:

- Moon rocks are **very old**, typically over 3 billion years old, whereas 90% of Earth rocks are younger than 600 million years old.
- Rocks from the surface of the Moon are **extremely dry**, containing less than a thousandth the water on average that Earth rocks have.
- Moon rocks contain few or no minerals (rock substances) that would have required the presence of water to form, such as clay, or any sediment.
- Moon rocks are loaded with tracks from high-energy cosmic rays, and are covered with tiny craters from micrometeorites, or "zap pits."
- Moon rocks are often made of minerals that are rare or unknown on Earth, which were made by the shocks of impacts. We know this from reproducing the impacts with high-speed guns.

Moon rocks are *similar* to Earth rocks in often being made of chemical elements that are common on Earth, including silicon, aluminum, iron, and titanium, all mixed with oxygen.

The Moon is lacking in volatile substances, such as water. There is little or no free oxygen, nitrogen, or organic material.

However, analysis in 2011 of samples collected by the Apollo 17 mission in 1972 suggests that there may be water in rocks from deep in the Moon's interior. There have also been indications of water, in the form of ice, at the poles of the Moon. The quest for lunar water continues.

Of the four primary geological processes on Earth's Moon,

1. Impact cratering dominates, by far!
2. The Moon has little tectonism. There are some faults and some scarps, or cliffs, from shrinkage of the Moon's crust when the lava cooled. There is no global system of plate tectonics, because the Moon has so little internal heat to drive it. Since the Moon is less than 1/80th of the mass of Earth, nearly all the heat it ever had leaked out long ago.
3. The Moon has no active volcanism. Most maria formed over 3 billion years ago, like the lava plains on Earth's ocean floor. Some activity was as recent as about 1 billion years ago.
4. The only source of geological gradation (or erosion) is *impacts,* mainly from micrometeorites and Solar wind. This is *very slow*. The astronauts' footprints should last 100,000 to 1 million years.

▶ **Figure 18-14** Lunar breccia is rock welded together by impacts. (Source: NASA/LPI)

▶ **Figure 18-15** Lunar glass is impact melt. (Source: NASA)

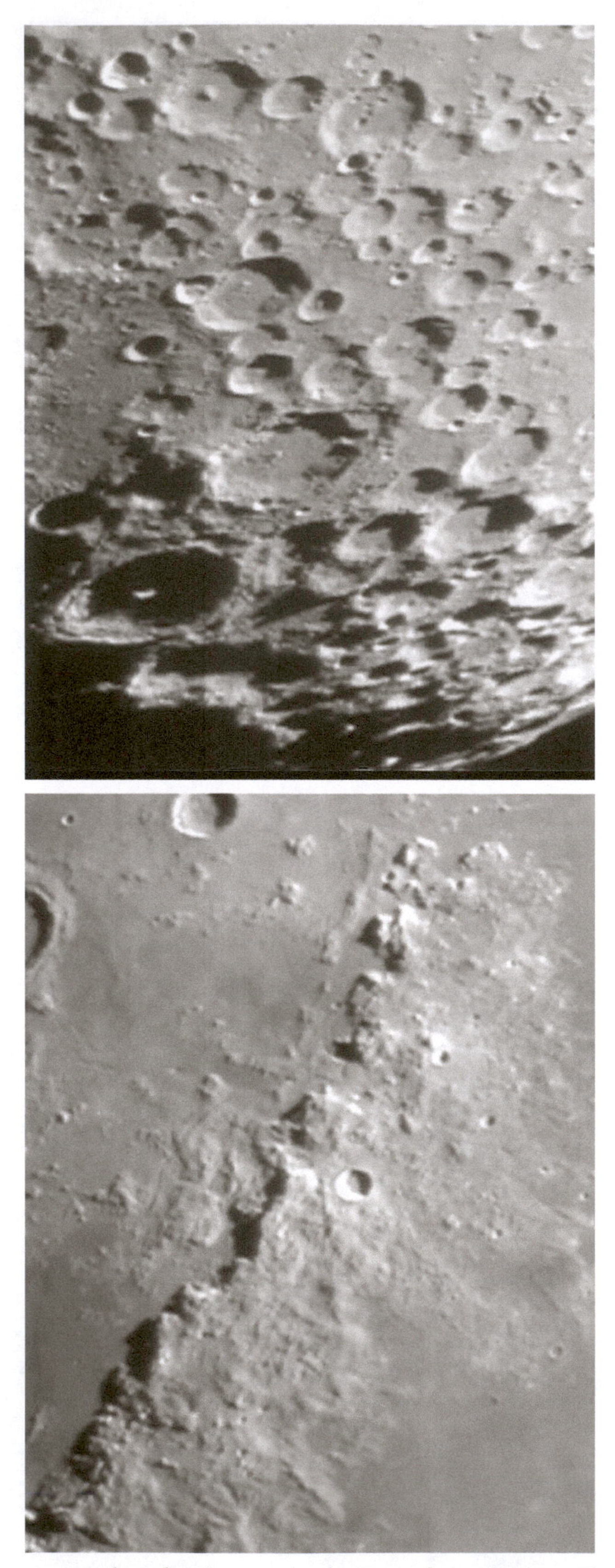

► **Figure 18-16** *Top:* Impact cratering dominates the lunar surface. *Bottom:* The mountains on the Moon are the rims of giant impact craters, unlike any mountains on Earth. (Images courtesy of Frederick Ringwald, at Fresno State's Campus Observatory)

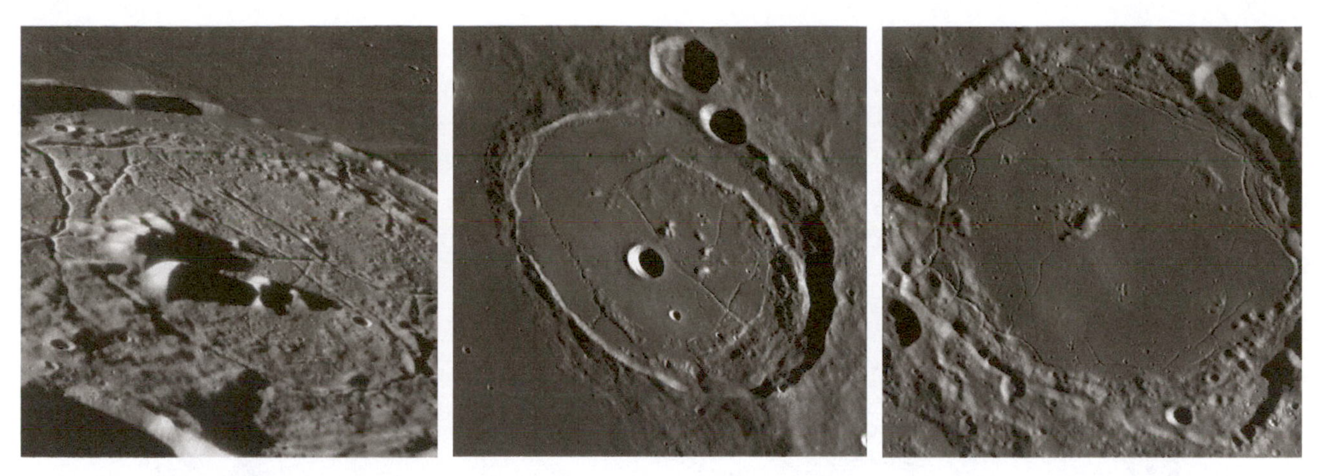

► **Figure 18-17** The Moon's surface has some ancient tectonism (cracking), but no global plate tectonics. This is because the Moon has only about 1/80th the mass of Earth, so that nearly all the heat the Moon ever had in its interior leaked out long ago, in the first billion years of its history. (*Left:* Source: NASA/Apollo 16 crew; *Middle and Right:* Source: NASA/GSFC/Arizona State University/Lunar Reconnaissance Orbiter)

► **Figure 18-18** Ancient lunar volcanism made this rille. Since nearly all the heat leaked out of the Moon long ago, there is no active volcanism today. (Source: NASA)

► **Figure 18-19** Lunar rilles are thought to have formed in a way similar to this lava tube in Hawai'i. (Source: NASA/LPI/S. Rowland)

► **Figure 18-20** Hadley Rille was explored by the *Apollo 15* crew. It was found to have been a lava tube 3.25 billion years ago. (Source: NASA/Apollo 15 crew/James Irwin)

Why haven't astronauts returned to the Moon since 1972?

The Golden Age of lunar exploration was between 1959 (with the first flyby of the Moon, Luna 1, by the Soviet Union) to 1976 (with the last sample-return robot, Luna 24, by the Soviet Union). The climax of this was in 1969–1972, with the six landings on the Moon by the Apollo project and a total of twelve astronauts, all from the U.S.A.

Why haven't we gone back?

- The Apollo hardware was so large, powerful, and complex, each mission cost about $2 billion (in 1969 dollars, which would be over $12 billion today). NASA has never had a budget big enough to cover this expense and also to carry out other programs, such as the Space Shuttle.
- The $24 billion spent to do it was resented by the American public, who lost interest after the first landing and were soured on all things governmental and military by the Vietnam war, which was raging out of control at exactly that time (1968–1972).
- The Soviet Union, after four failures of their own giant Moon rocket (the N-1), quit trying to go to the Moon, so the element of Cold War military competition was lost. No other country has ever tried to send humans to the Moon.
- Much of the technology needed to send astronauts to the Moon, such as the Saturn V rocket and the Lunar Module lander, was abandoned in the 1970s. Bringing them back will require a significant new development program. Do you have any idea how hard it is to get parts for a *car* that old?

  It is ***not true*** that we can't bring back the Saturn V because the plans were lost. This is an urban legend spread by *The New York Times*, who really ought to have known better.
- NASA's program to return astronauts to the Moon was cancelled, in favor of sending astronauts to asteroids. NASA is developing the Multi-Purpose Crew Vehicle (MPCV), also called the *Orion* capsule, and the Space Launch System (SLS) rocket for this.

  China is preparing to launch its first space station. They say they want to land Chinese astronauts (called *taikonauts*) on the Moon by 2025. Interest in the Moon is heating up: since 2005, Europe, Japan, China, India, and the United States have all sent spacecraft there.

  NASA retired its Space Shuttle fleet in 2011. NASA now sends astronauts to the International Space Station on Russian rockets. NASA is supporting development of private space launchers by several corporations.

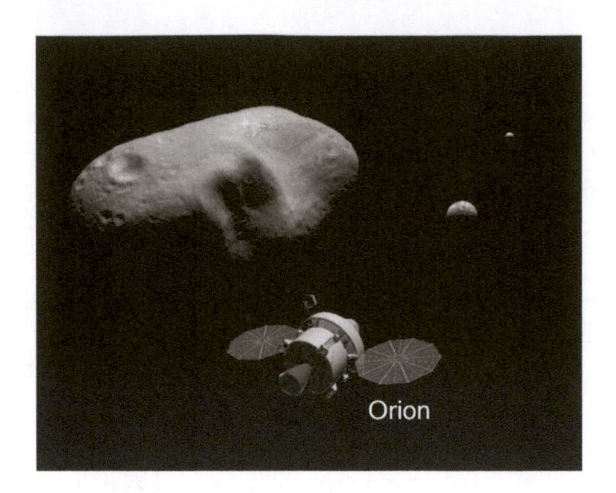

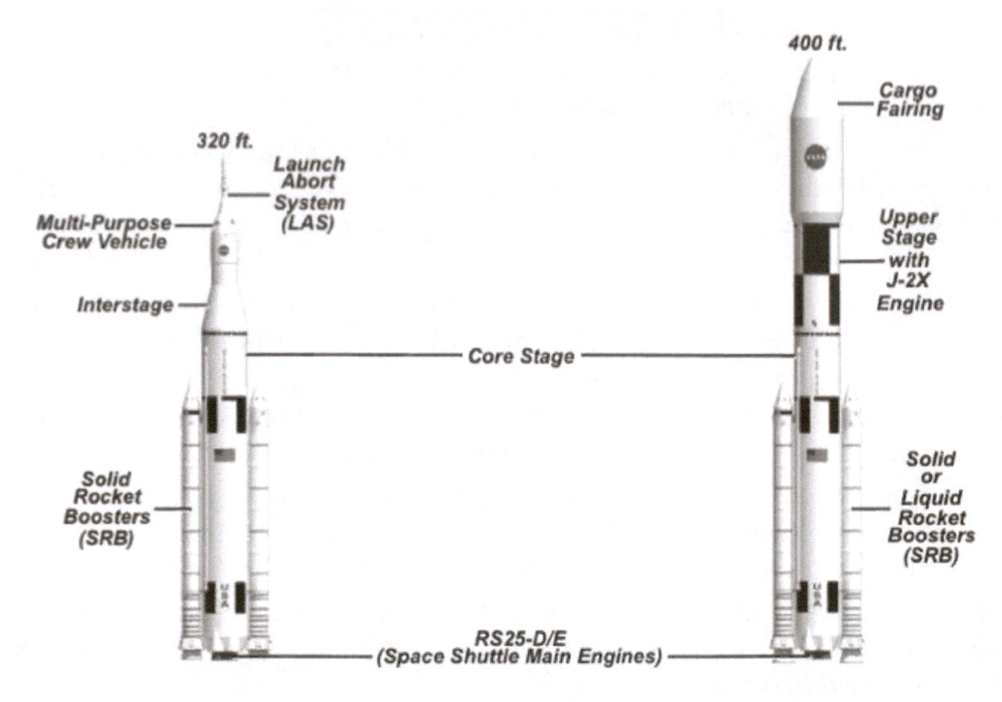

► **Figure 18-21** *Top:* In 2010, NASA's program to return astronauts to the Moon was cancelled, in favor of sending astronauts to asteroids. NASA is developing the Orion Multi-Purpose Crew Vehicle (MPCV) and the Space Launch System (SLS) rocket for this.

China is preparing to launch its first space station. They want to land humans on the Moon by 2025.

NASA retired the Space Shuttle in 2011. NASA now sends astronauts to the International Space Station on Russian rockets. NASA is supporting development of private space launchers. (Image courtesy of Frederick Ringwald, using NASA images)

Right: NASA is also developing the Space Launch System (SLS), a rocket with the lifting capability of the Saturn V Moon rocket. It will be essential for sending astronauts anywhere beyond low-Earth orbit. (Image courtesy of Frederick Ringwald, using NASA images)

CHAPTER

19

Mars

What Practical Good Is Space Exploration?

Direct Benefits

- Communications satellites make long-distance phone calls, TV, and the Internet possible.
- Weather satellites save lives, when the warnings are heeded.
- GPS and other navigation satellites also save lives and speed commerce.
- "Spy" satellites prevented World War III. They made it impossible for the Soviet Union to invade Western Europe by surprise.
- The pictures of Earth taken by the Apollo astronauts directly inspired the founding of the modern environmental movement.

Indirect Benefits

- Science began as an investigation of planetary motion, by Copernicus, Galileo, Kepler, and Newton. Without modern science, you'd probably never have heard of indoor plumbing, soap, antibiotics, or cell phones. It's likely you wouldn't be alive at all: smallpox, plague, cholera, and death during childbirth would be common. Superstition and witch hunts would also be common. You'd likely survive by subsistence farming as a serf, or as a slave. You probably wouldn't be a noble, since there weren't many of them.
- Astronomers kept records of sunspots for centuries before their nature was understood. We now have plenty of technology affected by solar activity, including radio, cell phones, and GPS. What was once an obscure curiosity is now practical business, because we can't always know all our wants or needs in advance.
- The greenhouse effect in the atmosphere of Venus alerted us to the hazard of allowing it to become a problem in the atmosphere of Earth. The real value of studying the planets is how they help us understand that most important planet, Earth.
- Mars was similar to Earth during its first billion years. Why is there no obvious life on Mars now? Knowing this might be useful for making sure that life survives on Earth.

- Fewer people worldwide are employed at learning about impacts by asteroids and comets than work at an average fast-food restaurant. As Larry Niven observed, "The dinosaurs became extinct because they didn't have a space program."
- The computer aboard the Apollo Lunar Module was the direct ancestor of the original personal computer. Its design was pushed by the need to be small and lightweight, unlike most computers of its day. Solar panels and fuel cells were developed for use in space flight. Teflon was invented for it. Velcro was popularized by it. Space flight creates economic opportunities because of the inventor's paradox: the more ambitious goal can have a better chance of success, because more possibilities are open. Wars also do this, but space flight is so much more fun.
- Politics does matter. The last time humans walked on the Moon, they were Americans racing Russians. Next time, they may be Americans racing Chinese—or they may be Chinese.
- Understanding our place in the Universe is important for the human spirit. So is fostering a spirit of adventure. This is why children are so often interested. Among the sciences, astronomy and space exploration have educational value rivaled only by paleontology.

"Space exploration is the carrot that incites people to become scientifically literate. So I view it as an economic development plan...The economic return is the scientists and technologists who invent the new tomorrow."

–Neil deGrasse Tyson

"We need art as we need dreams," [astronaut] Wally Schirra concluded.

"Dreams? Did you say dreams?"

"Without our dreams we wouldn't be where we are: dreaming of going to other planets, to other solar systems, and finding other Earths, our Earth, among billions of stars."

"Our Earth? Did you say our Earth?"

"Certainly. Because it's our Earth, it'll always be our Earth that we're looking for, it'll always be our Earth that we discover. I don't dream about the Moon. I know enough about the Moon to know how unpleasant and inhospitable it is. There's not one bit of Moon that's worth the Earth or what we could bring back to Earth as a trace of civilization. I don't dream about Mars . . . Mars and the Moon are two ugly islands. So then, you say, what's the point of going to them? The point is to be able to say I've been there, I've set foot on them and I can go further, to look for beautiful islands . . ."

–From *If the Sun Dies,* by Oriana Fallaci

Mars

Mars is the most Earth-like planet, with an atmosphere, weather, clouds, polar caps, ice, snow, seasons, and even a 24-hour day.

- 1 "sol" = 1 Mars day = 1 rotation period = 24 hours, 37 minutes.
- 1 Mars year = 687 Earth days = 1.88 Earth years.
- The radius of Mars ≈ 1/2 the radius of Earth.
- The total area of Mars is about equal to the land area of Earth. (Earth's Moon has about the same area as North America.)
- On Mars, the surface gravity is about 1/3 that of Earth's.
- On Earth's Moon, the surface gravity is about 1/6 that of Earth's.
- Mars has seasons, like Earth's. Before 1965, seasonal changes in the appearance of Mars were attributed to vegetation. They are now known to be caused by *global dust storms*.
- *Temperatures* on Mars get very cold (–100°C) every night, when heat escapes from the thin atmosphere.
- The *polar caps* of Mars are both seasonal (made of CO_2 ice) and permanent (made of water ice).
- Mars has an *atmosphere*, but we humans can't breathe it:
  - It is 95% CO_2 (carbon dioxide). At the surface of Mars, the pressure is 7 millibars, similar to outside a jet airliner window. (1 bar = Earth standard, 7 millibars = 0.007 bars.)
  - Mars can "snow" its own atmosphere, when CO_2 turns into dry ice.
  - Mars has icy water clouds, 20–30 km over its volcanoes.

Figure 19-1 Mars is the only planet for which we can see a solid surface (aside from Earth). Earth and Mars are shown together here to show that Mars is about half the diameter of Earth. They are never this close together: the closest Mars comes to Earth is 35 million miles. (Source: NASA)

► **Figure 19-2** Mars has an atmosphere, which has global dust storms about every Martian year. (Source: NASA/JPL/Viking 1 team)

► **Figure 19-3** Space enthusiasts dream about sending astronauts to Mars. It will be more difficult than sending them to the Moon: Mars is over 200 times farther away. Current plans for an expedition to Mars are for a round trip time of 2–3 years. Improved rocket technology may be able to shorten this time, but it isn't here—yet. (Source: NASA/Pat Rawlings)

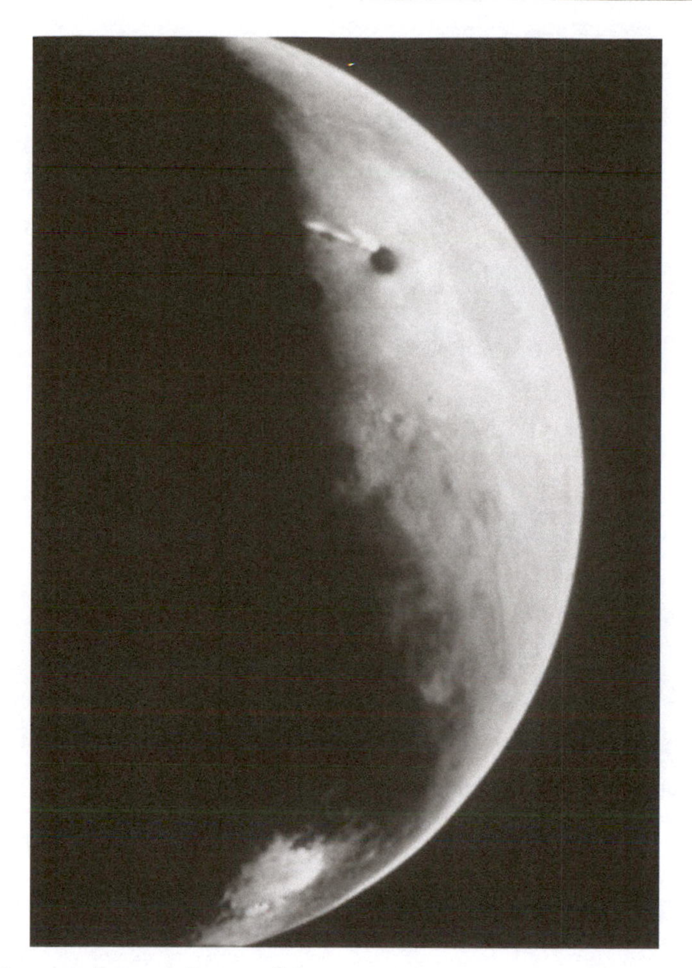

► **Figure 19-4** "What?, I sometimes ask myself in amazement: Our ancestors walked from East Africa to Novaya Zemlya and Ayers Rock and Patagonia, hunted elephants with stone spearpoints, traversed the polar seas in open boats 7,000 years ago, circumnavigated the Earth propelled by nothing but wind, walked the Moon a decade after entering space–and we're daunted by a voyage to Mars? But then I remind myself of the avoidable human suffering on Earth, how a few dollars can save the life of a child dying from dehydration, how many children we could save for the cost of a trip to Mars–and for the moment I change my mind. Is it unworthy to stay at home, or unworthy to go? Or have I posed a false dichotomy? Isn't it possible to make a better life for everyone on Earth *and* to reach for the planets and the stars?"

– From *Pale Blue Dot*, by Carl Sagan

(Source: NASA/JPL)

NASA Mars Odyssey orbiter since 2001

European Space Agency (ESA) Mars Express orbiter since 2003

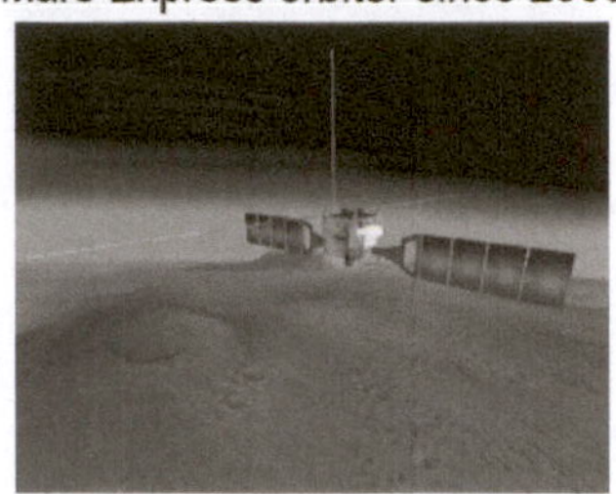

NASA Opportunity rover since 2003

NASA Mars Reconnaissance Orbiter since 2005

NASA Curiosity rover since 2012

NASA MAVEN (Mars atmosphere) orbiter 2014?

NASA InSight lander 2016?

ESA/Russia ExoMars 2016 and 2018?

NASA Sample Return 2020?

► **Figure 19-5** The real Mars invasion is happening now, and it is from Earth to Mars. NASA and ESA (the European Space Agency) have an active program of Mars exploration, by robots. (Source: NASA/ESA)

► **Figure 19-6** The classic tale of the invasion of Earth by Martians is *The War of the Worlds*, by H. G. Wells. It was published in 1898, just four years after the Sherlock Holmes stories, when automobiles and indoor plumbing were uncommon, and airplanes five years from being invented. Into this world, Wells story introduced the ideas of inhuman and utterly remorseless intelligent aliens armed with ray guns, giant war transport machines, and poison gas in an invasion of Earth. Read it, it's a good one! (Images © Bettmann/CORBIS)

▶ **Figure 19-7** In 1938, Orson Welles (no relation to H. G. Wells) broadcast a radio-show adaptation of *The War of the Worlds*. Because it was performed in a format similar to that of a news broadcast, and also because the actors used factual place names, many radio listeners mistakenly thought the show was an actual news report of an invasion from Mars. (© Bettmann/CORBIS)

In 1894, Percival Lowell established Lowell Observatory in Flagstaff, Arizona. Lowell mapped "canals" on Mars. He thought they were the engineering projects of a dying civilization.

Lowell popularized the idea of life on Mars. So did the 1898 science-fiction novel, *The War of the Worlds* by H. G. Wells, and the *John Carter of Mars* series by Edgar Rice Burroughs. The canals proved to be optical illusions, from chains of craters near his eye's resolution limit aided by Lowell's overactive imagination, and his unwillingness to subject his ideas to proper scientific testing.

Spacecraft Exploration of Mars

The *Mariner 4* (1965) and *Mariners* 6 and 7 (1969) spacecraft flybys of Mars were a real shock. They showed no canals or any sign of life, only ancient cratered terrain, "like the Moon." There were calls to end Mars exploration ("we understand Mars"), although less than half the planet had been seen.

In 1971, *Mariner 9* became the first spacecraft to orbit Mars, and was a stunning success. It would have been a failure if it had been only a flyby, since it arrived during a global dust storm. It found:

- The shield volcanoes in the Tharsis region. These are the tallest mountains known in the Solar System. The tallest, Olympus Mons, is 69,800 feet (21.3 km) high. (Mount Everest is 29,028 feet.)
- Valles Marineris, also called "Mariner Valley" or "the Big Valley." It's a *really* Grand Canyon, over 3000 miles long.
- Dendritic channels (dendritic = tree-like), resembling dry river beds, made by running liquid water.

The atmosphere of Mars is too thin for liquid water to exist on the surface of Mars today. Mars was therefore probably warmer and wetter in the past. Did life develop? If not, why not?

Also, if the atmosphere of Mars was once thicker than it is today, where did it go?

All four of the primary geological processes are important on Mars:

▶ **Figure 19-8** *Left:* Percival Lowell thought he could see "canals" on Mars. (Source: NASA) ***Right:*** One of Lowell's drawings of Mars, from his book *Mars as the Abode of Life*. (Science Source).

▶ **Figure 19-9** Your present author has always been impressed by all the detail Percival Lowell was able to see in Mars, because whenever I look through a telescope at Mars, I see something like this. Mars is a small, red planet, not really impressive to novice telescope observers. This is why, whenever I host a public night at the telescope, I go for the brightest, most spectacular objects, such as the Moon, Jupiter, Saturn, and colorful binary stars, and not Mars. Kids are always disappointed by Mars, since they expect the spacecraft images elsewhere in this chapter.

Percival Lowell was an example of how *not* to do science. Scientists need to be fair and objective. They must *not* let their imaginations run wild: speculation must be *disciplined*. Lowell was so in love with his idea that Mars was inhabited, he "cherry picked" his data. Any observation that supported his idea he loudly supported, but anything that did not he ignored. (Image courtesy of Frederick Ringwald, using the 16-inch Hans D. Isenberg Telescope that is now on Mars Hill, when it was still at Lindheimer Astronomical Research Center)

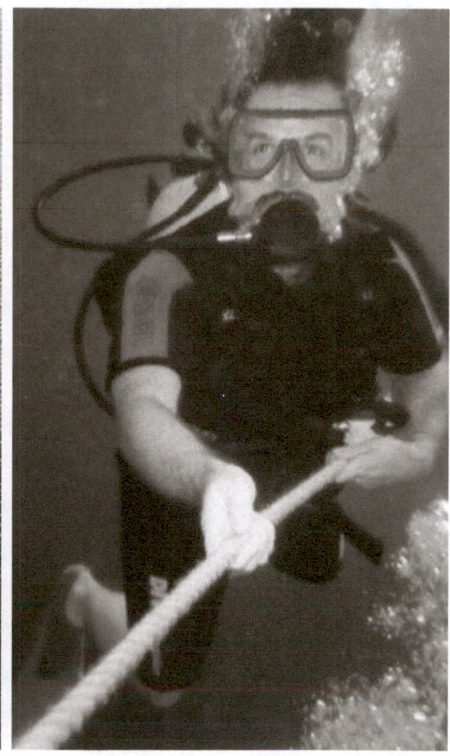

▶ **Figure 19-10** It was known in Percival Lowell's time that spectra show that Mars has a thin atmosphere. It is only about as dense as the air outside of a jet airliner's window. (It also contains no breathable oxygen.) The thin atmosphere of Mars is why Mars always gets so cold at night: all the heat of the day escapes into space. Because the atmosphere of Mars is so thin, liquid water can't exist on the surface of Mars. It would boil away into space, because the Martian atmosphere is so thin. Only "a race of madmen" would be building canals on Mars, wrote Alfred Russel Wallace.

Left: This is the main reason that astronauts need to wear space suits. If they didn't, in empty space, the absence of air would cause what is happening to the water in this glass to happen to their blood. (Image © Alexander Chelmodeev, 2013. Used under license from Shutterstock, Inc.)

Right: Lack of air pressure like this can cause a painful, debilitating, and often fatal condition known as "the bends." Deep-sea divers need to avoid it, by not coming from deep water (at high pressure) to the surface (at low pressure) too quickly. This is not the only reason that the author is shown here hanging on for dear life. (Image courtesy of Frederick Ringwald)

Humans have died in space because of lack of air pressure. In 1971, the *Soyuz 11* spacecraft sprung a leak while still in space. Nearly all the cabin air escaped into space. The crew were not wearing space suits, and blacked out from lack of oxygen a few seconds later. They died of pulmonary embolisms.

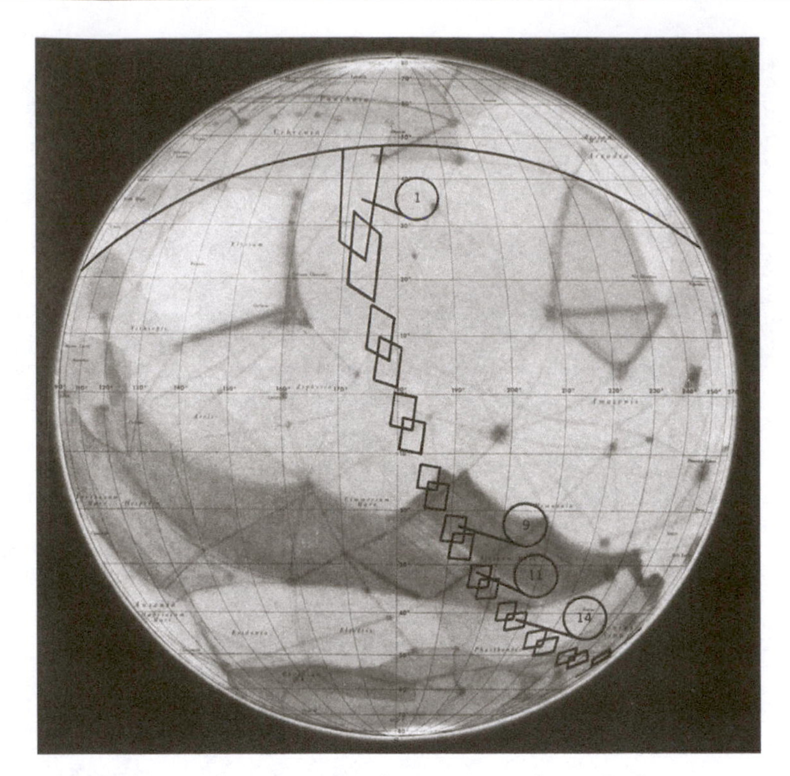

► **Figure 19-11** In 1965, *Mariner 4* became the first spacecraft to fly by Mars. It took 22 pictures and radioed them back to Earth. (Source: NASA)

► **Figure 19-12** *Mariner 4* found no canals, only ancient cratered terrain. (Source: NASA)

▶ **Figure 19-13** The canals are optical illusions, caused by the human eye and brain incorrectly "connecting the dots." *Left:* What Lowell thought he saw. (Public domain image by Giovanni Schiaparelli, 1877) *Right:* What another astronomer saw on Mars. (Public domain image by E. M. Antoniadi, 1909)

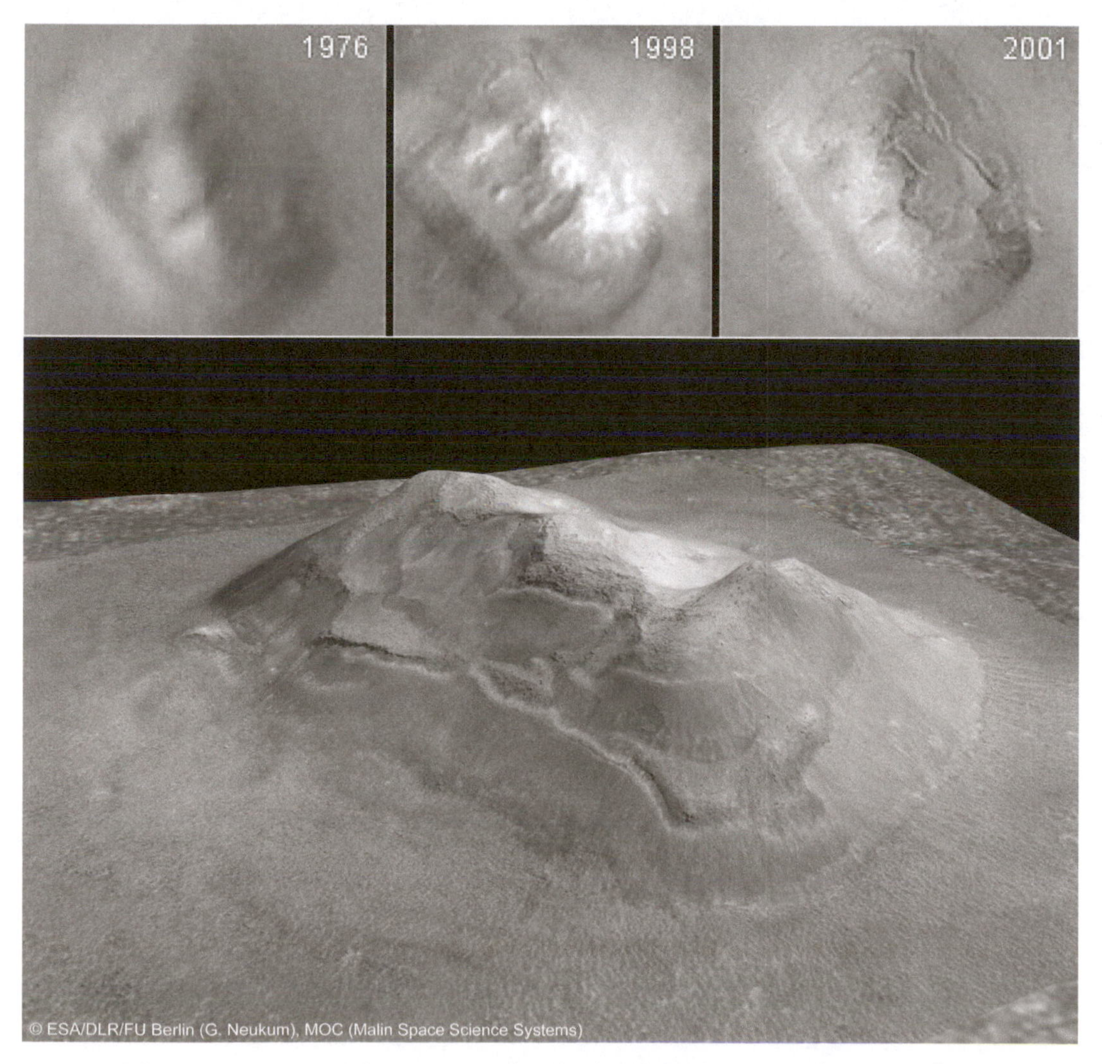

▶ **Figure 19-14** "The face on Mars" is an example of what happens when people "want to believe," too much. Close-ups show this hill on Mars really doesn't look like a face: the human eye and brain are good at picking out patterns, even when they aren't there. (Source: NASA)

► **Figure 19-15** If you "believe" in the face on Mars, do you also believe in the smiley-face on Mars? (Source: NASA)

► **Figure 19-16** There's really no need for pseudo-science to make up stories. The real Mars is a world of wonder. Ancient, cratered terrain dominates the southern hemisphere of Mars. (Source: NASA)

1. Impact cratering: the southern hemisphere of Mars is mostly ancient, cratered terrain. This is what Mariner 4 saw.
2. Volcanism: Olympus Mons the tallest of the giant shield volcanoes on Mars, is nearly three times taller than Mt. Everest. It's impossible for mountains on Earth to be this tall, because their great weight would make them sink into Earth. Since the surface gravity of Mars is 1/3 that of Earth's, the mountains of Mars can be taller. Also, the *lack* of plate tectonics on Mars allows its volcanoes to attain their huge size: Hawai'i moves, along the Pacific.
3. Tectonism is less important than on Earth. Mars has no global plate tectonics, because it's smaller than Earth, and had insufficient interior heat. Mars has no tectonic (folding) mountains, like Earth's.

   However, there is one spectacular feature, *Valles Marineris*, which is one great fault. The Valley is 5000 km long, 700 km wide at the widest, and has rims four miles high, with slumping from avalanches.
4. Gradation:

   Wind erosion is common, including active sand dunes and dust devils. Water erosion is also common, as shown by rampart craters, from splashes, dry runoff (dendritic) channels much like dry river beds, and outflow channels, which were made by massive outpourings in the past.

   Since 2012, the Mars *Curiosity* rover has been exploring the geological history of Mars, particularly the history of water on Mars. The Mars rover *Opportunity*, from 2004 to now, has made on-site measurements, including chemical analysis of rocks, which show clearly that there was a salty ocean there, at least twice over the history of Mars. In 1997, the Mars *Pathfinder* rover showed that where it explored had been flooded at least five times by channel outflow in the distant past.

   Terraforming: Might it be possible for humans to change the climate of Mars, to make it warmer and wetter? Robert Zubrin, in his book *The Case for Mars* (see also www.marssociety.org) discusses this. Planting plants at the poles would absorb sunlight and make the planet warmer. The details are left as an exercise for the student.

► **Figure 19-17** Mars has *big* volcanoes, the tallest mountains of any kind known in the Solar System. This is Olympus Mons, the tallest, and it is surrounded by clouds. (Source: NASA)

► **Figure 19-18** The volcanoes on Mars are as big across as Arizona. (Source: NASA)

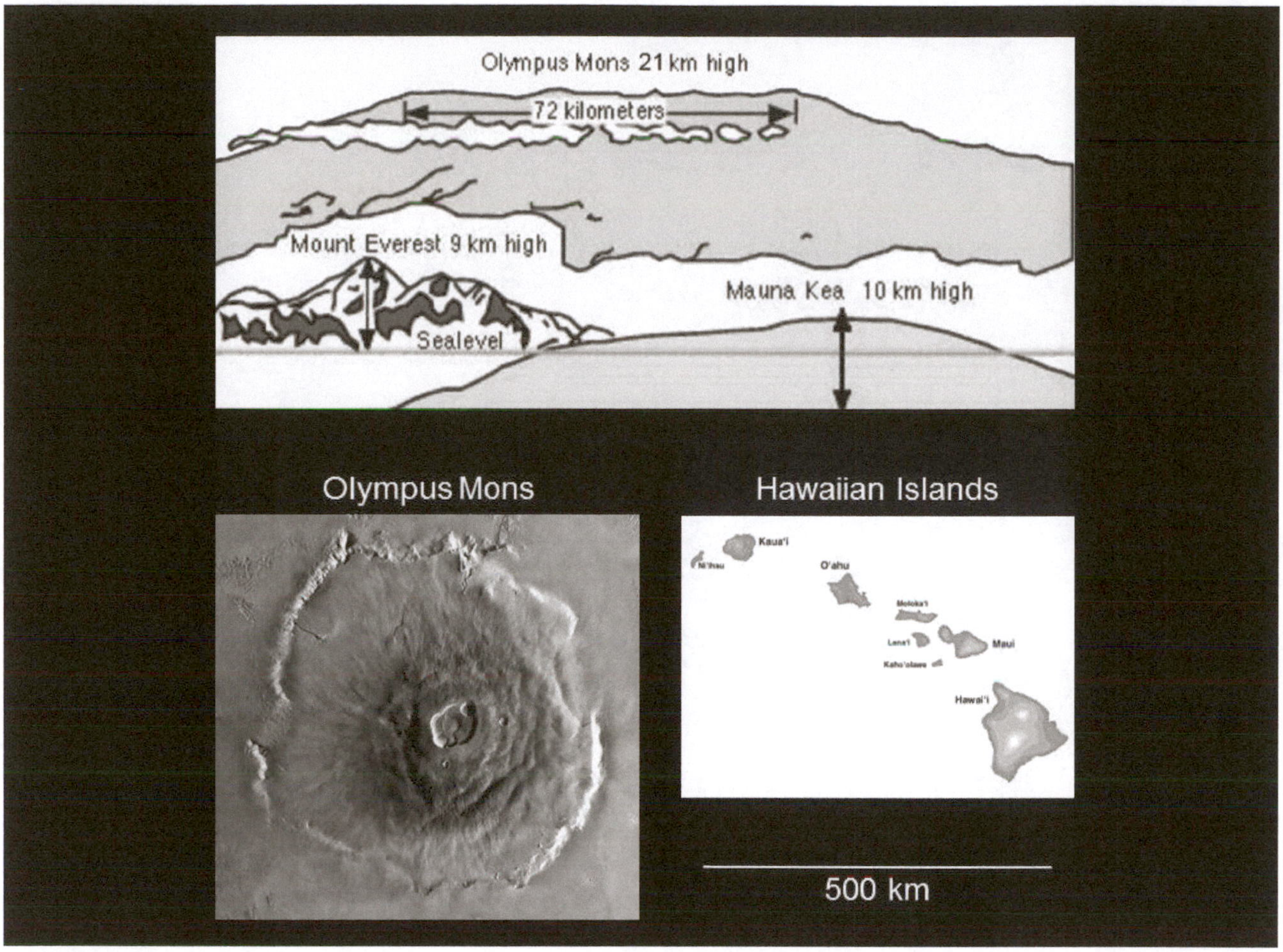

▶ **Figure 19-19** *Top:* Olympus Mons is 21,300 meters (69,800 feet) high, 2.4 times taller than Mount Everest. If a mountain as tall as Olympus Mons were on Earth, it would be so heavy it would sink into Earth. Mountains can be this tall on Mars because the surface gravity of Mars is only 1/3 of that on Earth. (Source: NASA) ***Bottom, left:*** Olympus Mons is a shield volcano, like Mauna Kea in Hawai'i. (Source: NASA/JPL) ***Bottom, right:*** Another reason Olympus Mons is so tall is the lack of plate tectonics on Mars. The shield volcanoes in Hawai'i are swept into a chain of islands as the hotspot under them is swept along by the motion of the Pacific Plate. (Source: USGS)

► **Figure 19-20** The Grand Canyon of Mars (also called Valles Marineris) dwarfs the Grand Canyon of the Colorado River, which would be about the size of the rectangle in the center of the image. (Source: NASA/LPI)

► **Figure 19-21** This is the Grand Canyon of the Colorado River in Arizona. The cliffs here are about 1.6 km (1 mile) deep. Imagine what Vallis Marineris would look like. (Image courtesy of Frederick Ringwald)

▶ **Figure 19-22** This is a landslide in Vallis Marineris. These cliffs are 6.4 km (4 miles) high. (Source: NASA)

▶ **Figure 19-23** In 1976, the *Viking 1* and *2* robots first searched for life on Mars. They did find weird soil chemistry, but the searches for life were *inconclusive*. (That's the trouble with robots: they're good at finding what they're programmed to find, but life is a complex phenomenon.) (Source: NASA)

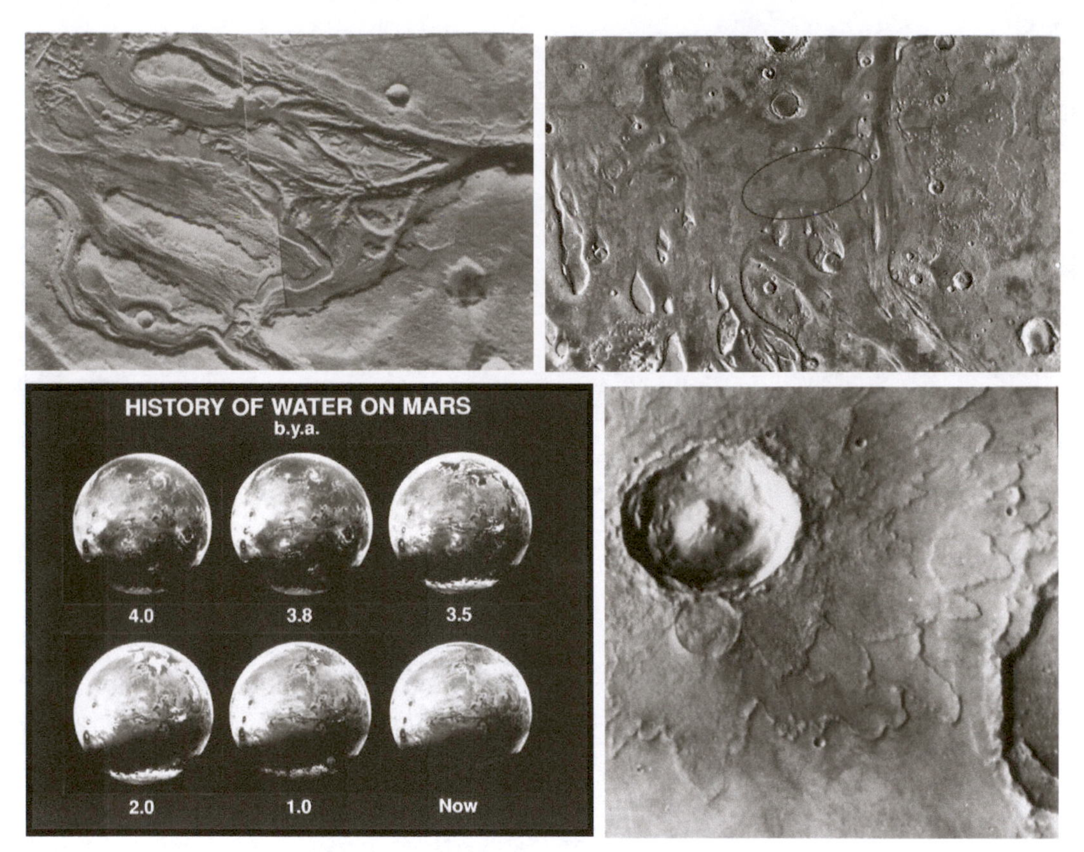

► **Figure 19-24** There is also plenty of evidence that large amounts of liquid water once existed on Mars. This couldn't happen today, because of the thin atmosphere of Mars. This implies that Mars once had a thicker atmosphere and was much warmer and wetter. Over 3 billion years ago, Mars may have been quite similar to Earth at that time. Since then, it has lost much of its atmosphere. (Source: NASA)

► **Figure 19-25** So, we keep dreaming about sending astronauts to Mars. Life is such a complex, unpredictable phenomenon, it may be necessary to send humans in order to discover it. (Source: NASA/Pat Rawlings)

CHAPTER

20

Cosmic Debris

Asteroids, Comets, and Kuiper-Belt Objects

Mars has two moons, Phobos and Deimos. They are thought to be captured asteroids, since Mars is close to the main asteroid belt, and since they have spectra similar to those of carbonaceous asteroids. Phobos and Deimos are about 10 km across, about the size of the asteroid that hit Earth 65 million years ago, which is thought to have caused the extinction of the dinosaurs. Since one never knows when or where a rock from space is going to hit, *serendipity* is important here.

Serendipity is a happy accident.

> "...Chance favors only the mind that is prepared."
>
> –Louis Pasteur (author of the germ theory of disease)

Popular media sometimes make this look bad, as if scientists who make serendipitous discoveries are bumbling along by trial and error. This is unfair: the scientists need to have the brains to recognize the lucky break, when they get it.

Examples of serendipity that this course has covered include:

- Wilhelm Roentgen's discovery of X-rays (1895)
- Henri Bequerel's discovery of radioactivity (1896)
- Ernest Rutherford's discovery of the atomic nucleus (1911)

A meteorite is a rock from space that has fallen to Earth. Serendipity is important for learning about meteorites, because one never knows in advance when and where a meteorite is going to fall:

- The L'Aigle meteorite shower (1803)

  Before this, many scientists ridiculed the idea that stones could fall from the sky. In 1803, in the French village of L'Aigle, there was a large meteorite fall, with over 3,000 stones heard and seen falling from the sky by hundreds of people.

  Now you know why this author is skeptical of claims that UFOs must be spacecraft from other worlds. If there were a clear case, like the meteorite shower at L'Aigle, I'd be convinced.

- The Allende meterorite (1969)

  The Lunar Receiving Lab was built in Houston, Texas in 1969, to examine the Moon rocks returned by the Apollo missions. By luck, barely a month later, there was a major meteorite fall in Allende, Mexico. By even better luck, this meteorite turned out to be unusually old, even for a meteorite. This is how we know the Solar System formed 4.57 ± 0.02 billion years ago.

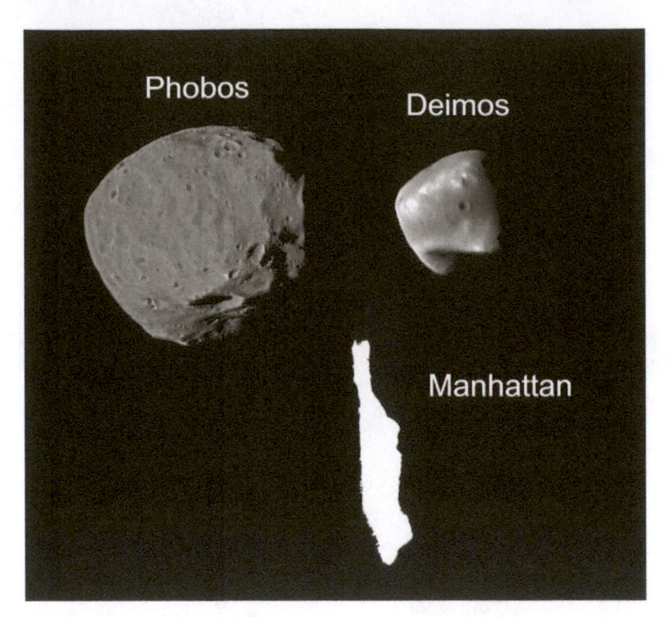

▶ **Figure 20-1** Phobos and Deimos compared to the island of Manhattan, in New York City. (Image courtesy of Fredrick Ringwald from NASA images)

▶ **Figure 20-2** *Left:* A carbonaceous chondrite meteorite, from the author's collection. Carbonaceous meteorites are stony meteorites rich in carbon and other organic material. This is one that fell in Allende, Mexico, in 1969. (Image courtesy of Fredrick Ringwald) *Right:* Essentially, it's a piece of stardust, direct from the Solar Nebula. Radiometric dating shows an age of 4.57 ± 0.02 billion years. (Source: NASA-Spitzer Science Center)

▶ **Figure 20-3** A stony-iron meteorite from the author's collection. (Image courtesy of Frederick Ringwald)

▶ **Figure 20-4** An iron-nickel meteorite (Image © MarcelClemens, 2013. Used under license from Shutterstock, Inc.)

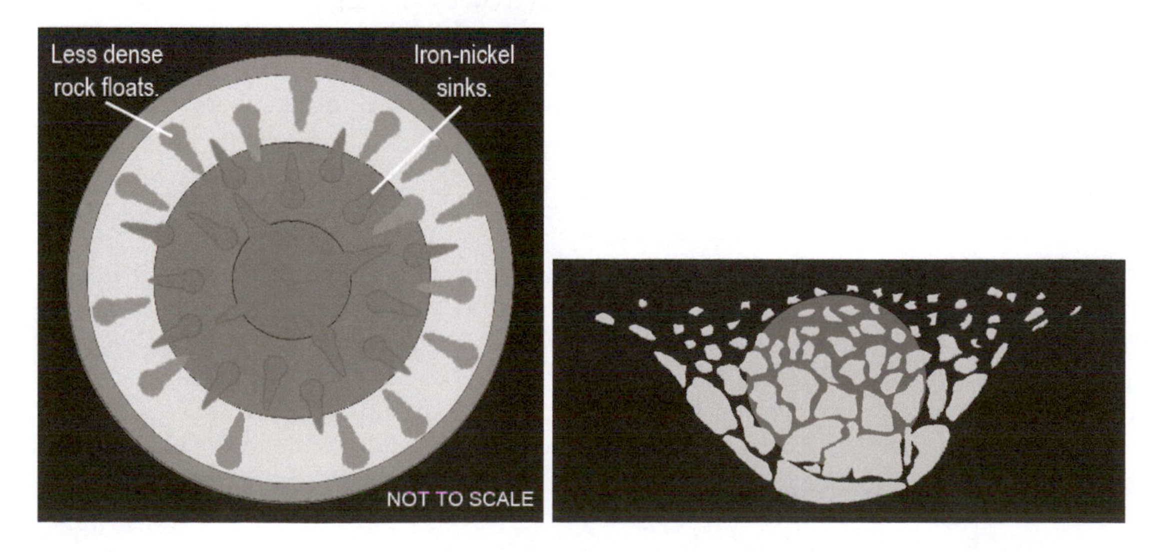

▶ **Figure 20-5** *Left:* Iron, stony-iron, and stony meteorites form in asteroids by differentiation by gravity. (Image courtesy of Fredrick Ringwald) ***Right:*** They then are spread back into space by fragmentation caused by collisions of asteroids. (Image courtesy of Frederick Ringwald)

Cosmic Debris: Asteroids, Comets, and Kuiper-Belt Objects

Meteoroids: are rocks (or most often, dust particles) in space.
Meteors: fall into Earth's atmosphere and burn up, since they move so fast.
Meteorites: hit Earth.

A typical meteor shower shows about 60 meteors per hour; for example, the Perseids in August. (The radiant, which is the place in the sky the meteors come from, is in the constellation Perseus.) The 2001 Leonid shower (from the constellation Leo) was like a fireworks show, with over 1000 meteors per hour.

Asteroids are rocky and from the Inner Solar System.
Comets are icy and from the Outer Solar System.

Otherwise, the distinction is blurring with the more we learn about both.

Meteoroids enter Earth's atmosphere undetected. Asteroids are discovered by telescope.

Comets are named after their discoverers. Asteroids are also named after people, as an honor. These are the only two kinds of natural objects in space that astronomers get their names on, officially. Star registries, in which one pays for star names, are *not* official.

Most, but by no means all, asteroids are in a belt between Mars and Jupiter. This was a planet that never formed, because it was stirred too often by Jupiter's gravity. It was *not* a planet that exploded.

Short-period comets orbit the Sun in less than 100 years. Halley's comet, which returns every 76 years, is an example. Short-period comets have orbits near the ecliptic plane. They therefore probably come from the *Kuiper Belt*, the remnant of the outermost part of the Solar Nebula, 40–50 AU from the Sun, predicted by Gerard Kuiper. Pluto, Eris, and over 1000 other icy bodies are now known, mostly beyond Neptune in the Kuiper Belt.

Long-period comets orbit the Sun in over 1000 years. They have randomly inclined orbits. They probably come from the *Oort Cloud*, a spherical shell of comets extending into interstellar space, over 10,000 AU from the Sun, predicted by Jan Oort.

Asteroids are often shown in science fiction movies to be hazards to space ships. None of the real spacecraft that passed through the asteroid belt suffered even one dust hit: the asteroid belt is *very* spread out. All the asteroids put together would have a mass <50% of Earth's Moon. Still, over 300,000 asteroids are known, with over 50,000 discovered every year, and over 10^6 are estimated to have diameters larger than 1 km. *Hundreds cross Earth's orbit, and so could hit Earth.*

Meteoroids smaller than a house (5–10 meters in diameter) almost entirely burn up in Earth's atmosphere. This happens about once per year somewhere in Earth's atmosphere.

Famous Impacts

Chelyabinsk, Russia in 2013 made a noise so loud it injured 1,500 people, mainly from glass from broken windows. Stony meteorites are still turning up.

Tunguska, Siberia in 1908 was probably by a comet. It exploded in the air, knocking down trees 30 miles away. It was about as powerful as the Nagasaki bomb (20 kilotons).

Meteor Crater, Arizona was formed in about 50,000 BC, by an asteroid 50 m across. It left the best-preserved crater on Earth, about 1 mile across and 500 feet deep. Its energy was comparable to the most powerful nuclear bombs (5 Megatons = 5,000 kilotons).

▶ **Figure 20-6** *Left:* This is an artist's depiction of the Leonid meteor storm of 1833. It is estimated there were over 100,000 meteors/hour. (Public domain image by Adolf Vollmy, 1888) *Top right:* The best meteor storm the author has ever seen was the Leonid meteor shower of 2001. It was like a good Fourth of July fireworks show, with over 3000 meteors/hour. (Comenius Univ. Bratislava, AGO Modra, Nov. 16, 1998 (Toth et al., 2000, EMP 82–83, 285)) *Bottom right:* This is a photo from the Russian space station Mir, taken during the 2001 Leonid meteor shower. Meteors are clearly shown burning up in Earth's atmosphere, below the station. (Source: Peter Jenniskens/APL/UVISI/MSX/BMDO)

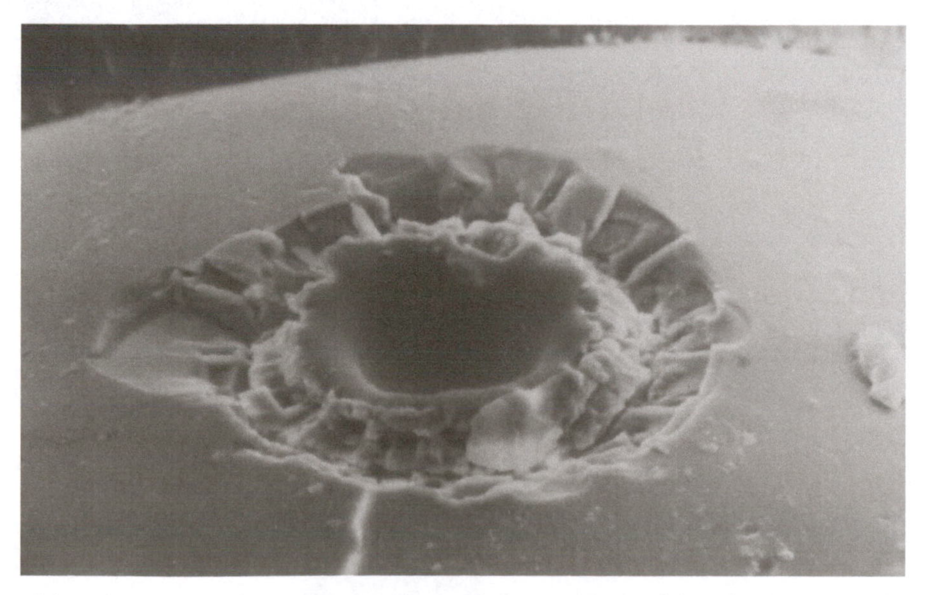

▶ **Figure 20-7** This micrometeorite crater, or "zap pit," was caused by the impact of a piece of dust. Spacecraft that have returned from space are often peppered with them. Moon rocks are full of them, on their parts that were exposed to space. (Source: NASA)

► **Figure 20-8** Copernicus crater on the Moon is 93 km (53 miles) across. The central peaks, from rebound from the impact, are 1 km (half a mile) high. (Source: NASA/JPL/USGS/Lunar Orbiter 2)

► **Figure 20-9** Copernicus crater is one of the freshest and youngest craters on the Moon, about 1 billion years old. It shows central peaks, terraced walls, an ejecta blanket, and secondary craters. (Source: NASA/GSFC/Arizona State University/Lunar Reconnaissance Orbiter)

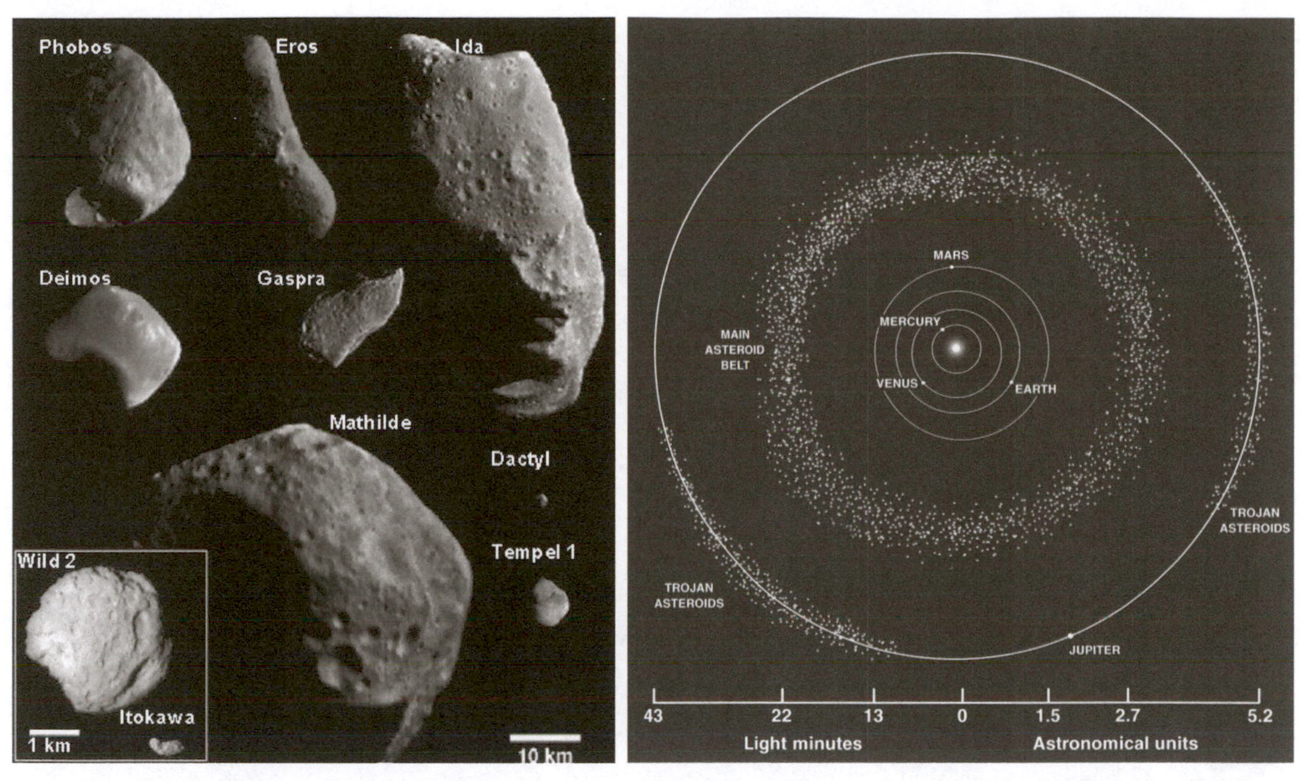

► **Figure 20-10** Asteroids are rocks, mainly in the inner Solar System in a belt between Mars and Jupiter. Many are loosely packed piles of gravel. (Source: NASA)

► **Figure 20-11** In 1996, Comet Hyakutake had a tail over 100 degrees long. (Image courtesy of Matt Mazurek)

► **Figure 20-12** Comets are dirty snowballs from the outer Solar System. *Left:* Edmund Halley predicted the return of Comet Halley in 1758. Its last apparition was in 1986. The next is predicted for 2061. (Image © Georgios Kollidas, 2013. Used under license from Shutterstock, Inc.) ***Top right:*** The orbit of Comet Halley. (Image courtesy of Frederick Ringwald) ***Bottom right:*** Comet Halley in 1986 (Source: NASA/NSSDC/William Liller)

▶ **Figure 20-13** Throughout history, people have reacted fearfully to comets. *Top:* Halley's comet is depicted in the Bayeux tapestry, which tells the story of the Norman conquest of England in 1066. "Isti mirant stella" means "they marvel at the star": in the next frame, King Harold looks worried. (© Gianni Dagli Orti/Corbis) *Bottom right:* A depiction of a really nasty-looking comet, from 1857 (© Bettmann/CORBIS) *Bottom left:* Today, we like to think we're modern and sophisticated. Nevertheless, the "Heaven's gate" mass suicide was a response to the 1997 apparition of Comet Hale-Bopp. (Image courtesy of Matt Mazurek)

▶ **Figure 20-14** The comet impact in 1908 in Tunguska, Siberia flattened trees over a 50 km (30 mile) radius. In 2013, a similar but smaller impact in Chelyabinsk, Russia, injured over 1000 people, mainly from flying glass from broken windows. (© Bettmann/CORBIS)

▶ **Figure 20-15** The Great Russian Meteor of 2013, also called the Chelyabinsk Superbolide, was the largest impact on Earth since the Tunguska event. Eyewitnesses reported the meteor was brighter than the Sun, and felt a pulse of heat from it. The meteor is estimated to have been 17 meters (56 feet) in diameter and exploded 23 kilometers above the surface of Earth. If it had come in at a steeper angle, it could have struck Earth and released there most of its 440 kilotons of energy (30 times larger than the Hiroshima nuclear bomb). (Image © Migel, 2013. Used under license from Shutterstock, Inc.)

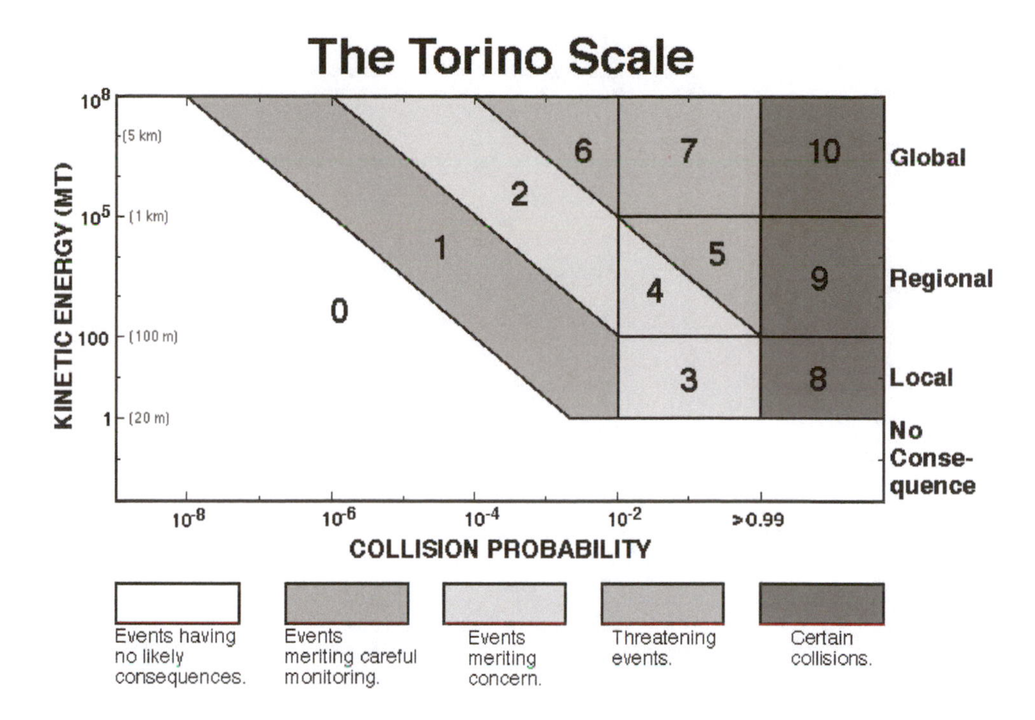

► **Figure 20-16** An asteroid scare: How do scientists communicate preliminary results, to encourage further observations, without panicking the press and the public—and *not* keep anything secret?

Several asteroid scares have happened since 1997, when an e-mail message among astronomers found its way to the front page of *The New York Times*, with a headline that the end of the world was nigh. The next day, the *Times* printed a retraction. It's not right to scare the public like that, needlessly. The press needs to understand these are only preliminary results. They're not definite: they need more observations. The trouble is that anyone can get e-mail, but then trying to keep them secret is trouble, and impossible anyway.

The Torino Impact Hazard Scale was devised soon after. It did not prevent several subsequent asteroid scares in subsequent years. (Used with permission of Richard P. Binzel (MIT))

CHAPTER

21

Mercury, Venus, and Atmospheres

Mercury

Mercury is hard to observe from Earth because it's never farther than 27 degrees from the Sun. (One hour on a clock is 30 degrees.) *Hubble Space Telescope* is not allowed to observe it because the spacecraft will heat up too much.

There have been over 40 missions to the Moon, 20 to Mars, and 20 to Venus. Mercury has been visited by only two spacecraft: *Mariner 10,* which made three flybys in 1974–75, and *MESSENGER,* which went into orbit around Mercury in 2011.

Mercury is the innermost and smallest of the terrestrial planets. It takes only 88 days for Mercury to orbit the Sun. (Earth, of course, takes 365.2422 days.)

The day side of Mercury is HOT, reaching temperatures of 800°F (= 427°C = 700 K).

Mercury's terrain is superficially similar to that of Earth's Moon with heavily cratered highlands and smoother maria (lava plains).

Images from *MESSENGER* show that Mercury was highly volcanically active between 3.8 and 4.0 billion years ago. Its ancient cratered terrain, like the Moon's, preserves a record of the history of the early Solar System.

► **Figure 21-1** Venus *(left)* and Mercury *(right),* just after sunset. (Image courtesy of Frederick Ringwald)

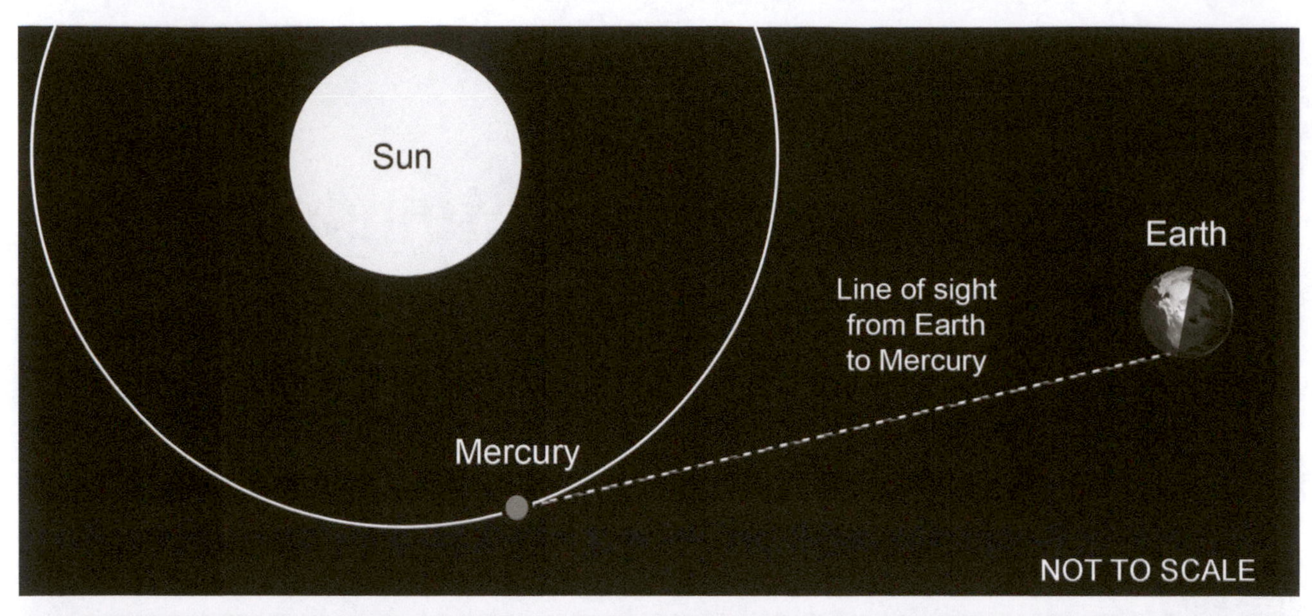

► **Figure 21-2** Mercury can be seen only just after sunset or just before sunrise because it is so close to the Sun. It's never visible at the zenith at midnight. (Images courtesy of Frederick Ringwald)

► **Figure 21-3** Mercury is the innermost planet to the Sun. As shown in this image by the *MESSENGER* spacecraft, Mercury looks much like Earth's Moon. (Source: NASA)

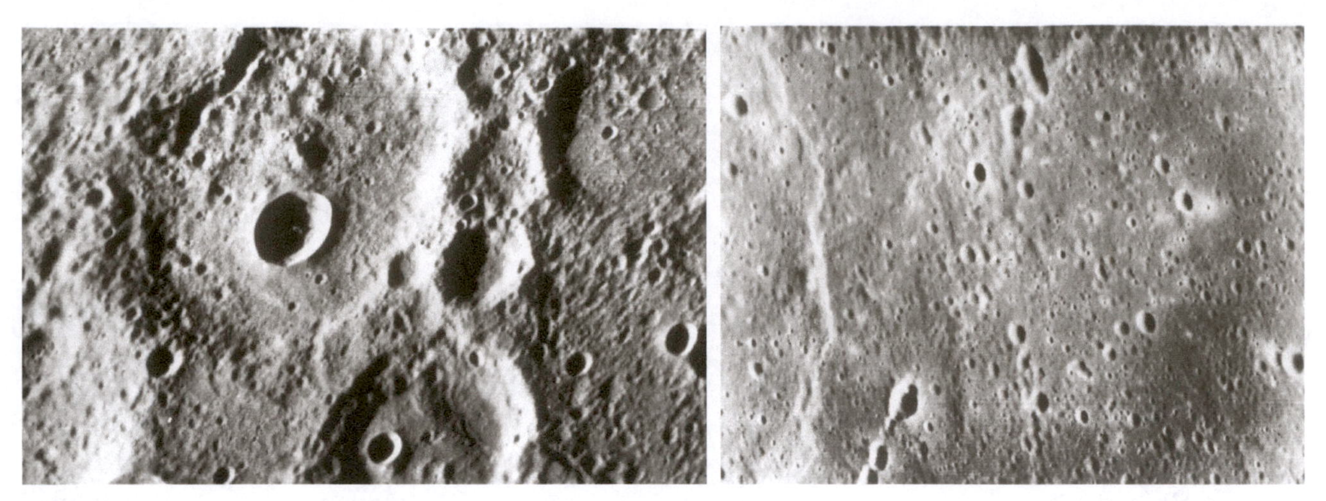

► **Figure 21-4** Mercury resembles Earth's Moon, with similar patterns of heavily cratered, ancient highland terrain *(left)* and smoother maria, or lava plains *(right)*. (Source: NASA)

Venus

The orbit of Venus is between those of Mercury and Earth. This is why Venus, like Mercury, is always seen either setting after the Sun at sunset or rising before the Sun at sunrise. Venus is the brightest object in the sky, after the Sun and Moon. Venus looks like a very bright star to the unaided eye, so it is sometimes called "the morning star" or "the evening star," even though it is a planet.

There have been many successful scientific missions to Venus since *Mariner 2* became the first successful flyby of another planet in 1962. Carl Sagan was on the *Mariner 2* team.

Venus is sometimes called Earth's "sister planet." Its mass, radius, density, and surface gravity are about the same as Earth's.

It's not easy to observe much from Earth. Venus has a featureless white appearance that goes through phases.

Observations with radio telescopes in the 1950s found temperatures of 735 K (860°F), hotter than Mercury. This is hot enough to melt tin and lead, and for surface rocks to glow. *Mariner 2* confirmed this high temperature.

The Soviet Union sent many spacecraft. *Venera 14* was the last lander, in 1982. It imaged the surface for 57 minutes. There has also been radar mapping, most recently by the United States by the *Magellan* orbiter, in 1990–1994.

The rotation of Venus is retrograde (backwards) and very slow, with a 243-day period. Its year is 225 Earth days long, longer than the Venusian day.

Venus has a *dense* atmosphere, made up of 96% CO_2, (carbon dioxide) *at 90 bars*, about the same as the pressure at the bottom of Earth's oceans. Earth's atmospheric pressure is 1 bar. Landing craft therefore need to be built like submarines to resist the crushing high pressure, as well as the high temperatures.

The clouds are mostly sulfuric acid. They never reveal the surface. One could never see the Universe beyond the clouds, from the surface of Venus.

▶ **Figure 21-5** Venus, shown here left of the Moon, is the brightest object in Earth's sky, after the Sun and Moon. Here it is shown rising before sunrise. (Image courtesy of Frederick Ringwald)

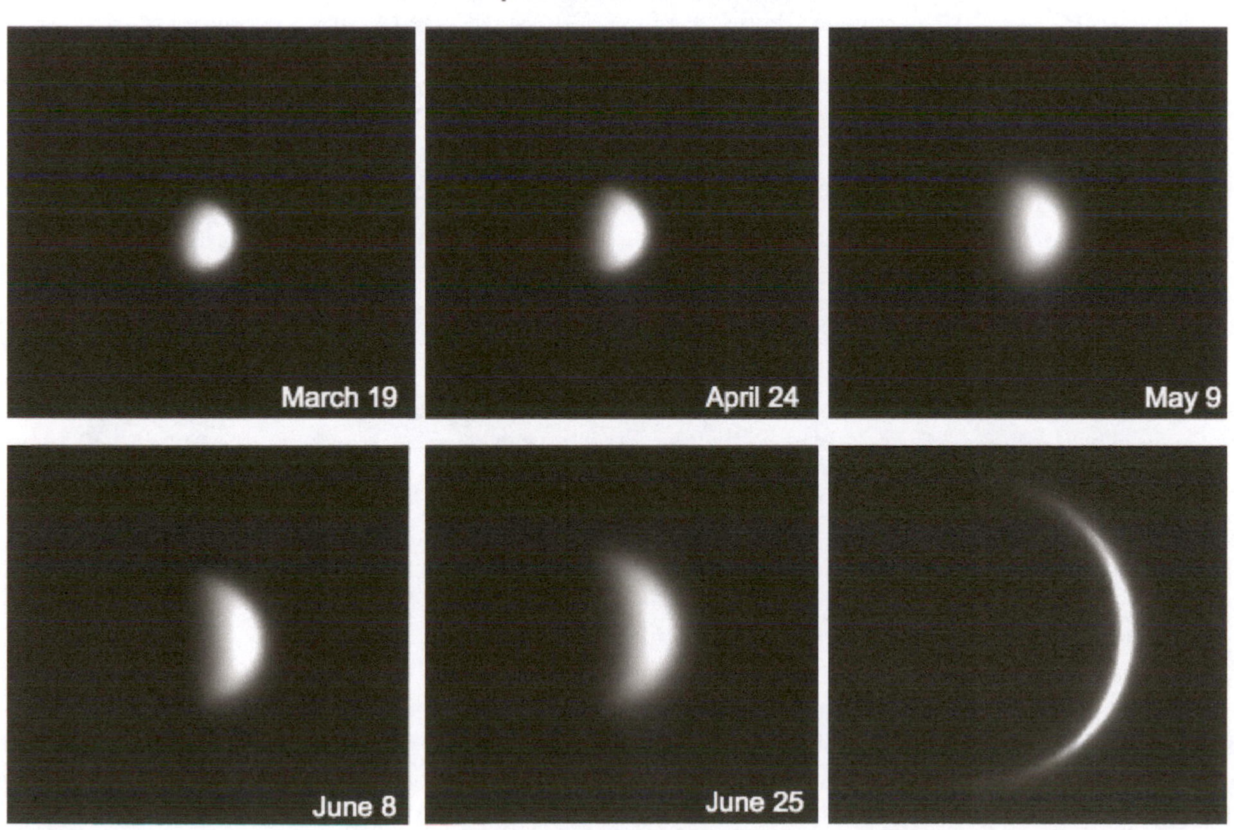

▶ **Figure 21-6** Observing the phases of Venus convinced Galileo that Copernicus was right. Mercury goes through a similar cycle of phases. (Image courtesy of Frederick Ringwald and Dr. Greg Morgan)

► **Figure 21-7** Venus is sometimes called "Earth's sister planet," since Venus has about the same mass, radius, density, and surface gravity as Earth does. Venus is unlike Earth in that its atmosphere makes it hotter than a blast furnace for steel making. (Source: NASA)

► **Figure 21-8** Before the spacecraft exploration of Venus, Venus was thought to be a steamy swamp, like Earth was during the age of the dinosaurs. The reasoning went that if Venus had clouds, these had to be clouds of water vapor, like Earth's. Since Venus was closer to the Sun, it had to be hot. Venus therefore had to be hot and humid like Earth was during the age of dinosaurs. Notice how the chain of reasoning went from "We can't see any detail" to "There must be dinosaurs." Don't you do that.

This is not a photo of Venus, but of a swamp in Florida. There are no dinosaurs on Venus. The clouds are sulfuric acid, not water vapor. There is not a trace of water on Venus. It *is* hot. (Image courtesy of Frederick Ringwald)

► **Figure 21-9** **The *Magellan* spacecraft used radar to map the surface of Venus. Radar works by bouncing radio signals off objects, so it can see through clouds.** (Source: NASA)

The four primary geological processes on Venus:

1. Impact craters are rare. They're hardly eroded at all, though. This shows that the whole surface had an episode of global resurfacing about 500 million years ago. No one knows why.
2. Tectonism: There are no plate tectonics, since the surface isn't broken into rigid plates, like Earth's crust. This may be because it's partially molten, from the atmosphere's high temperature.
3. Volcanism: The surface is *dominated* by lava flows. The atmosphere's high temperature partially melts the surface, making rocks soft, like plastic.
4. Gradation is primarily from lava flows.

In his Ph.D. thesis in 1957, Carl Sagan found the high temperatures are caused by *the greenhouse effect.*

"Greenhouse" here is a misnomer. It's not like a greenhouse, which lets in light and gets hot by preventing air from circulating. CO_2 is transparent to visible light radiated by the Sun. CO_2 isn't transparent to the infrared light radiated by Venus. Venus therefore heats up, but can't cool itself down by radiating heat into space. It therefore gets very hot!

Venus is very dry. All water probably broke down from the high temperature, and the hydrogen escaped from the planet aeons ago.

Venus has a runaway greenhouse effect. The hotter it gets, the drier it gets, and the drier it gets, the hotter it gets. This is an example of positive feedback, also known as "a vicious cycle."

Did this make the dense CO_2 atmosphere? Earth's CO_2 is mainly in limestone rocks. If all Earth's limestone rocks vaporized, Earth's atmosphere would become as dense as that of Venus.

► **Figure 21-10** The surface of Venus is dominated by lava flows, a volcanologist's paradise, with examples of nearly every type of hot volcanic structure known in the Solar System. This is because the atmosphere keeps the surface so hot that rocks aren't brittle the way they are on Earth: they're partially molten, and so are more like plastic. (Source: NASA/Magellan)

► **Figure 21-11** Basaltic lava plains, like the floors of Earth's oceans (under the sediment) and the maria of Earth's Moon, Mercury, and Mars, dominate the surface of Venus. (Source: NASA/Magellan)

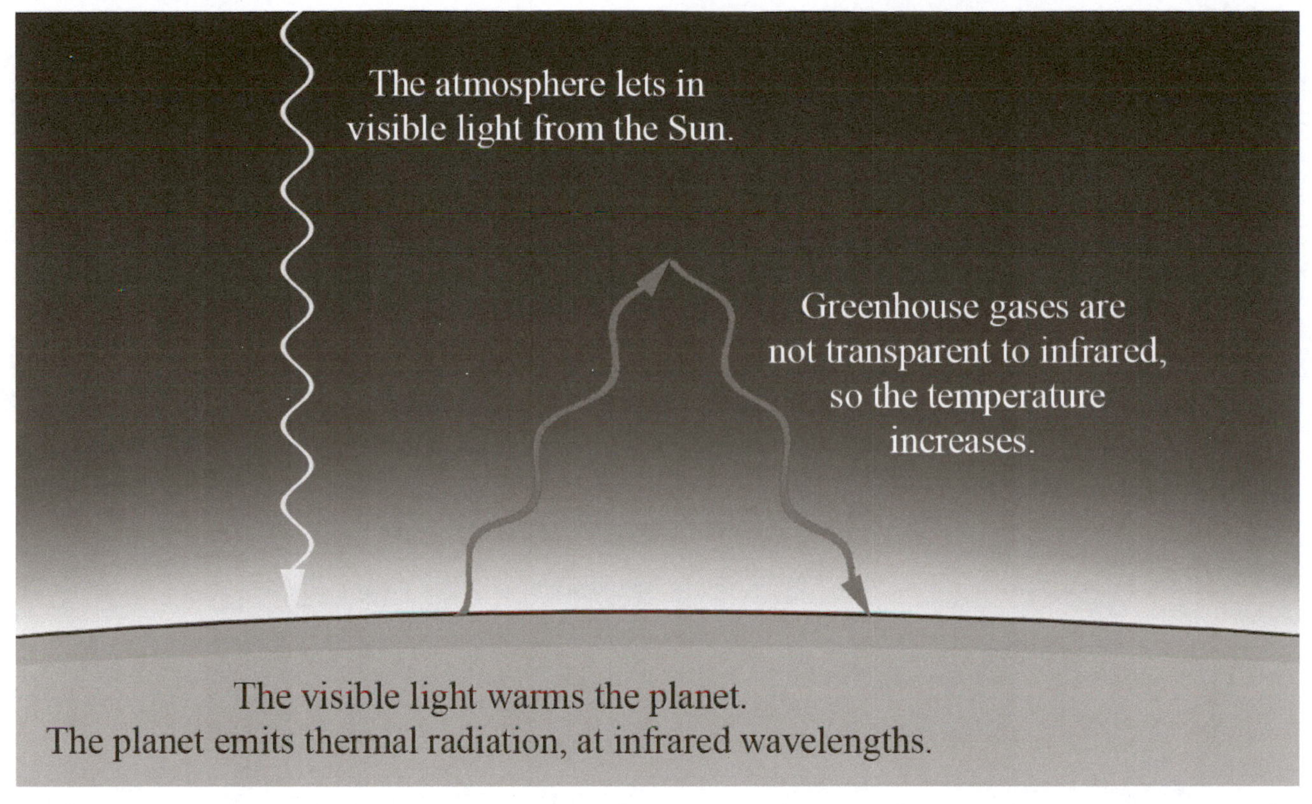

► **Figure 21-12** **Venus is hotter than Mercury because of *the greenhouse effect* (Sagan 1957). The atmosphere of Venus is mostly CO_2, which is transparent to visible light, but not to infrared radiation.** (Image courtesy of Frederick Ringwald)

Comparative Planetary Atmospheres

The real value of studying the planets is what this tells us about *Earth*. The atmosphere of Venus is about 90 times denser than Earth's atmosphere. It would squash you like a bug. Earth's atmosphere is about 150 times less dense than the atmosphere of Mars. Mercury and the Earth's Moon can't retain significant atmospheres because neither of them has enough mass or gravity. They're also too hot.

Venus and Mars have relatively inert, chemically reducing atmospheres:

Venus	96% CO_2
Mars	95% CO_2

In stark contrast, Earth has a highly reactive, oxidizing atmosphere. Life forms, such as blue-green algae, "poisoned" Earth's atmosphere 2.4 billion years ago in the Great Oxydation Event. The present composition of Earth's atmosphere is:

Earth	78% N_2
	21% O_2
	~0.03% CO_2

Earth's atmosphere has been completely remade over its long history, from the interplay of volcanism, water, Earth's crust, and especially, *life*.

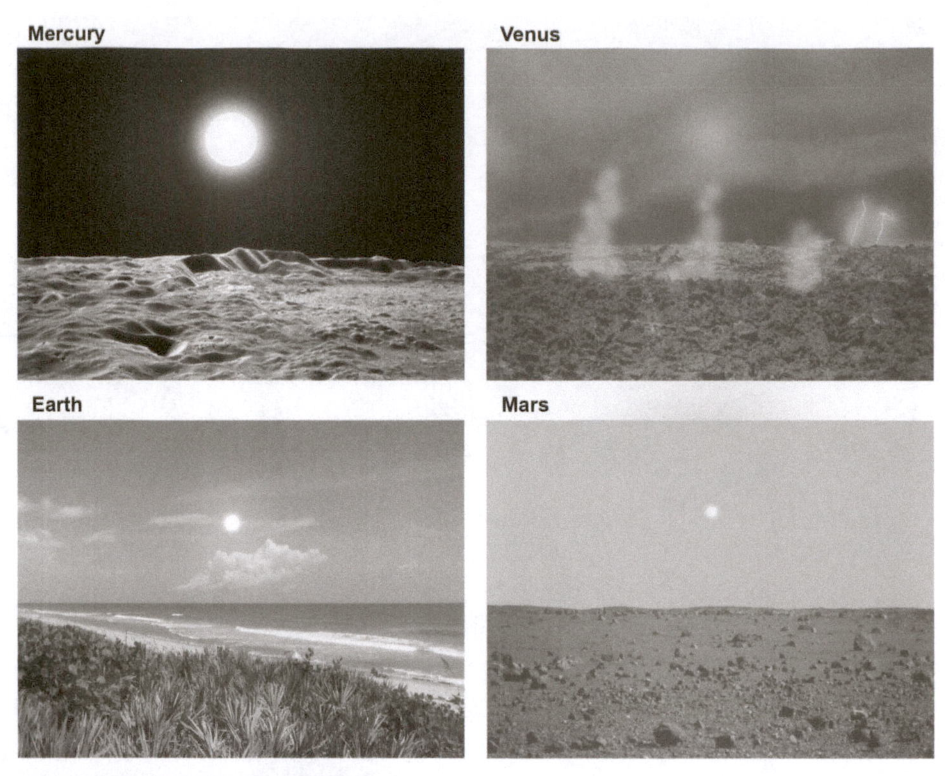

Average surface temperatures

Venus without greenhouse:	–30°F
Venus with:	870°F
Earth without greenhouse:	0°F
Earth with:	59°F
Mars without greenhouse:	–81°F
Mars with:	–71°F

▶ **Figure 21-13** *Top:* Comparative atmospheres: the real value of studying the planets is what they show us about Earth. Mercury is too hot, and has too little surface gravity from its small size and mass, to have much atmosphere. Venus and Earth have substantial atmospheres. Mars has a thin atmosphere because of its low surface gravity. Venus and Mars have atmospheres almost entirely composed of carbon dioxide. Earth's atmosphere is the only one rich in oxygen. (Image courtesy of Frederick Ringwald, using NASA images)

Bottom left: Most of the CO_2 that was once in Earth's atmosphere is now in limestone rocks, much of it from the action of life on Earth. (Image courtesy of Frederick Ringwald)

Bottom right: The greenhouse effect raises the temperatures of all planets with greenhouse gases, such as CO_2, in their atmospheres. (Image courtesy of Frederick Ringwald)

Carbon Dioxide (CO_2) in Earth's atmosphere

CO_2 is a greenhouse gas. It's transparent to visible light from the Sun, but it absorbs infrared radiation from Earth, so Earth's temperature increases.

The CO_2 content of Earth's atmosphere has increased by nearly 40% since the 1830s, about half of this since 1958. Measurements of the gases in air bubbles trapped in ice cores show there is now more CO_2 in Earth's atmosphere than at any time in the past 150,000 years.

This is because of human activity, especially burning fossil fuels such as oil and coal. Human CO_2 emissions are 130 times greater than those of natural sources, including volcanoes and hot spring

Earth is now the warmest it has been in 400 years, measured from gases trapped in ice cores, and from direct temperature measurements since 1855. Since 1855, the ten hottest years were, in order: 2010, 2005, 1998, 2003, 2002, 2009, 2006, 2007, 2004, and 2001. The 2000s were the warmest decade on record, with the 1990s second, and the 1980s third. Claims of "global cooling" since 1998 are false: there has been zero cooling.

Global average surface temperature has risen by 0.5–1.0°C since the 1880s. At present rates, it may rise by another 1.0–5.0°C, with a most likely rise of 2.5°C (4.5°F) by 2100, the warmest in two million years.

Average sea level around the world has risen by 10–25 cm since the 1880s. It may rise by another 10–90 cm by 2080.

Predicted effects of global warming include:

- Rising sea level and coastal erosion;
- Acidification of oceans—bleaching and death of coral reefs;
- Expanding deserts—fertile regions such as the U.S. Midwest becoming arid;
- Northward shift of agricultural areas and tropical diseases, such as malaria, yellow fever, dengue fever, and West Nile virus;
- Longer and more severe hay fever season;
- Severe economic impact—over 100 million people worldwide live within one meter of sea level, most in poor countries such as Bangladesh.

There may yet be hope. In 2012, CO_2 emissions in the U.S. dropped within 5% of their 1990 levels because of increased use of shale gas from hydraulic fracturing. This technique, also called fracking, is not without its own problems, since it is not always practiced safely.

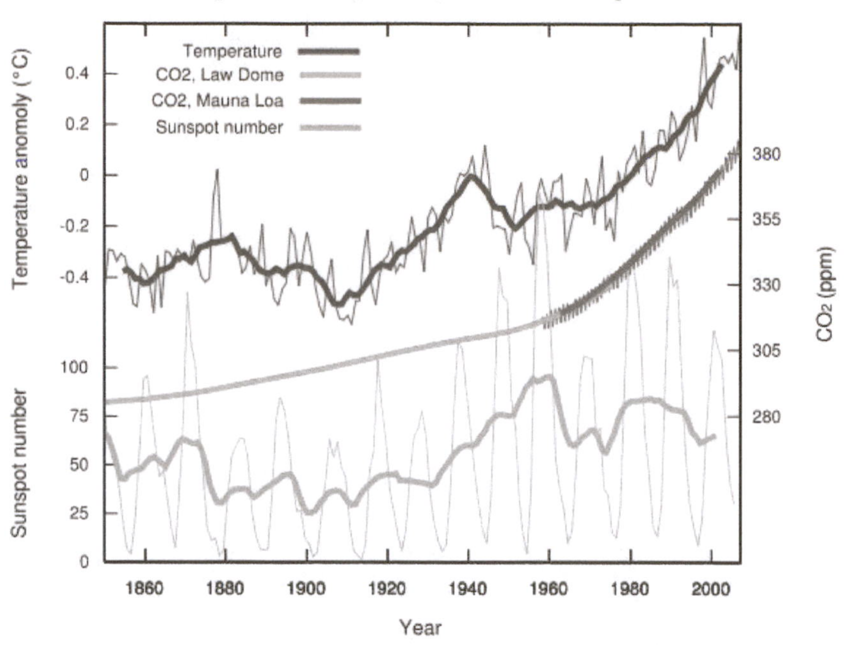

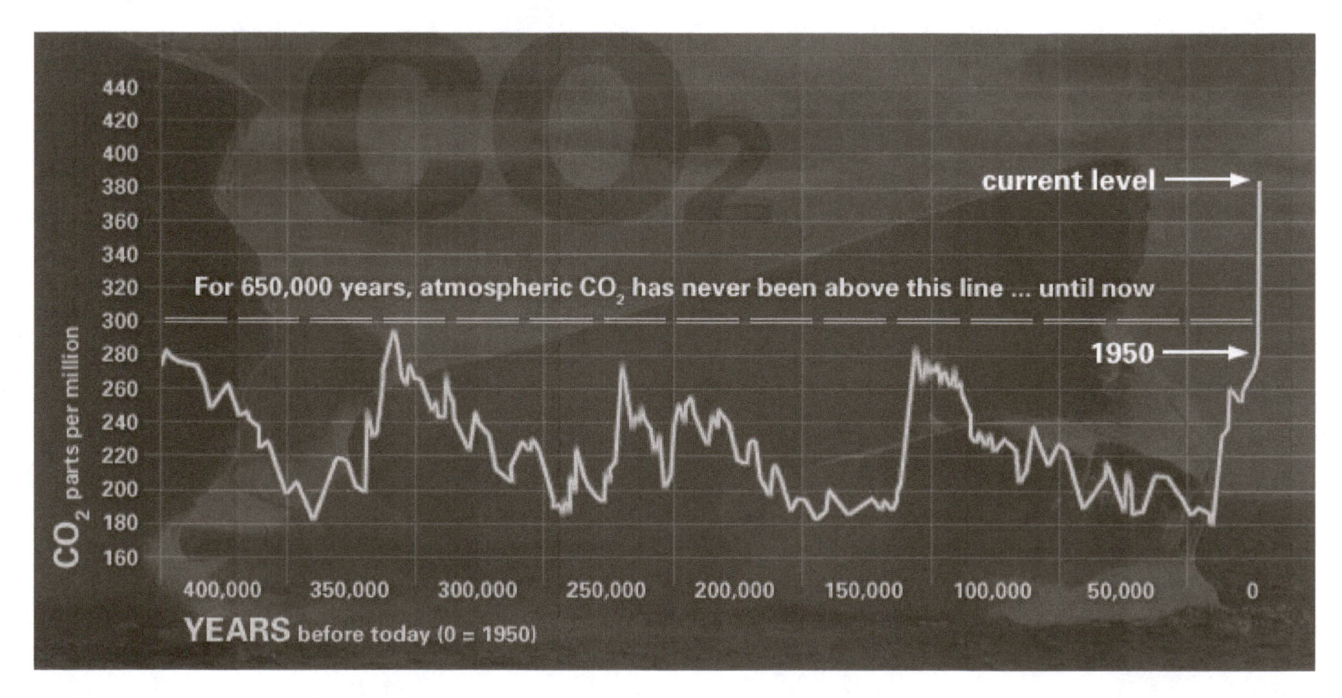

► **Figure 21-14** Global average temperature correlates with atmospheric CO_2 concentration. (*Top:* Source: NASA, *Bottom:* Source: NOAA)

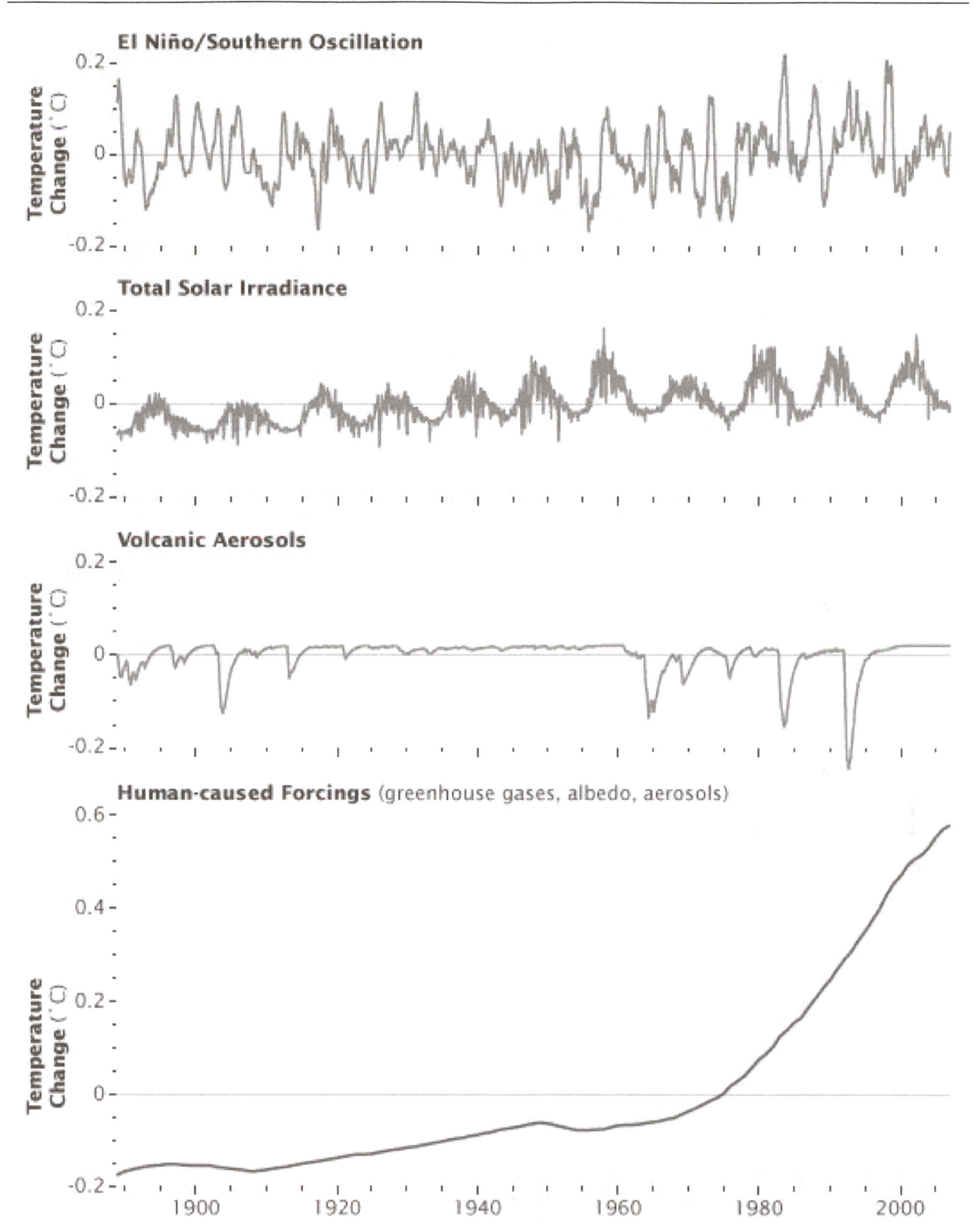

► **Figure 21-15** That global warming is real is shown in the bottom plot here. That it is caused by humans is shown in the other three plots. A total temperature variation of less than + 0.15°C comes from natural sources, such as variability in solar irradiance. A temperature change of greater than +0.8°C is anthropogenic: caused by humans. (Source: NASA Earth Observatory)

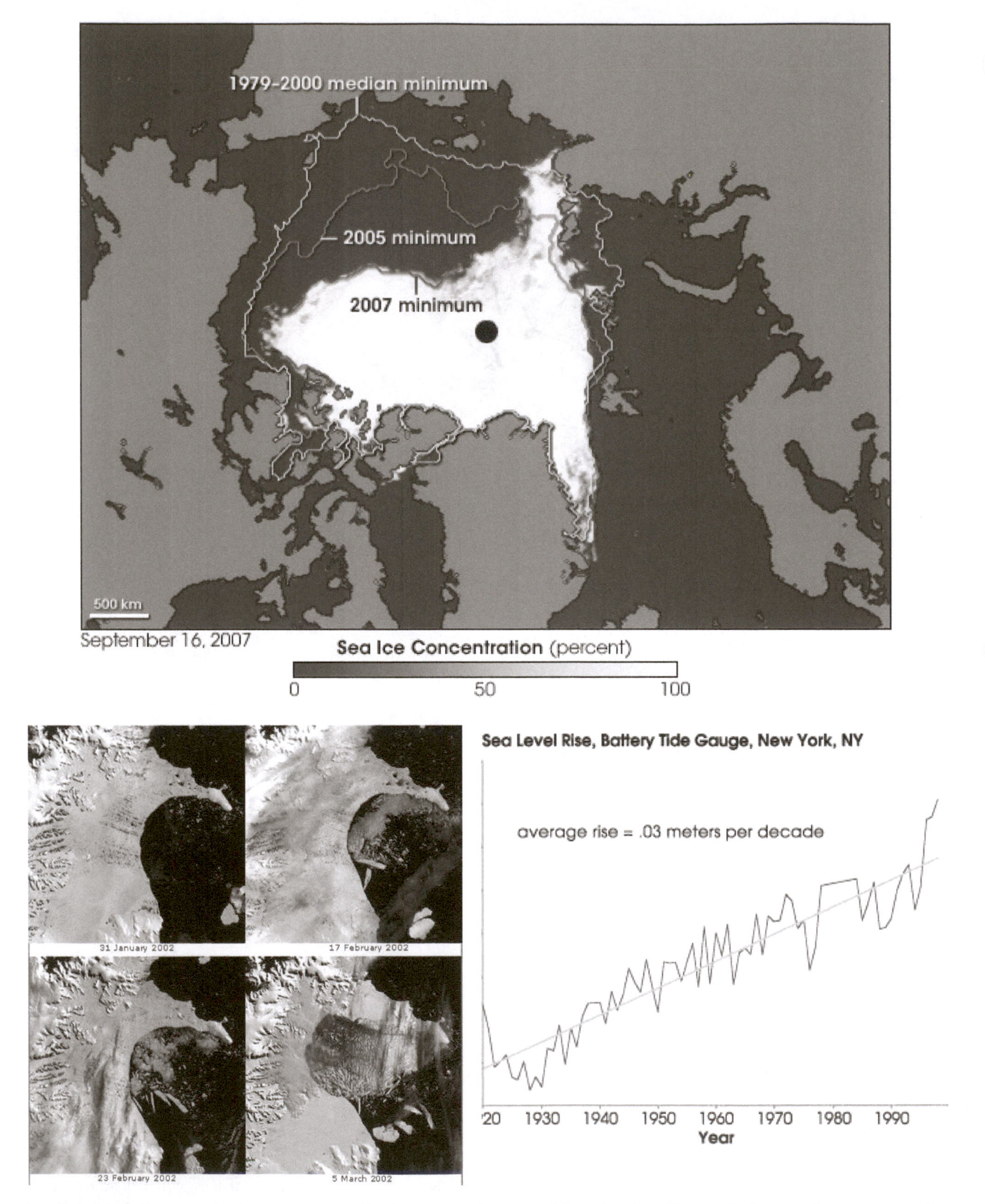

► **Figure 21-16** *Top:* Like any scientific theory, to be considered scientific, global warming must make predictions that can be tested by experiment. Here is a specific prediction: by 2045, the summer Arctic sea ice will be gone. (Source: NASA) *Bottom left:* The Larsen B ice sheet in Antarctica, 10,000-year-old ice, is receding at a rate unprecedented in modern history. (Source: NASA) *Bottom right:* Sea level is rising on the average of 0.03 meters per decade, as measured in New York. (Source: NASA Goddard Institute of Space Studies)

Why Was Earth's Climate So Warm during the Age of the Dinosaurs?

The question isn't why Earth was so warm then. The real question is why is it cooler now? The average global temperature of Earth now is about 10°C (18°F) cooler than during the age of the dinosaurs (the Mesozoic era), as well as throughout much of the rest of Earth's history.

The cool period we are in now is one of at least five that occurred throughout Earth's history, for which geological, chemical, and fossil evidence exists. The current cool period is called the Pliocene-Quaternary glaciation. It began about 2.58 million years ago, although conditions leading to it may have begun as much as 20 million years ago.

The other cool periods included the Carboniferous-Permian or Karoo (350 to 260 million years ago), the Andean-Saharan (460 to 430 million years ago), the late Proterozoic (850 to 630 million years ago, during which Earth may have frozen solid, the "snowball Earth" hypothesis), and the Huronian (2.4 to 2.1 billion years ago). These cool periods are thought to be caused by continental uplift. The current cool period may be from the uplift of the Himalayas and the Tibetan Plateau.

During these cool periods, glaciers advanced and retreated over much of the Northern and Southern Hemispheres cyclically, in ice ages. Over the last 750,000 years, ice sheets expanded into the U. S. Midwest at least eight times. The last was at maximum extent 20,000 years ago.

Between these ice ages were interglacial periods, in which the glaciers retreated. We are in one of these interglacial periods now, which began about 10,000 years ago and is called the Holocene epoch. These glacial-interglacial cycles during the cool periods are thought to be caused by variations in Earth's orbits. This is the Milankovitch hypothesis, which was proposed in the 1920s, but evidence for which accumulated only in the 1970s. The next glacial period is predicted to begin at least 50,000 years from now.

The Layers of Earth's Atmosphere

The air becomes thinner with altitude.

- The *exosphere* is at altitudes above 250 km. It's essentially empty space.
- The *thermosphere* is at altitudes of 90–250 km. It includes the *ionosphere*. The aurora is here. The air temperature increases with altitude to 2000 K, which is hot enough to ionize air. Short-wave radio waves reflect from the ionosphere. This is how they travel around the world.
- The *mesosphere* is at altitudes of 50–90 km. This is where most meteors burn up.
- The *stratosphere* is at altitudes of about 10–50 km. Because the air here is so thin, it's stratified (or layered), with little air circulation. Because of this, air temperature increases with altitude from the bottom of the stratosphere. Commercial airliners fly here, since this is above most air turbulence. The O_3 (ozone) layer is at about 20 km.
- The *troposphere* is from Earth's surface to altitudes of about 10 km. This is the "weather layer," in which air circulation transfers heat from the ground. This causes airplane turbulence and astronomical "seeing." Nearly all clouds are in the troposphere. Air temperature usually decreases with altitude in the troposphere.

The Two Ozone Problems

The ozone layer is at the bottom of the stratosphere. Ozone (O_3) absorbs ultraviolet radiation from the Sun. This protects Earth from the harmful effects of ultraviolet radiation, including skin cancer and cataracts in humans and damage to plants.

Chlorofluorocarbons (CFCs) from refrigerants and aerosols break down ozone in Earth's stratosphere. This makes a hole in the ozone layer over the Antarctic every spring. Its growth appears to have stopped: this is attributable to the international ban on CFCs that was enacted in 1987.

There's a second ozone problem: ozone is the component of smog that is most harmful to human health, and it can be a by-product of automobile exhaust. Ozone belongs in the stratosphere, not near the ground.

CHAPTER

22

The Outer Solar System

Before exploring the Outer Solar System (including the planets Jupiter, Saturn, Uranus, and Neptune), let's take a brief diversion, into the lab.

► **Figure 22-1** Jupiter, Saturn, Uranus, and Neptune are more like small stars than planets like Earth. None have solid surfaces, and all have extensive systems of moons and rings. (Image courtesy of Frederick Ringwald with images by NASA)

The Miller-Urey Experiment

In 1953, Stanley Miller was a student working in Harold Urey's lab. Both Miller and Urey were interested in the origin of life. To do this, they wanted to duplicate the early Earth's atmosphere, to see if chemicals similar to those in living things would form.

Miller filled a flask with the most primitive and common molecular gases in the Universe: water vapor (H_2O), ammonia (NH_3), and methane (CH_4). He then added energy, with electricity.

This *directly* made amino acids and proteins—the building blocks of life!

This has been repeated many times since. It's *easy* to make the building blocks of life—although we still don't know how to make life itself, because life is so complex.

A common misconception is that that life originated because of chance, or "just happened" (as Mark Twain wrote in *Huckleberry Finn*). Much of what we still don't understand about the Universe, including the origin of life, is clearly *not* due to chance: it's due to physical laws we don't yet understand.

One might therefore *expect* most new planets to be knee-deep in protein. Indeed, in space we often *do* see complex organic compounds. "Organic" in this course specifically means *containing carbon.*

Carbon is common throughout the Universe. All living things on Earth are made of organic compounds because carbon can bond to itself readily to make large, complex molecules that carry *energy* and *information.*

These can carry energy, as in hydrocarbons such as the octane in gasoline. They can also carry information as in DNA, the molecule that carries the genetic code in all life on Earth.

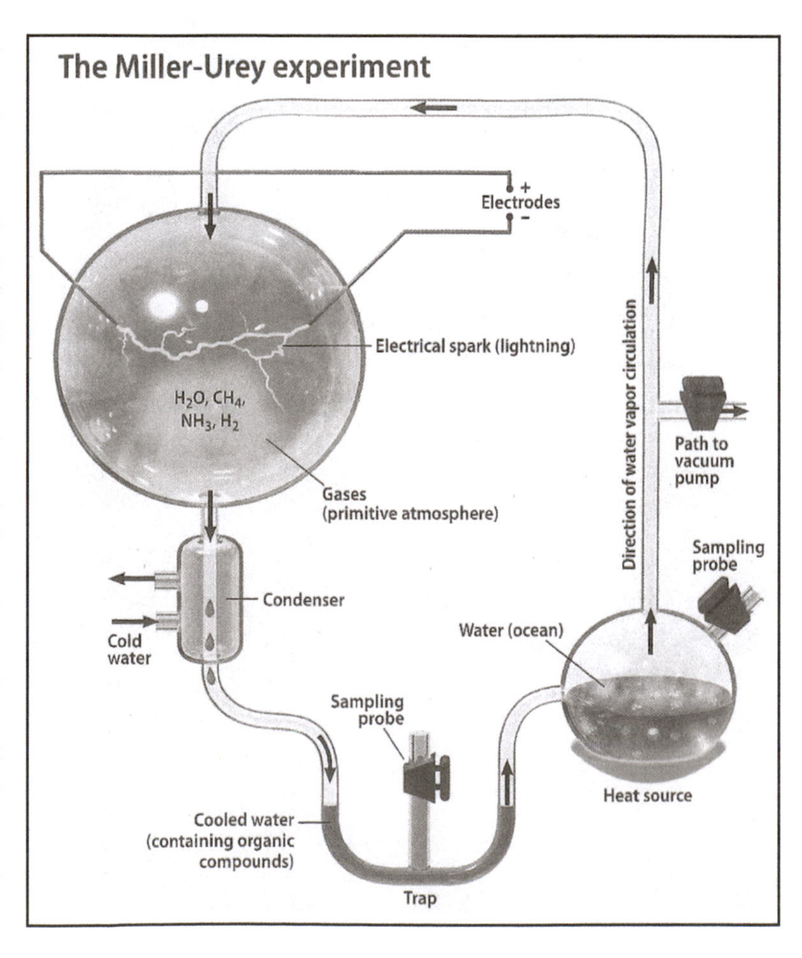

► **Figure 22-2** In the Miller-Urey experiment, simple gases that are known from spectra to be abundant in the Universe (water vapor, methane, ammonia, and molecular hydrogen) were zapped with electricity, a source of energy known to have been present in the early Earth's atmosphere (because minerals formed by lightning have been found in rocks dating back to that era). What this makes is amino acids: the building blocks of life.

No experiment yet done has produced life from non-living laboratory chemicals. Still, amino acids are the building blocks of proteins, and proteins are what living matter is made of, including you.

This prebiotic "brown organic tarry gunk" is present all over the Outer Solar System. This is because the Outer Solar System is cold, and so preserves it.

The implication is that any young planet should be covered in the building blocks of life. Perhaps this means that life is common throughout the Universe? (From Astronomy Magazine–2009 September p26 by Roen Kelly)

"Brown organic tarry gunk" in the Outer Solar System (more formally, Polycyclic Aromatic Hydrocarbons, or PAHs)

Organic compounds are seen all over the Universe, especially the Outer Solar System, where it's cold and so preserves these compounds.

We still haven't discovered life anywhere outside Earth's atmosphere (yet). This "brown organic tarry gunk" isn't necessarily living: it is to life what a pile of bricks is to a cathedral. A cathedral is more than just a pile of bricks, but to build a cathedral, one must start with a pile of bricks.

This "brown organic tarry gunk" isn't found on Mercury or Venus, because the high temperatures there break it down. It's not on Mars, because the thin atmosphere has no ozone layer, so ultraviolet radiation from the Sun sterilizes the surface. It's *all over* Earth, in the form of *living things*.

This "brown organic tarry gunk" is present in space seemingly everywhere it can exist. It has been found in:

- Carbonaceous asteroids (the most common type) and meteorites
- Comets: The nucleus of Halley's Comet is *black* with organic material.
  Comets are "dirty snowballs." They differ from asteroids mainly by being icier, since they're from the outer Solar System, which is colder.
- When Comet Shoemaker-Levy 9 collided with Jupiter in 1994, it made *Earth-sized* clouds of organic compounds in Jupiter's atmosphere!
- Planetary atmospheres: Jupiter and Saturn have brown, yellow, and red clouds.
- The atmospheres of moons: Titan, the largest moon of Saturn
  Triton, the largest moon of Neptune

► **Figure 22-3** The "brown organic tarry gunk" made in the Miller-Urey experiment is present all over the Outer Solar System. Seemingly everywhere it can exist, it does exist. Carbonaceous meteorites, such as this one here, are the most common type of asteroids in space. They are rich in organic compounds: over 50 amino acids, the building blocks of proteins, have been identified in them. (Source: NASA)

- The surfaces of Titan, Triton, and Iapetus, another moon of Saturn. Since Iapetus is icy, its organic matter looks like "cookies and ice cream."
- Planetary rings: Uranus and Neptune have rings made of black carbon dust.
  Jupiter's ring is silicate dust, like the ordinary dirt on the surface of Earth.
  Saturn's rings are by far the easiest to see because they're so bright, being made of ice.
- Outside the Solar System, in dusty dark nebulae where stars form, throughout the Milky Way, and in other galaxies.

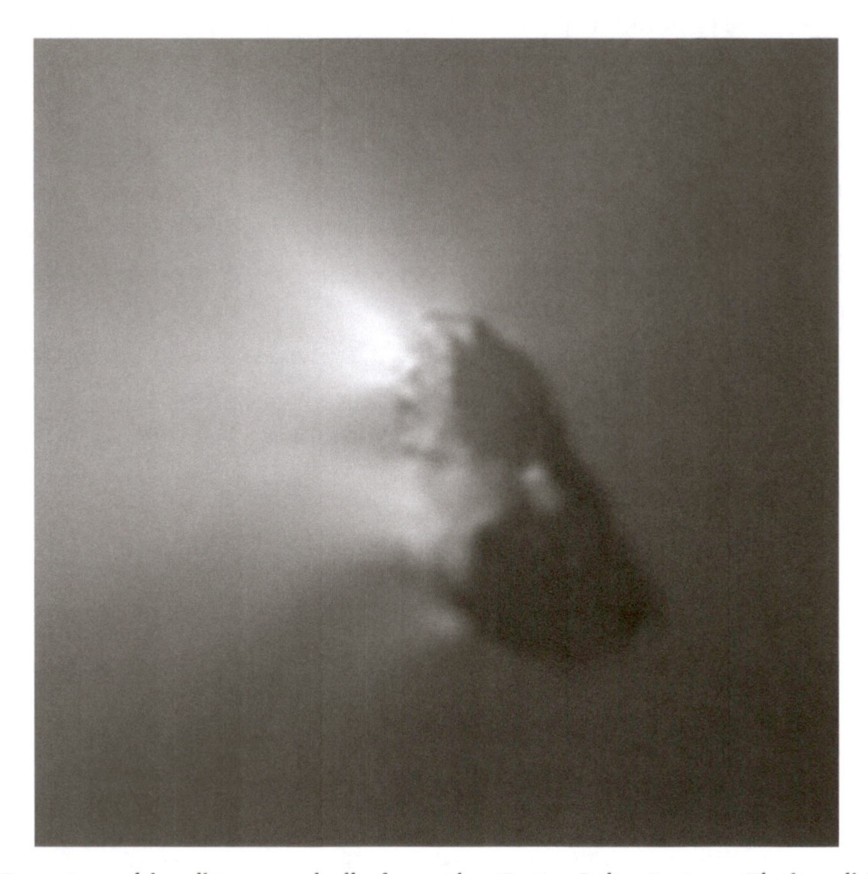

► **Figure 22-4** Comet are big, dirty snowballs from the Outer Solar System. Their solid, icy nuclei, from which their gaseous tails evaporate (actually sublimate, with solid ice turning directly into gas in the emptiness of space) turned out to be unexpectedly dirty. This image is of the nucleus of Halley's Comet, taken by a spacecraft during its most recent apparition in 1986. The nucleus is black with organic material, not white like a snowball. (Source: ESA/MPS)

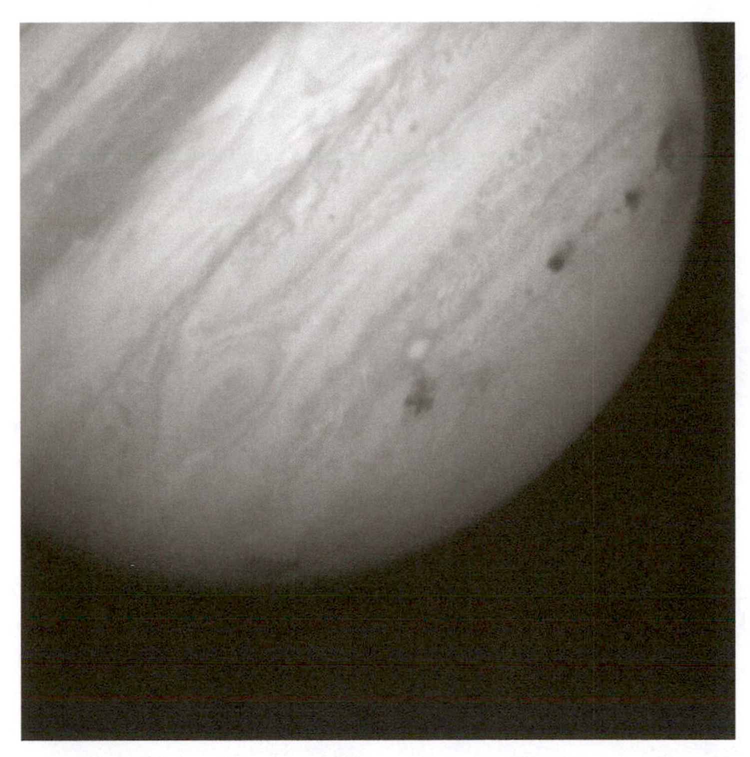

► **Figure 22-5** In 1994, Comet Shoemaker-Levy 9 broke up, and the pieces hit Jupiter. The impacts cooked up dark clouds of organic material, shown here. But then, the atmosphere of Jupiter is already full of organic material, with the browns, oranges, and reds of the clouds also shown here. (Source: NASA/STScl)

► **Figure 22-6** Titan is the largest moon of Saturn. It has an atmosphere, with a haze of hydrocarbons that obscures its surface. (Source: NASA/JPL)

► **Figure 22-7** *Top left:* Titan is the only moon or planet other than Earth to have liquid on its surface. Titan's lakes aren't water, though: they're liquid methane and ethane, which are hydrocarbons. (Source: NASA/JPL)

Top right: Titan's surface shows a cycle of evaporation, rain, and drainage of methane, similar to how the cycle with clouds and water (the hydrological cycle) works on Earth. What look like rocks there are H_2O ice, which at the −180°C temperatures, is as hard as rock. (Source: NASA)

Bottom: Around Titan's equator is a "desert," devoid of hydrocarbon lakes and rain. In it are dunes of organic material thought to look like coffee grounds. (Source: NASA/JPL)

► **Figure 22-8** Triton is the largest moon of Neptune. Like Titan, it has an atmosphere, and organic material on its surface. This may be what Pluto looks like up close. We'll know when the *New Horizons* spacecraft arrives at Pluto in 2015. (Source: NASA/JPL)

► **Figure 22-9** Iapetus is an icy moon of Saturn. It is as black as pitch on one side, and as white as snow on the other side. This is because one side is pitch—organic material—and the other side is snow. (Source: NASA/JPL)

The Outer Solar System

The Giant Planets, also called the Jovian Planets, since "Jove" is a nickname for Jupiter. They include:

Jupiter, Saturn, Uranus, Neptune

- Jupiter and Saturn are gas giants, mainly made of hydrogen gas, with deep atmospheres.
- Uranus and Neptune are ice giants, with a larger fraction of ice.
- All resemble miniature planetary systems, somewhat like the Solar System.
- All have extensive systems of moons:

  Jupiter has at least 67, Saturn 62, Uranus 27, and Neptune 13.
- All have rings. Saturn's are the largest and brightest, because they are ice.

  The others are dark dust.

► **Figure 22-10** The Solar System is only our local Universe, the Sun's family of planets, moon, and smaller objects. The Sun is one of hundreds of billions of stars in the Milky Way Galaxy, shown here. Far outside the Solar System, dust lanes in the Milky Way and other galaxies contain organic molecules. The "brown organic tarry gunk" is indeed spread throughout the Universe. (Image © MarcelClemens, 2013. Used under license from Shutterstock, Inc.)

- All have deep, stormy atmospheres, and no solid surfaces.

  The gas turns into liquid, at high pressure and temperature, deep in the interiors. There are probably solid cores in the centers of each, possibly made of diamond, but they're covered with thousands of kilometers of liquid and gas.

- All rotate rapidly, stretching the clouds into bands around the planets.

  Jupiter rotates in 9.9 hours. Earth rotates in 24 hours.

- All radiate more heat from their interiors than they get from the Sun.
- All have fierce, vicious radiation belts. These are similar to the Van Allen belts around Earth, but *much* larger.

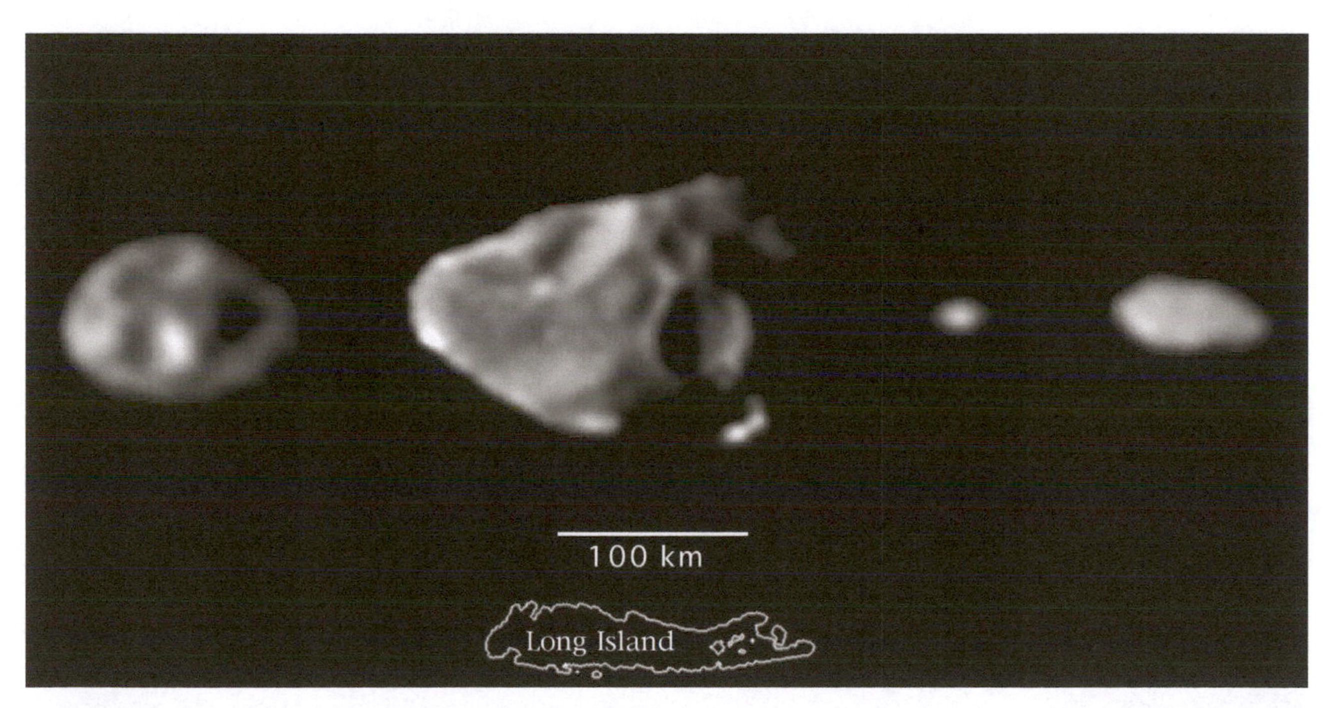

Figure 22-11 All the giant planets have extensive systems of moons. These are some of the smaller moons of Jupiter. (Source: NASA/JPL/Cornell University)

► **Figure 22-12** All the giant planets have rings. Saturn's are the most famous because they are the easiest to observe, but Jupiter, Uranus, and Neptune all also have them. These are the rings of Jupiter, which are made of dark dust. (Source: NASA/JPL)

► **Figure 22-13** Even a quite modest telescope will show that Saturn's rings look like this. Saturn's rings are so easy to see because they are made of bright, reflective ice particles. (Image courtesy of Frederick Ringwald)

► **Figure 22-14** Through *Hubble Space Telescope*, Saturn looks like this. One does expect more from a billion-dollar spacecraft. (Source: NASA/STScI)

► **Figure 22-15** Saturn looks like this from the *Cassini* spacecraft as it orbits Saturn. One should expect more from a 3.4 billion-dollar spacecraft. Saturn's rings have about 5000 ringlets, about as many as there are grooves on an LP record. They were totally unexpected, and are cleared by the gravity of the many moons. But then, the rings of Saturn are really just millions of tiny moons. Most of them are pieces of ordinary ice, about the size of the ones in an ordinary freezer. (Source: NASA/JPL)

► **Figure 22-16** *Left:* This is an artist's conception of the rings of Saturn close up. No spacecraft has been close enough to see this much detail. (Source: NASA/JPL/U. of Colorado/Marty Peterson) ***Right:*** We know that the rings of Saturn are made of particles of ordinary water ice because of their spectra. We know they are a few centimeters to 2 meters across because of how radio signals from spacecraft scatter, when the spacecraft are on the rings' opposite side. In this image, one can see the larger particles themselves, and the long shadows they cast at high Sun angle. Saturn's rings are thought to be only 10–20 meters thick. (Source: NASA/JPL/SSI)

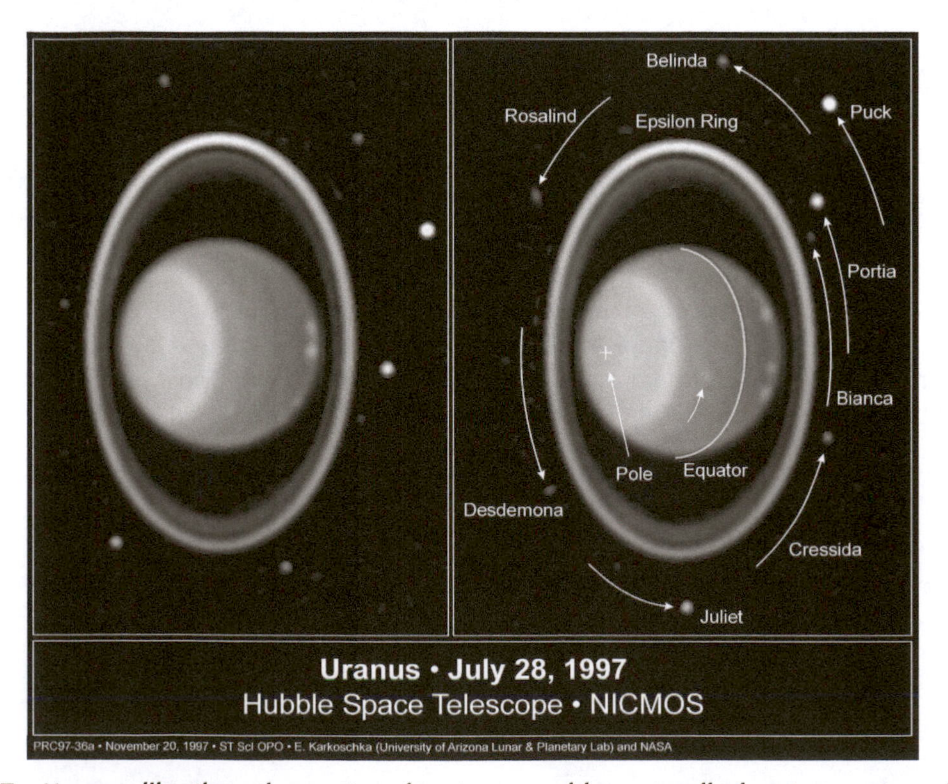

► **Figure 22-17** Uranus, like the other outer planets, resembles a small planetary system. It has a deep, storm atmosphere, rings (made of dark, carbon-rich dust), and many icy moons (named after characters by Shakespeare and Pope). It is unlike any of the other planets in that it's tilted on its side: no one knows why. (Source: NASA/STScI)

► **Figure 22-18** Neptune also has rings. They are difficult to observe since they are made of dark, carbon-rich dust, which is why these images avoid showing the planet Neptune: it is so much brighter its light would overpower the rings. Neptune's rings do not go all the way around Neptune: they are therefore more properly called ring arcs. (Source: NASA/JPL)

► **Figure 22-19** All the giant planets have deep, stormy atmospheres. Jupiter's Great Red Spot (at center right) is similar to a hurricane, except that it is three times larger than Earth (shown alongside) and has existed for at least 350 years.

The Great Red Spot had two new companions in 2008, shown here. These were called "Son of Red Spot" (at lower left) and "Son of Red Spot Junior" or "Baby Red Spot" (at center left), which appeared in 2008 and has since merged with the Great Red Spot. (Source: NASA/STScI)

▶ **Figure 22-20** *Top:* Could there be life floating in the atmosphere of Jupiter? There is water, heat (from Jupiter's gravity), and organic chemistry (which we know from spectra). (Image courtesy of Frederick Ringwald) *Bottom:* The Galileo probe didn't see any life as it parachuted into Jupiter's atmosphere in 1995, but it had no camera because it had to withstand Jupiter's gravity. A more modern, all-digital design, on a more long-lived balloon, might be able to search for life. (Source: NASA)

► **Figure 22-21** Saturn also has a deep, stormy atmosphere, as shown by the storm at top. (Source: NASA/JPL)

► **Figure 22-22** Neptune also has a deep, stormy atmosphere. Its Great Dark Spot, shown here in 1989, was a storm similar to a hurricane about the size of Earth. It disappeared by 1997. (Source: NASA/JPL)

▶ **Figure 22-23** All the giant planets are more like small stars than planets like Earth, in that they emit at least as much infrared radiation as they get from the Sun. The heat shown here flowing out of Jupiter shows that planets *don't* shine entirely by reflected sunlight: that's only for visible light. (Source: NASA/STScI)

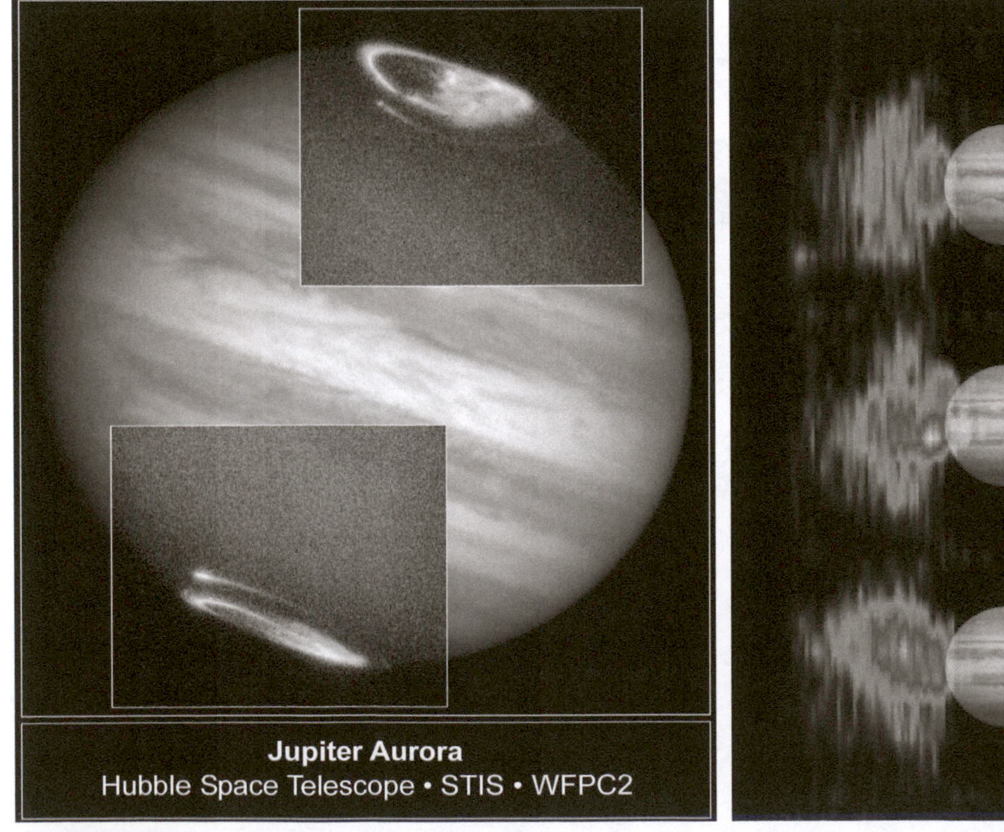

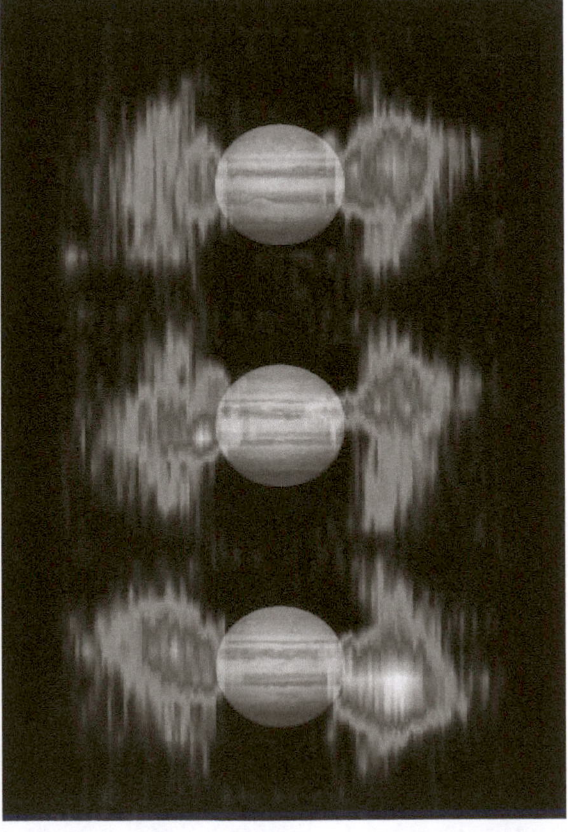

▶ **Figure 22-24** *Left:* Jupiter has a powerful magnetic field and fierce, vicious radiation belts thousands of times more extensive than Earth's. The magnetic fields can channel protons and electrons from the Sun into Jupiter's magnetic poles. This can light up the top of Jupiter's atmosphere there, causing aurora borealis (northern lights) and aurora australis (southern lights). Earth has auroras too, but Jupiter's are much larger. (Source: NASA/STScI) *Right:* Jupiter's radiation belts emit radio waves, which are among the most powerful in the Solar System. (Source: NASA/JPL)

The Galilean Moons of Jupiter

The Galilean Moons of Jupiter were discovered in 1610 by Galileo and are visible in binoculars. (Memory aid: I Eat Green Carrots.)

	Io	Europa	Ganymede	Callisto
Impact Cratering	None!	Rare	Common	Dominant: saturation cratered
Tectonism	*Very* Active	Common: cracks in icy surface.	Active plate tectonics, in ice. Folding mountains, like Earth, but icy.	None
Volcanism	Dominant!	Cryo- (cold) volcanism only: water seeps through the icy surface.	Cryovolcanism only	None
Gradation	By lava flows: extremely active	Dominant: very active flowing ice, surface is *very* flat	Slow; some ancient cratered terrain	Very slow: ancient cratered terrain

Io (pronounced "EYE-oh," although "EE-oh" is also acceptable) is by far the most active volcanic body in the Solar System. It has hundreds of volcanic features and dozens of active volcanoes.

Europa is thought to have an **ocean of liquid water** beneath its icy surface, because:

1. Europa is *so* flat—by far the flattest body in the Solar System.
2. Europa's surface is icy (known because of spectra), and Europa's interior is heated by Jupiter's tides, the way Io is.
3. Liquid water seeps through Europa's ice in a process called *cryovolcanism* (cold volcanism: not of molten rock, but of water).
4. Europa's magnetic field is similar to what one might expect from a salty ocean, since salt is a good conductor of electricity, and electricity is always connected with magnetism. Ganymede and Callisto are also suspected to have salty, underground oceans, because of measurements of their magnetic fields by the Galileo spacecraft.

Saturn's moon **Enceladus** (pronounced like "enchiladas") also shows evidence of warm liquid water under its icy surface, since it has geysers of liquid water shooting into space.

There are two other large planetary moons, both of which have *atmospheres*:

- **Titan** (moon of Saturn) has a hazy atmosphere, loaded with organic chemicals
- **Triton** (moon of Neptune) has an atmosphere, too, although it's even colder

▶ **Figure 22-25** In 1610, Galileo discovered the four largest moons of Jupiter ("the Galilean satellites"): Io, Europa, Ganymede, and Callisto. Io *(left)* and Europa *(center)* are shown here. (Source: NASA/JPL)

▶ **Figure 22-26** Io, pronounced "EYE-oh," is the innermost of the Galilean satellites. It is sometimes called "the Cosmic Pizza," because of its yellow and orange colors, which are from sulfur. The black spots aren't anchovies, however: they're volcanoes. Notice the lack of impact craters, because the surface of Io is resurfaced constantly by volcanism. Io is slightly larger than Earth's Moon. It has essentially no atmosphere. (Source: NASA/JPL)

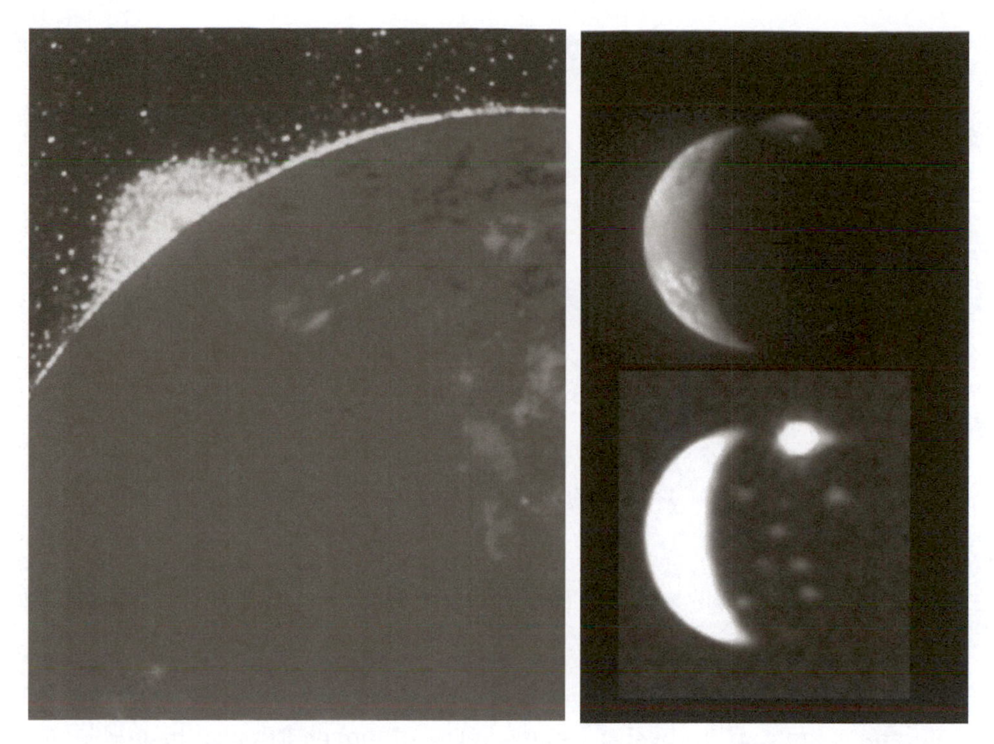

► **Figure 22-27** *Left:* Io is the most volcanically active body in the Solar System, with over a dozen volcanoes active at any given time. (Source: NASA/JPL) *Right:* The heat flowing out of Io that powers the volcanism comes from the gravity of Jupiter and its other moons. Whereas the Moon's gravity on Earth causes tides of water in Earth's oceans that lift up a few feet, the gravity of Jupiter and its other moons on Io causes tides that lift up the rocky surface of Io by as much as 15 km (9 miles) twice every 42 hours, the orbital period of Io around Jupiter. (Source: NASA/Johns Hopkins University/APL)

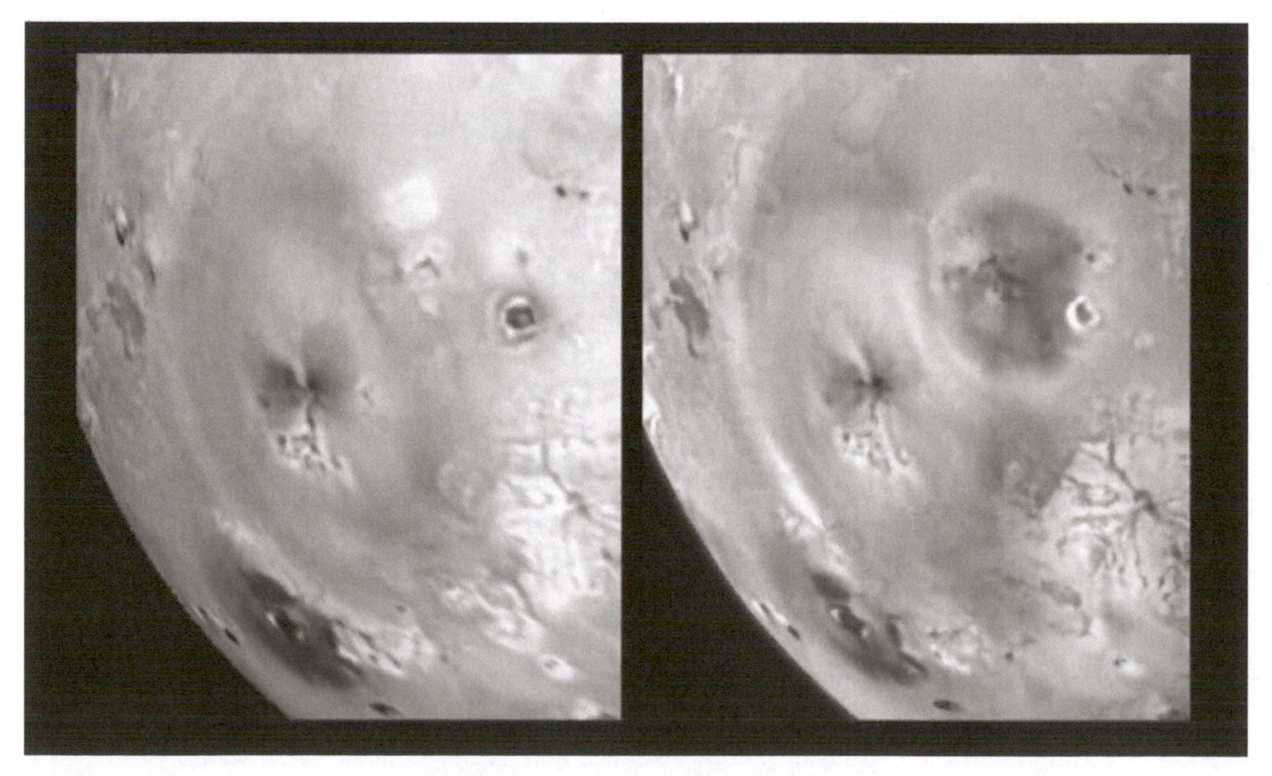

► **Figure 22-28** *Left:* Io in April 1997. (Source: NASA/JPL) *Right:* Io in August 1997. See the gray lava flow at upper right, which is about the size of Arizona? It wasn't there four months earlier. (Source: NASA/JPL)

► **Figure 22-29** Europa is the next furthest Galilean satellite of Jupiter. It is also heated by Jupiter's gravity, but not as much as Io. Spectra show that its surface is covered in ice. Europa is very flat, with the highest point on Europa being barely 100 meters higher than the lowest point. Notice the lack of impact craters: the surface is young and actively resurfaced. Europa is slightly smaller than Earth's Moon. It has essentially no atmosphere. (Source: NASA/JPL)

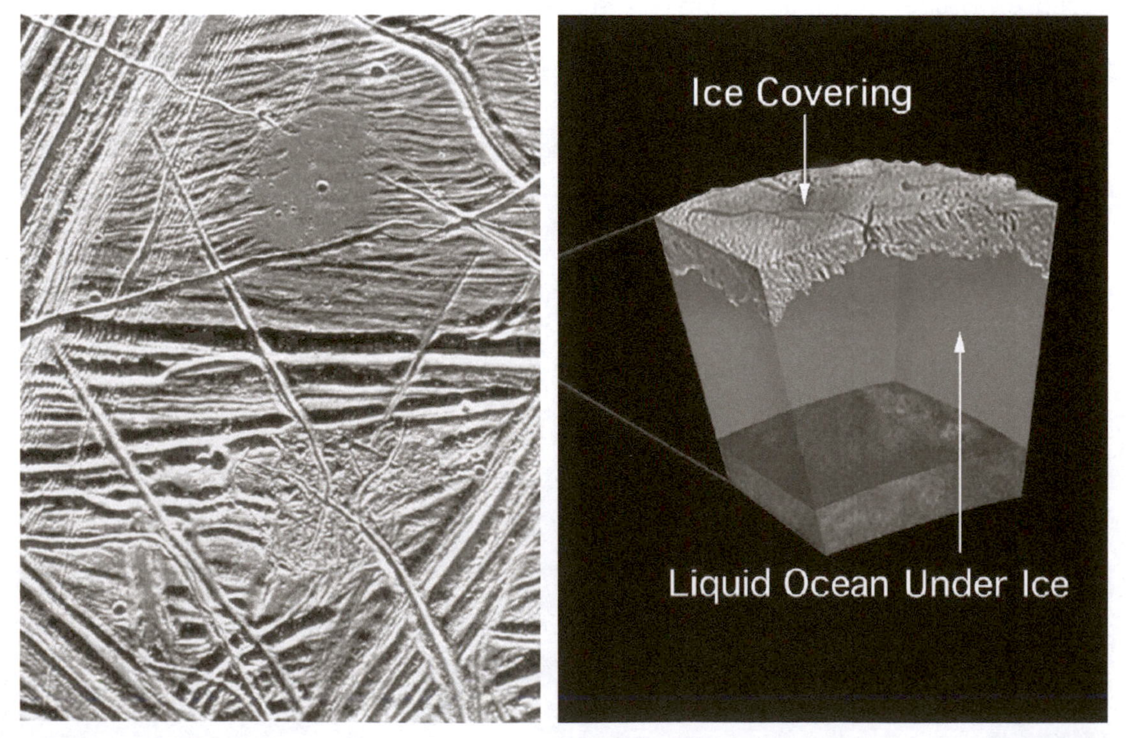

► **Figure 22-30** *Left:* Europa has liquid water seeping through its icy surface. (Source: NASA/JPL) ***Right:*** This, as well as the flatness and the youth of Europa's surface and its magnetic field, all support the idea that under the icy surface may be an ocean of liquid water, which is kept warm by Jupiter's gravity. (Source: NASA/JPL)

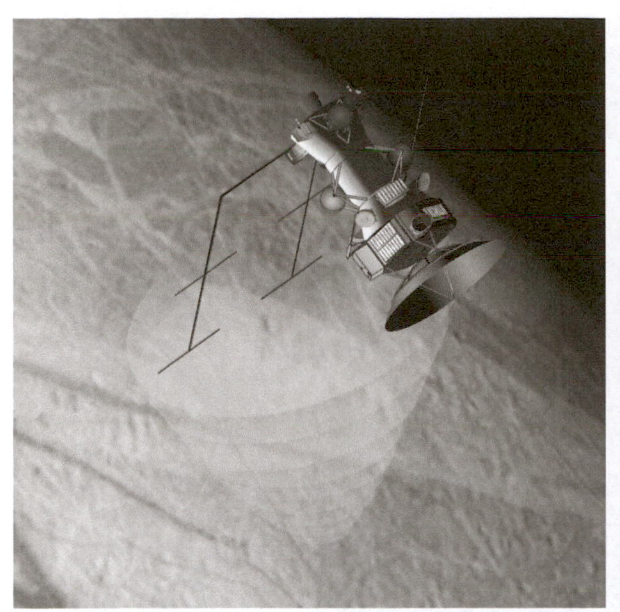

► **Figure 22-31** With liquid water, heat, and organic chemistry (known from spectra), what more could one want for life? It may be more likely that there is life on Europa than on Mars. *Left:* One spacecraft that has been planned to search for life on Europa would use radar that could sense through the ice, to confirm that there really is an ocean. It may conceivably find beds of coral or seaweed, if they're down there. (Source: NASA/JPL) *Right:* There are dreams of sending a lander with a probe to melt through the ice and explore the ocean, similar to robots used in Antarctica. (Source: NASA/JPL)

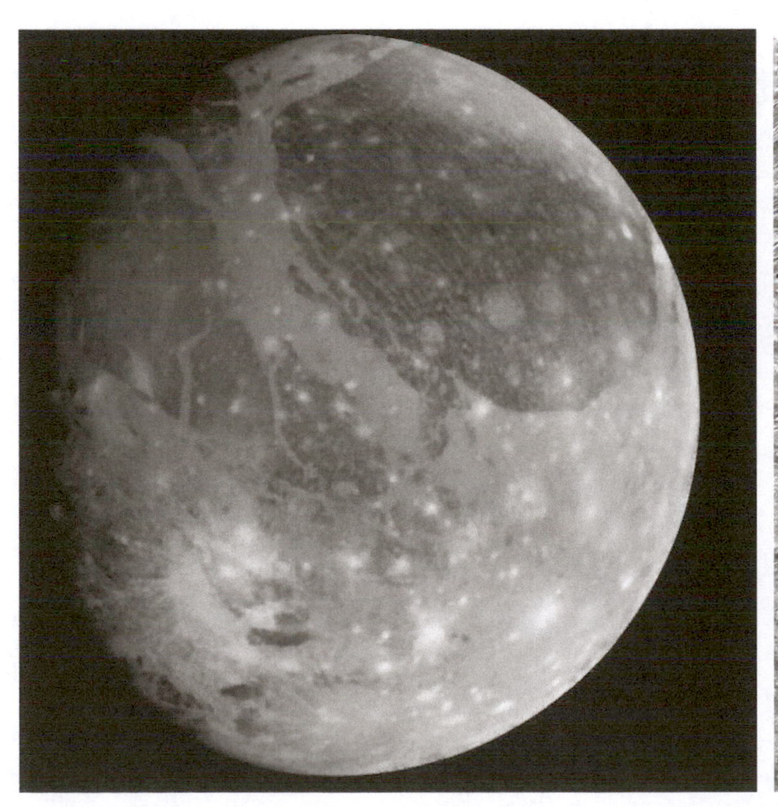

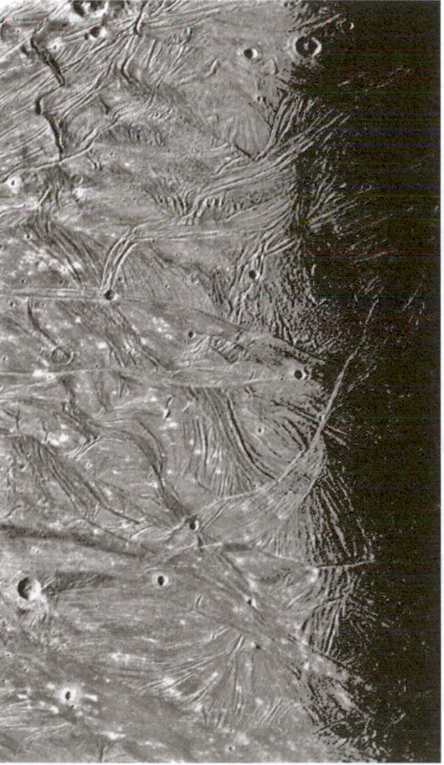

► **Figure 22-32** *Left:* Ganymede is the largest moon in the Solar System. It's a dirty snowball bigger than the planet Mercury. (Source: NASA/JPL) *Right:* Ganymede is about half rock and about half ice. It is the only solid body known in the Solar System other than Earth to have a form of plate tectonics, its grooved terrain shown here. (Source: NASA/JPL)

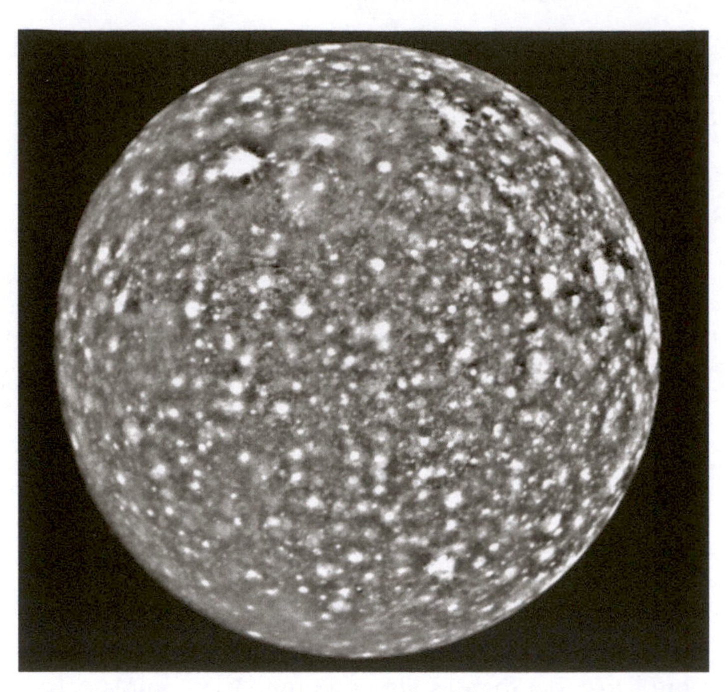

► **Figure 22-33** Callisto is the biggest snowball you may ever see. It is larger than Earth's Moon and slightly smaller than Mercury. It is also made of mostly ice, with an ancient, heavily cratered surface. A common theme in science fiction is that the aliens are interested in Earth because of its water. This is absurd. Aliens could help themselves to Callisto, with all the water anyone could ever need, in easily transportable solid form, and not have to deal with ornery humans and their nuclear weapons. (Source: NASA/JPL)

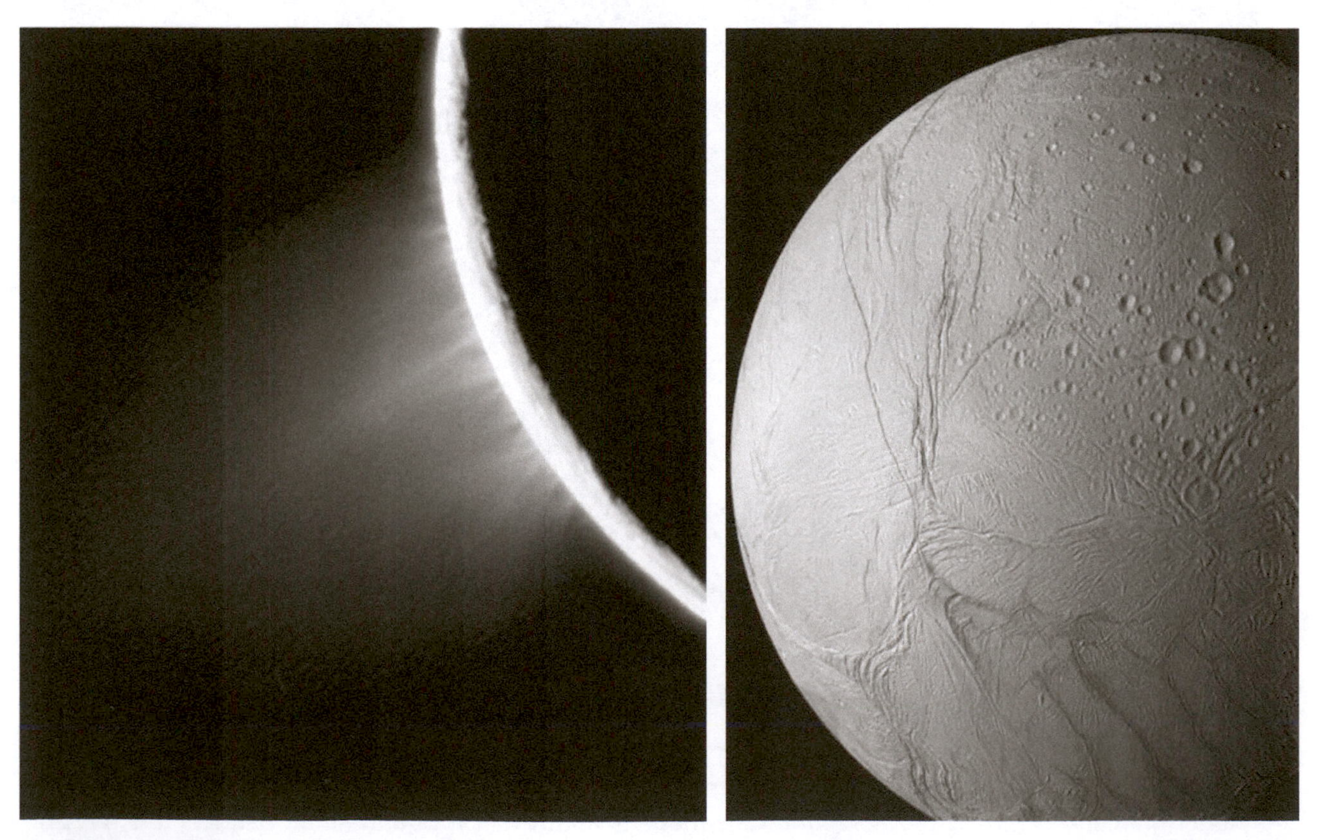

► **Figure 22-34** Enceladus (pronounced like "enchiladas") is a small, icy moon of Saturn. It has geysers of liquid water. Might it have life, under the ice? (Source: NASA/JPL)

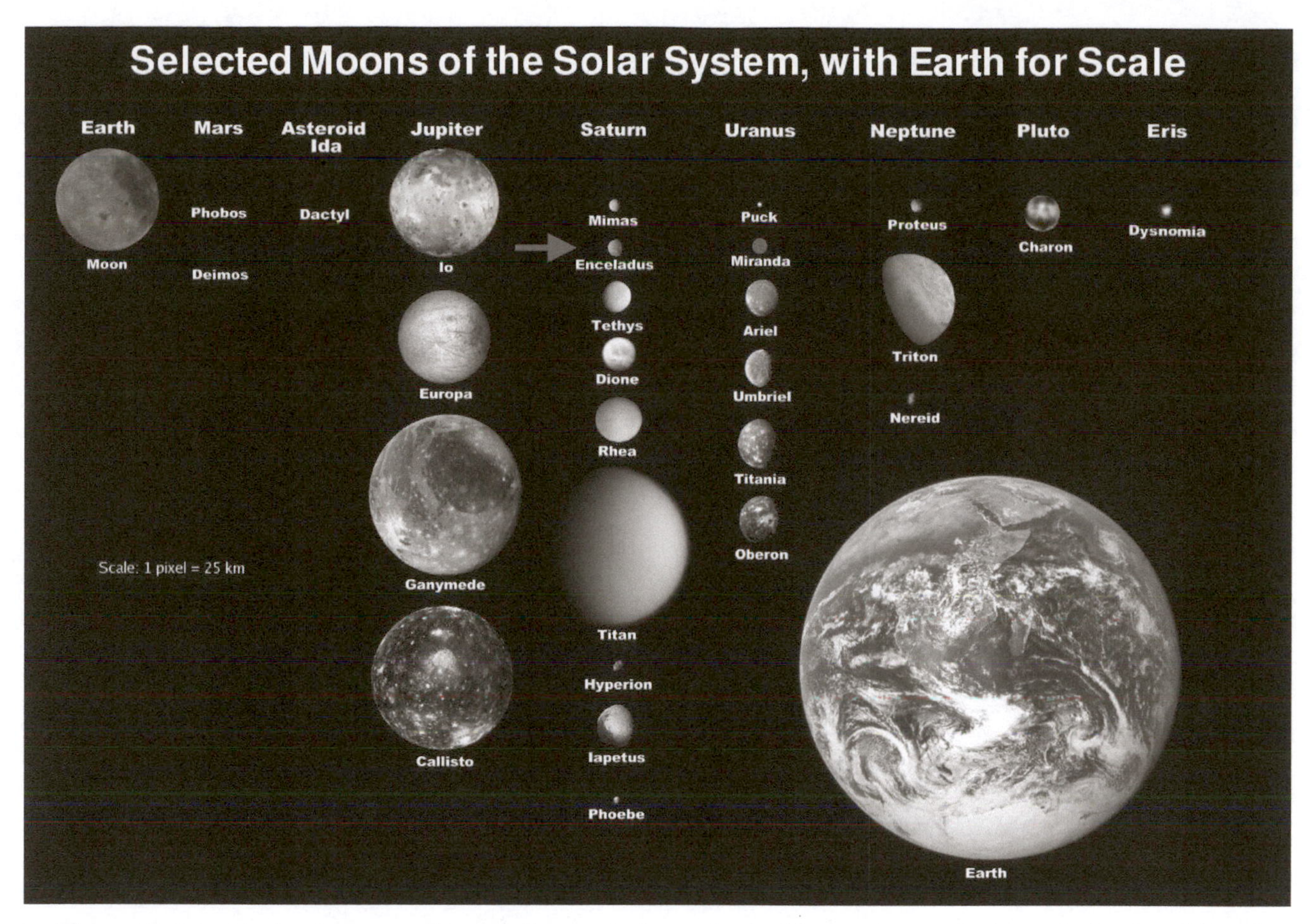

► **Figure 22-35** Notice how small Enceladus is. *(It is marked with a red arrow.)* If Enceladus has life, what other small, icy bodies in the Solar System contain life? If it turns out that Pluto has life in its icy interior, this author would think this was terribly funny. (Source: NASA)

CHAPTER

23

The Sun and Nuclear Energy

Proper Name

Sol (as in "the Solar System")

Size

1 solar radius = 109 Earth radii
(Jupiter radius = 11 Earth radii)

Temperature

Temperature at surface = 5,800 K ≈ 10,000°F, about as hot as a lightning bolt
Temperature at center (core) = 15 million K

Observing

Solar observing can be done safely, but one mistake and you're blind!

- Solar filters always look like mirrors, and always go in the *front* of the telescope tube.
- They NEVER go in front of the eyepiece. These are DANGEROUS and should be destroyed!
- Solar observing need not be a risk to one's eyesight, because of $100 webcams. These can be put at the eyepiece of a telescope—with a safe solar filter over the front of the telescope's tube, to prevent burning out the webcam!—and can display the image of the Sun on a laptop computer.

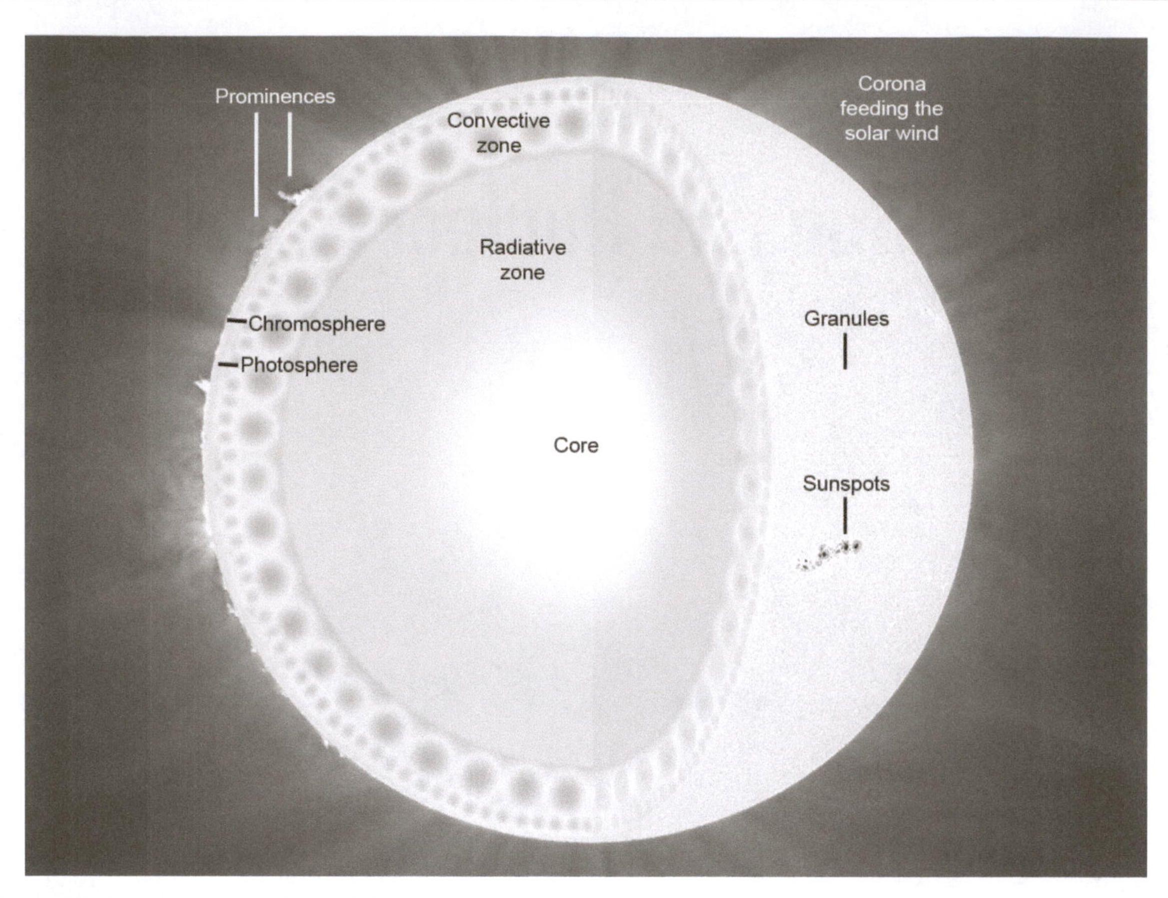

► **Figure 23-1** A cross-section of the Sun (Image courtesy of Frederick Ringwald, using images courtesy of Dr. Greg Morgan and by NASA)

► **Figure 23-2** *Left:* Safe solar filters must fit securely over the *front* of a telescope's tube. I like to duct tape them on, to prevent fools from removing them. Beware of anything that is fool-proof: fools can be very clever. Never use solar filters that go over the eyepiece only: these are dangerous and should be destroyed. *Right:* Safe solar filters look like a mirror, so that less than 1/1000th of 1% of the light gets into the telescope in the first place. Every time before using a solar filter, check it for scratches or holes in the reflective coating, and do not observe the Sun with it if there are. (Images courtesy of Frederick Ringwald)

▶ **Figure 23-3** Notice the webcam in this dedicated solar telescope's eyepiece holder. There is really no reason to risk harm to anyone's eyes these days, because of inexpensive webcams. These can fit into the eyepiece holder of a telescope, and can project an image of the Sun onto the screen of the laptop computer at left. One advantage of this is that more than one person can observe the Sun at the same time, by looking at the laptop's screen. Another advantage is that, even if you screw up, you burn up a webcam—and not a human eye. (Image courtesy of Frederick Ringwald)

▶ **Figure 23-4** Observing the Sun at the Downing Planetarium with the Central Valley Astronomers, Fresno's astronomy club. Notice that none of them are looking through telescopes at the Sun: the one at left is looking at the Moon. The others are projecting images of the Sun onto laptop computers and portable TVs. (Image courtesy of Frederick Ringwald)

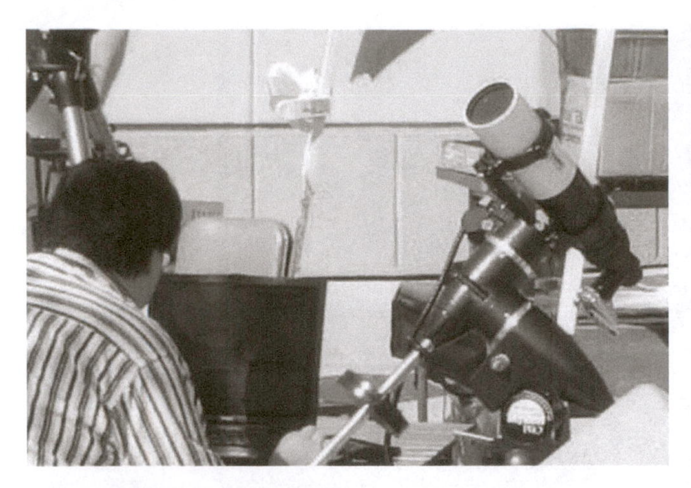

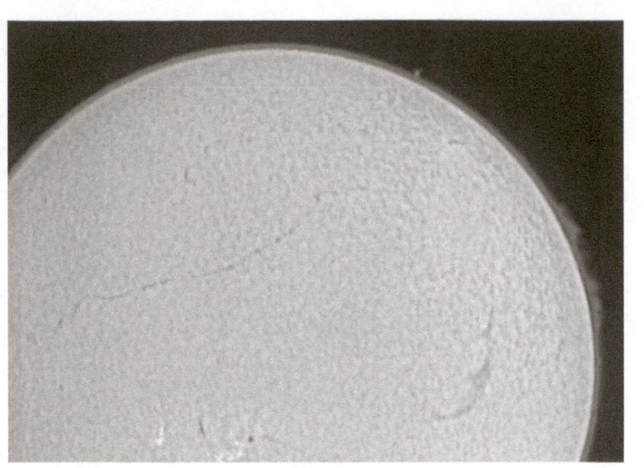

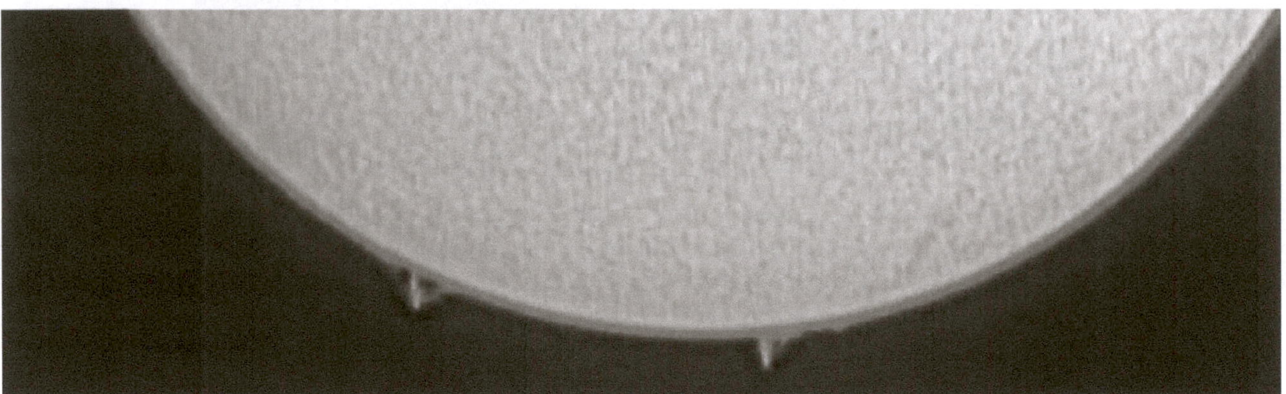

► **Figure 23-5** Safe solar observing (with no eyepiece!) can be done with a webcam and a dedicated solar telescope. The images at top right and at bottom were taken with the dedicated H-alpha solar telescope, webcam, and laptop computer shown at top left, about two minutes before the picture at top left was taken. (Images courtesy of Frederick Ringwald)

► **Figure 23-6** Professional solar observers have their own specialized instruments. This is the McMath-Pierce Solar Telescope on Kitt Peak, Arizona. It extends deep underground, so that the sunlight it catches is spread out. This gives a detailed solar image and also reduces the effects of heating from so much sunlight. (Image courtesy of Frederick Ringwald)

▶ **Figure 23-7** The McMath-Pierce Solar Telescope is shown here just before sunrise. This was the photo that convinced the author that he really could do good photography, since it is just a snapshot done with a $50 camera. It helps a lot to be able to see something that will make a good photograph. (Image courtesy of Frederick Ringwald)

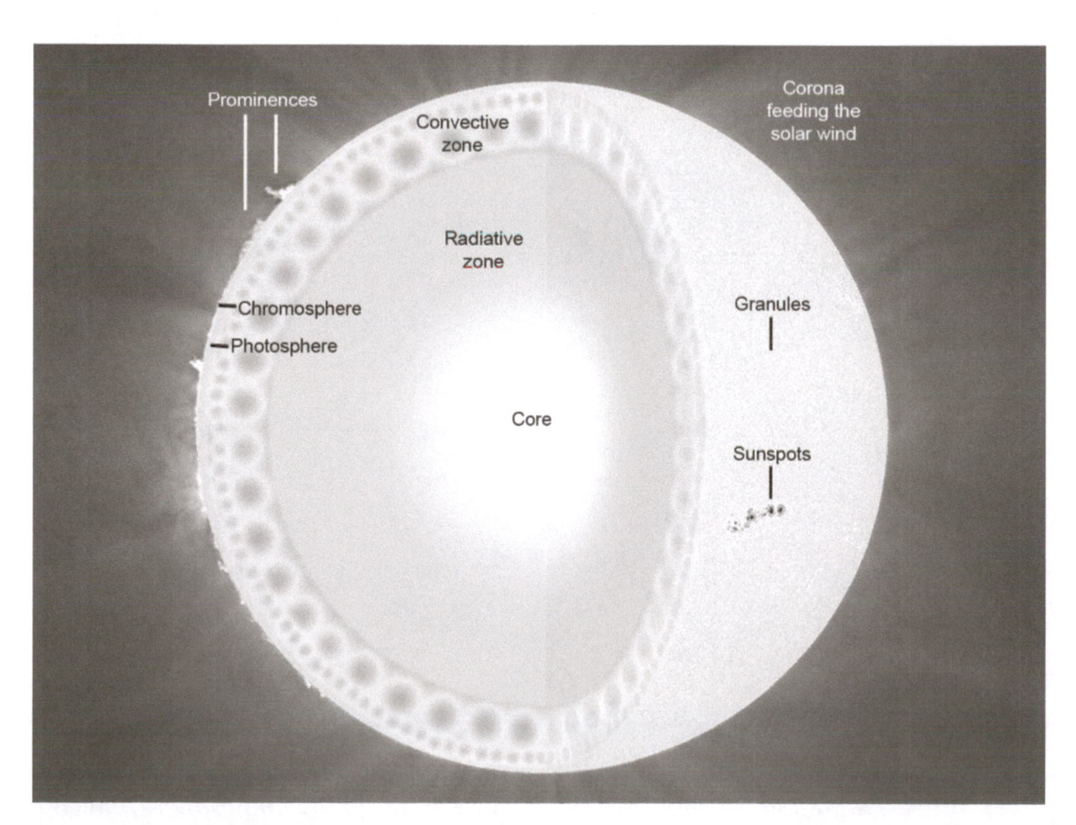

▶ **Figure 23-8** Let us now take a tour through the Sun, from its core to its outer layers. (Image courtesy of Frederick Ringwald, using images courtesy of Dr. Greg Morgan and NASA)

▶ **Figure 23-9** How do we know what's inside the Sun? We have ways of observing it, both directly and indirectly. *Neutrinos* are ghostly particles made by nuclear reactions. One trillion neutrinos pass through every square centimeter of you every second. This image was made with solar neutrinos. That's right, this is a picture of the fires in the heart of the Sun. (R. Svoboda, Univ. of California, Davis [Super-Kamiokande Collaboration]) *Indirectly: Sound waves* are observed at the Sun's surface. They carry information from its interior, similar to how seismic waves show what's inside Earth. This is called *helioseismology.* The Sun's interior has now been mapped all the way to its core.

▶ **Figure 23-10** These are granules in the Sun's visible surface, called the *photosphere*, which means "light sphere." They are bubbles that are rising up from the Sun's interior, like bubbles in boiling water. Each one is about 1500 km (1000 miles) across, about the size of the great state of Texas. They come and go in about 15 minutes. (Source: NASA/Solar Dynamics Observatory)

► **Figure 23-11** The red *chromosphere* (color sphere) is the Sun's atmosphere. An atmosphere is a layer of transparent gas held to a surface by gravity. Earth's atmosphere (which we are breathing) is *much* cooler than the Sun's atmosphere! (Image courtesy of Greg Morgan)

► **Figure 23-12** The solar corona is the hot gas escaping from the Sun. Some texts say that the corona is the Sun's atmosphere. This is incorrect, because the corona is not bound to the Sun by its gravity. The corona is more like a wind, that is escaping from the Sun. The corona is a million times fainter than the photosphere, so it can be seen only during a total solar eclipse (and through special instruments). It changes in shape over an 11-year cycle: at left it is shown on 1980 February 16, and at right it is shown on 1988 March 18. (Image © Vladimir Wrangel, 2013. Used under license from Shutterstock, Inc.)

Sunspots

Sunspots are caused by strong magnetic fields in the solar surface, or photosphere. Sunspots come and go over an 11-year cycle.

The last solar maximum was in 1999–2001. This was "the real Y2K problem," as solar flares are most common during Solar maximum.

Earth's atmosphere and magnetic field protect people on Earth from many of these effects, but not spacecraft.

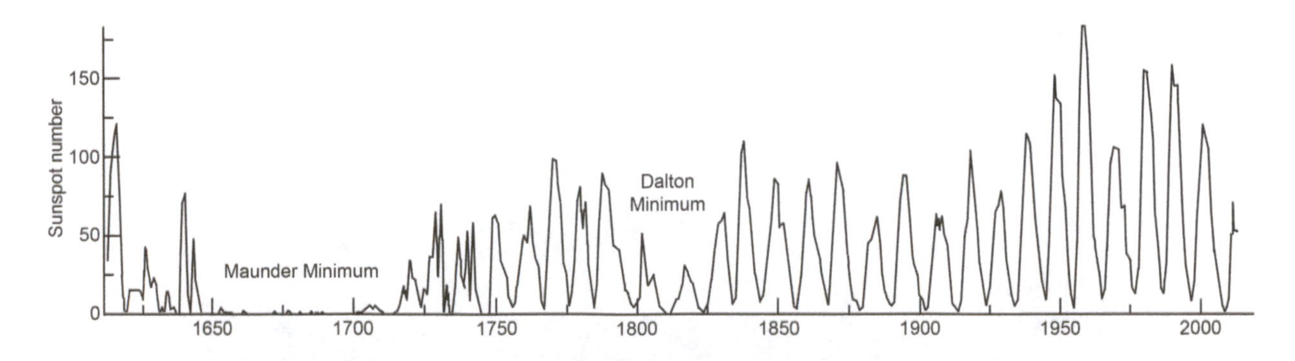

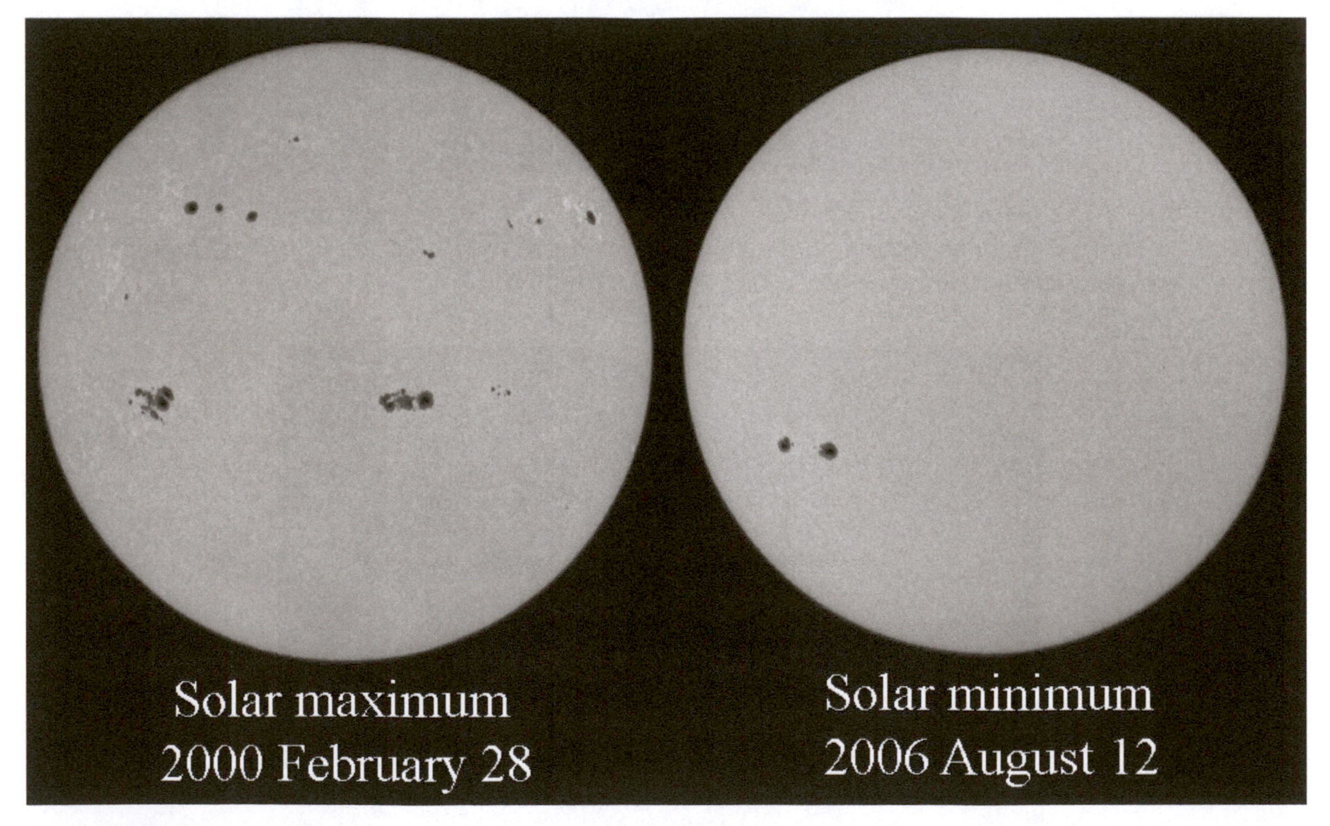

► **Figure 23-13** *Top:* Sunspots also become common and uncommon over the 11-year solar cycle. The cycle is actually 22 years long, because it is caused by the Sun's magnetic field reversing itself. During the Maunder Minimum from 1645 to 1715, the solar cycle apparently shut down, with almost no sunspots observed. At the same time was a period of unusually cold weather, globally. The Dalton minimum from 1790 to 1830 also coincided with cold weather globally. In 1816, "the Year Without a Summer" led to famine. (Image courtesy of Frederick Ringwald with data from the U.S. Naval Research Laboratory and NOAA-NWS Space Weather Prediction Center) *Bottom:* The image at left shows how common sunspots are during solar maximum. The image at right shows the Sun at solar minimum, with few sunspots halfway through the 11-year cycle. (Source: NOAO/ESA/SOHO)

The Sun in X-rays shows active regions. These are storms of million-degree gas, which appear directly above the sunspots. The Sun has a "Dr. Jekyll and Mr. Hyde" personality. In the story, "The Strange Case of Dr Jekyll and Mr. Hyde," by Robert Louis Stevenson, Dr. Jekyll takes a drug and becomes Mr. Hyde, his evil part, which he names "Mr. Hyde" since Dr. Jekyll feels obliged to hide him. Mr. Hyde is much smaller than Dr. Jekyll, because he is only part of Dr. Jekyll.

The Sun radiates 97% of its energy in visible light. Images of the solar photosphere, look calm. The Sun radiates only a small part of its energy in high-energy X-rays, and it's violent.

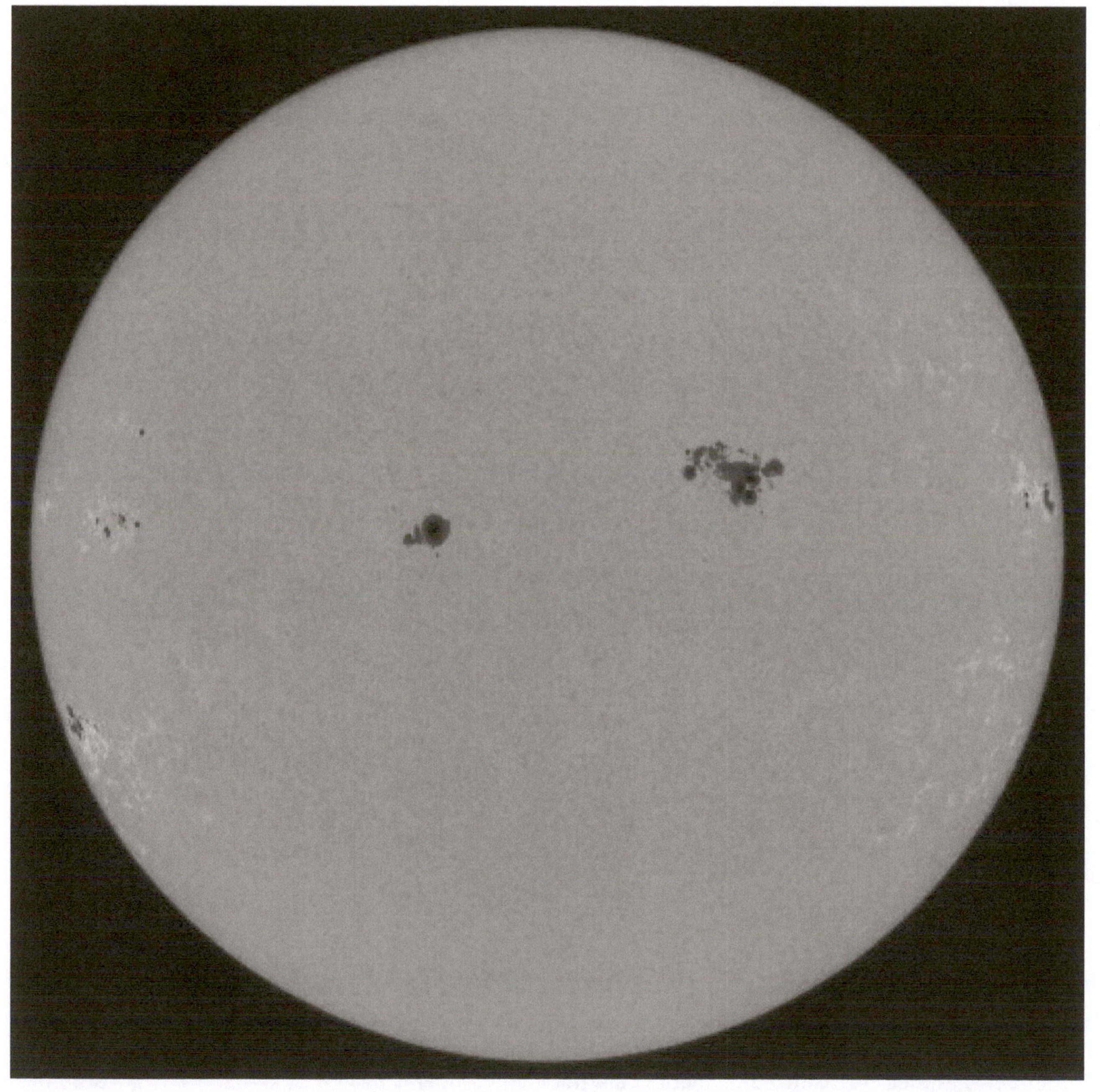

► **Figure 23-14** The solar photosphere shown here in visible light shows sunspots. (Source: NOAO/ESA/SOHO)

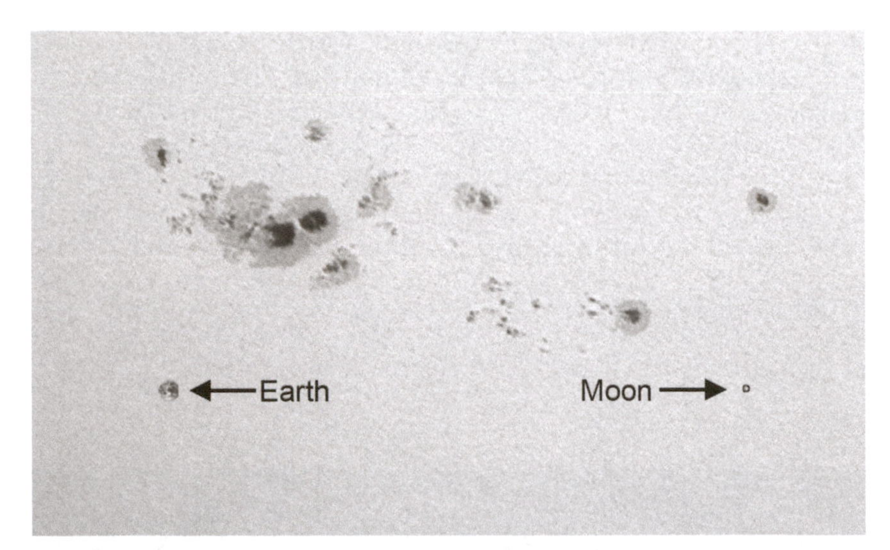

► **Figure 23-15** Sunspots are like Moses parting the Red Sea. Sunspots are places on the Sun's photosphere where the Sun's magnetic field bunches up and pushes away gas in the Sun's surface. This makes sunspots a few hundred degrees cooler than their surroundings. Sunspots therefore look dark because they are silhouetted: they are still quite hot. To show how large sunspots can be, here is a picture of a big sunspot group alongside of images of Earth and the Moon. The sizes and the distance between Earth and the Moon are to scale with the sunspots. Earth and the Moon are never this close to the Sun, of course. (Image courtesy of Frederick Ringwald)

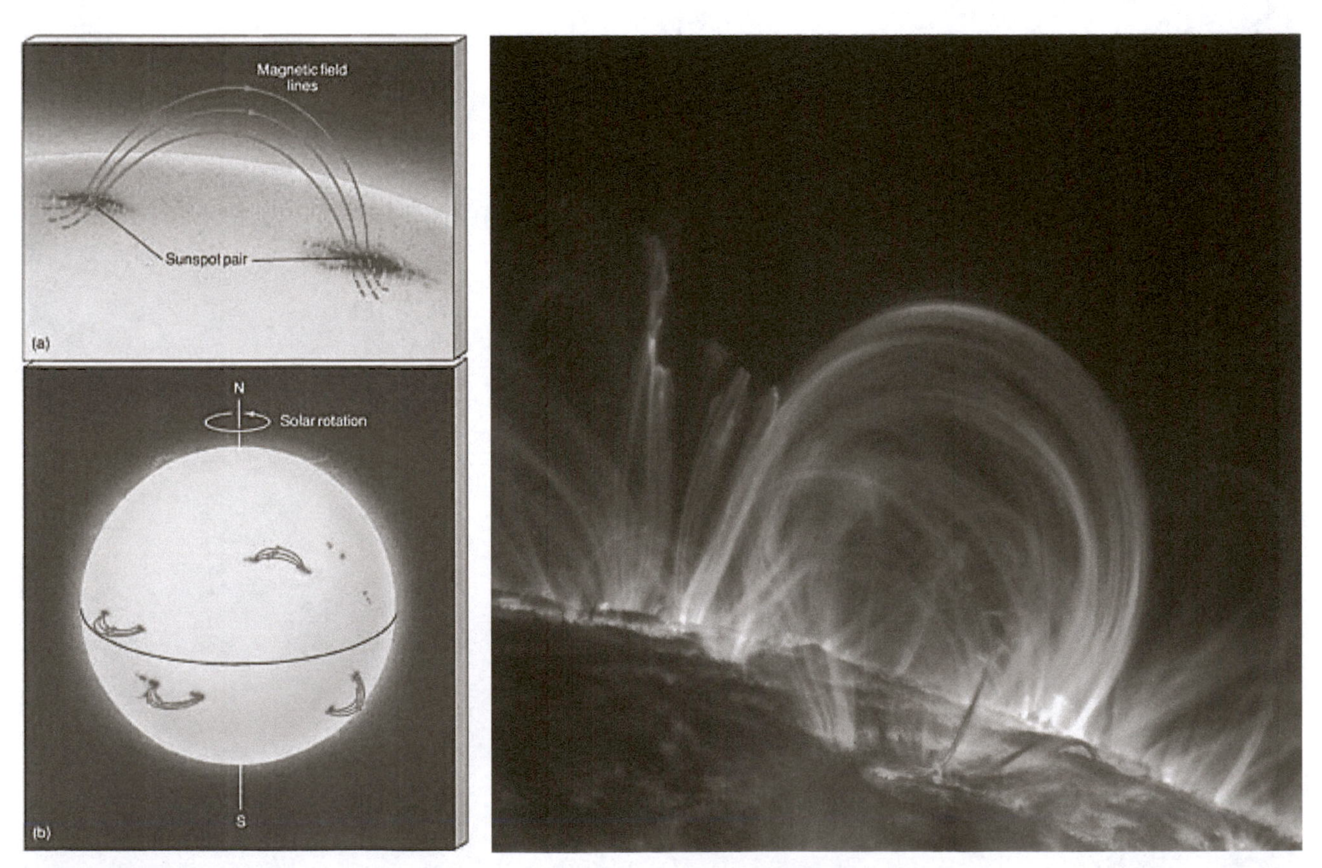

► **Figure 23-16** Sunspots are related to *prominences,* in the Sun's magnetic field. Prominences look like loops, or flames, of hot gas. Prominences often form with pairs of sunspots, with the sunspots' north and south poles at their feet. Since prominences are hot, thin gas that is heated by the Sun's magnetism, they are buoyant, and can rise over a million miles into space. (Source: NASA)

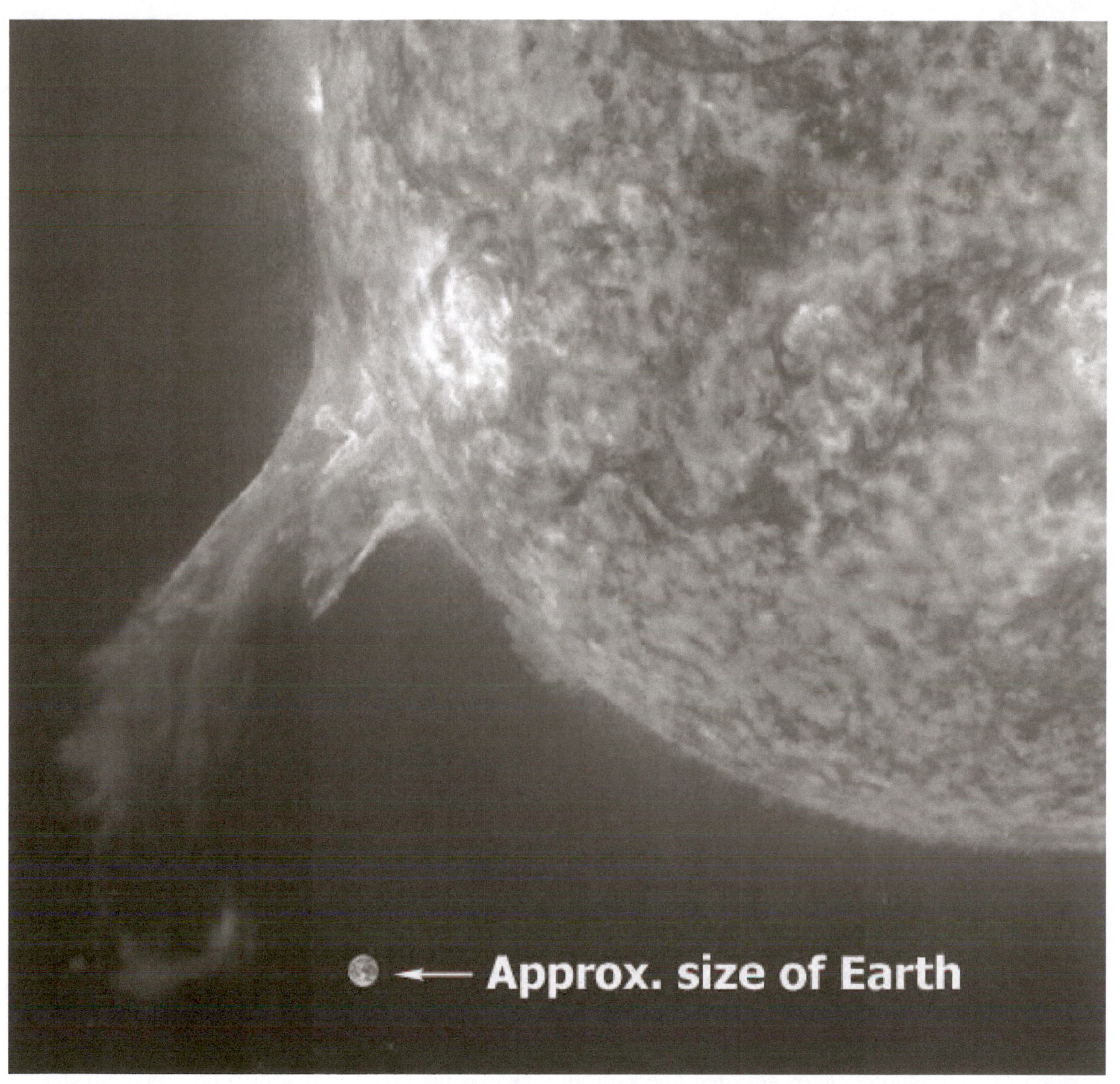

▶ **Figure 23-17** Prominences are streams of gas. They rise into space over several hours. Don't confuse solar prominences, shown here, with solar flares. Prominences are leisurely, and fun to watch over the course of an afternoon. Flares are explosive, and last only a few minutes. (Source: NASA/ESA/SOHO)

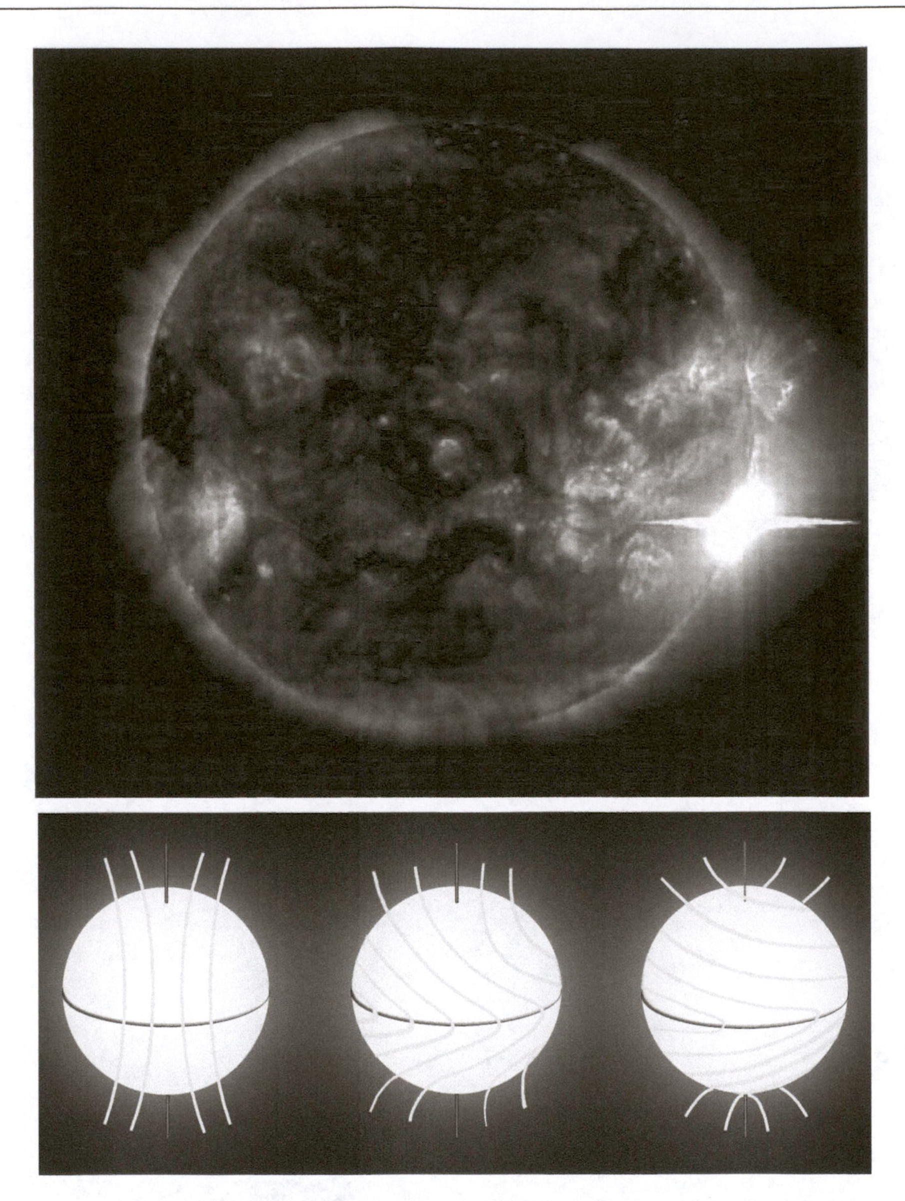

► **Figure 23-18** *Top:* A solar flare is explosive. Don't confuse them with solar prominences: flares are ten times more energetic and ten times faster, so they are 100 times more powerful than prominences. (Source: NASA/ESA/SOHO) *Bottom:* Flares get their energy from magnetic field lines that stretch until they break. The Sun is not solid but gas, so it rotates faster around its equator (about once per 25 days) than around its poles (about once per 30 days). This causes its magnetic field lines to stretch and eventually break, like rubber bands. (Image courtesy of Frederick Ringwald)

► **Figure 23-19** Solar flares cause *coronal mass ejections* (CMEs). (Source: NASA/ESA/SOHO)

Effects of Radiation from Solar Flares

- It makes aurorae. (The upper atmosphere glows like a fluorescent lamp.)
- It's a health danger to astronauts, airplane flight crews, and frequent flyers. Pregnant women should avoid flying, particularly over the poles.
- It ages spacecraft electronics: satellites have failed after solar flares.
- It charges up high-voltage power lines on Earth, causing electric blackouts.
- It makes Earth's uppermost atmosphere, called the ionosphere, reflective.
- This can do weird things to radio signals. Local taxi drivers may become unable to talk to their dispatcher, but able to talk to the one in Kansas City. Cell phones, pagers, and GPS can also be affected.

Radiation can't be seen, heard, smelled, or tasted. But what *is* it, exactly?

1. Charged particles, including protons, electrons, and alphas;
2. Neutrons;
3. High-energy electromagnetic radiation, including gamma rays, X-rays, and ultraviolet radiation.

All are so energetic they *ionize* matter, including *your cells*.

High-energy radiation passing through a human body is like a stream of little bullets, killing one cell at a time. This is what causes radiation sickness.

Radiation can cause cancer, because it disrupts the normal chemical processes in cells, turning them into cancer cells. Ironically, radiation used carefully can cure many cancers. This is because tumors, which are masses of cancer cells, are denser than healthy cells. Tumors therefore absorb more radiation, and so can be burned away by it.

Microwaves are *not* ionizing radiation: they're not from nuclear reactions. Microwave ovens are sometimes called "nukers." This is incorrect, since they have nothing to do with nuclear energy. Microwaves *are* radiation, but so is visible light. So is the infrared radiation that a conventional oven uses. Still, don't use a microwave oven that has obviously been damaged, because the microwaves can burn you!

Microwaves are absorbed by water, which makes them useful for cooking food. They don't heat the plate the food is on, because there is little water in the plate. Microwaves are useful for cooking because they heat food so efficiently: little of their energy is wasted, because most foods are full of water. Microwaves can't cause cancer any more than radio waves from cell phones can, because they just don't have enough energy: microwaves aren't *ionizing* radiation.

A problem with ionizing radiation is that *human eyes can't see it*. Objects can be dangerously radioactive well below being radioactive enough to "glow in the dark." Health problems can begin well before this happens: no human being has ever been radioactive enough to glow in the dark, not even a dead one. People don't turn into mutants, as in bad science fiction. Ionizing radiation can cause stillbirths and mental defects in children born after the radiation exposure.

Radiation doses:

0.1 rem: is a normal yearly sea-level dose, from cosmic rays and natural radioactivity in rocks. Twice as much is normal for Denver, "the mile-high city."

1 rem: is a typical dose from crossing Earth's radiation belts—for example, by astronauts on their way to the Moon. This is also the maximum dose allowed for nuclear workers for a year. Alan Shepard, commander of Apollo 14 in 1971, died of leukemia (blood cancer) in 1998. He never denied that it might have been caused by the radiation exposure he got during his trip to the Moon.

40 rem: is a typical dose for a 2.5-year Mars expedition, outside Earth's magnetic field. A human body can heal itself over time: this level of radiation elevates the risk of cancer about 1% per year, about as much as if the astronauts had stayed on Earth and smoked cigarettes for 2.5 years.

75–200 rem within 30 days: At such high radiation doses, a human body can't repair cell damage fast enough. Radiation sickness results: symptoms include nausea, vomiting, chronic fatigue, diarrhea, hair loss, and for heavy doses, skin sores, unconsciousness, and death.

500 rem, promptly, is a fatal dose. Over 6000 rem is an instantly fatal dose.

Coronal Mass Ejections can emit over 2000 rem/hour. One happened two weeks after Apollo 16, in 1972. It would have killed the astronauts if they'd been on the Moon when it happened. Any spacecraft to take astronauts to Mars must protect them from radiation.

▶ **Figure 23-20** Radiation storms travel out into *the solar wind*. There is no clear physical boundary between the Sun's corona and the solar wind: it's only a semantic difference. The solar wind is mainly made of protons and electrons. This radiation mostly passes around Earth because of Earth's magnetic field. Some protons and electrons from the solar wind leaks into Earth's north and south magnetic poles. This causes auroras. (Source: NASA)

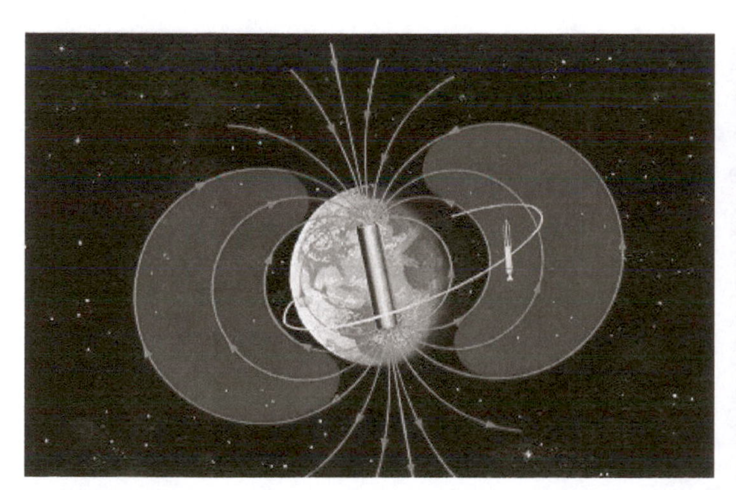

▶ **Figure 23-21** The Van Allen radiation belts are around Earth. *Left:* This radiation is mostly protons from the solar wind, trapped by Earth's magnetic field. (Source: NASA) *Right:* They were discovered in 1958 by America's first successful satellite, *Explorer 1*. (Source: NASA) Jupiter and the other outer planets also have trapped radiation belts, but which are much more powerful.

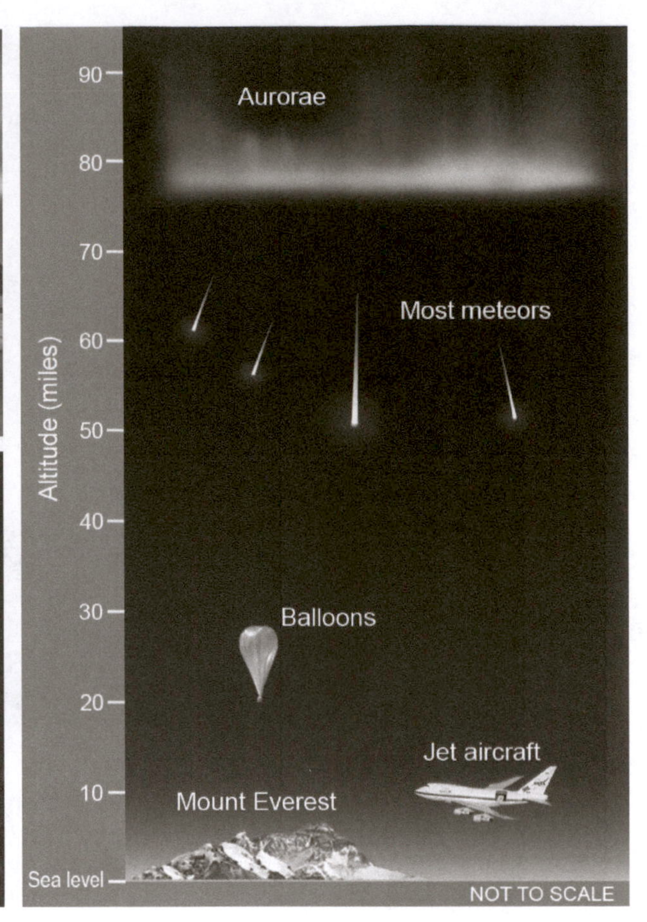

▶ **Figure 23-22** *Top left:* Solar radiation storms light up the top of Earth's atmosphere, making the *northern lights*, also called the *aurora borealis*… (Image © John A. Davis, 2013. Used under license from Shutterstock, Inc.)

Bottom left:…or the *southern lights*, also called the *aurora australis*. This was taken by an astronaut on a Space Shuttle, looking down on it from space, since the shuttles typically orbited 400 km (300 miles) above Earth's surface. Relatively few people have seen the Southern lights. This is because fewer people live in Antarctica than in Norway, Sweden, Russia, Canada, and Alaska.

The farther north one lives, the more often one sees an aurora borealis. Your present author used to see them often, sometimes nightly, when going to grad school in New Hampshire, during the solar maximum of 1988–1992. From the author's present address in Fresno, California, auroras are visible maybe once per decade, but I did see a red curtain aurora during a class observing session in the country about 50 km north of Fresno in 2001. (Source: NASA)

Right: Aurorae are higher than most of Earth's atmosphere. (Image courtesy of Frederick Ringwald, using NASA images)

How Do the Sun and the Stars Shine?

The four known forces of nature:

1. Gravity —Always attracts
2. Electromagnetism (electricity + magnetism + light) —Can attract or repel (Opposites attract)
3. Weak nuclear force (accumulates in heavy nuclei; e.g., uranium or plutonium, making them split, causing *radioactivity*) —Repels, but is weak
4. Strong nuclear force (binds nuclei together despite the like charges of their protons) —Attracts, but only inside nuclei

Nuclear physics for beginners (pronounced "NEW-clear")
The Sun generates energy by *nuclear fusion*: by joining atomic nuclei together at high temperature, in its core.

Hydrogen nucleus: mass = 1.007276470 amu
Helium nucleus: mass = 4.00150618 amu
(1 amu = 1 atomic mass unit = $1.6605402 \times 10^{-27}$ kg)

Notice that: $4 \times 1.007276470 = 4.02910588 > 4.00150618$

So, in other words: $4 \times \text{mass(hydrogen)} > \text{m(helium)}$

So: 4 hydrogens become 1 helium **+ energy**, since $E = mc^2$

We know this because we can reproduce fusion in laboratories—and in nuclear weapons.

This is *unlike* how a commercial nuclear reactor works, by nuclear *fission*.
Fission liberates energy by splitting large nuclei, such as uranium or plutonium.

Nuclear weapons

- "Atom" bombs, used on Hiroshima and Nagasaki in 1945, used *fission*.
- "Hydrogen" bombs, or thermonuclear weapons
  These, first tested in 1952, are 1000 times more powerful. They use *fusion*, with a fission bomb to trigger fusion reactions in hydrogen.

The primary effect of a nuclear weapon is the blast, which can knock down buildings for miles. The secondary effect is the tremendous heat of the fireball: one weapon can produce tens of thousands of severe burn injuries. The tertiary (third) effect is radiation, which can last for years.

Let us hope they will never be used again!

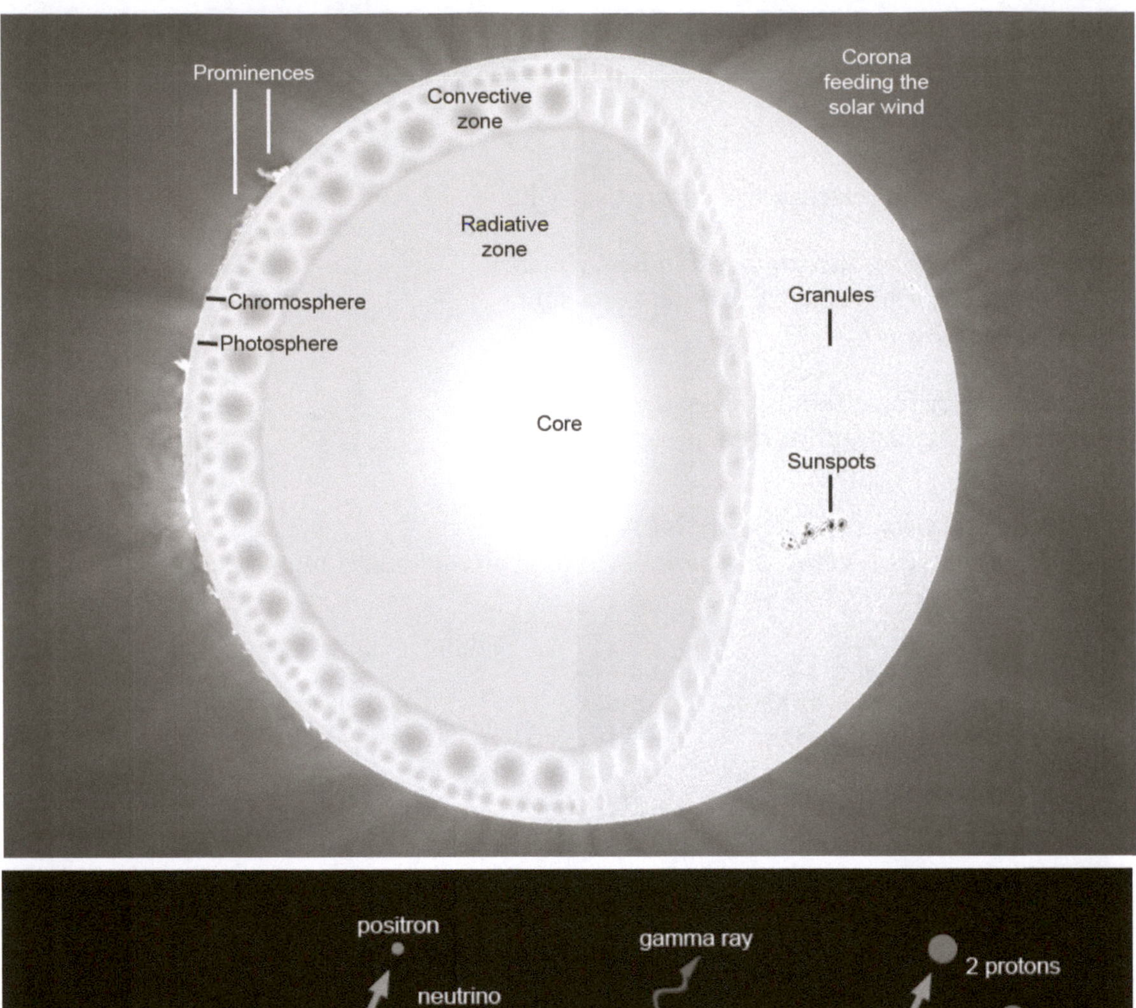

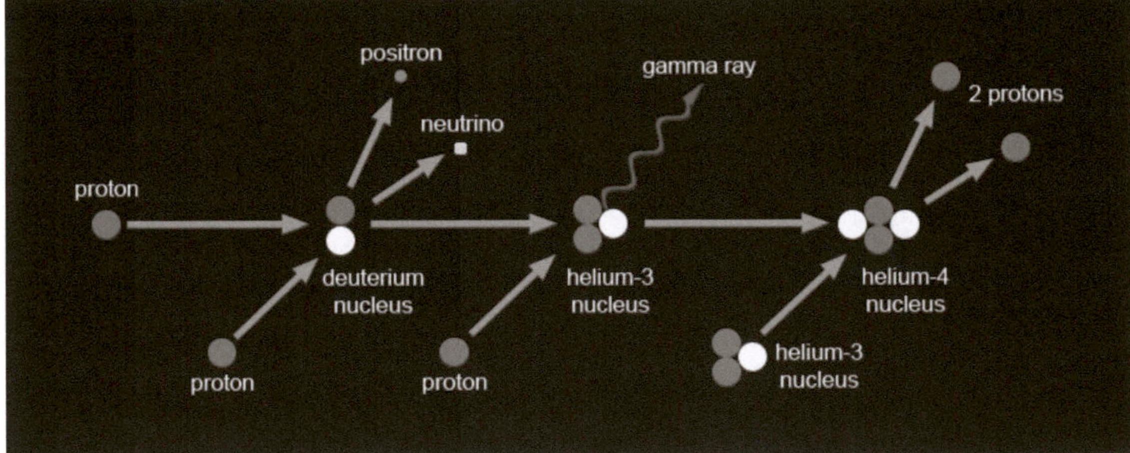

► **Figure 23-23** **Nuclear fusion** is how the Sun and the stars shine. The Sun's core is hot enough so that nuclear fusion reactions can happen. In each of many reactions per second, four hydrogen nuclei join (or *fuse*) into one helium nucleus, which releases energy. Similar, more complicated reactions are how stars cook up all the elements heavier than helium, including the carbon, nitrogen, and oxygen that largely makes up you. (Images courtesy of Frederick Ringwald)

► **Figure 23-24** I wish we could use fusion practically, to make electricity. This is a photo of one experiment. (Source: DOE/Sandia National Labs)

► **Figure 23-25** *Left:* Einstein's most famous equation, $E = mc^2$, means that energy equals matter times the speed of light, squared. The speed of light is a large number: its square is an even larger number. *Since* $E = mc^2$, matter can turn into a huge amount of energy. This is what powers the stars for billions of years. Sadly, it is also how nuclear bombs can destroy whole cities. (Source: DOE) *Right:* Energy can turn into matter, too. A high-energy gamma ray can turn into an electron and its anti-matter counterpart, a positron. (Source: Brookhaven National Laboratory)

▶ **Figure 23-26** Einstein was horrified at how his work was put to use to develop nuclear weapons. (Left: © National Geographic Society/Corbis, Right: Source: DOE)

"If only I had known, I should have become a watchmaker."—Albert Einstein
"Human history becomes more and more a race between education and catastrophe."—H. G. Wells
"If knowledge can create problems, it is not through ignorance that we can solve them."—Isaac Asimov

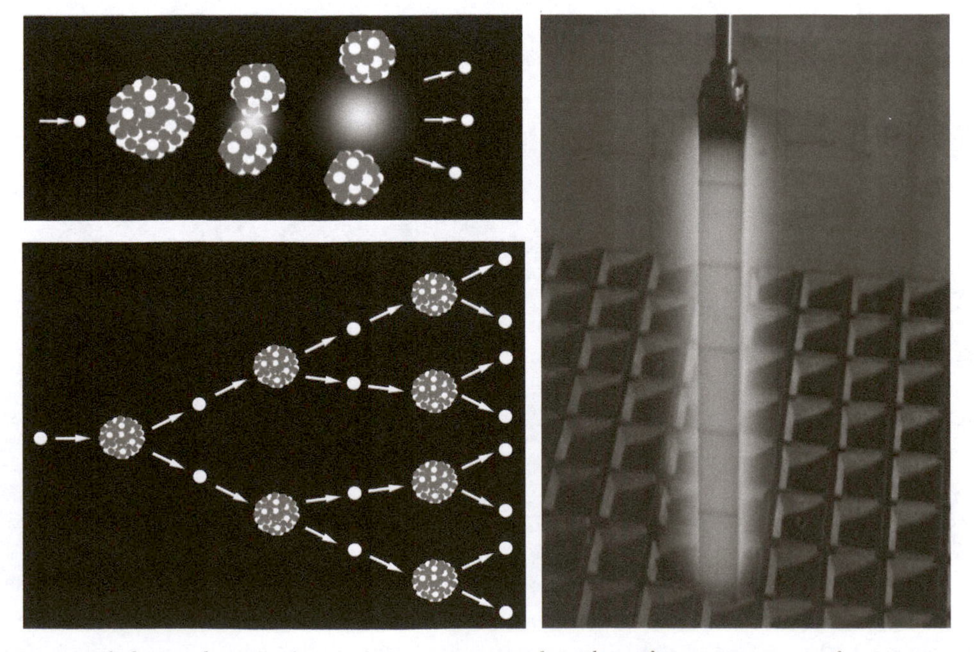

▶ **Figure 23-27** *Top left:* Nuclear fission is how a conventional nuclear reactor works. Heavy nuclei are split by neutrons. (Image courtesy of Frederick Ringwald) *Bottom left:* Uranium-235 or plutonium-239 can sustain a chain reaction, so that 1 kg of U-235 has the energy of 14,000 barrels of oil or 3,000 tons of coal. (Image courtesy of Frederick Ringwald) *Right:* Nuclear waste is radioactive, and needs to be stored safely for over 10,000 years. (© Tim Wright/CORBIS)

► **Figure 23-28** This is your present author at age 22, while serving in the U. S. Navy, at the conn of a nuclear submarine. Yes, that's really me. Yes, that's a genuine nuclear submarine. Uncle Sam is noted for his keen sense of humor, you know!

Movies sometimes show old-fashioned diesel submarines that run out of air. Nuclear submarines, which have been in use since 1954, are not like this. The air in a nuclear submarine is fresh and clean. It is kept that way by the electricity made by the nuclear reactor.

The reactor also provides all the power needed for the submarine to take over 100 sailors around the world dozens of times, with enough electricity to make fresh water for drinking and even showers. What limits how long a nuclear submarine can stay underwater is food for the crew, which has physical volume and can fill up the submarine's interior. Nuclear submarines can stay submerged for 6 months or longer, if necessary.

The nuclear reactor that provided the power for all of this was barely 60 centimeters (2 feet) in diameter, and 1 meter (3 feet) long. Since $E = mc^2$, the uranium fuel in the reactor didn't need to be very massive to generate a large amount of energy. (Image courtesy of the crew of the USS Ethan Allen SSN 608)

► **Figure 23-29** Thermonuclear ("hydrogen") bombs use fusion, triggered by fission. The "atom" bombs used on Hiroshima and Nagasaki were fission weapons. H-bombs are 1000 times more powerful. (Image courtesy of Frederick Ringwald using some DOE photos)

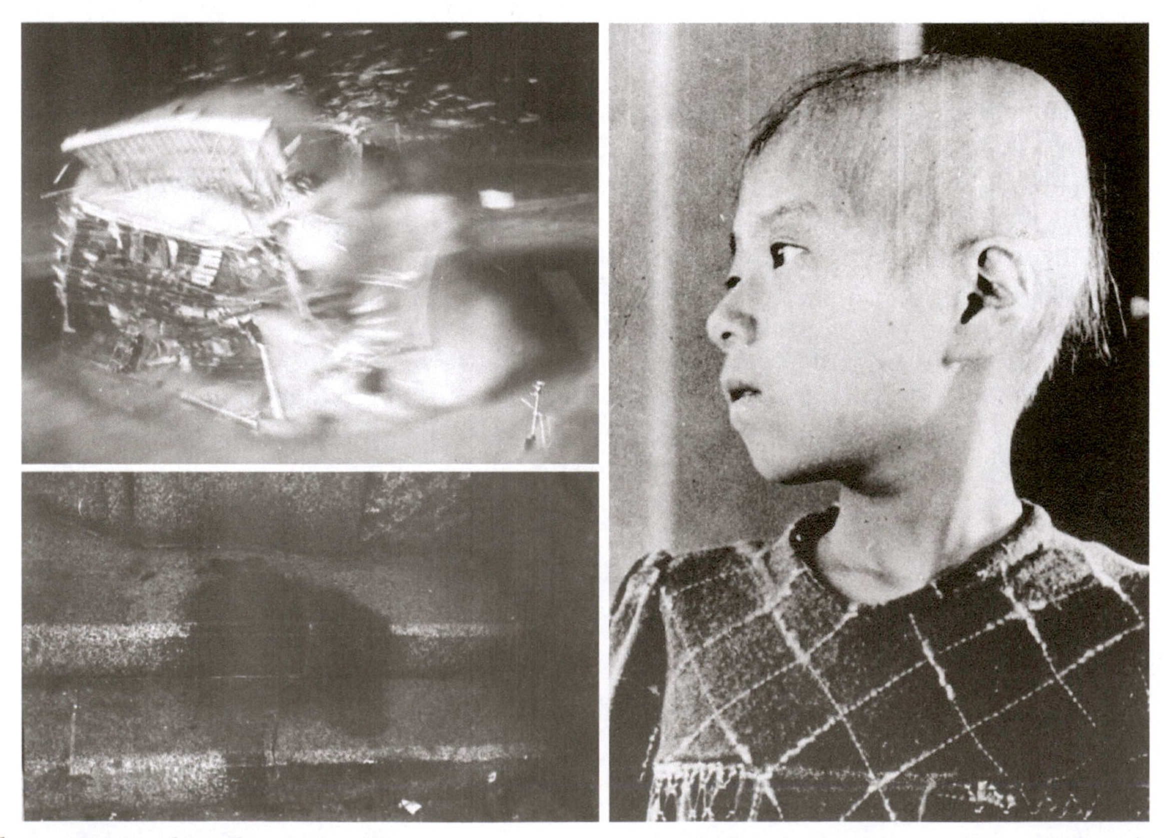

► **Figure 23-30** The effects of nuclear weapons: *Top left:* The first is the **blast,** which can knock down buildings miles away, (Source: DOE/NNSA/NSO) *Bottom left:* Second are the **burns,** which can set fires for miles (here all that is left of a person is a shadow on the pavement), (Source: DOE) *Right:* Third is the **radiation.** About half the people who died in the Hiroshima and Nagasaki bombings in 1945 died of radiation because they didn't know to get away from it. (Source: DOE)

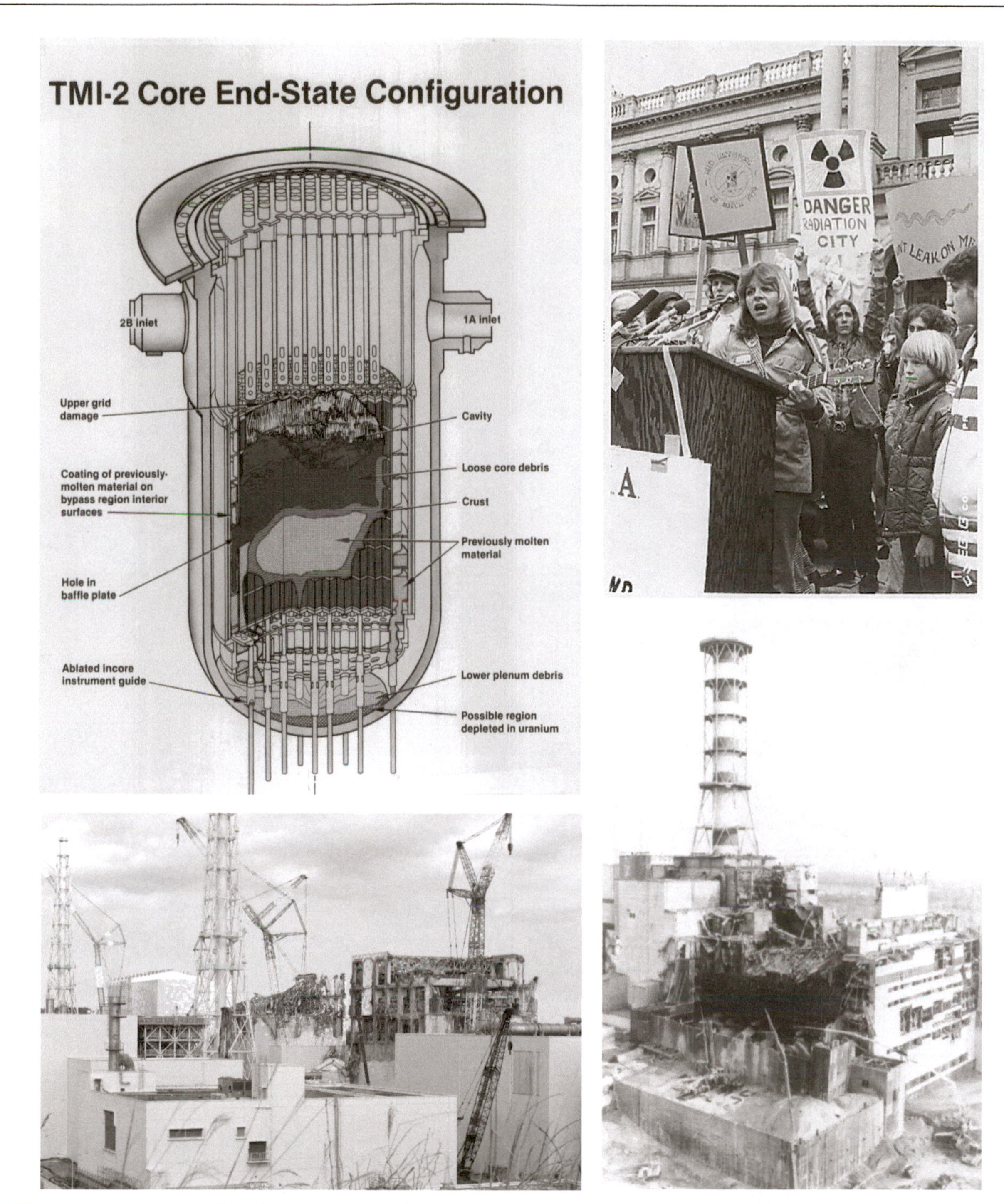

► **Figure 23-31** Safety has been a particular concern for commercial nuclear energy. Both Three Mile Island (in 1979 in Pennsylvania) and Chernobyl (in 1986 in the Ukraine) had loss-of-coolant accidents due to operator error. *Top left:* Three Mile Island partly melted down but had no fatalities, because its containment building held. (Source: NRC) *Top right:* There was a loss of public trust for the Three Mile Island operators. In particular, in response to an aggressive line of questioning by a news reporter, a utility vice president said, "I don't know why we need to tell you each and every thing that we do." The U.S. civilian nuclear power industry still hasn't recovered from this. (Source: NARA) *Bottom left:* Chernobyl had no containment vessel. It had a large radiation release over Europe. Cancers and deaths continue to this day. (© Mainichi Newspaper/AFLO/AFLO/Nippon News/Corbis) *Bottom right:* The Fukushima accident in 2011 in Japan was four loss-of-coolant accidents, but due to a natural disaster of an earthquake and tsunami. It had explosions and fires with 7 deaths, and is ongoing. (Source: International Nuclear Safety Program/DOE)

► **Figure 23-32** Nuclear fission power has had four problems since its inception in the 1950s: 1. Safety; 2. Waste disposal; 3. Security and weapons proliferation; 4. Cost, which has been the real killer. Civilian nuclear power reactors cost $6–12 billion.

No new civilian nuclear power plants have been completed in the U.S. since the Three Mile Island accident, in 1979. (Image © axyse, 2013. Used under license from Shutterstock, Inc.)

Chapter 24
The Stars

Apparent magnitude (m) is how bright a star *looks*.

Star	m
Sun	– 26.8
Sirius	– 1.43
Vega	0.0
Rigel, Betelgeuse	0.3, 0.7 (variable)
Polaris, Orion's Belt, Big Dipper	2–3
Unaided eye limit in a dark, country sky	≈ 6.5
8-inch telescope limit	13
Hubble Space Telescope limit	31–32

This is a logarithmic scale: +5 magnitudes = 100 times fainter,
+1 magnitude ≈ 2.5 times fainter

Absolute magnitude (M) is the apparent magnitude a star *would* look, if it were a standard distance of *10 parsecs* (32.6 light-years) away. Therefore,

$$m - M = 5 \log d - 5 \quad \text{or} \quad d = 10^{\left(\frac{m-M+5}{5}\right)}$$

Absolute magnitude therefore measures how powerful a star *really is*, not just how bright it *looks*.

It therefore measures a star's **luminosity**, which is *its total power output*.

1 parsec = 3.26 light-years. This unit of distance comes from how astronomers measure stellar distances, by *parallax*. Parallax proves that Earth really does move around the Sun.

Telescopes accurate enough to measure parallax didn't appear until 1838. This was only three years after Galileo's books were taken off the *Index of Prohibited Books*, and over 200 years after his trial.

► **Figure 24-1** Hipparchos of Nicaea, in the 2nd century BC, made the earliest-known map of the stars. On it, he defined *apparent magnitudes*. The brightest stars he defined as 1st magnitude. The faintest stars (that he could see) he called 6th magnitude. (Image © rook76, 2013. Used under license from Shutterstock, Inc.)

► **Figure 24-2** Proxima Centauri is 4.2 light-years from Earth (about 25 trillion miles). (Recall that *one light-year* is the *distance* that light travels in one year, about 6 trillion miles.) Although light-years are useful for measuring distances between stars, astronomers prefer to use different units: **parsecs**. One parsec = 3.26 light-years. Proxima Centauri is therefore 1.3 parsecs from Earth. Astronomers prefer parsecs because they're based on how astronomers measure distances, by stellar parallaxes. (Source: 2MASS/IPAC/Caltech/University of Massachusetts)

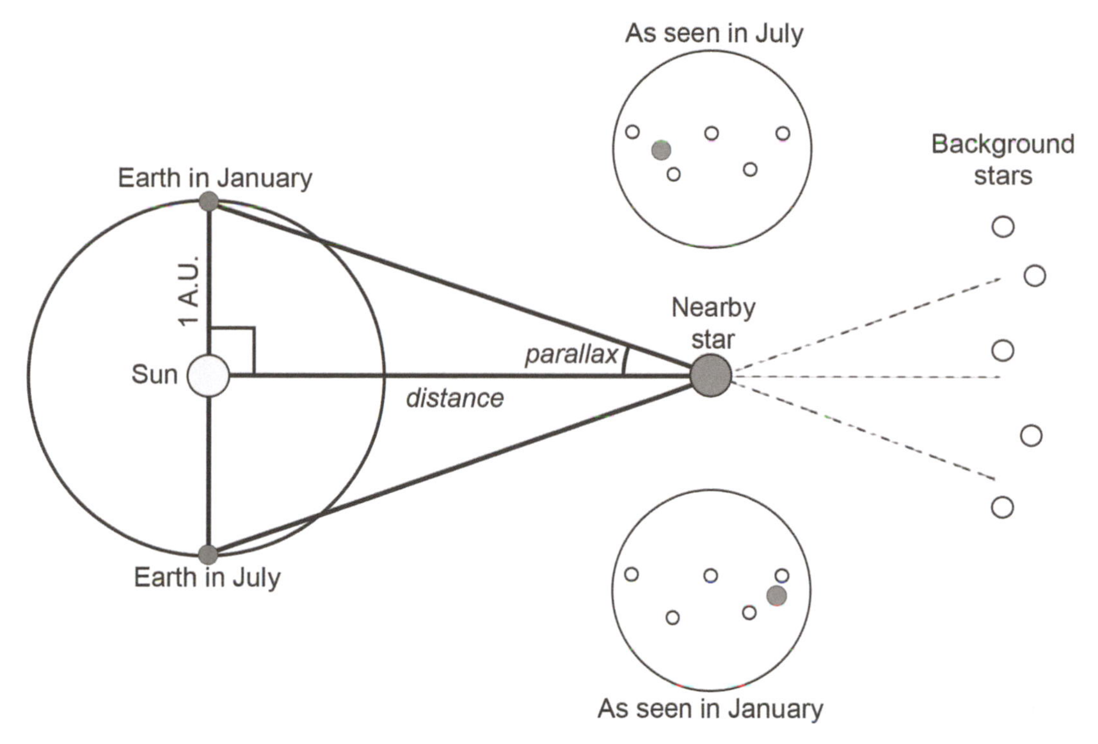

▶ **Figure 24-3** Distances between stars clearly can't be measured with a tape measure. Astronomers use parallax, which is the angle between two images of a star that are seen from different angles. This is also how human depth perception works: two eyes look at different angles, and the brain measures these angles and perceives depth. (Image courtesy of Frederick Ringwald)

Spectral Types

Different stars show different lines in their spectra. The spectral types were originally classified by the strength of the hydrogen lines: A, B, C, etc. Annie Jump Cannon led a major effort at Harvard to classify the spectra of nearly 250,000 stars, from 1880 to 1940.

Cecilia Payne-Gaposchkin figured out how to interpret these spectra quantitatively. In 1925, she found that:

- Stars are made of dense gas. The Sun is 1.4 times denser than water, on average.
- Stars are about 71% hydrogen (H), with about 27% helium, and 1–2% all other, heavier elements (which astronomers call "metals," even though many of them are not metals), by mass, or about 90% H, 9% He, and 1–2% metals, by number of atoms.

In other words, 90% of the atoms in the Universe are hydrogen.

In 1929, she found that spectral types change with *temperature*. This mixed them up:

O	B	A	F	G	K	M				
							(R	N	S	turned out to be just chemically peculiar M types)
								L	T	introduced in 1999, for brown dwarfs.
Now:										
O	B	A	F	G	K	M	L	T	Y	

Temperatures, in Kelvins:

30,000		10,000		6000	3000	900	600 K

Sub-types:
O3, O4, O5, … , O9, B0, B1, B2, …, B9, A0, A1, A2, … (etc.)
(Each sub-class is a few hundred Kelvins.)

The Sun is a G2 star, since T = 5800 K.

▶ **Figure 24-4** Different stars have different colors, from red to yellow to white to blue. (Image courtesy of Frederick Ringwald)

► **Figure 24-5** Remember from Chapter 11, "More Tricks of the Light," that flames work by *thermal radiation*. They give off light because they are hot. Hotter flames make more high-energy light, which has short wavelengths. As flames get hotter, they change in color, from red hot (in hot charcoal), to orange (in a wood fire), to yellow (in a bonfire), to white (in a blast furnace for steelmaking). Even hotter flames are blue. Stars work in exactly the same way, because physical law is *universal*. The same physical laws work everywhere in the Universe. (Image courtesy of Frederick Ringwald)

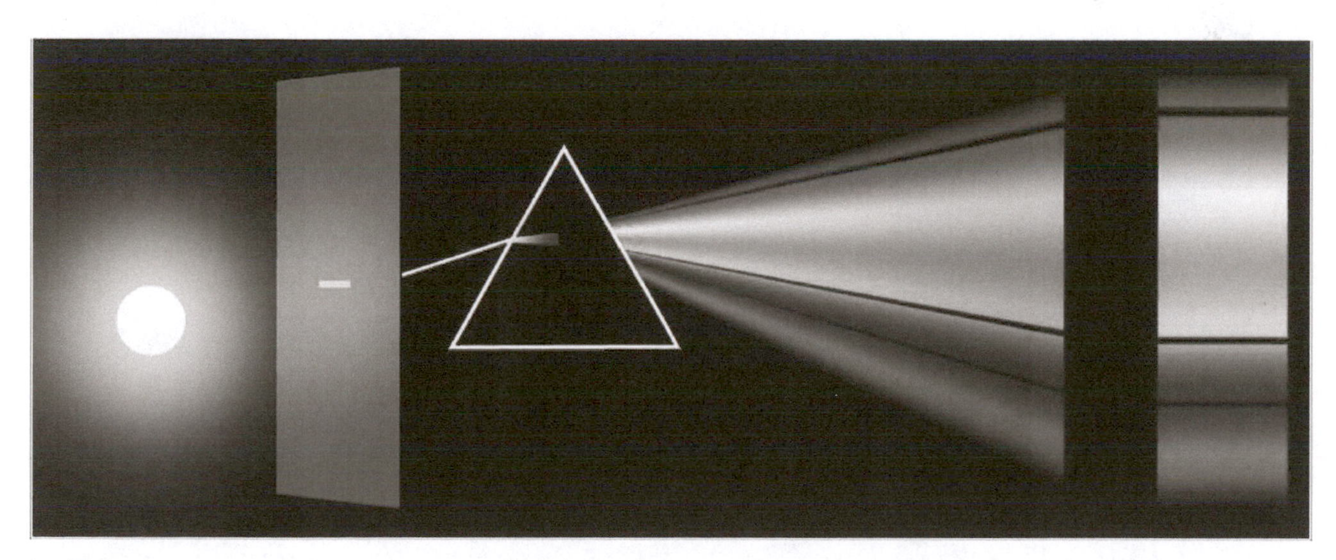

► **Figure 24-6** Recall from Chapter 10, "Light and Spectra," how astronomers can use a prism or a grating to break starlight into its colors, also called its spectrum. In the late 1800s, it was noticed that the spectra of stars follow a distinctive pattern of *spectral types*. (Image courtesy of Frederick Ringwald)

▶ **Figure 24-7** Harvard College Observatory began hiring women as "computers" in the 1880s. Annie Jump Cannon led a nearly 40-year campaign at Harvard to photograph and to classify the spectra of nearly quarter of a million stars. From this came the spectral types. (© Hulton-Deutsch Collection/CORBIS)

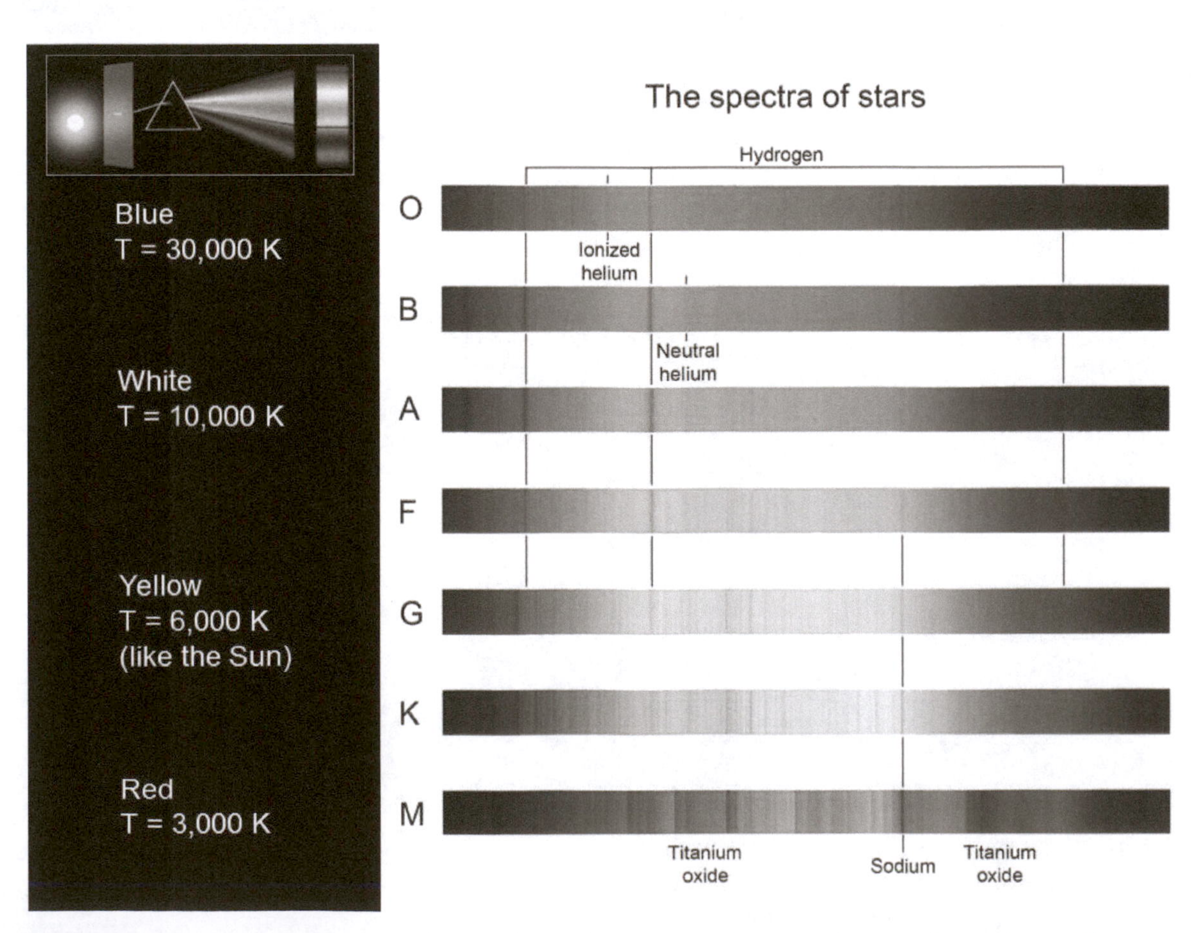

▶ **Figure 24-8** The spectra of stars follow a distinctive pattern of *spectral types*, shown here. Blue stars have lines of ionized helium, which can only exist at high temperatures. These are the O types. Red stars have lines of molecules, which can only exist at relatively low temperatures. These are the M types. Yellow stars like the Sun, are in-between. These are the G types. (Images courtesy of Frederick Ringwald with spectra from NOAO/AURA/NSF)

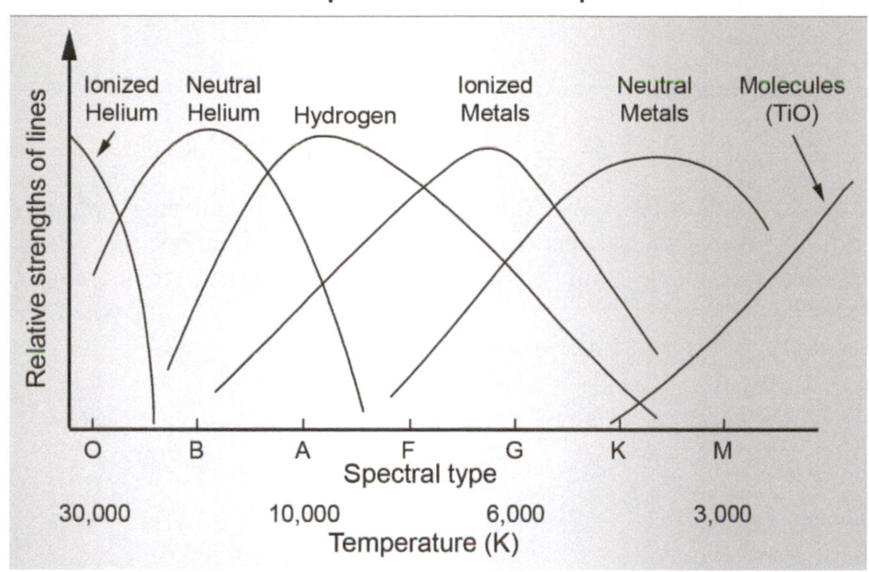

► **Figure 24-9** *Left:* In 1925, Cecilia Payne-Gaposchkin got the first Ph.D. in astronomy in the United States, at Harvard. She used the then-new science of quantum mechanics to interpret stellar spectra quantitatively. By doing this, she found that *most stars are mainly made of* **hydrogen**. For her postdoctoral research, she found that *spectral type depends on* **temperature** *only*. A-type stars, for example, don't have the strongest hydrogen lines because they are the hottest stars. At higher temperature, hydrogen ionizes, so that the hydrogen lines become weaker. O types are hotter, since they have ionized helium in their spectra; B types are between O and A types in temperature, showing neutral helium. This put the spectral types out of alphabetical order once and for all. (Image courtesy of Frederick Ringwald)

Right: "The reward of the old scientist is the sense of having seen a vague sketch grow into a masterly landscape."—Cecilia Payne-Gaposchkin (© Bettmann/CORBIS)

Memory Aids for the Spectral Types

OBAFGKM

> Oh, Be A Fine Girl, Kiss Me.
>
> –Henry Norris Russell, 1915

> Only Boys Accepting Feminism Get Kissed Meaningfully.
>
> - Geoff Marcy, 1980

Since the addition in 1999 of the L and T types, for brown dwarfs, there is now a need for new memory aids, for OBAFGKMLT.

Y types were added in 2012. These are essentially super-planets, 2–13 times more massive than Jupiter. H types still haven't been formally defined, but Jupiter-sized planets are known to exist, so astronomers may eventually get spectra of enough of them to define a class.

The spectral types may therefore become:

OBAFGKMLTYH.

> A memory aid for LTYH is **Listen To Your Heart**.
>
> – Roger Key, Fresno State, 2005

Science majors please note: in your own scientific work, give some careful thought to systems of names for things. Try to avoid saddling future generations with clunky systems of names, *like this*.

A problem here is that once people start using your work, your ability to change it has ended. People often resist adopting new standards: look at the argument about Pluto. Admitting you were wrong is essential to science, but humans often don't like to do it.

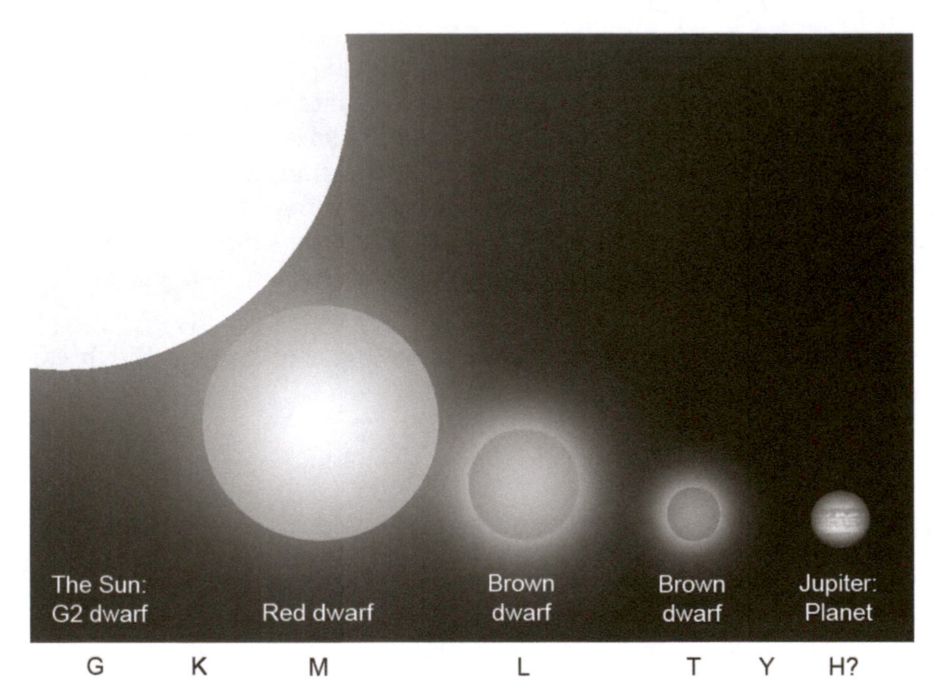

▶ **Figure 24-10** If an object is less than 80 times the mass of Jupiter, it doesn't have enough temperature and pressure in its core to have nuclear reactions, like a star. These are called *brown dwarfs*. Their heat is left over from their formation. Brown dwarfs dribble this heat into space, the way that big tanks of water cool down slowly, or like Jupiter and other giant planets do.

In 1999, two new spectral types were added for the first time in over 100 years. These are the L and the T types. L-type objects are about half low-mass, cool stars, and about half brown dwarfs. T types are all brown dwarfs.

In 2012, Y types were added. These are super-planets, 2–13 times more massive than Jupiter. The only letter that is unused for spectral types is H, so H may eventually be used for giant planets, such as Jupiter. (Image courtesy of Frederick Ringwald)

CHAPTER

25

The H-R Diagram and the Lives of the Stars

Spectral Types and Luminosity Classes

Stars of the same spectral type or color or temperature can have greatly different luminosities, and so must have greatly different *radii*.

For example: Capella is a star in the constellation Auriga. It is almost the same temperature as the Sun, so it's a G type.

Capella, however, is 160 times more luminous than the Sun. This is because Capella has 160 times more area than the Sun. Capella must therefore have a radius $\sqrt{160} = 12.7$ times larger than the Sun's radius.

Therefore, Capella is a G giant, or G8III. The Sun is a G2 dwarf, or G2V.

Luminosity Classes

Ia, Ib	Extreme supergiants	
II	Supergiants	(like Betelgeuse or Rigel)
III	Giants	(like Capella or Arcturus, a K)
IV	Sub-giants	
V	Dwarfs	(like the Sun or Sirius, an A1V)
VI	Sub-dwarfs	(metal-poor)
VII	White dwarfs	(burnt-out stellar remnants)

"Dwarfs" are small stars. "Dwarves" are small people.

A plot of luminosity (or absolute magnitude) versus spectral type (or temperature) is called **the Hertzsprung-Russell (H-R) diagram.** It shows how the stars change over time, or in other words, **evolve.**

▶ **Figure 25-1** The Hertzsprung-Russell diagram shows how stars work, and how they change over time. It is also an example of what scientists like: one diagram that brings together essentially everything that stars do. *Left:* Ejnar Hertzsprung (Image from the AIP Emilo Segre Visual Archive) ***Right:*** Henry Norris Russell (Image from the AIP Emilo Segre Visual Archive)

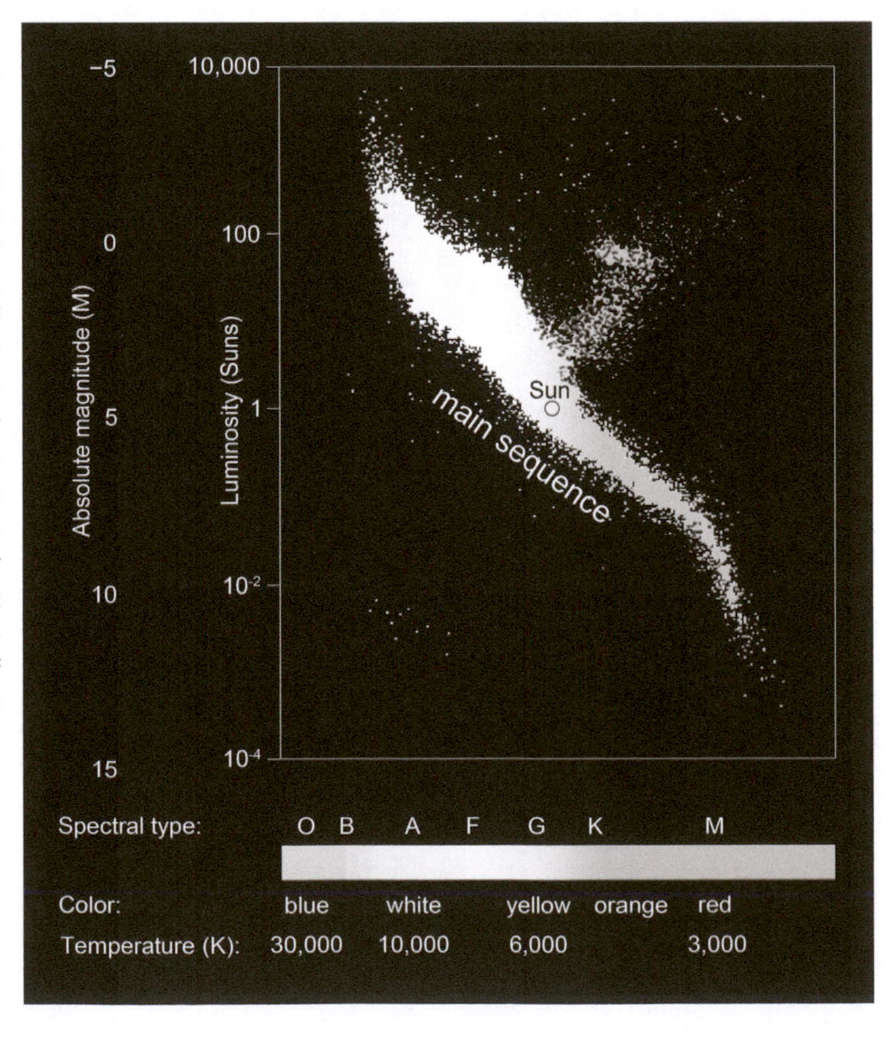

▶ **Figure 25-2** The Hertzsprung-Russell diagram is a plot of the spectral types (which depend on temperature only, and which also are equivalent to colors) of stars against their luminosities, which are the same as the stars' power outputs (or absolute magnitudes, which are a logarithmic way of expressing luminosities). In other words, it's a plot of spectral type versus luminosity for stars. In other words, it's a plot of spectral type versus power output for stars. In other words, it's a plot of (decreasing) temperature versus luminosity for stars. In other words, it's a plot of (decreasing) temperature versus color for stars. (Image courtesy of Frederick Ringwald, using data from ESA/Hipparcos)

► **Figure 25-3 Thermal radiation** is the light any hot object gives off because it's hot. Examples include a red-hot poker in a fire, the Sun, and the stars. Notice that the H-R diagram shows that the hot stars are also the powerful (or *luminous*) stars. Flames do this, too. A big, roaring bonfire is much more powerful *and* much hotter than a little match. Stars therefore act like flames, because the laws of nature are *universal*. (Image courtesy of Frederick Ringwald)

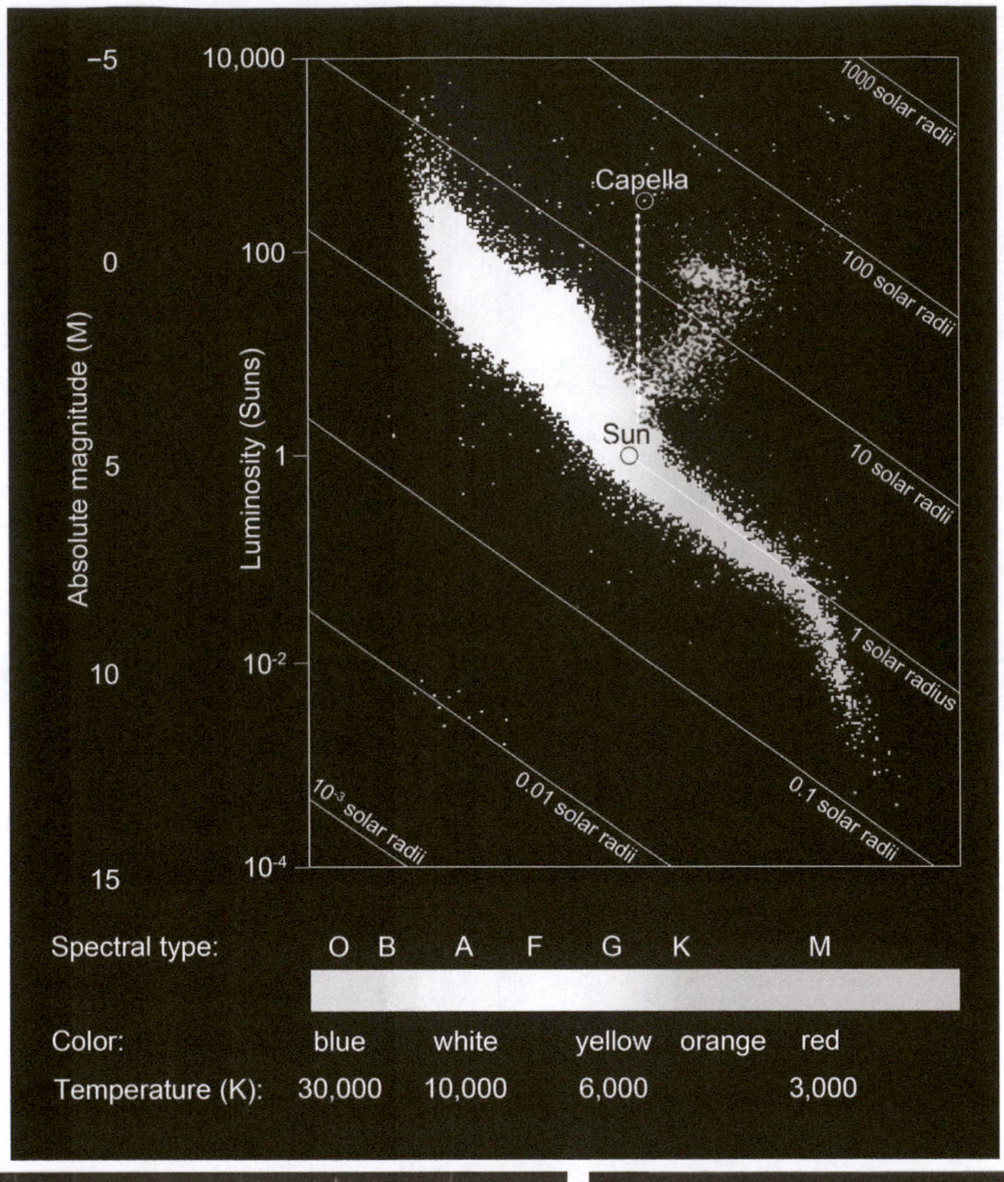

► **Figure 25-4** Capella is a star with a pleasant yellow color, because it has a temperature nearly the same as the Sun's temperature. Plotted on an H-R diagram that shows stellar radii (at top), Capella is 160 times more luminous than the Sun. How can Capella have the same temperature as the Sun, and yet be 160 times more powerful? The answer is that Capella is a giant star. Capella has 160 times the area of the Sun, because it has a radius 13 times larger than the Sun's radius (and since $13^2 \approx 160$). (Images courtesy of Frederick Ringwald)

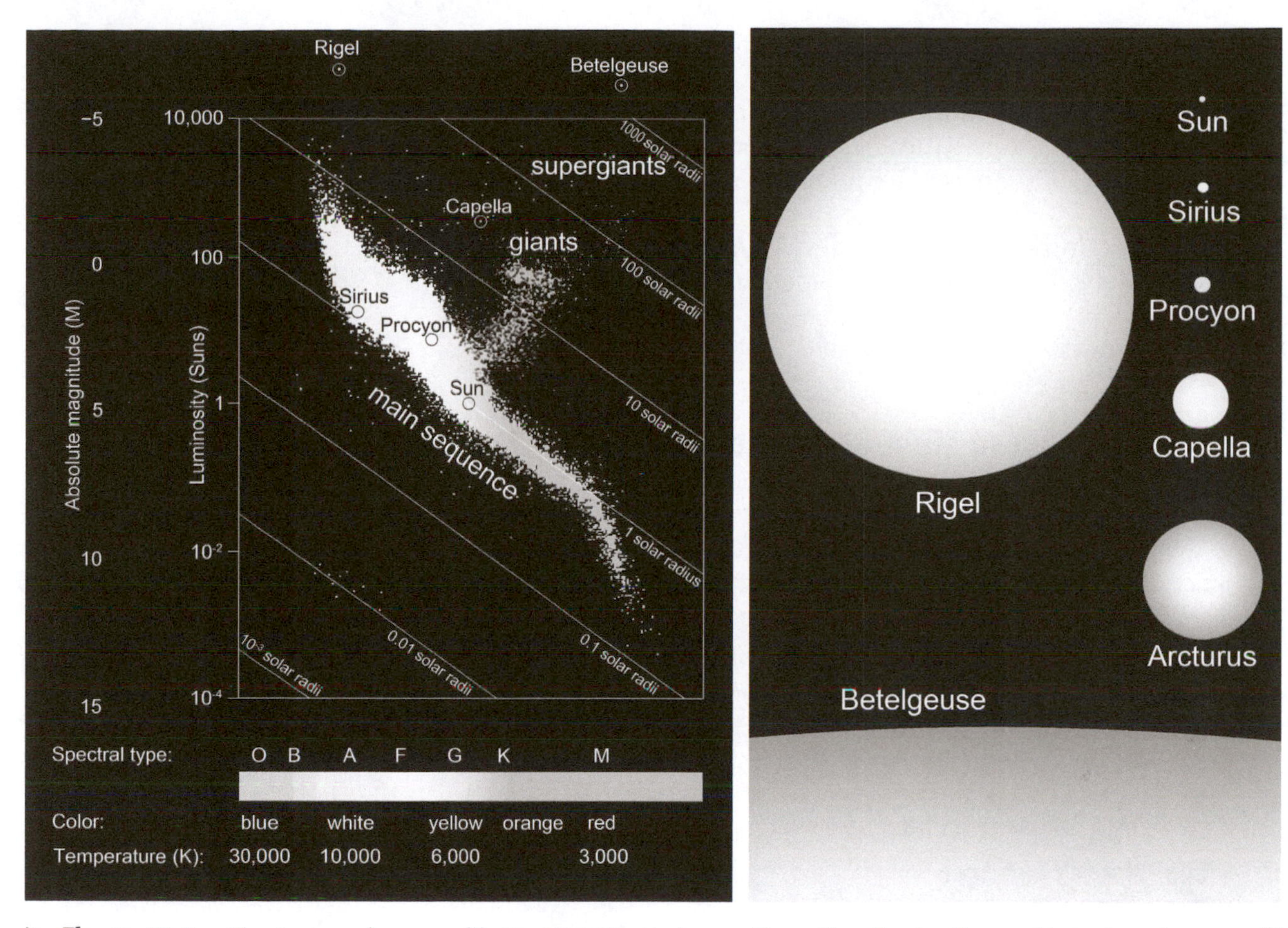

► **Figure 25-5** Giant stars have radii 10–100 times larger than the Sun's. Supergiant stars have radii 100–1000 times larger than the Sun's. Both giants and supergiants come in a wide variety of temperatures, from hot, blue supergiants such as Rigel, to cool, red supergiants such as Betelgeuse, and nearly everything in between. Notice that giants and supergiants are plotted on the H-R diagram above the main sequence, because they are more luminous than main-sequence stars like the Sun. (Images courtesy of Frederick Ringwald)

The Sun is a **G2V** star. It has an absolute magnitude M = 4.86.

This is not particularly luminous (or powerful).
The most luminous supergiants, such as Betelgeuse and Rigel, have M ≈ –7.
This is about 60,000 times more luminous than the Sun.

But then, the faintest M dwarfs and white dwarfs have M ≈ 15.
This is about 10,000 times less luminous than the Sun.

The Sun is more luminous than 80–90% of all stars.

Observed Number of Main-Sequence Stars Versus Spectral Type

About 30% of all stars are burnt-out cinders of what used to be main-sequence stars. These are called white dwarfs. About 10% of all stars are giants, or supergiants. The rest of the stars are on the main sequence.

The most massive stars on the main sequence are also the most luminous stars, and they are rare: less than 1% of main-sequence stars are O types.

Stars of intermediate mass and luminosity, like the Sun, are more common: about 10% of main-sequence stars are between F, G, and K dwarfs. The least massive stars on the main sequence are also the least luminous stars, and they are by far the most common: about 80% of main-sequence stars are M dwarfs.

This makes sense, when one considers:

Stellar Masses and Ages

Type	Mass	Lifetime, on the Main Sequence
O	60 Suns	<3 million years
A	3 Suns	300 million years
G	1 Sun	10 billion years
M	0.1 Suns	>1 trillion years

Massive stars are more luminous, because they have more nuclear fuel (the hydrogen in their cores). Massive stars are also so luminous because their cores are hotter since the pressure there is greater because they're so massive.

More luminous stars expend their nuclear fuel faster. They don't live for very long, so there aren't many of them.

Stars more massive than 60 Suns are so luminous, they come apart because of radiation pressure. They're so bright, their light literally tears them apart.

Notice also: the Sun, at 4.57 billion years old, is now about halfway through its main-sequence lifetime.

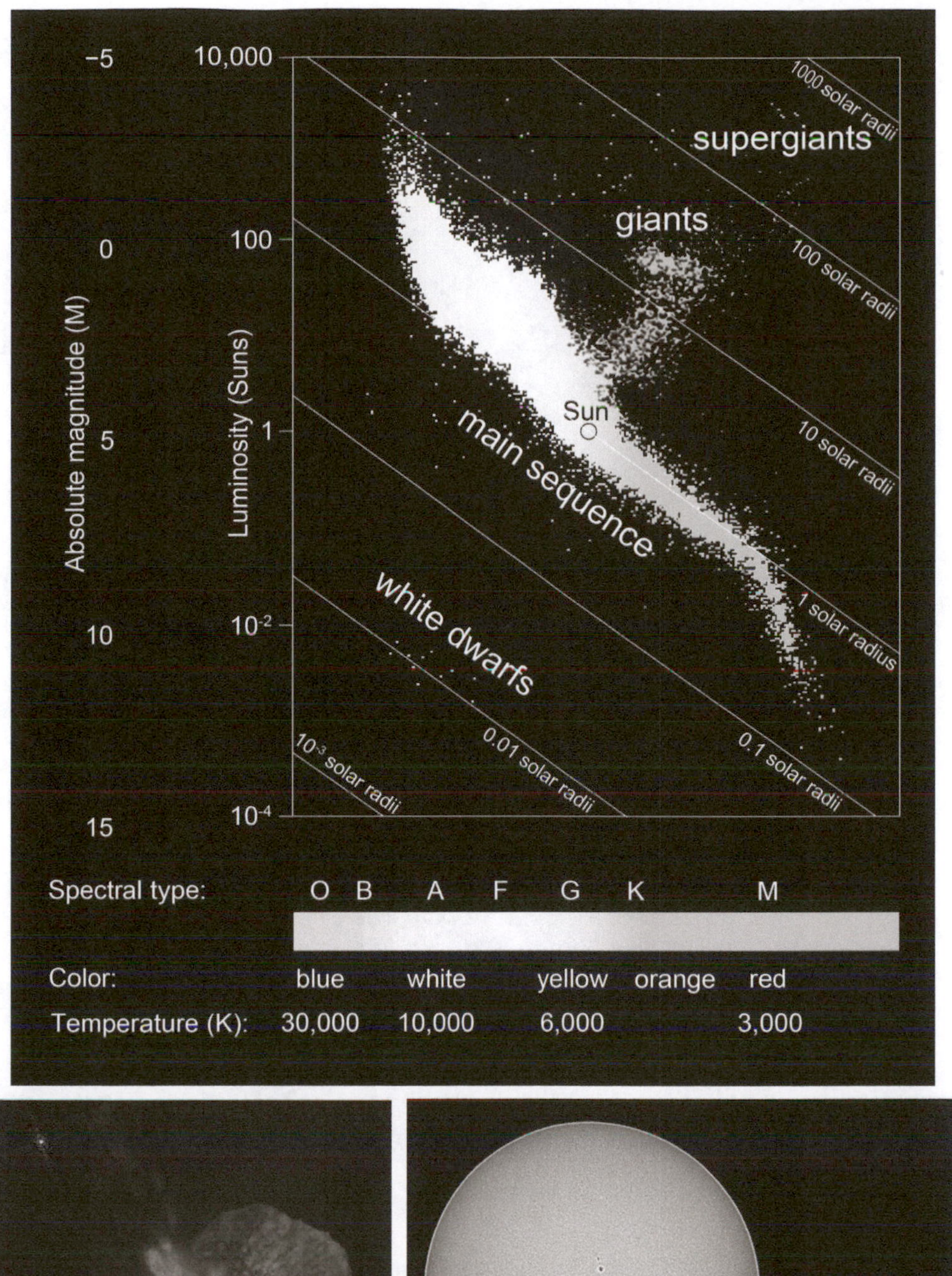

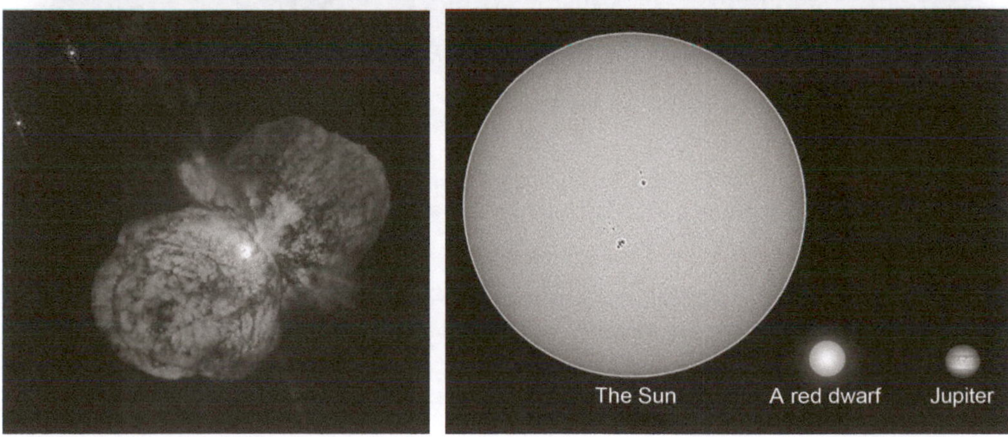

► **Figure 25-6** On the main sequence:

- The most massive stars have masses of ~60 M_{Sun}, such as Eta Carinae, shown at bottom left.
- The most massive stars are OB types. They have the shortest main-sequence lifetimes, since they are very luminous and burn themselves out quickly.
- M dwarfs are the least massive, live the longest, and are by far the most common stars. The Sun is more luminous than 80–90% of stars, nearly all of them K and M dwarfs.

(Images courtesy of Frederick Ringwald, except the image of Eta Carinae, which is by NASA/STScI)

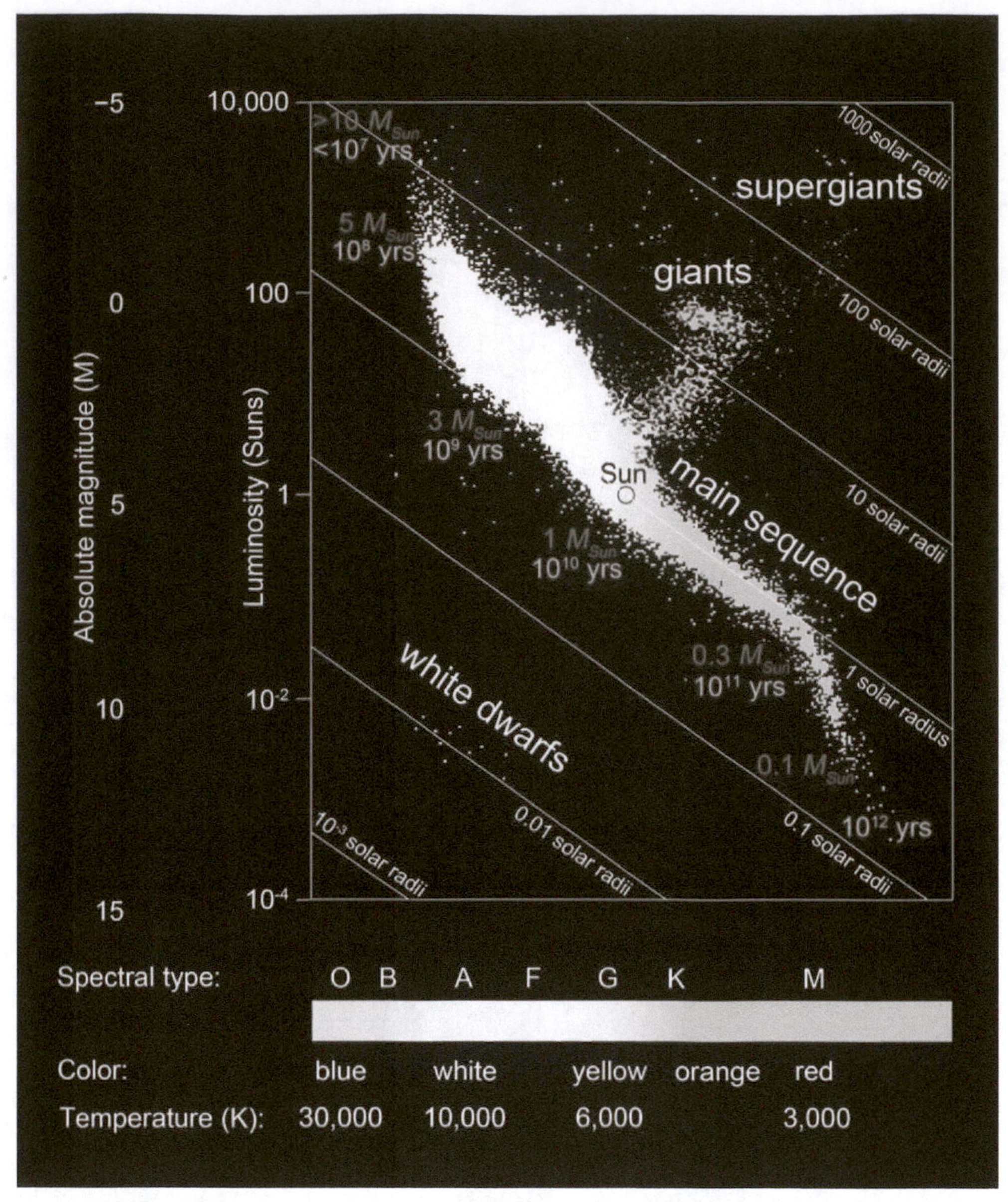

► **Figure 25-7** This H-R diagram summarizes everything discussed so far about H-R diagrams. It includes:

- stellar radii,
- masses for stars on the main sequence *(in magenta),* and
- main sequence lifetimes *(in green).*

The stars' main-sequence lifetimes are easy to calculate:

Main sequence lifetime = energy in the star/observed luminosity,
where energy in the star $E = (0.007)\ (0.1)\ m_{star}c^2$.

This is because $E = mc^2$, nuclear reactions are 0.7% efficient (which is observed in laboratory experiments), and 10% of the star's mass is turned into energy. (Image courtesy of Frederick Ringwald, using data from ESA/Hipparcos)

Binary Stars

M dwarf stars are the most common stars. Except for M dwarf stars, *most stars* are double, or binary, stars, with companions orbiting them.

30% are single stars
50% are binaries
15% are triples
5% are multiple-star systems, with four or more stars orbiting each other.

We can measure stellar **masses** by observing their orbits, and then using Kepler's Third Law and Newton's Law of Gravity. **Eclipsing** binaries show us stellar **radii** (sizes).

Star Clusters

There are two general types of star clusters.

Open Clusters (e.g., M45 in Taurus, also called "the Pleiades" or "the Seven Sisters")
These have hundreds to thousands of stars. They're metal-rich because they've been enriched by new generations of stars, which cook up metals.

Globular Clusters (e.g., M13 in Hercules)
These have 10^5 to millions of stars. They're metal-poor because they are very old, having formed before many stars cooked up much metal.

Star clusters show us how stars change over time, or evolve. This is because all the stars in a cluster are the **same age**, since they had to form together, because they're still bound to the cluster by gravity. It's also because all the stars in a cluster are at nearly the **same distance** from Earth, so if one star in a cluster looks twice as bright as another star in the cluster, it's because it's genuinely twice as luminous (powerful).

We can see how stars of different luminosities, and masses, change over the age of a cluster. The cluster's H-R diagram shows this.

Very young open clusters can contain O or B stars, which are short-lived. Older clusters don't, since luminous O and B stars burn out so quickly.

Ancient globular clusters have only G-M main-sequence stars, which age slowly. They also have many red giants and supergiants, which used to be the O and B stars.

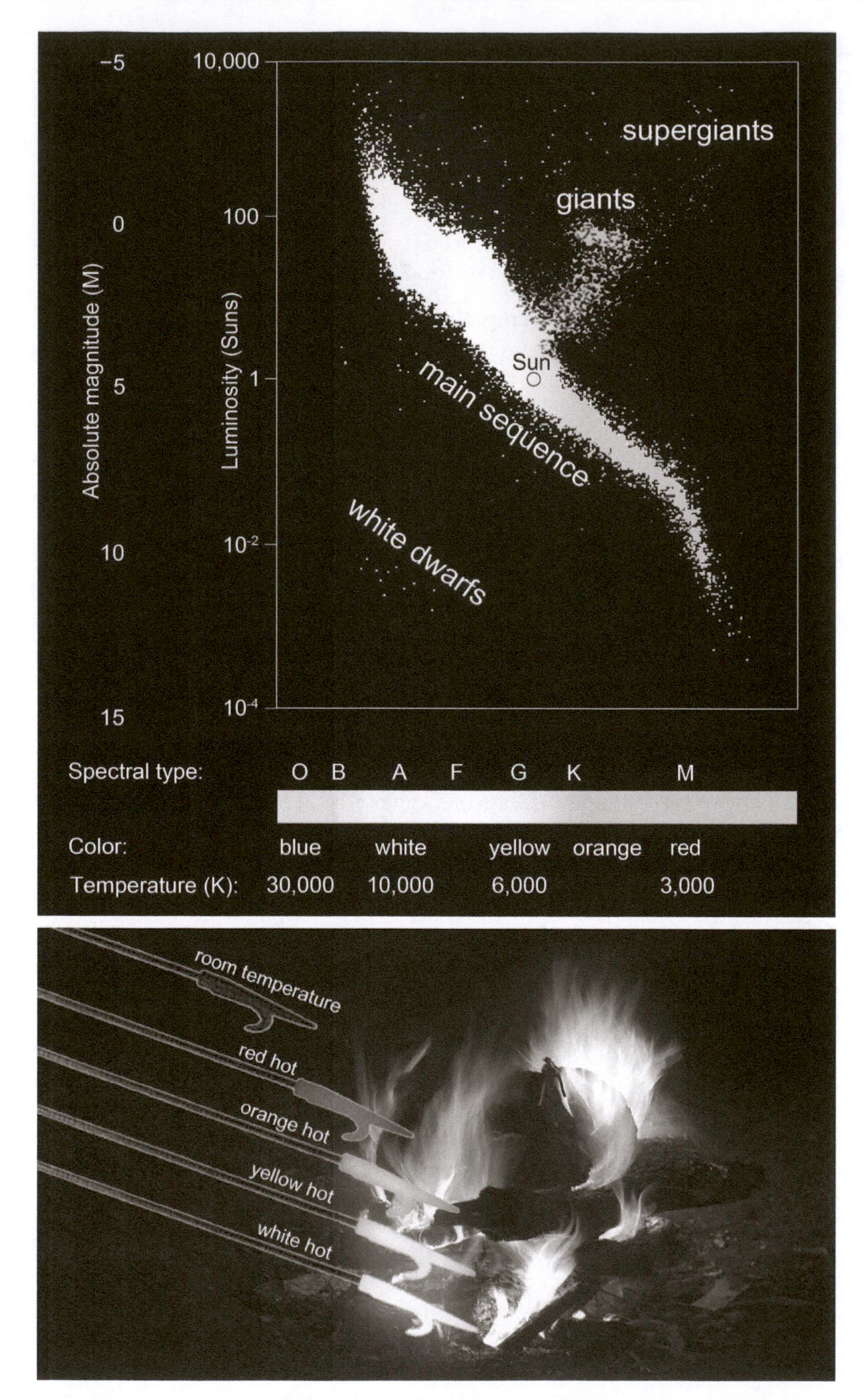

▶ **Figure 25-8** Recall that **thermal radiation** is the light any hot object gives off, because it's hot. Examples include a red-hot poker in a fire, the Sun, and the stars.

Notice that the hot stars are also the powerful (or *luminous*) stars. Flames do this, too.

If all stars are thermal radiators (in other words, they act like flames), why aren't all stars on the main sequence? Why do stars become red giants and supergiants? To answer these questions, we need to know how the masses and radii of stars are measured. (Images courtesy of Frederick Ringwald, using data from ESA/Hipparcos)

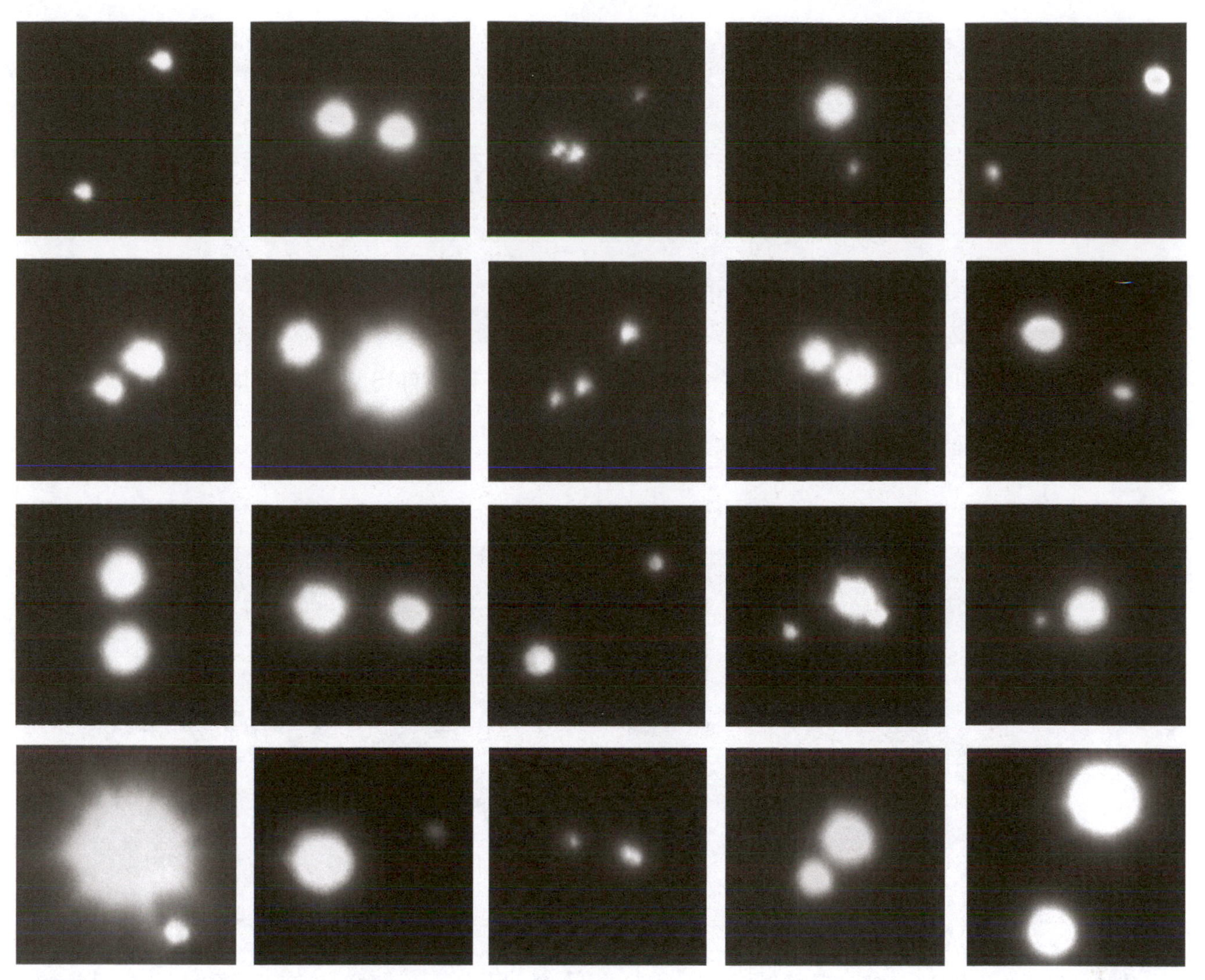

► **Figure 25-9** In much the way the Sun has planets orbiting it, most other stars have other stars orbiting them. Binary and multiple stars tell us stellar **masses**, by using Kepler's Third Law. If one can observe stars (or planets) orbiting another star, one can measure how strong the gravity is, and can use this to tell the masses of the stars or planets. (Images courtesy of Frederick Ringwald, at Fresno State's Campus Observatory)

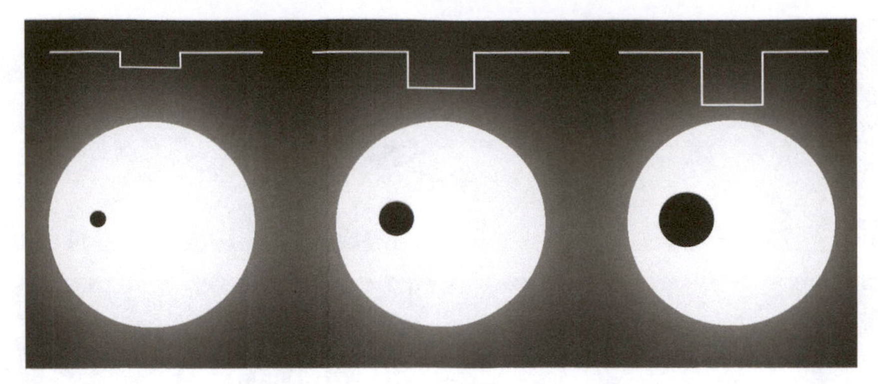

▶ **Figure 25-10** Eclipsing binary stars tell us stellar **radii**, from the depth of the eclipses. Not all star systems eclipse. Among those that do, the larger the stars, the more they cause each other to get dimmer when they do eclipse. (Image courtesy of Frederick Ringwald)

▶ **Figure 25-11** We can measure main-sequence lifetimes using star clusters. Star clusters tell us how stars change over time, or **evolve.** *Left:* An **open cluster** is metal-rich and young. (This is the open cluster M45, also called the Seven Sisters, or the Pleiades.) (Source: NASA/ESA/AURA/Caltech) *Right:* A **globular cluster** is metal-poor and ancient. (Image courtesy of Frederick Ringwald)

Stellar Evolution for Low-Mass Stars

Low-mass stars, like the Sun, spend about 10 billion years on the main sequence. They run out of nuclear fuel and swell up, becoming **red giant** stars, such as the stars Aldebaran or Arcturus.

They then literally come apart, throwing off **planetary nebulae.** Their cores, which were once their nuclear furnaces, are left behind. These cool down, turning into **white dwarf stars.**

We have now observed every stage of this process, and these observations confirm quantitative predictions for how common most classes of stars are.

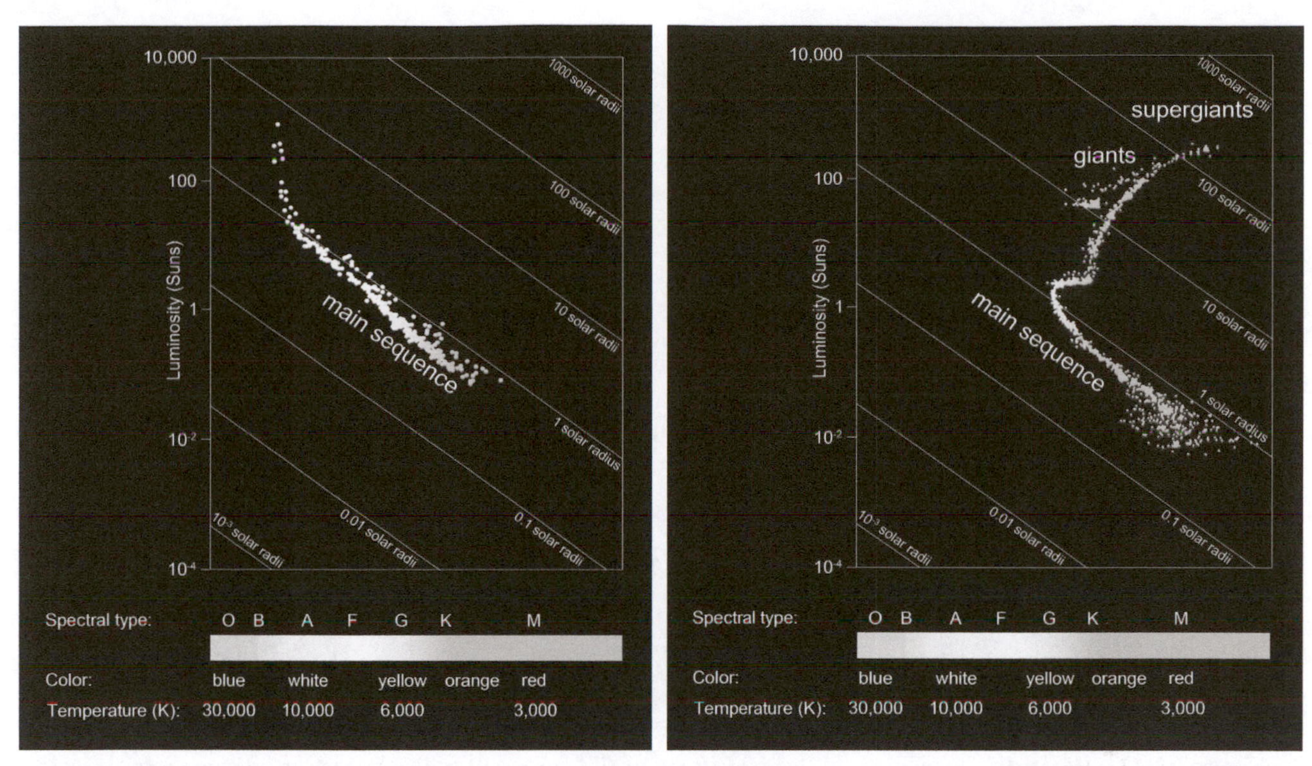

▶ **Figure 25-12** How star clusters evolve, and how fast it takes them to do it, is shown by their H-R diagrams. *Left:* This open cluster is 50 million years old. O and B stars are absent, because they don't last that long. *Right:* This globular cluster is 12 billion years old. It has no O, B, A, or F stars. The OBAF stars have all evolved into red giants (at upper right). (Images courtesy of Frederick Ringwald)

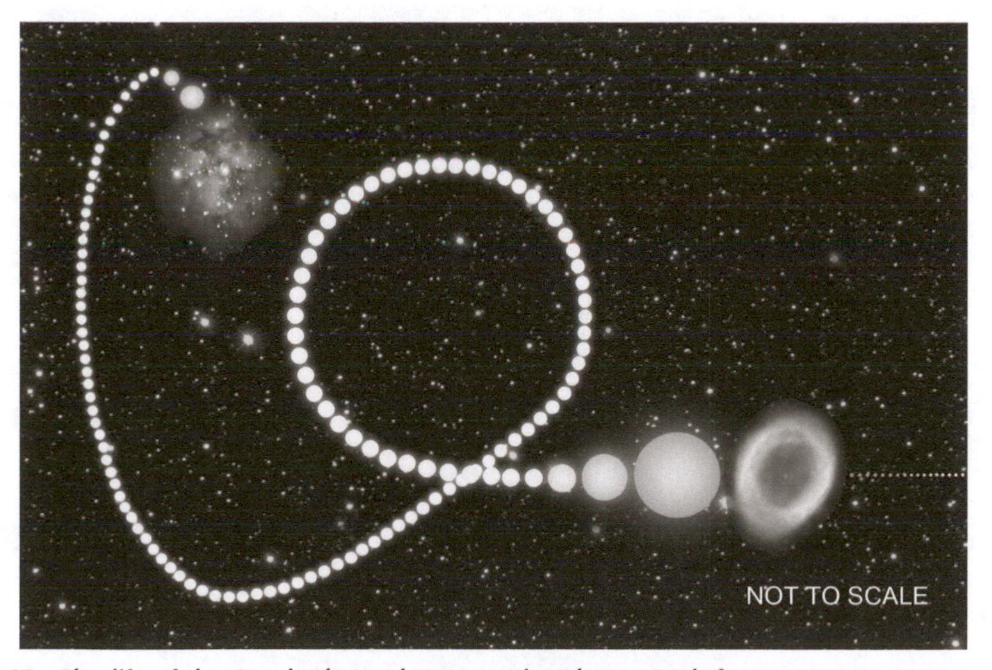

▶ **Figure 25-13** The life of the Sun is shown here as a time-lapse. Each frame represents 100 million years. The kink at lower left is where the Sun is now, 4.6 billion years after it formed by gravity in a nebula. The Sun is expected to last another 7.6 billion years (through the loop, which is meant to keep the images on the page), and then will leave the main sequence when it runs out of hydrogen nuclear fuel in its core. It will then come apart as a planetary nebula (not explode as a nova or supernovae), and leave behind a white dwarf, a hot cinder that used to be its core (at lower right). One reason scientists think this is that we've seen examples of every step of the process. (Image courtesy of Frederick Ringwald, using images from Mount Laguna Observatory)

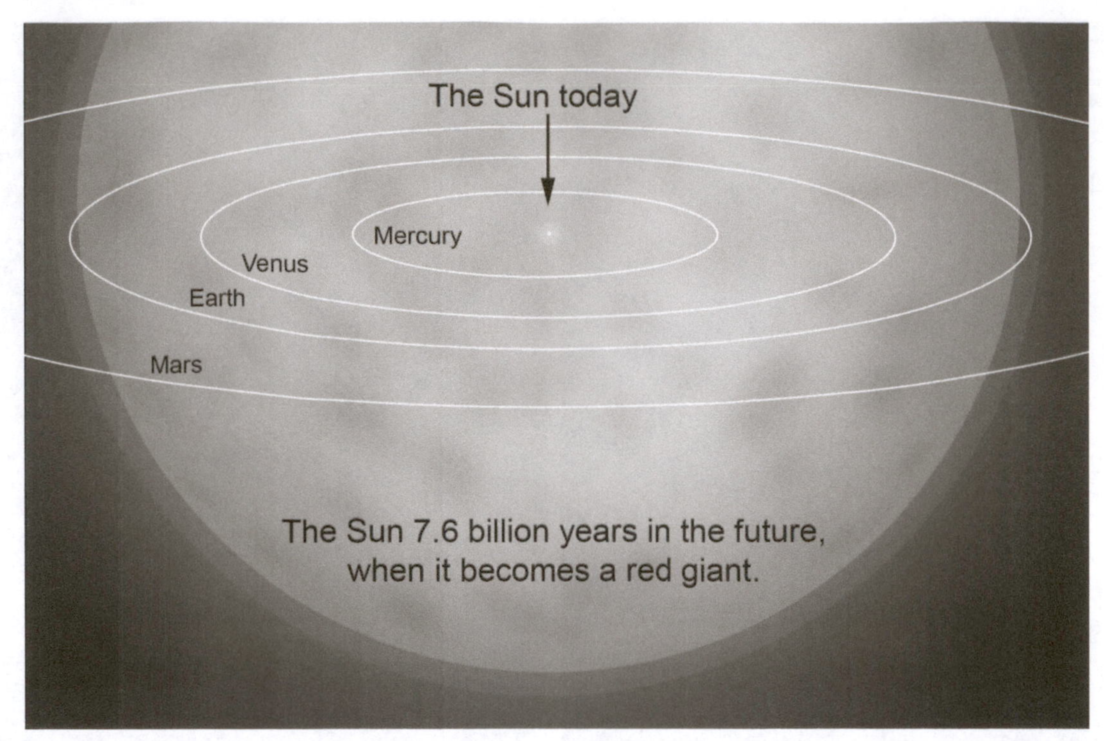

► **Figure 25-14** When the Sun dies, billions of years in the future, it will run out of hydrogen fuel for the nuclear fusion reactions in its core. This will make it expand and become a red giant. It will expand to a radius of about 1 Astronomical Unit, engulfing Mercury and Venus, and probably also Earth. Humankind had better move before this. (Image courtesy of Frederick Ringwald)

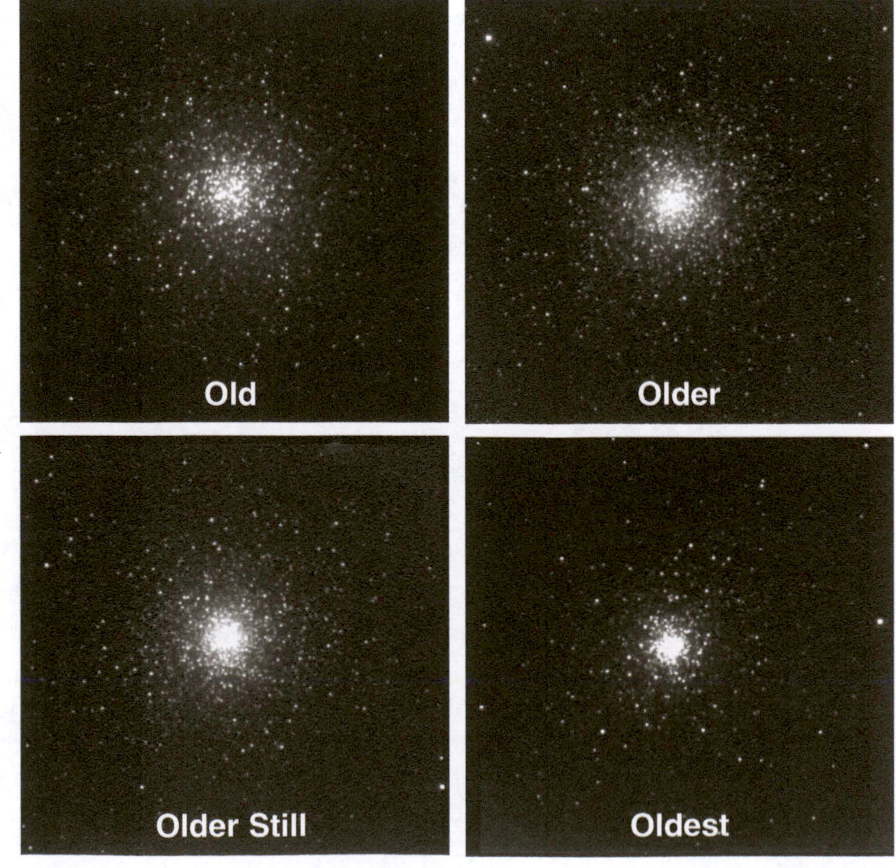

► **Figure 25-15** Why do stars become **red giants** when they die? It is because *gravity always attracts*, so mass sinks to the center over time. A star's outer parts therefore *expand*: these star clusters also do this.

This is called the Virial Theorem. It provides an independent check on the ages of star clusters, which confirms the ages found from the clusters' H-R diagrams. (Images courtesy of Frederick Ringwald)

► **Figure 25-16** Why stars become red giants (continued):

- As stars age, they cook up heavy elements. This heavy stuff sinks to the bottom.
- As stars become more centrally condensed, their cores become hotter.
- Their outer layers therefore expand, as in a hot air balloon.

When their cores run out of hydrogen (nuclear fuel), this expansion *accelerates*. (Left: Image © topseller, 2013. Used under license from Shutterstock, Inc. Right: Image © mack2happy, 2013. Used under license from Shutterstock, Inc.)

► **Figure 25-17** When the Sun becomes a red giant, it is expected to engulf several of the planets. One might wonder what this might do to the Sun. Little planets can harm a big, puffed-up red giant in a way similar to how little bullets can harm a big human body, if they are traveling fast, like planets. (Source: NASA/STScI/James Gitlin)

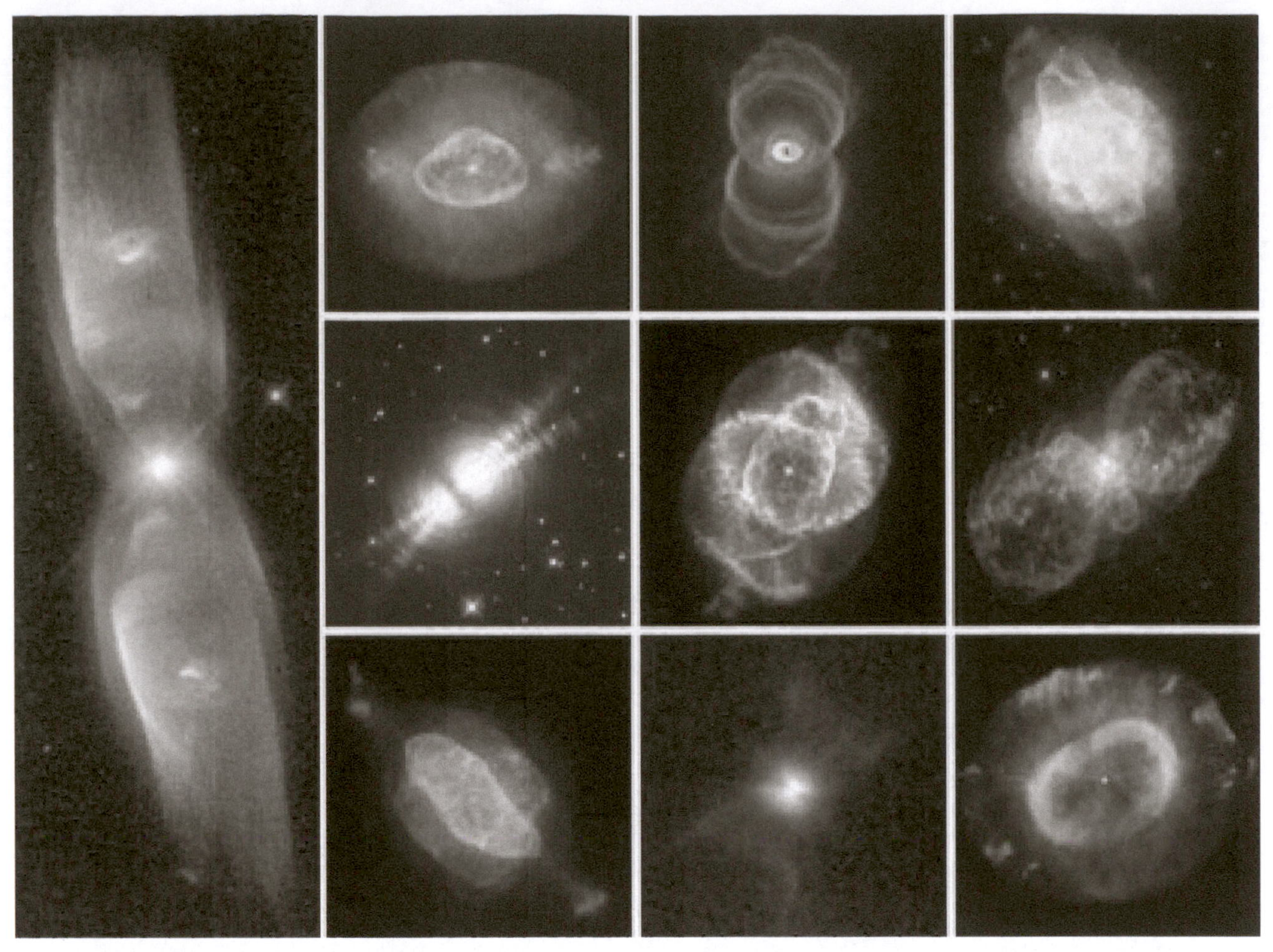

► **Figure 25-18** A red giant eventually comes apart as a **planetary nebula**, like these. (Source: NASA/STScI)

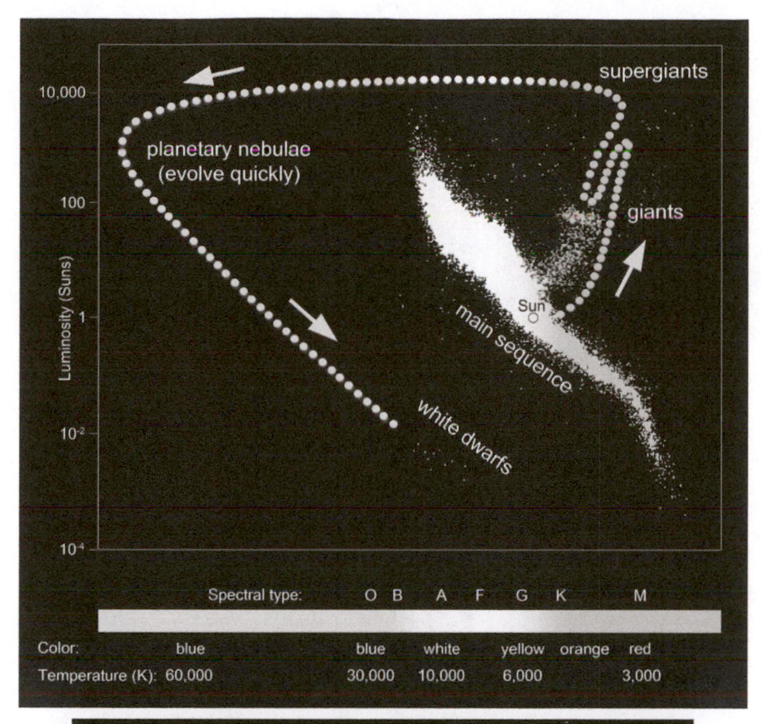

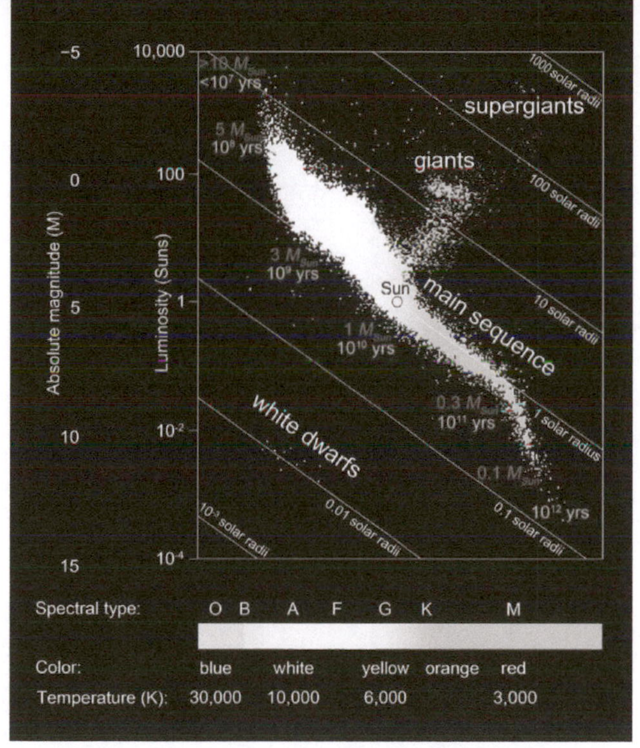

► **Figure 25-19**

Top:
Theory
A low-mass star like the Sun is expected to:
- Spend most of its life on the main sequence,
- Expand into a red giant,
- Eject a planetary nebula,
- Cool and shrink, and become a white dwarf.

Bottom:
Observations
Every step is observed!
- Stellar masses are measured in binary stars.
- Radii are measured in eclipsing binaries.
- Stellar lifetimes, from $E = mc^2$, are verified by star clusters' H-R diagrams and dynamics.
- Stellar populations are correctly predicted.

(Images courtesy of Frederick Ringwald; data from ESA/Hipparcos)

Chapter

26

Nebulae, Star Birth, and Star Death

Nebulae are clouds of gas and dust in interstellar space. In some, stars are forming. Others are the gas from stars that died. Whether they're stellar nurseries or stellar remnants, nebulae come in three distinctive types:

1. **Dark clouds**, which are full of dust. This dust comes in a wide variety of types, including *silicates* (rocks) and *carbon* (graphite, soot, and diamonds). There are also hundreds of known kinds of molecules, including the "brown organic tarry gunk" found all over the Outer Solar System.

   Dark clouds, and especially dusty "molecular clouds," are cold and dense, and are where stars begin to form by condensing under gravity. This makes observing how stars are born difficult, since very young stars are surrounded by dense cocoons of dust.

► **Figure 26-1** The Trifid Nebula has cold dark clouds, hot emission nebulosity (the pink parts at left), and warm reflection nebulosity (the blue parts at right) all close to each other. (Image courtesy of Frederick Ringwald, at Mount Laguna Observatory)

2. **Emission nebulae** are the glowing *red* clouds. They are lit up by high-energy ultraviolet radiation from stars, much like how electricity lights up the gas in fluorescent lights. Many nebulae where stars have just formed are emission nebulae, because their hot, young O and B stars illuminate them. Nebulae thrown off by stars that died, such as planetary nebulae, can also be emission nebulae. The central star of a planetary nebula can light up the nebula because the central star can be very hot, because it was once a star's nuclear furnace.

3. **Reflection nebulae** are the *blue* clouds in space. They shine by reflected starlight, much as how planets do. Reflection nebulae are blue because blue light from stars *scatters* more than any other color of light. This works in a way similar to how light from the Sun scatters in Earth's atmosphere, which is why Earth's daytime sky is blue.

Star formation is among the major unsolved problems of astrophysics. One reason for this is because, until recently, our ability to observe star formation was quite limited. We still understand few of the details, such as why low-mass stars like the Sun are common but massive stars are rare, or why so many stars are in binaries. We understand only the basics: gravity pulls gas clouds together into stars. Since the clouds that collapse into stars are rotating, they flatten into disks. Planets form in these disks. This is why the planets of the Solar System are all near the ecliptic plane.

Proplyds (pronounced PRO-plids) are these protoplanetary disks. Over 150 were discovered in the Orion Nebula with *Hubble Space Telescope*.

Our ability to observe star formation is improving, because instruments that detect radio and infrared radiation are improving. Radio and infrared wavelengths are longer than the dust grains that obscure our view, so they can go through clouds of dust.

Dust between the stars therefore causes *reddening* of starlight. Since only red light can get through the dust, stars obscured by dust look redder than they really are.

Dust *in Earth's atmosphere* (not in space) is **why sunsets are red.** When the Sun is low in the sky, we look through a great deal of dust in Earth's atmosphere. (See Figure 26-5.) Red light has the longest wavelengths that unaided human eyes can see. Only the red light can get through the dust, because of its long wavelengths.

Earth's daytime sky is blue because air molecules in Earth's atmosphere *scatter* light from the Sun. This is because the air molecules are about the same size as the wavelength of blue light. Other colors don't scatter as much, because their wavelengths are longer, so Earth's sky is blue.

Scattering of blue light from the Sun is also why oceans and glaciers are blue. Molecules of water, ice, and air are all about the same size as the wavelength of blue light, so they all scatter blue light more than any other color of light. (See Figure 26-6.)

▶ **Figure 26-2** This is the Rosette Nebula. Notice the stars in the center, which condensed by gravity from the gas. (Image courtesy of Frederick Ringwald, at Fresno State's station at Sierra Remote Observatories)

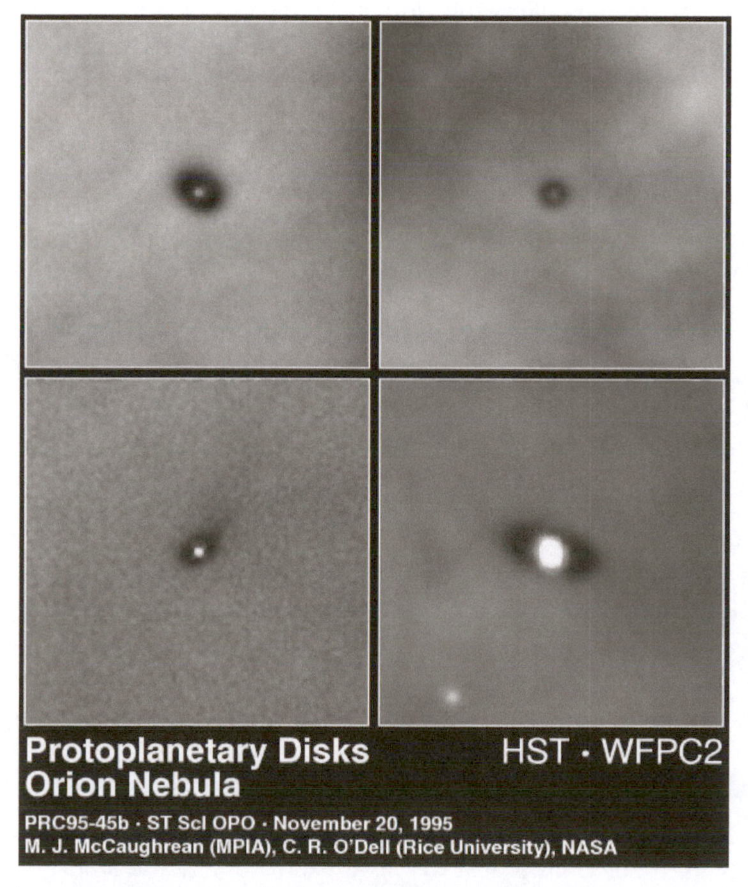

▶ **Figure 26-3** Protoplanetary disks are called **proplyds**. These are stars caught in the act of forming by gravity, with planets forming in the disks of gas and dust that are thrown out because they spin. (Source: NASA/STScI)

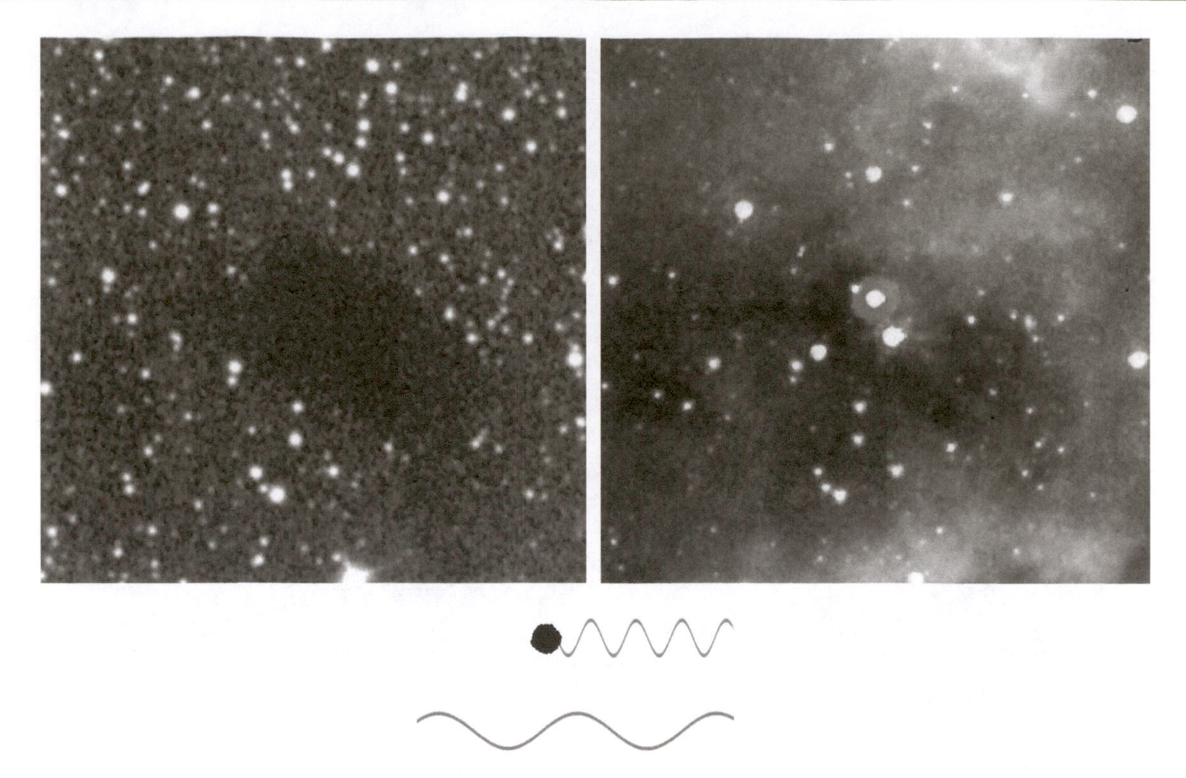

► **Figure 26-4** Stars form in dark nebulae, where dust cools the gas enough so its gravity can condense it into stars. Astronomers used to be unable to see inside dark clouds, though, because they're so dusty. *Infrared* telescopes and cameras now allow this. *Top left:* A dark cloud seen in *visible light* (Source: NASA) *Top right:* The same cloud is shown in *infrared radiation.* Notice the star forming inside it. (Source: NASA) *Bottom:* Dust grains absorb visible light more than infrared radiation, and violet light more than red light. (Image courtesy of Frederick Ringwald)

► **Figure 26-5** Toward the center of the Milky Way, our galaxy, one can see the *extinction* and *reddening* of background stars. At left, there is little dust to cause extinction, so the stars show clear. At center, dust clouds make the stars in back of them look yellow and red, from left to right: this is interstellar *reddening.* At right, there is so much extinction, stars no longer show through the dust lanes: this is interstellar *extinction.* (Image courtesy of Frederick Ringwald)

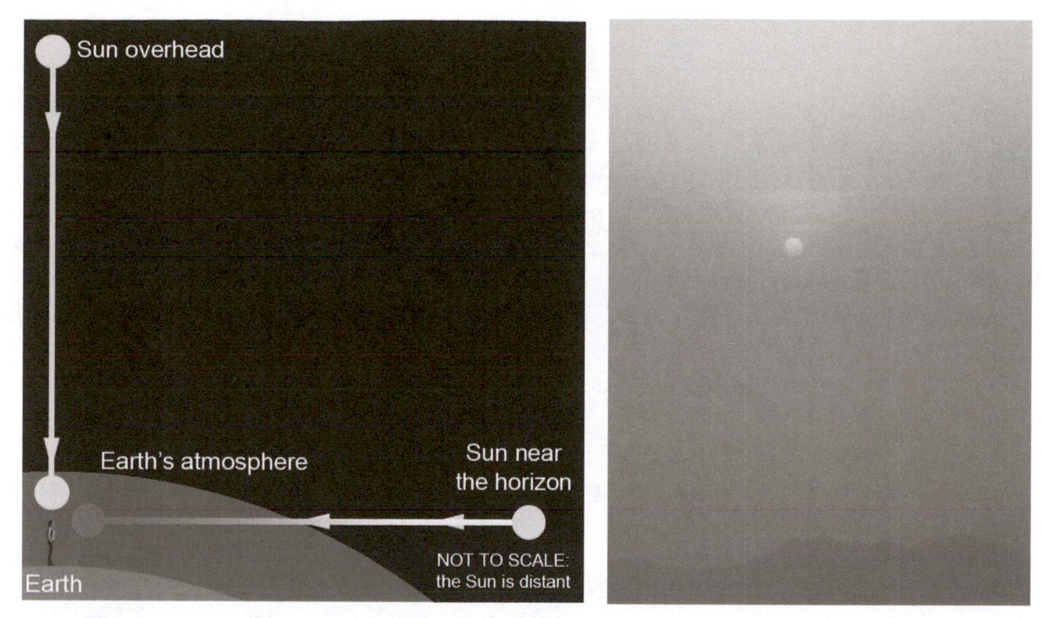

► **Figure 26-6** Sunsets are red because of *dust* in Earth's atmosphere. The dust absorbs all but the red light. This happens more when the Sun is near the horizon. This is because the Sun's light passes through a longer thickness of Earth's atmosphere, which contains more dust than when overhead. (Images courtesy of Frederick Ringwald)

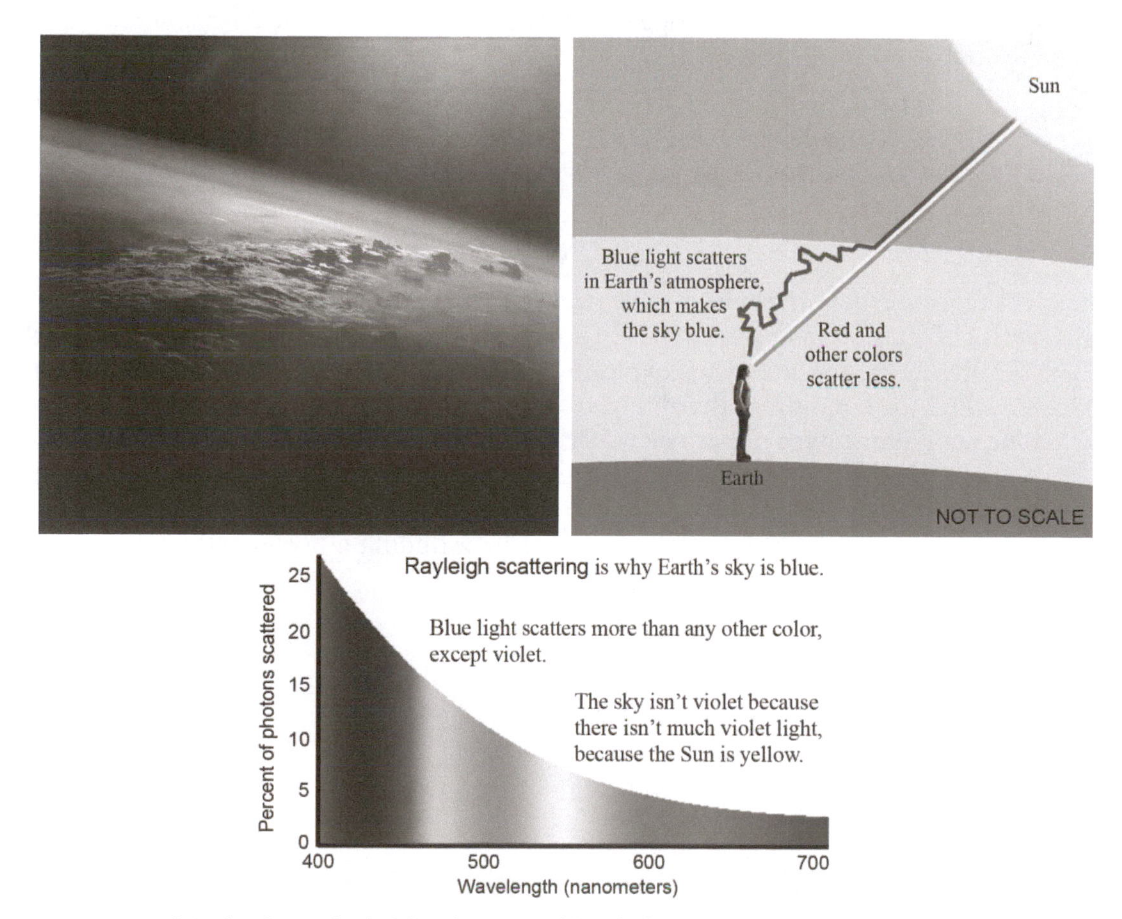

► **Figure 26-7** Earth's daytime sky is blue because blue light is *scattered* by *air molecules*. In the daytime, most light comes from the Sun. The air molecules that *scatter* this light are in Earth's atmosphere. Blue light scatters more because it has a shorter wavelength, which is smaller than the air molecules. Water and ice also scatter blue light. This is why oceans and glaciers are blue. (*Top left:* Source: NASA; *Top right* and *bottom:* Images courtesy of Frederick Ringwald)

White Dwarfs

Again, "dwarves" are small people, "dwarfs" are small stars. White dwarfs are thought to be burned-out cinders of stars once like the Sun. Recall that, 7–8 billion years in the future, the Sun should run out of the hydrogen that fuels the nuclear reactions in its core. When this happens, the Sun is expected to become a **red giant**, which should eject a **planetary nebula.**

A planetary nebula has a hot star in its center. The nuclear reactions that used to power the star will now be extinguished, though, so the central star should cool down. As it cools, it becomes a white dwarf. As it contracts under its own gravity, it becomes so dense it resembles a solid. A teaspoon of white dwarf material would have a mass of over a ton! (A teaspoon of lead has a mass of 11 grams.) Since the nuclear reactions that once powered the star made carbon, a white dwarf is essentially a *very* dense diamond.

The best evidence that white dwarfs exist is that we see them. The companion of Sirius has a mass of 1.0 Suns, measured by tracking its orbit and using Kepler's Third Law. Spectra show it's hot (27,000 K), but its luminosity is tiny (0.03 Suns). It must therefore be small, at 1% the radius of the Sun, or about the size of Earth.

Stellar Evolution for Massive Stars

Stars with masses over 11 Suns die more spectacularly. They run out of fuel, become red giants, and then become **red supergiants**, such as Betelgeuse or Antares. They burn heavier and heavier elements in the nuclear fusion reactions in their cores:

Hydrogen	→	Helium
Helium	→	Carbon
Carbon	→	Oxygen, Magnesium, and Neon
OMgNe	→	Silicon
Silicon	→	Iron

Nuclear fusion reactions that convert iron to heavier elements, though, require *more* energy than they release. This cools the massive star's core, which collapses and explodes in a Type II **supernova.**

(In a Type Ia supernova, a white dwarf explodes when it's pushed over the maximum mass a white dwarf can have, called the Chandrasekhar limit, from gas spilled on it from another star. In a Type Ib supernova, a blue supergiant with a mass of > 50 Suns explodes, as with Supernova 1987A.)

The best evidence that all of this happens is that we see these supernovae, and that their spectra show they are rich in heavy elements. When a supernova explodes, it seeds the Galaxy with heavy elements—*the stuff you're made of.* If you don't believe it, try finding a hotter place.

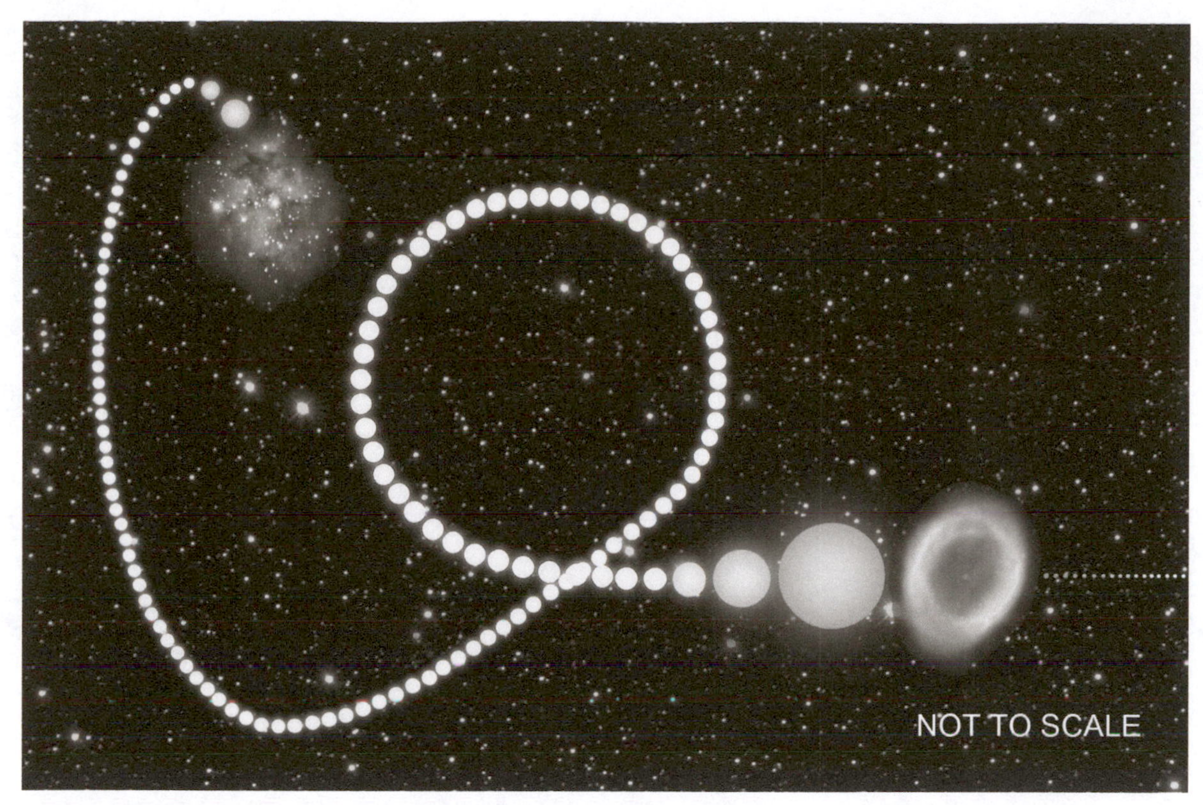

► **Figure 26-8** Let's consider again the end of a star's life. (Image courtesy of Frederick Ringwald)

► **Figure 26-9** When the Sun dies, 7–8 billion years from now, it should become a **red giant**. (Image courtesy of Frederick Ringwald)

▶ **Figure 26-10** The red giant should eject a **planetary nebula**. Notice the central star. (Image courtesy of Frederick Ringwald, using Fresno State's station at Sierra Remote Observatories)

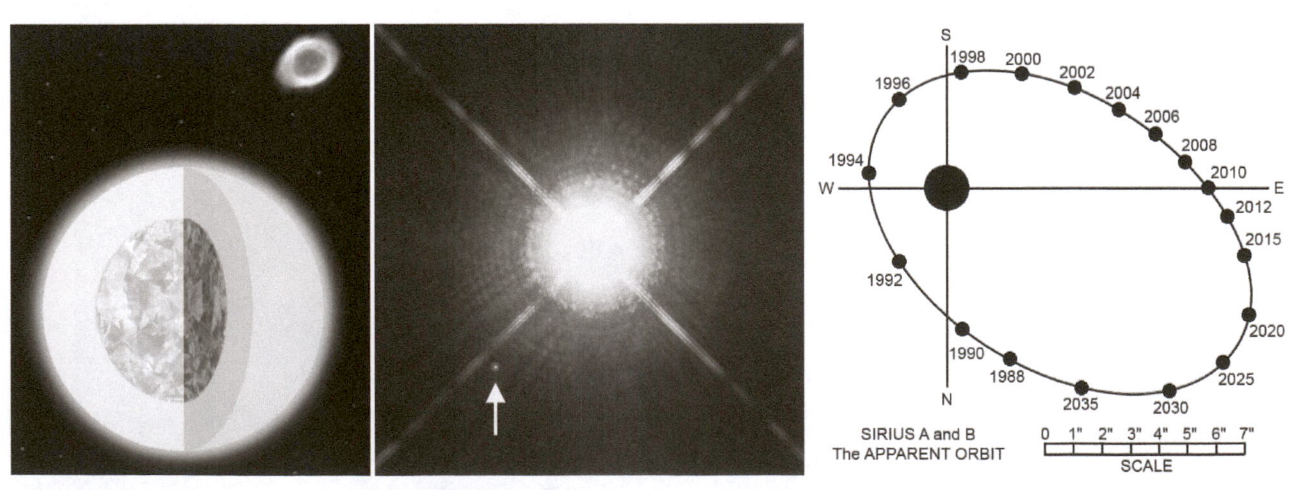

▶ **Figure 26-11** *Left:* The central star of a planetary nebula cools, becoming a **white dwarf**. (Image courtesy of Frederick Ringwald) *Middle:* The best evidence that white dwarfs exist is that we see them. Sirius, the brightest star in the night sky, has a white dwarf orbiting it, called Sirius B (visible here at lower left of Sirius A, which is much brighter). Sirius B is hot (27,000 K), but its luminosity is only 3% of the Sun's. It must therefore be small, at 1% the Sun's radius, or about the radius of Earth. With a mass of 1.0 Suns (measured from the orbit), Sirius B is a *very* dense diamond. (Source: NASA/STScI/Howard Bond/Martin Barstow) *Right:* The mass of Sirius B was measured from its 50-year orbit around Sirius A. (Image courtesy of Frederick Ringwald)

► **Figure 26-12** Evolution of **massive stars** (with masses > 11 Suns) is shorter and more eventful than for low-mass stars, like the Sun. Betelgeuse, shown here, is an M2 supergiant, with a mass of 14 Suns and a radius of 630 Suns. (Source: NASA/STScI)

► **Figure 26-13** The core of a massive red supergiant is thought to build up layers, like an "onion skin." (Image courtesy of Frederick Ringwald)

► **Figure 26-14** If Betelgeuse became a supernova, it would visible in the daytime for months. (Image courtesy of Frederick Ringwald)

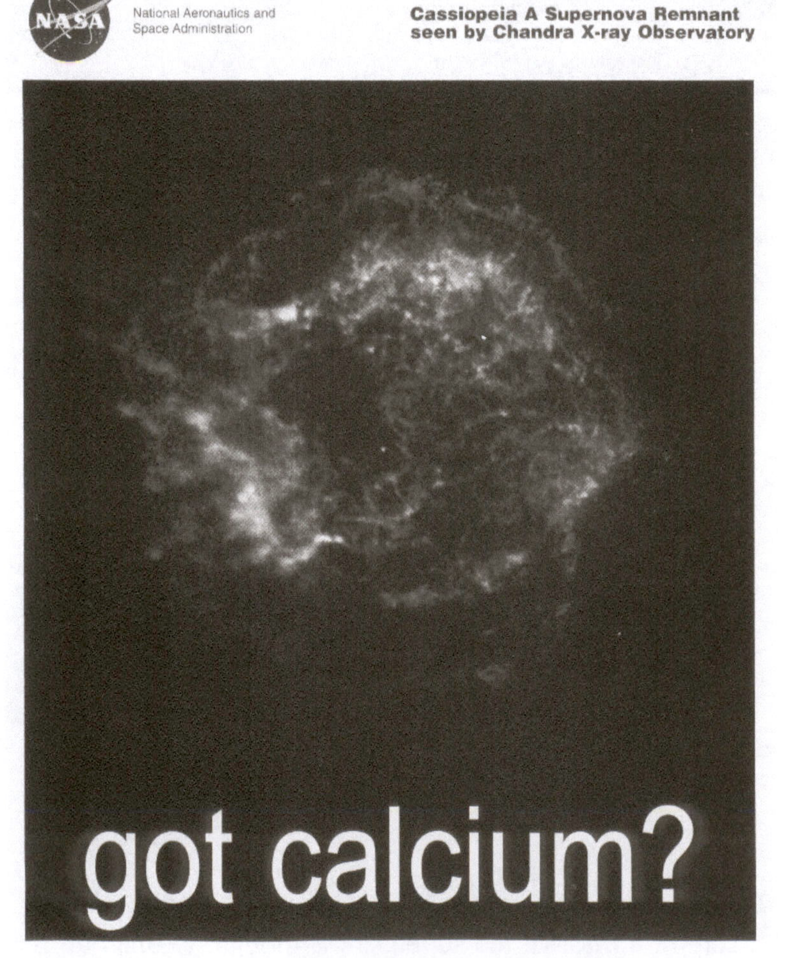

► **Figure 26-15** **Supernovae** are examples of stellar evolution occurring on a *human* timescale of just a few months. We can *see* them change. (Source: NASA/Chandra X-ray Observatory)

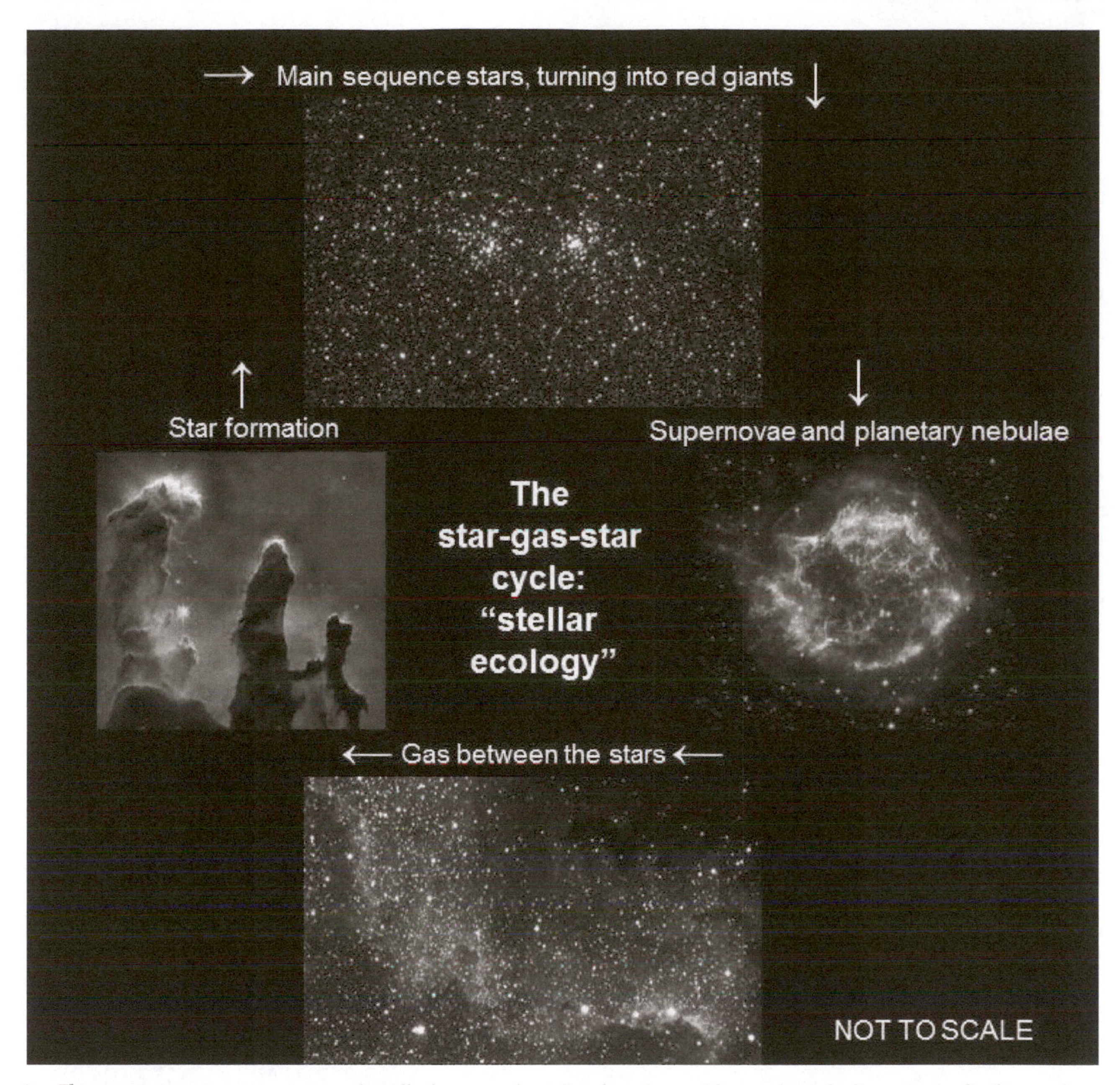

► **Figure 26-16** Supernovae make all elements heavier than iron. *(If you don't believe it, try finding a hotter place.)* Supernovae also blast large amounts of *all* elements into space, which form new stars. We are made of 3rd- or 4th-generation elements. (Source: NASA/STScl/Greg Morgan/Frederick Ringwald)

Notice that the more extraordinary the claims we make, the more extraordinary and detailed is the evidence we need to provide.

Pulsars

A **pulsar** is a rotating, magnetic *neutron star*, or a ball of nearly pure neutrons, packed together so tightly, they touch each other. There is a pulsar in the center of the *Crab Nebula*, which is a cloud of gas expanding outward from where a supernova exploded. Chinese astronomers observed this supernova when it exploded, in the year 1054 AD, and kept records that are still useful today. (Imagine doing something in your job that colleagues of yours still use, a thousand years in the future.)

We know pulsars are magnetic, because their radio waves show a distinctive spectrum (called synchrotron radiation), which we can reproduce on Earth, using electrons moving through magnetic fields.

Still, why are we so sure that neutron stars exist? And how do we know they're so small?

- Pulsars are the most precise clocks known in the Universe. Their pulses are so repeatable, and the times between pulses are so regular, it's hard to explain—unless it's from something that rotates, then it's easy to explain.
- They pulse *so* rapidly, some are as fast as thousands of times per second. To do this, and not have to travel faster than light (not allowed by relativity), they must be small, no more than a few kilometers across.
- Any object rotating this fast would fly apart, unless its gravity were very strong and it were made of a very strong solid.
- From neutron stars in binary star systems, Kepler's Third Law shows us that their masses are typically about 1.4 Suns.
- Since density = mass/volume, anything this massive and this small is *as dense as an atomic nucleus:* therefore, a neutron star.

A neutron star is so dense, there is no space between atoms or even between nuclei. The protons and electrons in atoms are so compressed into each other, they make a solid mass of neutrons, with no space between them.

Still, even in this extreme state, neutron star matter isn't infinitely strong. If a neutron star had a mass greater than 3 Suns, it would gravitationally collapse—into a black hole.

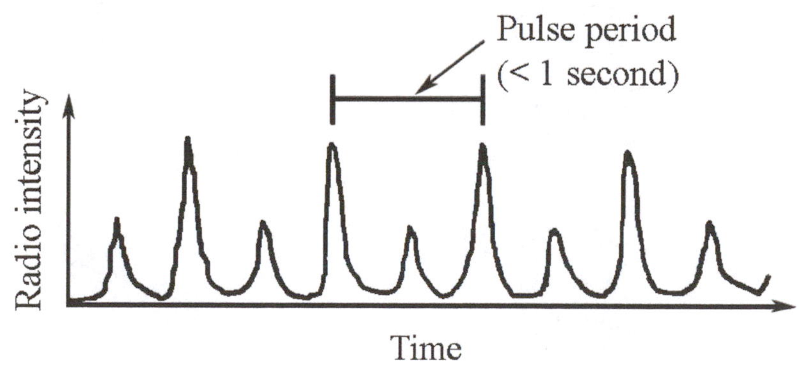

► **Figure 26-17** In 1967, Jocelyn Bell discovered *pulsars*. These radio sources repeated many times per second and *very* regularly. At first, they were called LGMs (Little Green Men). (Top: Image from Jocelyn Bell Burnell; Bottom: Image courtesy of Frederick Ringwald)

▶ **Figure 26-18** A **pulsar** is a rotating magnetic neutron star. (Sourc: NASA/Fermi Gamma-Ray Space Telescope)

▶ **Figure 26-19** The **Crab Nebula** is the remnant of a supernova, recorded by Chinese astronomers in 1054 AD. It is powered by a pulsar in its center. (Image courtesy of Frederick Ringwald, using Fresno State's station at Sierra Remote Observatories)

We can therefore summarize stellar evolution surprisingly easily:

1. Stars form by *gravitational collapse*, in a dark cloud.
2. When they start releasing energy by nuclear fusion in their cores, they settle onto the *main sequence*.
3. Low-mass stars like the Sun evolve slowly, with enough hydrogen in their cores to fuel them for billions of years. When this runs out, they become *red giants*, which come apart into *planetary nebulae*. The left-over cores turn into *white dwarf stars*.

   In other words: **Low-mass star → red giant → planetary nebula + white dwarf.**
4. Massive stars, with more mass than 11 Suns, evolve much faster, shining much hotter and with more luminosity. After "only" a few million years, they become *red supergiants*, which explode in *supernovae*, leaving *supernova remnants*.

   Their cores become superdense *neutron stars*, spheres of nuclear matter.

   In other words: **Massive star → red supergiant → supernova + neutron star.**
5. The very most massive stars, which are rare, have the mass of 50–60 Suns. These evolve even faster, exploding suddenly when they're still blue supergiants (as Supernova 1987A did). Do these leave behind black holes?

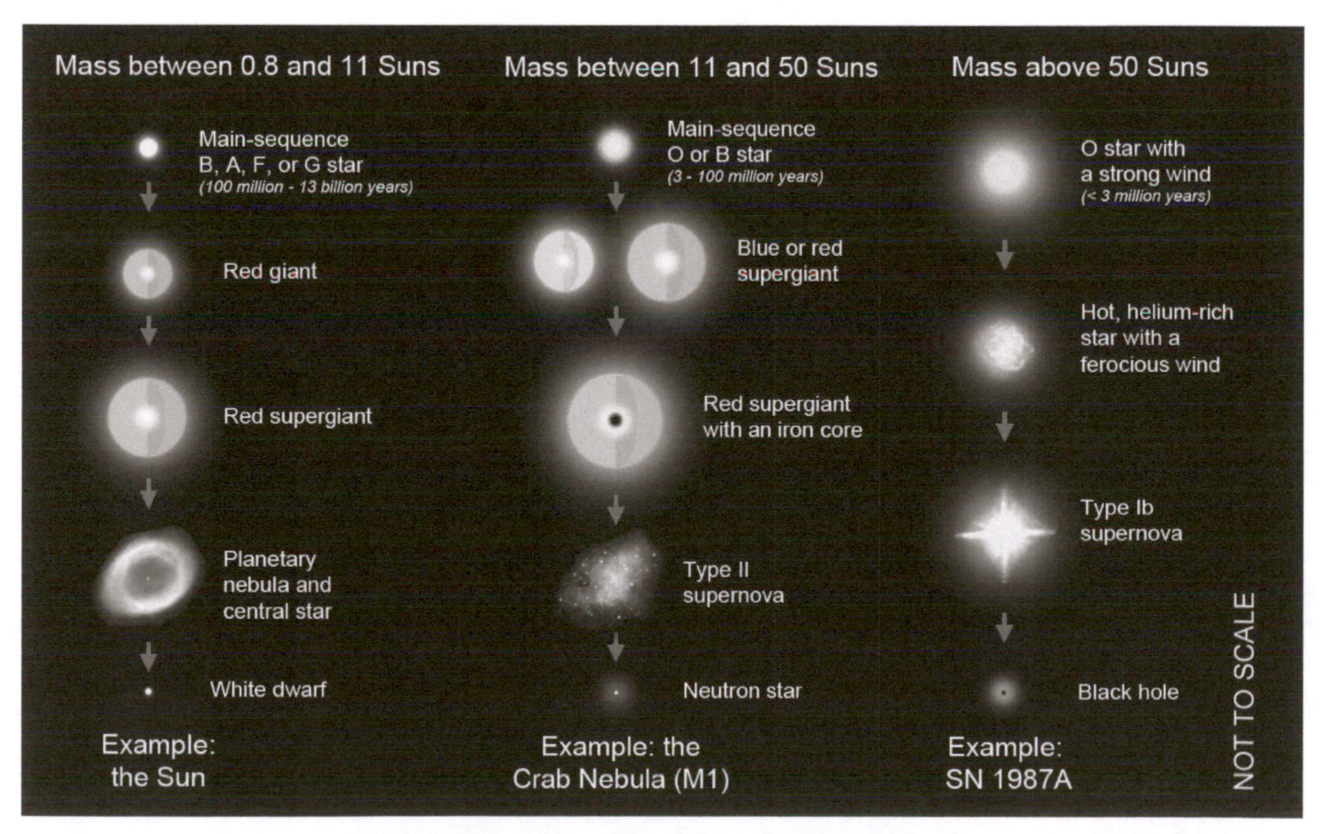

► **Figure 26-20** **We therefore can understand stellar evolution surprisingly easily. The differences between stars are almost entirely due to their *mass* and *age*.** (Image courtesy of Frederick Ringwald, based on an illustration in *Sky & Telescope,* 1997 December issue, page 38)

Chapter

27

Black Holes and Relativity

Modern Physics in a Nutshell

Black holes are one of the stranger predictions of modern physics. To understand black holes better, we must understand modern physics better.

Physics is the study of how things work. Modern physics, or physics since 1900, has a different character from classical physics, which came before 1900. Classical physics is often called Newtonian physics, since much of it comes from Newton's laws of motion and gravity.

Classical physics makes specific, deterministic predictions for how individual things move or behave. According to classical physics, if one knew all the laws of physics, and knew exactly how the Universe began ("the initial conditions"), one could calculate how everything would happen in the future, exactly. Modern physics shows that it's not so simple.

Modern physics has three main branches:

1. Quantum Mechanics
2. The Special Theory of Relativity
3. The General Theory of Relativity

Special Relativity and General Relativity were almost entirely the work of Albert Einstein, who also contributed to Quantum Mechanics.

1. **Quantum Mechanics** (Niels Bohr, Werner Heisenberg, Erwin Schrödinger, et al. 1900–1930)

   On the scale of atoms or smaller, reality becomes ruled by probability and randomness. It's not possible to calculate what any one atom will do, just what many are likely to do.

   Einstein was famously *wrong* when he said, "God doesn't play dice with the Universe." Niels Bohr, who was sitting next to him, said "Stop telling God what to do!"

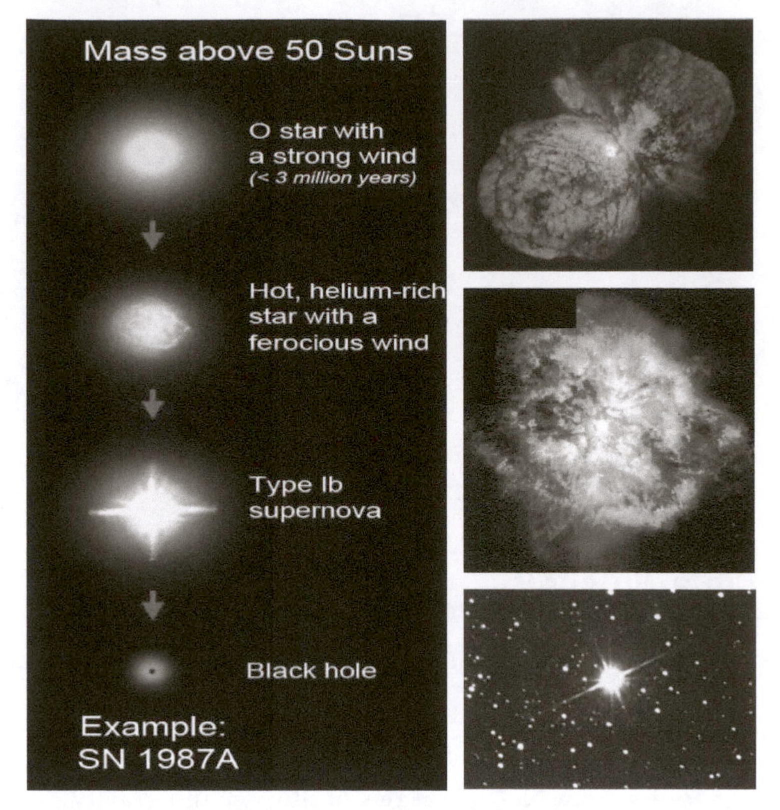

▶ **Figure 27-1** "There are more things in heaven and earth, Horatio, Than are dreamt of in your philosophy."—*Hamlet,* Act 1, Scene 5

Left: The most massive stars have lives that are nasty, brutish, and short. An O-type star with over 50 times the mass of the Sun is nearly a million times more powerful than the Sun, and has a main-sequence lifetime of less than 3 million years. (Image courtesy of Frederick Ringwald) ***Right, top:*** Again, we see examples of every step of the evolution. Fierce winds turn them into luminous blue variables, such as Eta Carinae. (Source: NASA/STScI) ***Right, middle:*** The wind drives off nearly all the hydrogen, which we know since spectra show strong helium lines but little hydrogen. (Source: NASA/STScI) ***Right, bottom:*** They then explode as Type Ib supernovae, such as SN 1987A. This may literally tear a hole in space and time. Did SN 1987A make a black hole? Astronomers are still observing what it left behind. (Image courtesy of Frederick Ringwald)

▶ **Figure 27-2** If we are to understand black holes further, we must become familiar with the works of Einstein.

Nature and Nature's laws lay hid in night:
God said, "Let Newton be!" and all was light.
– Alexander Pope

It did not last: the devil howling "Ho!
Let Einstein be!" restored the status quo.
– J.C. Squire

Einstein made scientists realize we aren't close to knowing it all. This is healthy for science: what remains to be learned about the Universe is much larger than what has been learned. (© Bettmann/CORBIS)

▶ **Figure 27-3** Quantum mechanics (QM) is how atoms work. (BSIP/Science Source) Einstein helped found QM (and got his Nobel prize for it) by showing that light isn't just a wave, it's also a particle. By analogy, it was realized that electrons and atoms aren't just particles: they also have wave properties. Electron orbits in a hydrogen atom illustrate how the properties of atoms are set by their wave nature. The wave properties of atoms give them an element of randomness. The randomness is averaged out for large numbers of atoms, as in the everyday world, but individual atoms can do strange things because of their random nature.

▶ **Figure 27-4** Although Einstein helped to create quantum mechanics, he found it "unsatisfactory." He did not like the random nature of atoms. He wanted physics to be deterministic, the way Newton's laws of motion are.

Einstein said, "God does not play dice with the Universe." Niels Bohr said to him, "Stop telling God what to do!"

This was a famous example of where Einstein was wrong. Nobody's perfect. The random properties of quantum mechanics are now well tested by experiment, if still a little strange. (Paul Ehrenfest, courtesy AIP Emilio Segrè Visual Archives/Science Source)

Chaos

Surprisingly, complex systems in the everyday world (in other words, things much larger than atoms) can *also* display seemingly random behavior. They can be very sensitive to initial conditions. Weather systems, for example, are so complex that a butterfly flapping its wings might affect the weather next week, which is why this is called "the butterfly effect." Of course, most small effects do *not* affect the whole system, but some can and do.

This was surprising, since Newton's laws, which everyday objects obey, were thought to predict an orderly, "clockwork" Universe. The study of complex systems is therefore often called **chaos** (or *chaotic dynamics*, or *non-linear dynamics*, or *complexity theory*). Chaos is a new branch of classical physics, since its importance was realized only with computer simulations of weather and other systems in the 1960s.

Also, in 1931, Kurt Gödel, an Austrian mathematician and logician, proved that for any set of assumptions, there exist a number of true statements that cannot be proved from these assumptions. (More precisely, that for any finite set of axioms, there exist a number of formally undecidable propositions.) In other words, even logic has its limits. (An excellent book on this is *Gödel, Escher, Bach,* by Douglas Hofstadter.)

Quantum indeterminacy, chaos, and Gödel's incompleteness theorems are three reasons to think that it may be possible to reconcile how free will and personal responsibility can exist in a deterministic Universe. It turns out that the Universe *isn't* strictly deterministic, or in other words perfectly predictable like a clock, as some followers of Newton insisted. Free will isn't necessarily an illusion: the illusion is the extent to which the Universe is deterministic. Despite what anyone might say, you are much more than a machine. Human beings are more than robots, or glands, or genes, or chemicals. Act accordingly!

► **Figure 27-5** The idea of a Universe that runs like clockwork, in which it's possible to deduce everything forever from the laws of nature, runs into trouble not only from the randomness of quantum mechanics.

Turbulence and chaos in a cloud are non-linear. Even though they follow Newton's laws of motion, small differences in conditions can lead to large differences in their overall behavior over time. This is called the butterfly effect: a butterfly flapping its wings in Brazil might cause a tornado in Texas. Of course, most small effects do *not* affect the whole system, but some can and do.

Even in principle, though, the Universe is still not completely predictable. Kurt Gödel's incompleteness theorems show that even in pure mathematics, it's still not possible to know everything. (Image courtesy of Frederick Ringwald)

2. **The Special Theory of Relativity** (Einstein 1905)
   Time *slows down* at speeds close to the speed of light (c).

   Fantastic as it sounds, this is now a well-observed effect. One example is cosmic rays. So is everything done in a high-energy physics lab.

   Another example is that astronauts age more slowly than they would have if they had stayed on Earth, at a rate of 9 milliseconds per orbit of Earth. This may not sound like much, but if the astronauts stay in space for a week, they can orbit Earth over a hundred times. Their total time dilation will add up to over a second, easily measured with precise clocks.

   As one's speed approaches the speed of light, time slows down. This means that time would stop at the speed of light. (It would also take an infinite amount of energy to accelerate to the speed of light. It's difficult to get an infinite amount of energy.)

   This is why we don't know how to travel faster than light. This is not a welcome prospect for anyone interested in flying to the stars. Einstein's General Theory of Relativity *might* provide a way around this, though.

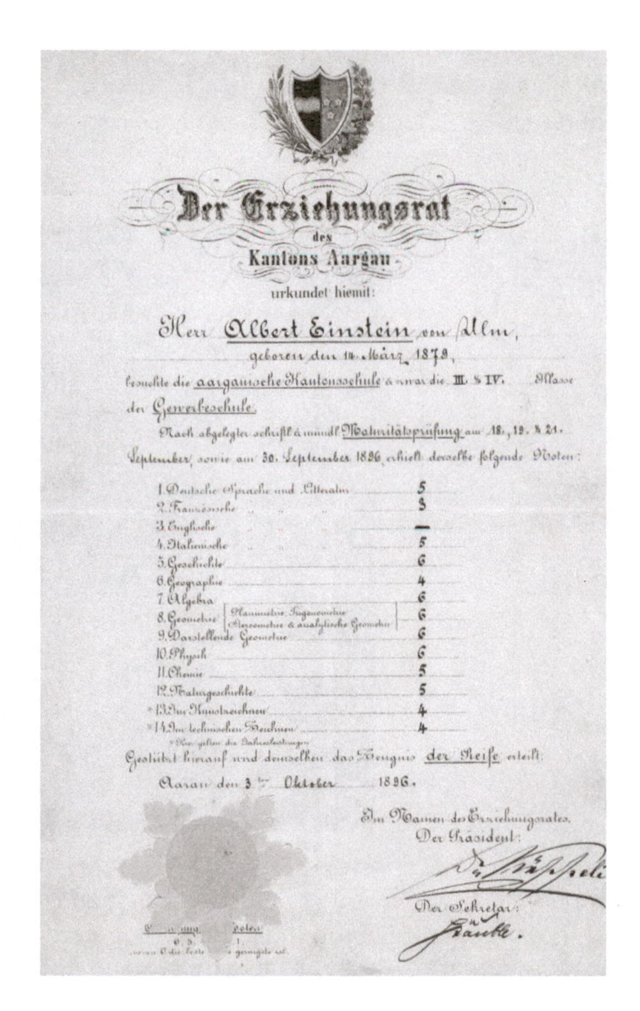

Der Erziehungsrat
des
Kantons Aargau
urkundet hiemit:

Herr Albert Einstein von Ulm,
geboren den 14. März 1879,
besuchte die aargauische Kantonsschule & zwar die III. & IV. Klasse
der Gewerbeschule.

Nach abgelegter schriftl. & mündl. Maturitätsprüfung am 18., 19. & 21. September, sowie am 30. September 1896 erhielt derselbe folgende Noten:

1. Deutsche Sprache und Litteratur	5
2. Französische	3
3. Englische	—
4. Italienische	5
5. Geschichte	6
6. Geographie	4
7. Algebra	6
8. Geometrie (Planimetrie, Trigonometrie, Stereometrie & analytische Geometrie)	6
9. Darstellende Geometrie	6
10. Physik	6
11. Chemie	5
12. Naturgeschichte	5
*13. Im Kunstzeichnen	4
*14. Im technischen Zeichnen	4

*Hier gelten die Jahresleistungen.

Gestützt hierauf wird demselben das Zeugnis der Reife erteilt.

Aarau den 3ten Oktober 1896.

Im Namen des Erziehungsrates,
Der Präsident:

Der Sekretär:

► **Figure 27-6** There are many myths about Einstein, and the worst is that he was a poor mathematician. The truth is that he was a fine mathematician. They do come better, but not by much. (He got help on his General Theory of Relativity from David Hilbert, the greatest mathematician in the world at the time, since Hilbert had an office down the hall from him.)

Einstein was *not* a poor mathematician. This is his high-school transcript. He got the highest grade (6/6) in math and physics (and history, too). He did get a C (3/6) in French.

He was also booted from school at age 15 essentially because he asked the teachers too many questions they couldn't answer. He later returned and graduated, and got into a top technical university. He mastered calculus at 15, a full year before your present author did. (Public domain image)

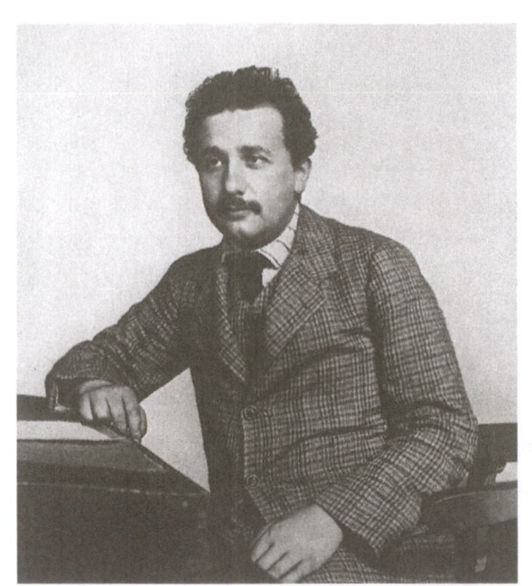

▶ **Figure 27-7** Albert Einstein's miracle year was 1905. He was working as a clerk in the patent office in Bern, Switzerland, because he couldn't get a teaching job since he loved to argue, and it did not endear him to his teachers. During his spare time, in 1905, he published five scientific papers. Four of them deserved the Nobel prize, and one of them did get him the Nobel prize, 16 years later. It wasn't the one on relativity. *Left:* Albert Einstein at the patent office, c. 1905. (© Bettmann/Corbis.) *Right:* The street that Einstein lived on, in Bern. (Image © TonyV3112, 2013. Used under license from Shutterstock, Inc.)

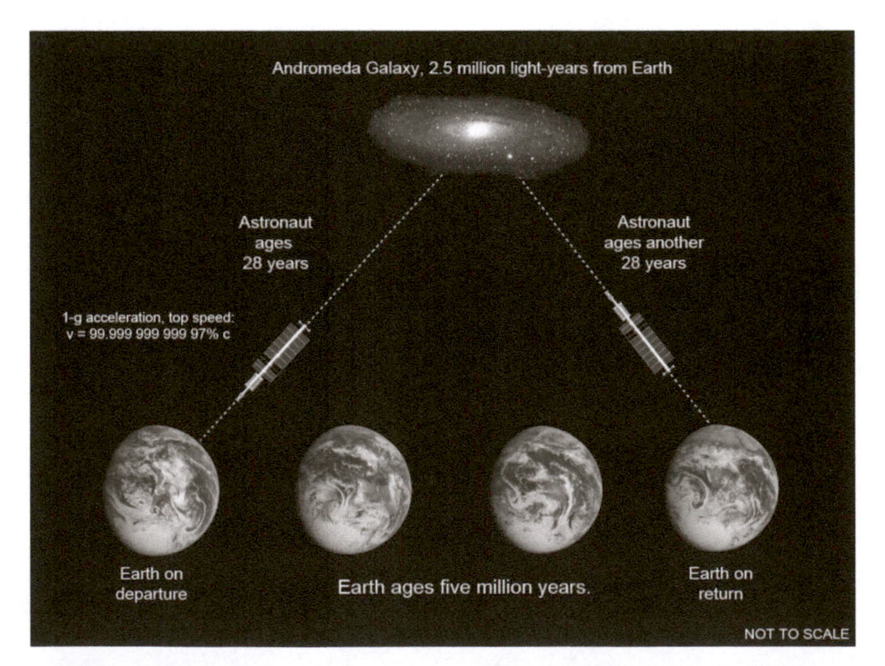

▶ **Figure 27-8** Einstein's special theory of relativity says the passage of time is not fixed: it depends on your motion. As your speed approaches the speed of light, time slows down. You don't notice anything strange. That happens when you look at anything that is not moving. If you got in a powerful starship, and accelerated to 99.999 999 999 97% of the speed of light in empty space (abbreviated as *c*), it would take you 28 years to get to the Andromeda Galaxy. If you then turned around and came back in the same way, you'd age another 28 years, so you'd age a total of 56 years before you returned to Earth. While you were away, Earth would have aged 5 million years. Since time slows down as one approaches the speed of light, it would stop at the speed of light. The speed of light is therefore the speed limit of the Universe. (Image courtesy of Frederick Ringwald, using NASA images)

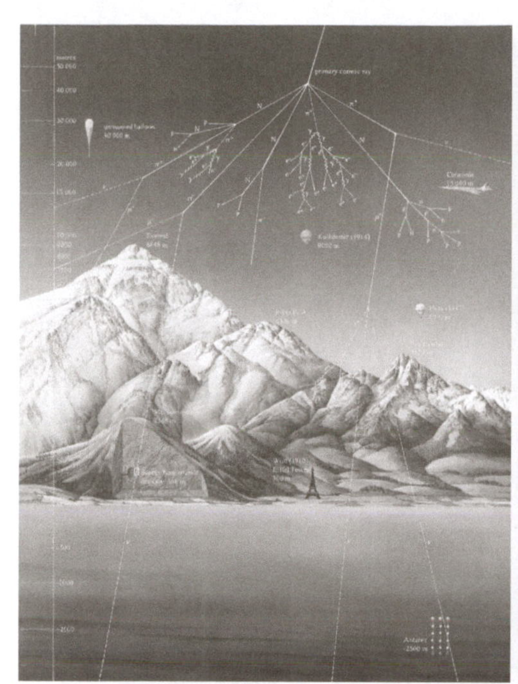

▶ **Figure 27-9** Einstein's Special Theory of Relativity may sound like pure science fiction, but it has been well observed in cosmic rays and in high-energy physics experiments. *Left:* Cosmic rays are particles from space, mostly protons, that travel at nearly the speed of light (*c*). The ones that we detect at sea level are the products of reactions the protons have with the nuclei of air molecules, high up in Earth's atmosphere. We wouldn't detect these at sea level, if they weren't traveling nearly at the speed of light, because time has slowed down for them. (Gary Hincks/Science Source) *Right:* High-energy physics experiments like this one observe special relativity many times every day. These experiments use magnets, in the big white buildings around the rings, to speed protons up to nearly the speed of light, where the effects of relativity affects everything they do. (Science Source)

3. **The General Theory of Relativity** (Einstein 1916)

   Time slows down in a gravitational field, which also *curves space*.
   The idea that space can curve isn't hard to understand: the surface of Earth is round.

   "Matter tells space how to curve, and curved space tells matter how to move."
   – John Archibald Wheeler

   Fantastic as it sounds, this is also well observed. One example is gravitational lensing, in which stars shift position during eclipses. This was the result that made Einstein famous to the general public, in 1919. (From 1905 to 1919, he'd been famous only among other scientists.)

   A **black hole** is called a hole because it really is a hole—in space, and in time. Time *stops* in a black hole, where the curvature of space becomes *infinite*. Both reasons are why a black hole is called a *hole*.

   ▶ No force at all holds a black hole up against gravity. Because of this, all the mass is in a point in the center, the *singularity*.

Recall the concept of escape velocity. The escape velocity of Earth is 11 kilometers/second (which is about 25,000 miles per hour). A rocket must go faster than this to break free of Earth's gravity: if it goes slower, it will either orbit or fall back to Earth.

- The *event horizon* is where the escape velocity becomes greater than the speed of light. Nothing can escape from inside a black hole's event horizon, not even light: that's why a black hole is *black*.
- In 1916, Karl Schwarzschild used Einstein's General Theory of Relativity to predict the existence of black holes. This is why the event horizon is also called the Schwarzschild radius. It's also called "the point of no return," since that's what it is.

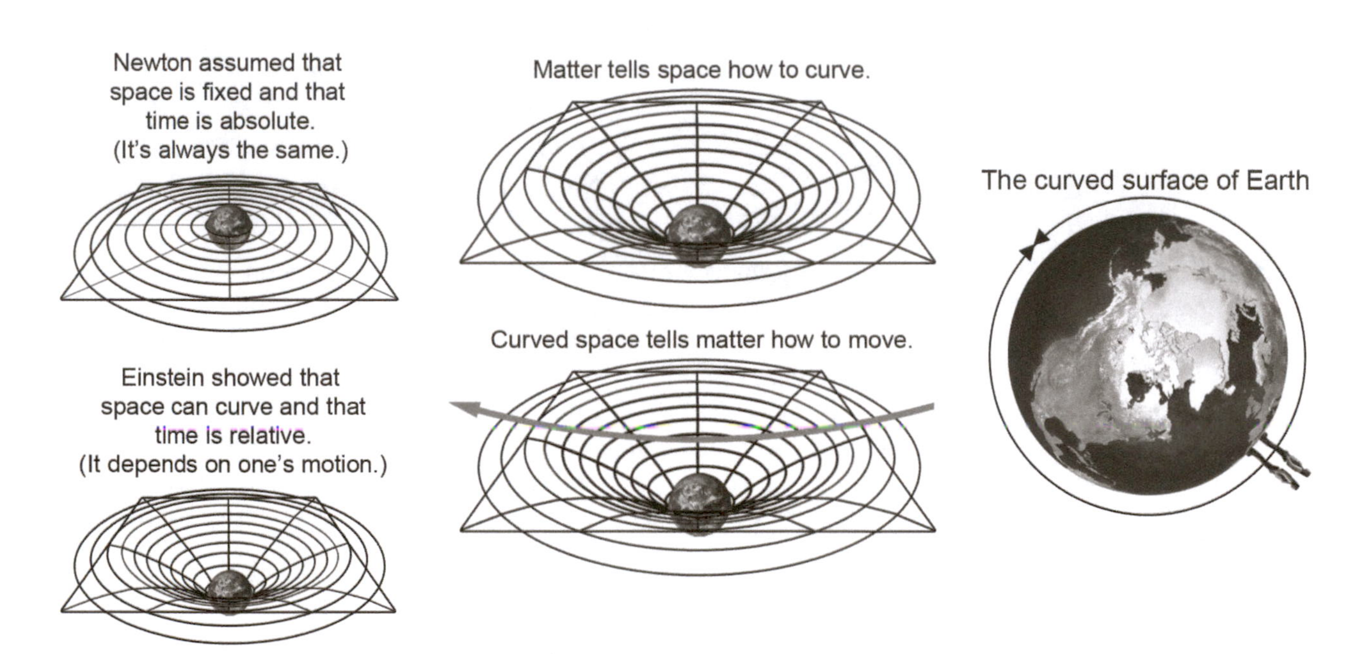

Figure 27-10 Einstein's General Theory of Relativity (1916) explains gravity as *curvature* of space and time. The idea that space can curve is not so crazy: the surface of Earth is curved. As Einstein noted, "When the blind beetle crawls over the surface of a globe, he doesn't realize that the track he has covered is curved. I was lucky enough to have spotted it." (Images courtesy of Frederick Ringwald; Earth image: NASA)

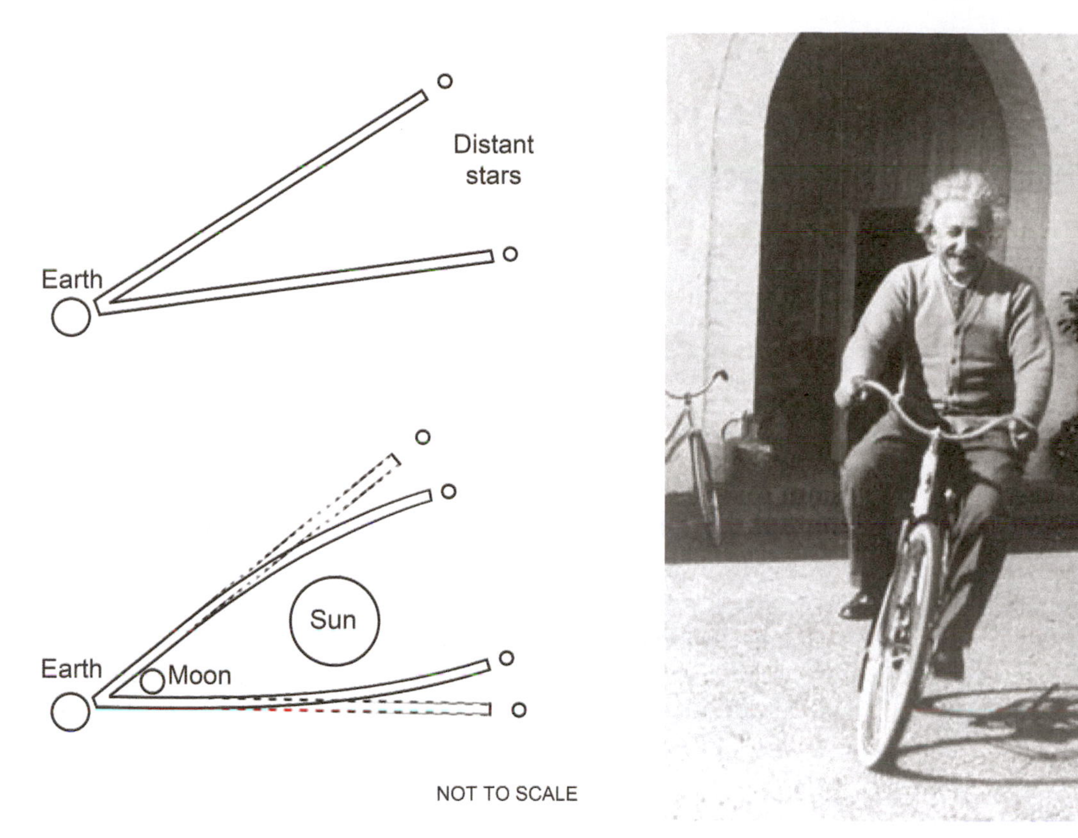

▶ **Figure 27-11** *Left:* Einstein's General Theory of Relativity predicts that gravity can *bend* starlight. This was first *observed* during a total solar eclipse in 1919. (Image courtesy of Frederick Ringwald) *Right:* This was the result that caused Einstein to become the most famous scientist of the 20th century. Before this, he'd been famous only to other scientists. After this, he was as famous as a movie star. (Archives of the California Institute of Technology)

▶ **Figure 27-12** Opinions differ about how Einstein handled his fame. Some observers say he didn't handle his fame well. Others say he clearly enjoyed being famous, and cultivated it. What's certain is that he didn't hesitate to use his celebrity to speak out against Hitler and the Nazis. Hitler and the Nazis hated him for it, and wanted to kill him. This caused him to flee from Germany to the United States. He wound up at the Institute for Advanced Study at Princeton University. (© Bettmann/CORBIS)

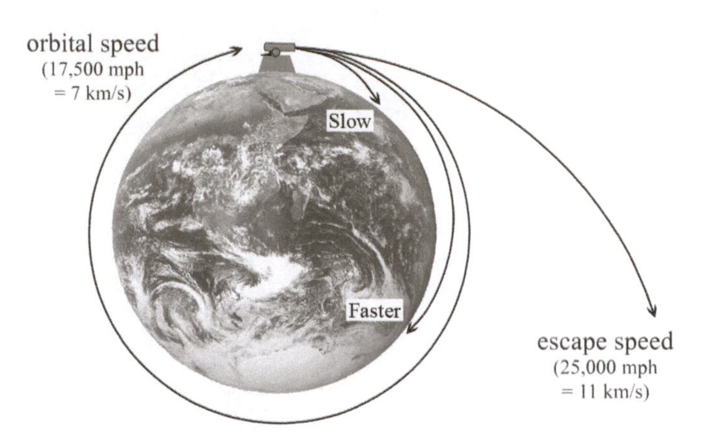

▶ **Figure 27-13** Why a black hole is black. *Left:* In 1916, Karl Schwarzschild realized that enough mass in a volume small enough would have an escape speed greater than the speed of light.

Recall the concept of *escape speed*. To reach orbit, a rocket must reach a speed of 7 kilometers/second (17,500 miles/hour), or else it will fall back to Earth. If a rocket reaches a speed of 11 kilometers/second (25,000 mph), it will reach escape speed and break free of Earth's gravity. A black hole has an escape speed greater than the speed of light. This means that nothing can escape from a black hole, not even light. A black hole is therefore black. (Image courtesy of Frederick Ringwald with an Earth image by NASA)

Einstein never accepted black holes. He thought they seemed too fantastic, even for him.

In Einstein's early career, he made advances remarkable for their originality and imagination. In his late career, he repeatedly found it hard to accept the implications of those same ideas. He never thought quantum mechanics was correct, even though he helped invent it, and there is no doubt about it now: without quantum mechanics, the digital electronics in your cell phone wouldn't work. Einstein did come to accept the expansion of the Universe, because Edwin Hubble could show it to him.

John Archibald Wheeler was credited with having made up the name "black hole," but he didn't really. In 1967, he was giving a talk at a scientific conference on black holes. He said, "One can say 'totally collapsed, self-gravitating object' only so many times before one wonders if there isn't a better name for them." An unknown audience member suggested, "How about 'black hole'?"

If no one had thought of black hole, today they might be called "frozen stars." This refers to how time stops in a black hole, because the curvature of space-time becomes infinite. It's also because black holes are thought to be made in supernova explosions, when very massive stars collapse when they die.

A common misconception is that black holes "suck." They don't—no more than any other gravity field does. A black hole's gravity field is no different from the gravity field of any other object, except that it's bottomless.

"Common sense is a set of prejudices one acquires before age 16."

—Einstein

"Nothing is too wonderful to be true, if it be consistent with the laws of nature."

—Michael Faraday

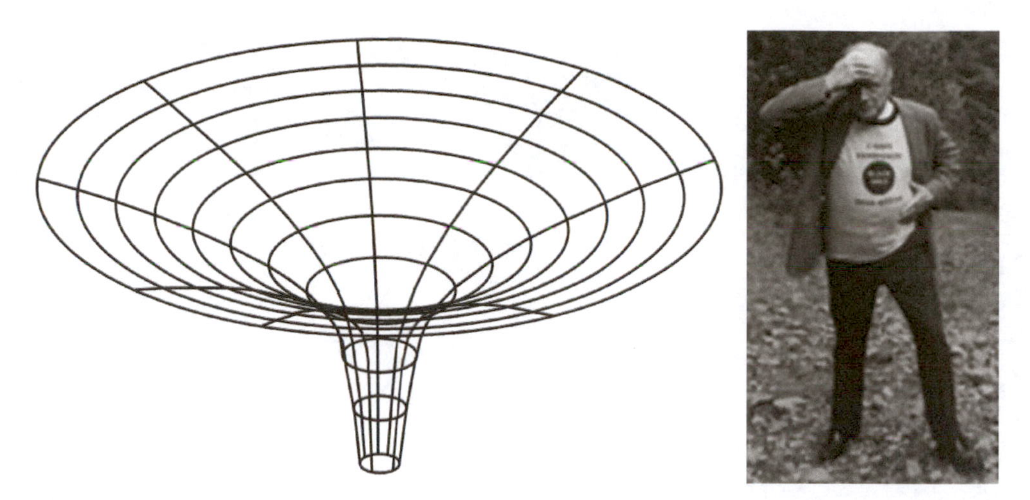

► **Figure 27-14** Why a black hole is a hole. *Left:* So much mass in such a small volume makes the curvature of space approach infinity. It tears *a hole* in space. Also: time *stops* in a black hole. (Image courtesy of Frederick Ringwald) ***Right:*** John Archibald Wheeler is widely credited with inventing the term, *black hole*, but an unknown member of the audience of a talk of his did it. (Emilio Segre Visual Archives/American Institute of Physics/Science Photo Library)

► **Figure 27-15** Contrary to popular myth, black holes don't "suck." (Neither does real quicksand.) (Image © Mediagram, 2013. Used under license from Shutterstock, Inc.)

► **Figure 27-16** *Left:* Steven Hawking first became famous by showing that black holes aren't so black, since they can evaporate. (Source: NASA) *Right:* Might some advanced civilization be able to put this to use? This artist's conception shows starships using a black hole to replenish their energy. (Image courtesy of Mark Garlick)

If a black hole is a hole in space and time, where does it lead?

No one knows. Some solutions of Einstein's equations in General Relativity suggest that they lead to "other Universes"—quite what "other Universe" means, this author doesn't know—or that they might be shortcuts to other places in our own Universe.

If so, they might offer a way to travel through the Universe faster than is allowed by Special Relativity, which of course limits speeds to the speed of light. Such shortcuts are called *wormholes*, which refers to a worm being able to travel through a hole it's eaten in an apple faster than it can walk around the outside surface of an apple. It also pays tribute to Isaac Newton, who was inspired to think of gravity by the fall of an apple.

Wormholes are a popular staple of science fiction. Some stories call it "warp drive"; some other stories call it "hyperspace," or a "star gate." It should be emphasized that, as of now, they are ***fiction***: scientists today don't know whether they exist, or not.

Astronomers and physicists have searched, but have so far still never found any material object or information that was genuinely traveling faster than the speed of light, despite some false alarms. Jets seen coming from quasars (to be covered later in the course), called superluminal sources, appeared to move faster than the speed of light, but this turned out to be an illusion.

It still hasn't stopped some scientists from speculating that a starship might be able to manipulate space-time and dive through wormholes it generates in the Universe. It sounds like a wonderful idea, but this author says: You first!

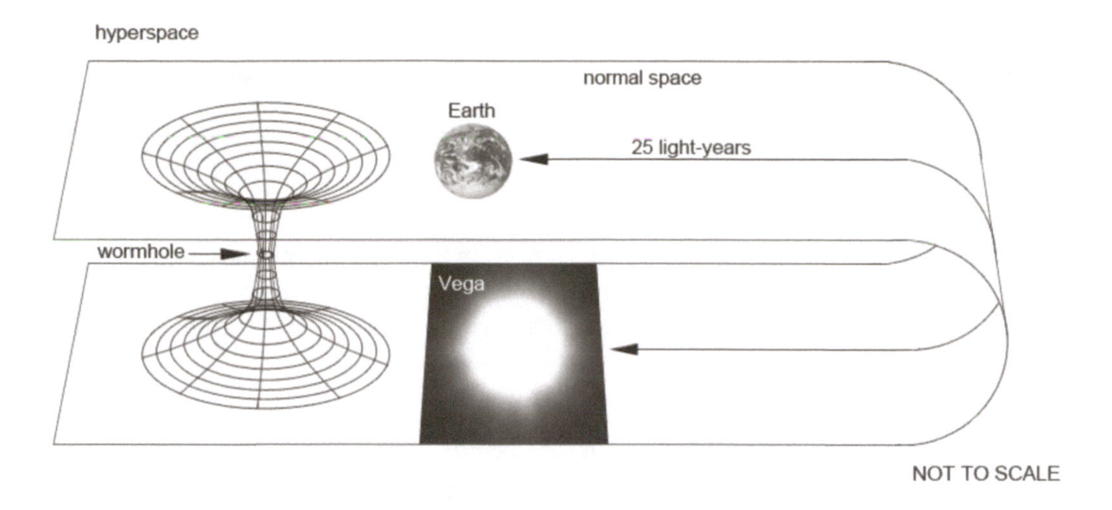

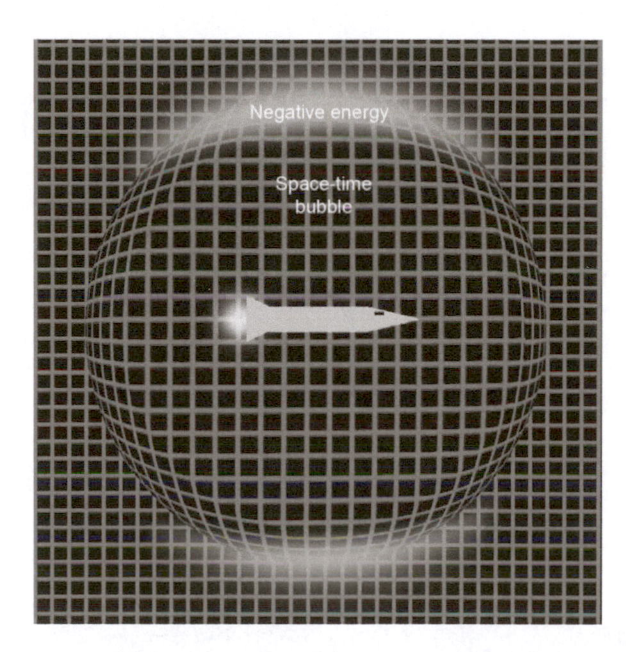

► **Figure 27-17** Wormholes may provide a way to travel faster than light by taking shortcuts through higher dimensions ("hyperspace" or "warp drive"). This is still science fiction. (Images courtesy of Frederick Ringwald)

Why are we so sure that black holes are real, and that we have found them?

1. "We know that neutron stars exist, and black holes are pretty close!"

   This isn't a strong argument. Just because it's plausible doesn't mean it must exist. It's like saying, "We know that horses exist, and so unicorns must exist, since unicorns are pretty close to horses."

2. Cygnus X-1 is the brightest source of X-rays in the constellation Cygnus the Swan. It was discovered in 1971. In Cygnus X-1, an OB supergiant star spills gas into the source of the X-rays. The spectrum of the X-rays fits very well to predictions of what X-rays from gas going down a black hole should be like.

   So, why didn't astronomers proclaim they'd discovered a black hole? Some did, but as Carl Sagan liked to say, "Extraordinary claims require extraordinary evidence." Discovering a black hole is an extraordinary claim: many astronomers wondered whether there might be some *other, independent* evidence that supports this claim?

3. To hold up against collapse, a neutron star with a mass of greater than 3 Suns would have to be made of material that's infinitely strong. Indeed, nuclear theory predicts how strong neutron star material should be: a realistic neutron star should collapse, if it had a mass greater than 2.7 ± 0.3 Suns.

   Can we find a compact object with a mass we can prove is greater than 3 Suns?

   We measure mass of stars and black holes with Kepler's Third Law, since if one can observe one object orbiting another object, it shows one how strong the gravity between the two objects must be. This is done by getting spectra all around the orbit, and measuring how much each of the two objects' gravity pull each other around by measuring the Doppler shift. This is similar to how we detect planets around other stars, by measuring how their gravity pulls their parent stars around.

► **Figure 27-18** Cygnus X-1 was discovered in 1971, because of its X-rays. (Source: NASA/CXC/MWeiss)

From 1971 to 1994, the compact star in Cygnus X-1 was hailed as a likely black hole candidate, but the case that we really had discovered black holes wasn't decisive. Mass estimates for Cygnus X-1 were in the range of 10 Suns, but they were only estimates. The problem was we couldn't measure the other properties of the system well enough to be sure, because Cygnus X-1 is unluckily nearly face-on to us.

4. V404 Cygni: In 1994, Jorge Casares, Phil Charles, and Tim Naylor (when all of them were at Oxford) got spectra to measure the masses of both stars in V404 Cygni, a binary system in which there had been a nova eruption. They found that one star is a K giant. The other, which emits X-rays, is compact, and has a mass of 6.0 ± 0.4 Suns—definitely a black hole.

   Now, over 20 similar black holes in binary star systems are known.

   Still, this is only indirect evidence. Can we show that black holes exist, more directly?

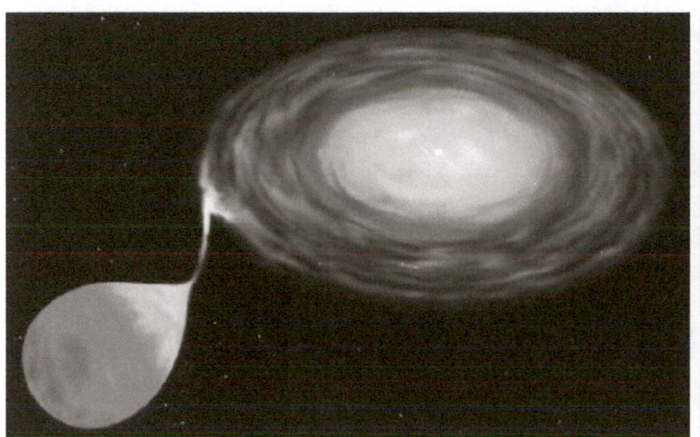

► **Figure 27-19** *Left:* **Since 1994, at least 20 black holes have been identified by how their gravity affects companion stars.** (Source: NASA/STScI) *Right:* **Jorge Casares in 1993, around the time he measured that the gravity of the compact object in V404 Cygni (GRS 1915+105/Nova Cyg 1938) pulled around its companion star so much, the compact object had to have a mass of at least 6.0 solar masses, and so is definitely a black hole. I was asking him, "What are you drinking in the middle of the day, Jorge?" It was water.** (Image courtesy of Elena Pavlenko)

5. Event horizons not seen: In 1997, Ramesh Narayan and Mike Garcia (both of whom are at Harvard) showed that binary star systems containing black holes are systematically fainter than similar star systems containing neutron stars. Why? Because they have something *black* in them—the event horizon.

   Still, a black cup of coffee is also black, and it isn't a black hole. Is there direct evidence that black holes exist?

6. Yes, there is. If one dropped a clock into a black hole, one would see the second hand go slower and slower, because of how time slows down, as predicted by Einstein's General Theory of Relativity.

This observation has now been done: a NASA spacecraft, called *Rossi X-ray Timing Explorer (RXTE)*, has seen the effect of space and time being dragged around a rotating black hole.

As gas falls into a black hole, waves in the gas get longer in wavelength. The observations with *RXTE* show this to happen in the way predicted by Einstein's General Theory of Relativity, because of *time slowing down*.

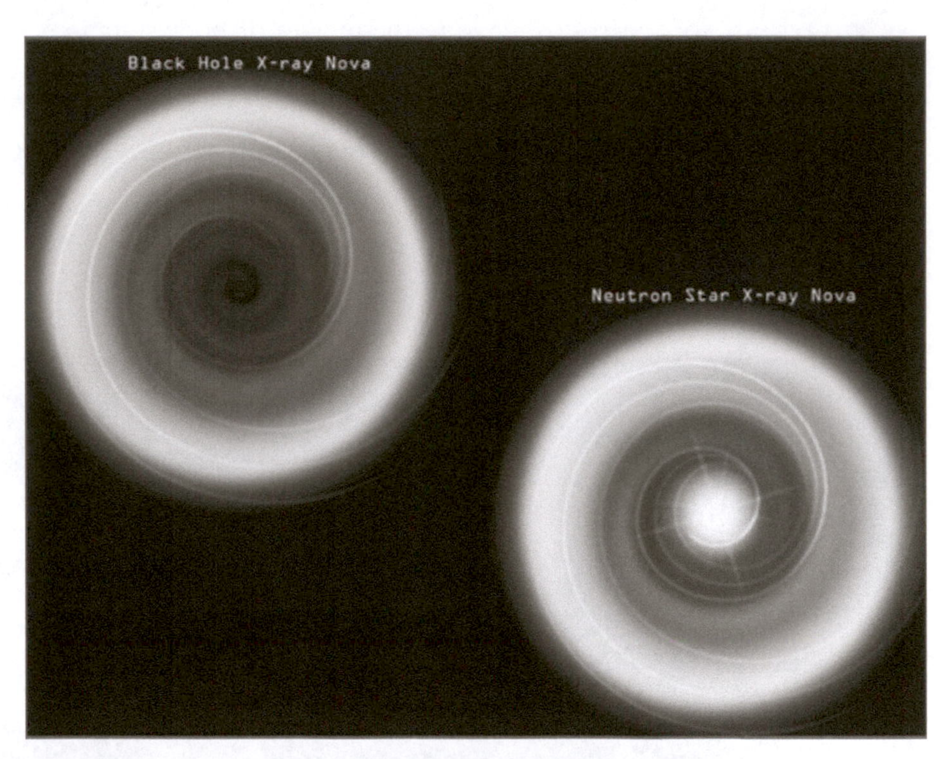

► **Figure 27-20** Black holes are black: binary star systems that contain black holes are systematically fainter than ones that contain neutron stars. Apparently, there is something black in the black-hole systems. It is reassuring to find that black holes are black, but it doesn't prove that they are black holes in the first place: a black cat and a black cup of coffee are also black, and they are not black holes. (Source: NASA/CXC/M. Weiss)

► **Figure 27-21** Black holes act like black holes: direct evidence that black holes exist comes from how their gravity follows predictions from Einstein's General Theory of Relativity. General Relativity describes gravity of curvature of space, and of time. This means that if one were to drop a clock into a black hole, one would see the clock's second hand going slower and slower, as the clock fell into the black hole. Exactly this has been observed. Waves in gas going down a black hole (in the binary star system GRO J1655-40) were observed by a NASA spacecraft (*Rossi X-ray Timing Explorer*) to spread out in time, just as predicted by General Relativity. (Source: NASA/CXC/April Hobart)

Chapter 28

Ultimate Address and the Large-Scale Structure of the Universe

California and the History of Astronomy

1888: Lick Observatory, near San José, became the first permanent mountaintop observatory, with the completion of its 36-inch refracting telescope. It demonstrated the usefulness of such facilities, with its weather, and more importantly, with its superior *seeing* (the lack of atmospheric turbulence, resulting in clear, detailed, high-resolution images). Edward Emerson Barnard used it to pioneer astronomical photography.

1896: The 36-inch Crossley reflector at Lick Observatory demonstrated the superiority of reflecting optics for large telescopes. Nearly all later large telescopes would be reflectors.

1908: The 60-inch telescope on Mt. Wilson, in the San Gabriel Mountains near Pasadena (and highly recommended as a day trip from L.A.), became the world's largest telescope.

1917: The 100-inch telescope on Mt. Wilson became the world's largest, until 1948.

1920s: Edwin Hubble used the Mt. Wilson 100-inch to discover the nature of galaxies (that they are "island Universes" of billions of stars outside our galaxy, the Milky Way).

1929: Edwin Hubble announced his discovery of the expansion of the Universe, based on spectra from the 100-inch, taken mainly by Milton Humason.

Humason came to Mt. Wilson as a mule driver during its construction. He fell in love with the director's daughter and stayed on as the facility's first janitor. Held back by his eighth-grade education, he sought out tutoring in math by several astronomers. The story goes that he was called in one night to substitute for the telescope operator, who was ill, and proved superbly capable with the instruments (although he had shown care and skill with instruments during the observatory construction). He became telescope operator and an accomplished scientist in his own right, eventually receiving an honorary doctorate.

1930s: Caltech astronomers Fritz Zwicky and Walter Baade discovered supernovae. Zwicky also discovered galaxy clusters and evidence for dark matter.

1943–44: Walter Baade discovered stellar populations. This was because he got every night on the 100-inch telescope for over a year, since his German nationality made him ineligible for the war work most other astronomers were doing at the time.

1948: The 200-inch (5-meter) telescope on Mt. Palomar, near San Diego, became the world's largest telescope, until 1991. In 1962, it was used to discover quasars. In 1995, it was used to discover brown dwarfs.

1958: The NASA/Caltech Jet Propulsion Laboratory, founded in 1936, built and operated America's first satellite, *Explorer 1*. JPL would become the world center for planetary exploration by robotic spacecraft, operating the *Viking* orbiters and landers to Mars (the first spacecraft purpose-built to search for life on another world), the *Voyager* probes to the outer planets, the *Galileo* Jupiter orbiter, the *Cassini* Saturn orbiter, and many more.

► **Figure 28-1** "There is a theory which states that if ever anyone discovers exactly what the Universe is for and why it is here, it will instantly disappear and be replaced by something even more bizarre and inexplicable.

"There is another theory which states this has already happened..."

—Douglas Adams, from *The Restaurant at the End of the Universe,* the sequel to *The Hitchhiker's Guide to the Galaxy* (Source: NASA/STScI)

Once upon a time, a long, long time ago, when your present author was six years old, my dear, sweet Auntie Agnes gave me a National Geographic globe of the world. It is a beautiful thing—I still have it—and my Mom insisted that I write a letter to Auntie Agnes to thank her for such a fine present.

So I sat down to write the letter. Auntie Agnes lived in England, and I noticed that I had to write "United Kingdom" at the end of her address. Likewise, it's necessary to write "U.S.A." at the end of an address when sending a letter to the United States from outside the country.

Notice how addresses go from local to global. Typically postal mail addresses list street number, street, city, state, postal code, and country.

This got me to wondering: What is my ultimate address? How would someone anywhere in the Universe address a letter to get it to me here?

(Of course, you probably wouldn't send a postal mail letter across the Universe. E-mail would be much better, one reason being that e-mail travels at the speed of light. Notice, however, that e-mail addresses also go from local to global coordinates, from user to machine to site to domain, such as .com or .edu.)

Here is my "Ultimate Address":

2345 E. San Ramon Ave.
Fresno, CA 93740-8031
U.S.A.
Earth
Solar System
Orion Spur
Perseus Arm (formerly Orion Arm)
Milky Way Galaxy
Local Group (of galaxies)
Local Supercluster
~~Local Supercluster Complex~~
Observable Universe

The United States is on Earth, which is a planet in the Solar System. The Solar System is the planetary system of the Sun, which is in the Orion Spur of the Perseus Arm of the Milky Way Galaxy, the local "island Universe" of 600 billion stars. (Bill Morgan's original 1951 map of the arms of the Milky Way listed it as the "Orion Arm," but modern observations are done by computer and much more thorough.)

The Milky Way is in the Local Group of Galaxies, which is on the outskirts of the Virgo Cluster of about 2,000 galaxies, which is the nearest large cluster of galaxies. The Local Group, the Virgo Cluster, and about 100 other galaxy groups make up the Local Supercluster of galaxies, which includes about 10,000 galaxies.

In 1989, astronomer Brent Tully thought he had evidence that superclusters of galaxies cluster into even bigger things, which he called "supercluster complexes" (and mercifully not "super-duper clusters"). This has turned out to be incorrect: recent deep surveys of the Universe, done largely by computer, show that the largest things in the Universe apparently are *superclusters* of about 10,000 galaxies. They are separated by roughly spherical *voids*, which are comparable in size to the superclusters, about 100 million light-years across. The voids are almost, but never entirely, devoid of galaxies.

If there were bigger things, such as supercluster complexes, the recent surveys are sensitive enough to have seen them, and they aren't there. So, in my Ultimate Address above, I have crossed out "Local Supercluster Complex," since apparently it doesn't exist.

For completeness, "Observable Universe" is included at the end of the Ultimate Address. Don't make the mistake of thinking that "Observable Universe" necessarily means the entire Universe. Since the speed of light is finite, it is possible to see only so far. The Universe may well be *much* larger than what can be seen. It may even be infinite in volume. Whether the Universe is finite or infinite, either is mind-boggling.

▶ **Figure 28-2** Earth... (Source: NASA/Apollo 8 crew/William A. Anders)

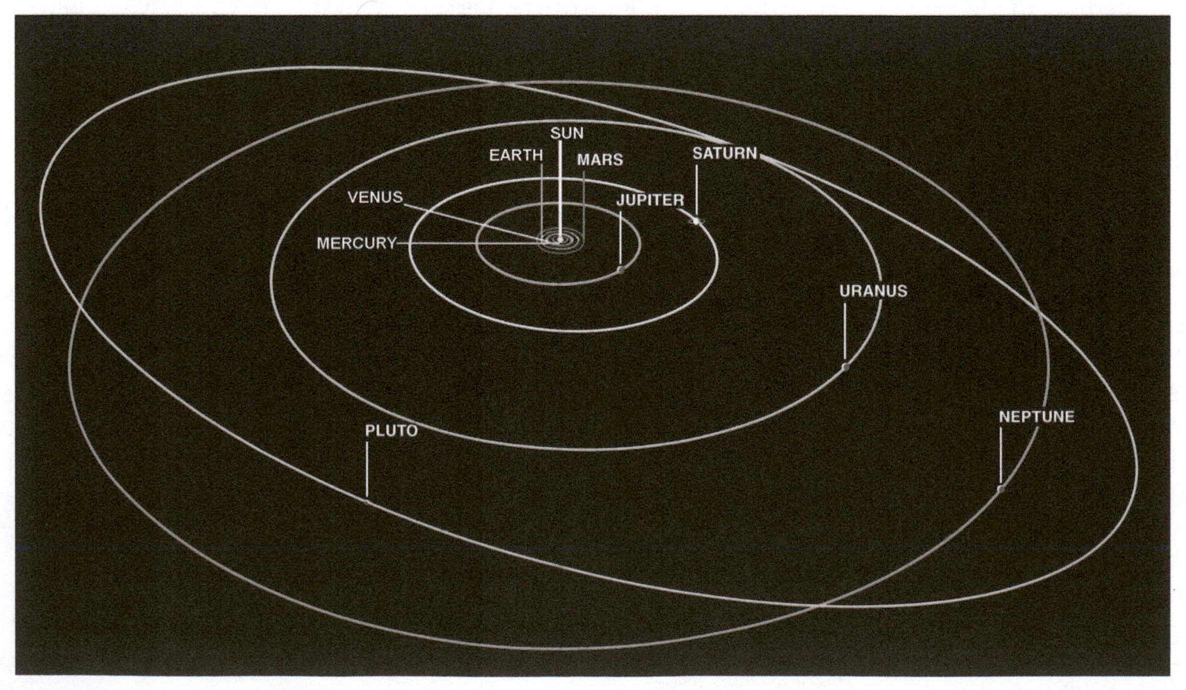

▶ **Figure 28-3** ...Solar System... (Source: NASA/LPI)

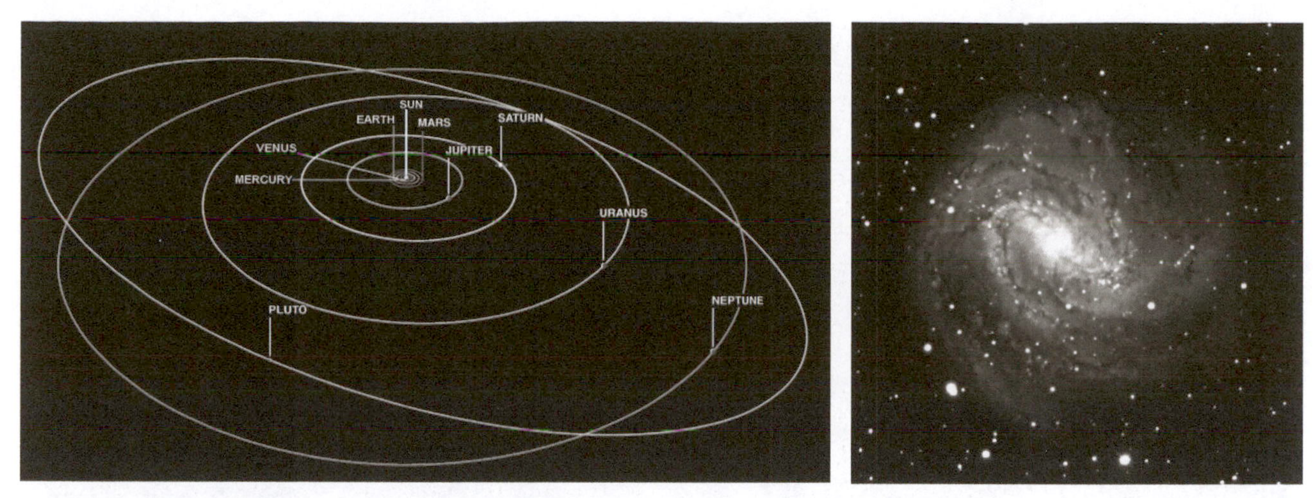

► **Figure 28-4** Don't confuse *galaxies* with *planetary systems*, or *systems of planets*. *Left:* The Solar System is our planetary system. It has one star (the Sun) and its system of planets. (Source: NASA/LPI) *Right:* This is a whole galaxy, with hundreds of billions of stars. (Image courtesy of Frederick Ringwald)

► **Figure 28-5** ...Orion Spur of the Perseus Arm... (Source: NASA/JPL/Caltech)

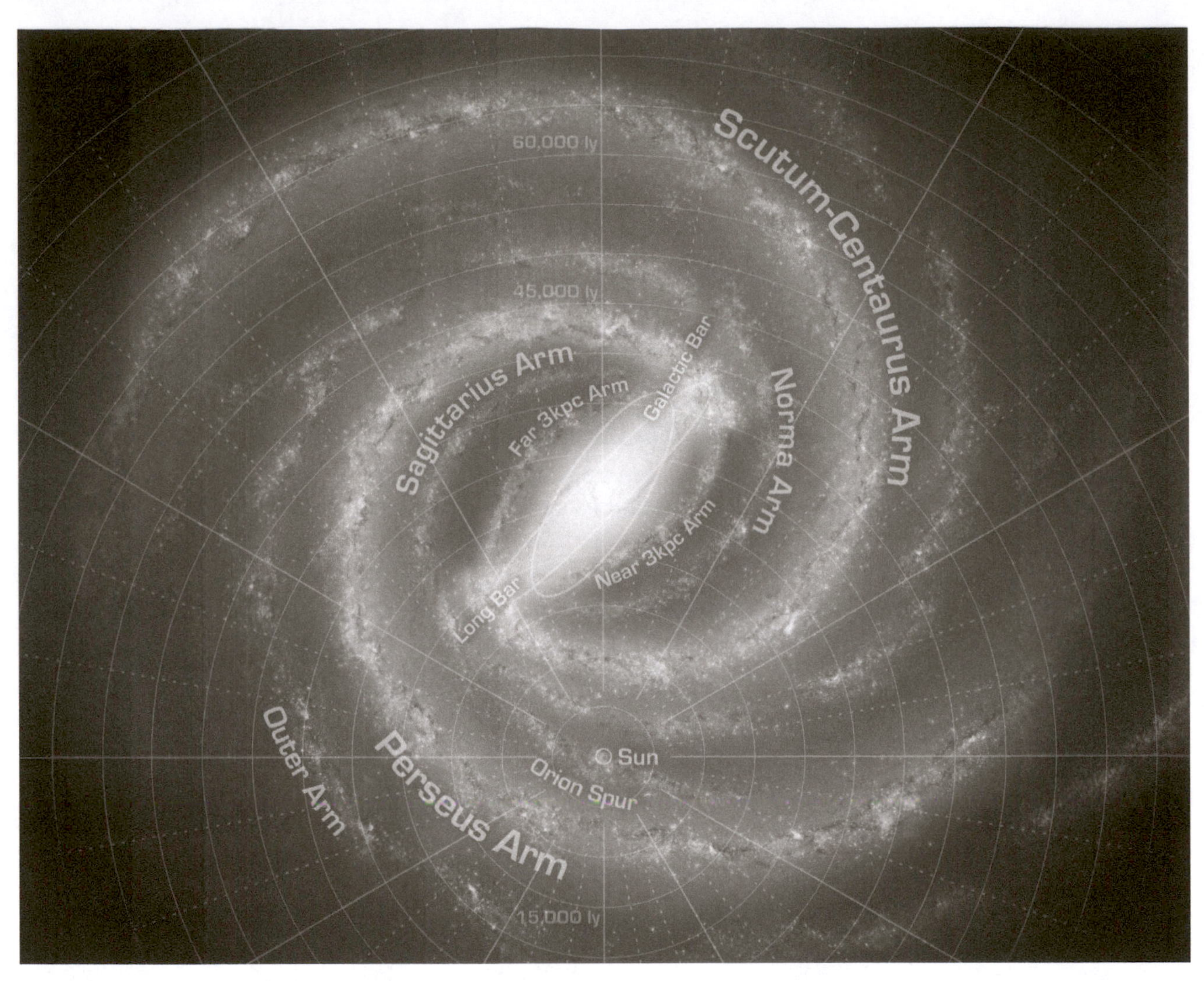

▶ **Figure 28-6** …Milky Way Galaxy… (Source: NASA/JPL/Caltech)

► **Figure 28-7** Here is an image of the Milky Way Galaxy without the lines centered on the Solar System, which is just one of hundreds of billions of other planetary systems orbiting their parent stars.

One might wonder, how was this picture taken? Humans don't have spacecraft that can fly outside the Milky Way, turn around, and take a picture of it (yet). This picture is a map, not a photo. It was built up tediously, of measurements of the positions and motions of thousands of stars, and then assembled like a jigsaw puzzle.

Notice how the center of the Milky Way Galaxy is elongated into a bar. Astronomers didn't catch onto this until recently, since by luck Earth happens to be looking approximately down the bar. Barred spiral galaxies like the Milky Way are not unusual. The bars come from tides from the gravity of neighboring galaxies. (Source: NASA/JPL/Caltech)

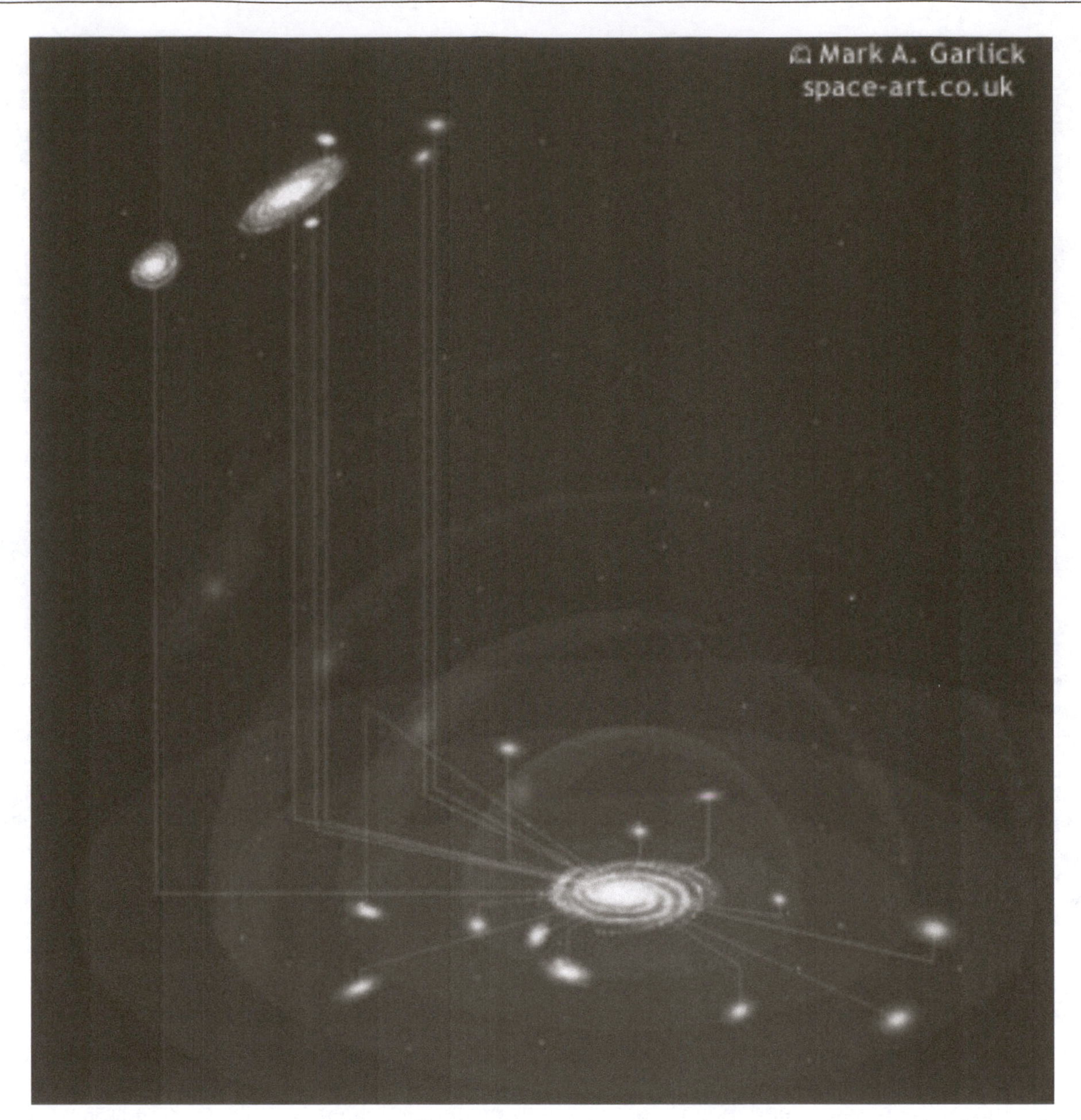

► **Figure 28-8** . . . Local Group (of galaxies). . . The Local Group includes two giant spiral galaxies, the Milky Way Galaxy (the one we're in, shown at lower right), and the Andromeda Galaxy, also called the Great Galaxy in Andromeda or M31 (shown at upper left). The Milky Way has about 600 billion stars; the Andromeda Galaxy has about 800 billion stars. Both the Milky Way Galaxy and the Andromeda Galaxy have about 1–2 dozen satellite galaxies. Each has 1–10 billion stars each. (Image courtesy of Mark Garlick)

▶ **Figure 28-9** ...Local Supercluster... The Virgo Cluster of galaxies is the nearest cluster of galaxies. It is 55–65 million light-years away, and we are on the outskirts of it. The Local Supercluster (of galaxies) is centered in it and contains about 100 smaller galaxy groups, one of which is the Local Group (of galaxies). (Image © AZSTARMAN, 2013. Used under license from Shutterstock, Inc.)

▶ **Figure 28-10** Superclusters of galaxies appear to be the largest things in the Universe. We don't know what the smallest things in the Universe are. (Quarks? Strings?)

Optical map showing the large-scale distribution of galaxies in a region of sky near the south galactic pole. The image, which has been colored blue, shows the clumpy nature of the universe. On scales of tens of millions of light years, galaxies seem to dot the surfaces of giant interconnected bubbles surrounding immense voids. The map covers 4,300 square degrees, or about 10% of the sky, and shows the distribution of some two million galaxies. Each pixel represents a patch of sky 8 arc minutes on a side; pixels are black where there are no galaxies, bright blue where there are more than 20 and darker blue elsewhere. (Maddox, Sutherland, Efstathiou & Loveday/Science Source).

► **Figure 28-11** Observable Universe. There are over 100 billion (10^{11}) galaxies in the Observable Universe. How many eyes are looking back, and is it an odd or an even number? (Source: NASA/STScI)

Chapter

29

Cosmology

Cosmology is the study of the Universe as a whole, including its history and origin. (It has nothing to do with haircuts or makeup: that's cos*met*ology.)

What evidence is there that the Big Bang really happened?

1. **The observed expansion of the Universe**, discovered in 1929 by Edwin Hubble: All but the nearest galaxies have redshifts in their spectra. Hubble's law is the farther the galaxy, the greater the redshift (so $v = H_0 D$).
2. Gravity requires the Universe to be either expanding or contracting: it can't be static. In 1916, Einstein discovered that **his General Theory of Relativity predicts this**, but he didn't call attention to it until 1929, when Edwin Hubble first observed it.
3. **The cosmic background radiation** was discovered in 1964 by Arno Penzias and Bob Wilson. It is exactly the thermal spectrum predicted for a Universe that was once hot and dense, and has since expanded and cooled.
4. **The Olbers paradox**: If the Universe were infinitely large and infinitely old, the night sky should not be dark. It should be as bright as the Sun, because in no matter which direction one looked, one would eventually see a star. The solution to the paradox is: the Universe is not infinitely old. Interestingly, since galaxies are so smoothly spread out, the Universe may be infinitely large, but it has a finite age.
5. **The abundances of helium and other light elements**: During the first three minutes of the existence of the Universe, the whole Universe was as dense and as hot as the center of a star. The same nuclear reactions that happen in stars were happening everywhere. We understand stars (and nuclear weapons) well enough to make predictions for the early Universe. Measurements from spectra, starting in 1972, agree with these predictions.
6. Recall the concept of look-back time: as we look deep into space, we look back in time. This is because light has a finite speed, and took time to get here.

   Looking deep into the Universe, **we see change over time**, or in other words, **evolution**:

   a. In the distant past, galaxies of all kinds were much bluer, with many hot, short-lived stars (OB types), after accounting for the effect of redshift.
   b. Galaxies now are more often observed in clusters than they were in the past. This clustering has progressed over the age of the Universe because of gravity.
   c. Irregular galaxies were much more common during the first 1–3 billion years of the existence of the Universe. Spiral galaxies have become more common since then, because of collisions and mergers between galaxies.

d. Giant elliptical galaxies, sometimes called "cannibal galaxies," have become more common over the age of the Universe, and in the centers of galaxy clusters. This is because they are observed to be eating other galaxies, mainly spirals.
e. Quasars are the most luminous (powerful) galaxies. Quasars are agitated galaxies, which are in the process of forming. They were much more common in the distant past than they are now. Normal spiral galaxies (like the Milky Way) have become more relaxed, having settled into stable orbits because of gravity.

▶ **Figure 29-1** In the 1920s, a debate gripped astronomy. One side, led by Harlow Shapley, said that the Milky Way *(top)* was the entire Universe. The other side, which included Edwin Hubble, said that there were other galaxies, or "island universes," of hundreds of billions of stars *(bottom)*. Shapley thought that these were nearby objects in the Milky Way, namely planetary systems in formation (which we now call proplyds, and were discovered by *Hubble Space Telescope* in 1995). Edwin Hubble won the argument, which was settled by his observations, made throughout the 1920s. (Top: Image courtesy of Frederick Ringwald at Fresno State's station at Sierra Remote Observatories; Bottom: Image courtesy of Frederick Ringwald at Mount Laguna Observatory)

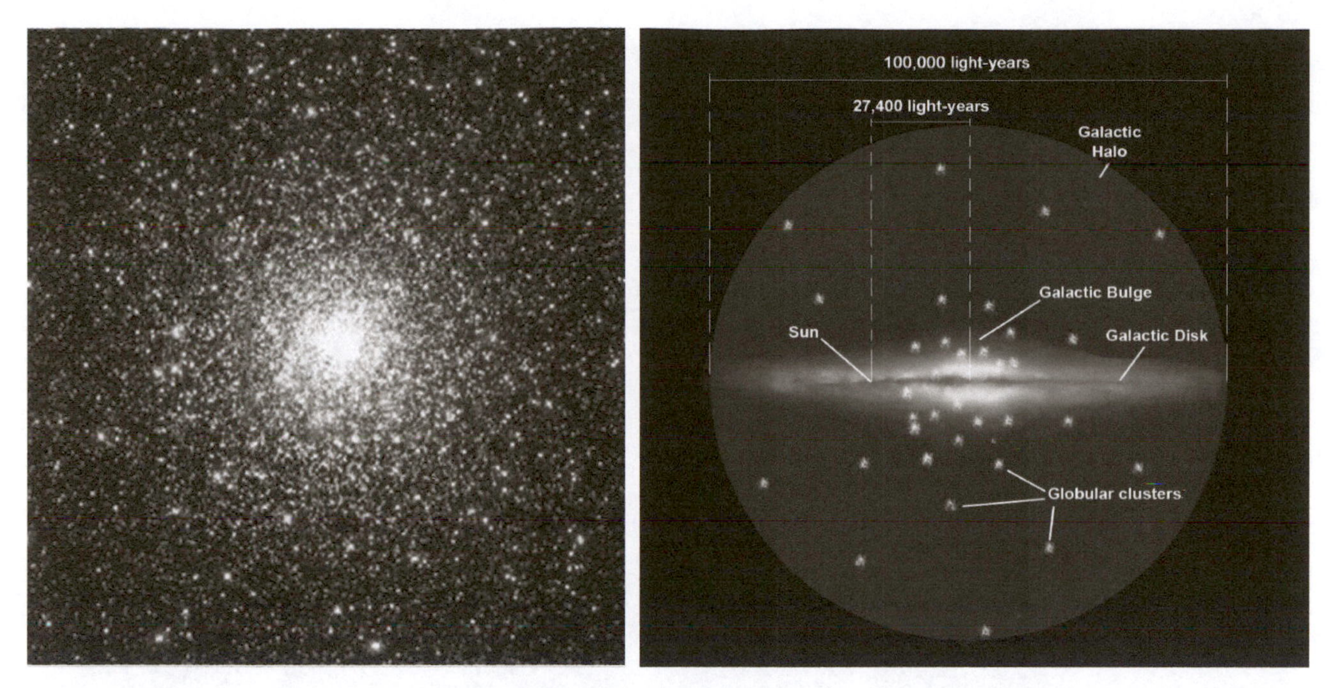

▶ **Figure 29-2** In hindsight, Shapley perhaps can be forgiven for this, because he had recently shown that the Universe was 100 times larger than was previously known. In 1915, Shapley showed the Solar System is not at the Milky Way's center. He did this by noticing that globular clusters (*left*), now known to be ancient clusters of stars, are not evenly spread out over the sky: nearly all of them are observable in boreal summer. Shapley measured the distances to many globular clusters, and found they are centered on a point in space about 30,000 light-years away, in the constellation Sagittarius. The Milky Way passes through Sagittarius, and is brightest and thickest there. Shapley realized that he had discovered the center of the Milky Way Galaxy (*right*), and that we are not close to it. (Left: Source: NASA/STScI; Right: Image courtesy of Frederick Ringwald)

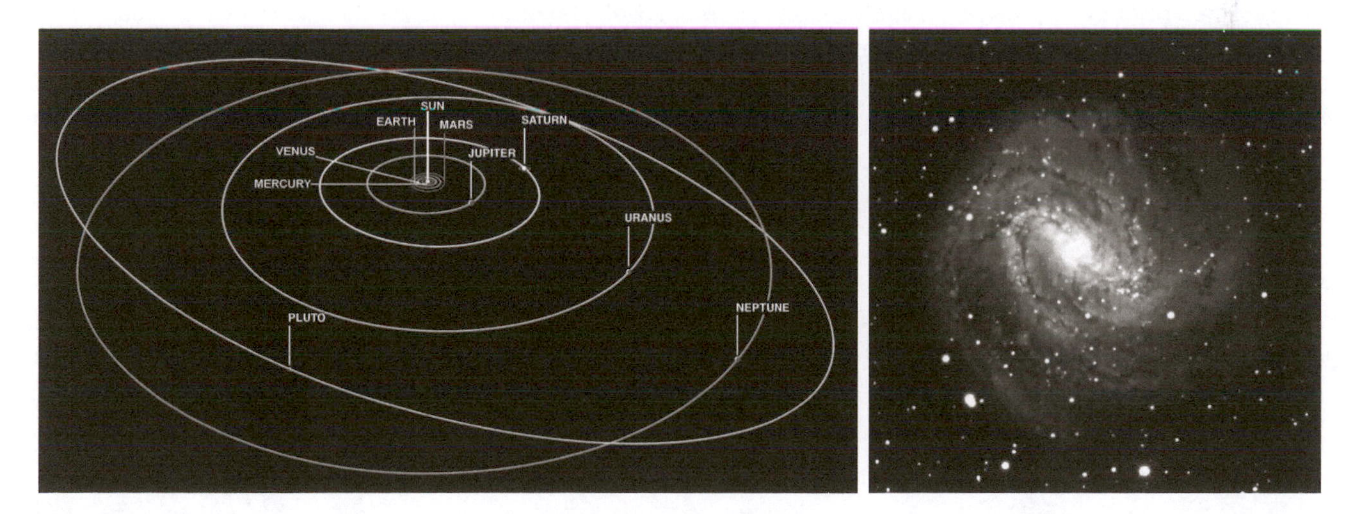

▶ **Figure 29-3** Again, don't confuse *galaxies* with *planetary systems*, or *systems of planets*. *Left:* The Solar System is our planetary system. It has one star (the Sun) and its system of planets. (Source: NASA/LPI) ***Right:*** This is a whole galaxy, with hundreds of billions of stars. (Image courtesy of Frederick Ringwald at Mount Laguna Observatory)

▶ **Figure 29-4** *Left:* In 1925, Edwin Hubble showed that galaxies are islands of billions of stars, outside the Milky Way. (Jean-Leon Huens/ National Geographic Creative) ***Right:*** M31, the Great Galaxy in Andromeda, is our nearest neighbor galaxy. Edwin Hubble obtained the first observations of individual stars in this galaxy that were detailed enough to show that they were similar to stars in the Milky Way—only much fainter, because they are much farther away. (Image courtesy of Frank S. Barnes III)

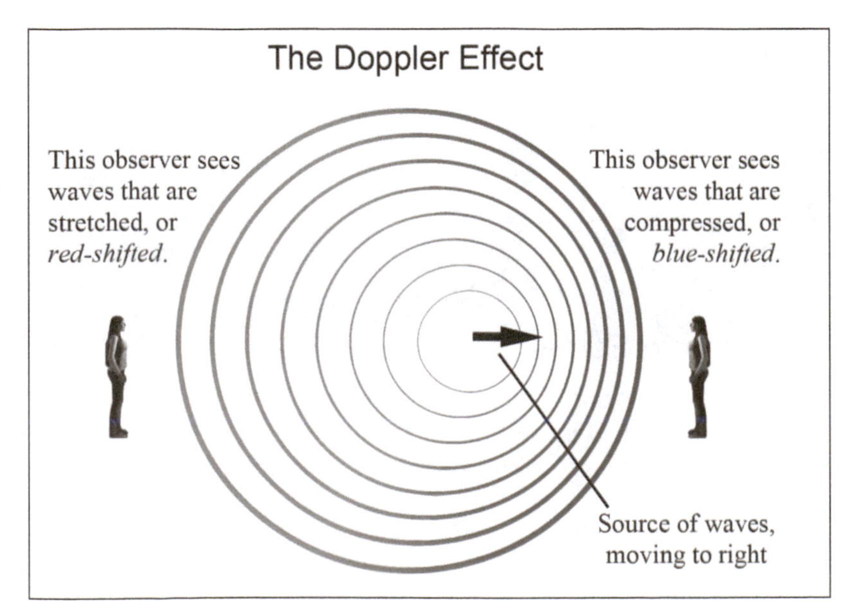

▶ **Figure 29-5** In 1929, Edwin Hubble discovered that *the Universe is expanding*. The spectra of nearly all galaxies are *redshifted*. Hubble's law is: $v = H_0 D$. The farther a galaxy is from us (D), the greater its redshift, and the faster it moves away from us (v). (Image courtesy of Frederick Ringwald)

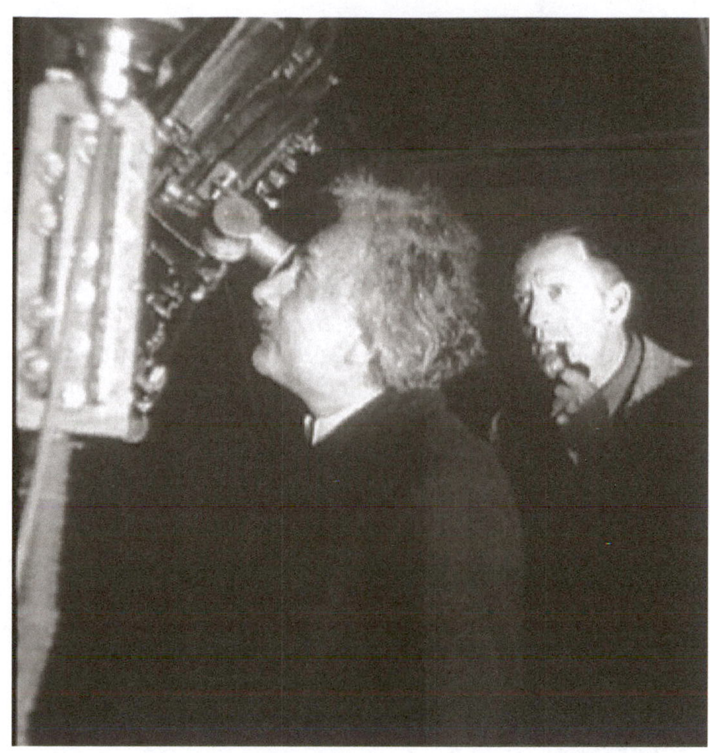

▶ **Figure 29-6** *Top left:* If the Universe is expanding, this implies that sometime in the past, everything was much closer together, or in other words, *denser*, than it is now. This implies that it began in a hot, dense state, widely referred to today as "the Big Bang." Recent measurements with *Hubble Space Telescope* and other instruments of the rate of expansion imply that the Big Bang happened 13.80 ± 0.04 billion years ago. (Image courtesy of Frederick Ringwald)

Bottom left: One way to understand the expansion of the Universe is to make an analogy to an expanding balloon, with galaxies painted on it. As the balloon expands, the galaxies move apart. This illustrates Hubble's law, $v = H_o D$, since the farther apart the galaxies are (D), the faster they move away from each other (v).

The expanding balloon analogy also illustrates that there *is* no center of the Universe. From the point of view of any galaxy, all the other galaxies are moving away from it. This is because the expansion of the Universe isn't an explosion of particles into space, like a firecracker: it's an expansion of space itself. (Image courtesy of Frederick Ringwald)

Right: Einstein visited Hubble at Mount Wilson in 1931, shown here. Einstein predicted the expansion of space over time, in his General Theory of Relativity in 1916. Einstein did not call attention to this, though, since it seemed too fantastic, even for him. He did come around to the idea, though, when Edwin Hubble was able to show him the expansion.

In hindsight (which can be so cruel), that the Universe can expand is not such a crazy idea. Since gravity always attracts, the Universe can't possibly be static. It must either expand or contract, since if it were static, gravity would make it contract. (Image from the Archives of the California Institute of Technology Photo ID 1.6-16)

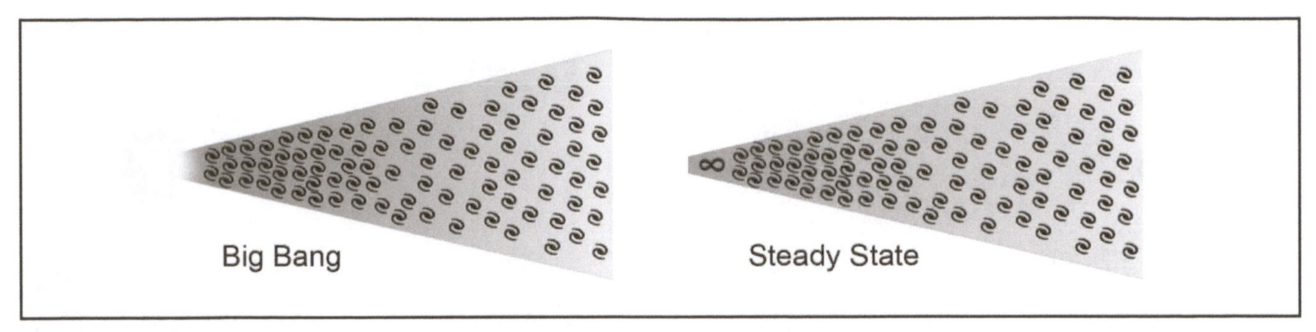

► **Figure 29-7** "Extraordinary claims require extraordinary evidence." It is a lot to expect anyone to accept that the Universe, and time itself, had a beginning, isn't it? What other evidence can support this? Fred Hoyle was an author of the Steady State theory, an early competitor to the Big Bang. In 1949, he made up the name "Big Bang" to make fun of it. This was on a BBC radio show, in which he pointed out that, in 1949, there was no other evidence that the Universe had a hot, dense origin: only the observed expansion. The Steady State theory supposed that the Universe is infinitely old, and expands because matter is created continuously. Before dismissing this as a crazy idea, remember that it's no crazier than supposing that all matter came into existence all at once, in the Big Bang. And so the matter sat, until 1964... (Image courtesy of Frederick Ringwald)

► **Figure 29-8** ...in 1964, Arno Penzias (right) and Bob Wilson (left) were two engineers, not astronomers. They were working for the phone company (Bell Labs, since broken into AT&T and subsequently renamed Lucent Technologies). They were putting together the first satellite communications system for long-distance telephone. Since satellites were crude and not powerful in 1964, Penzias and Wilson used an antenna much large than the ones in common use today, shown behind them. This antenna was also very sensitive. (Courtesy of AT&T Archives and History Center)

► **Figure 29-9** Because their antenna was so sensitive, Penzias and Wilson discovered that it was detecting a faint hiss of microwaves, coming from everywhere in the sky. Today, this is called the Cosmic Microwave Background, also known as the Cosmic Background Radiation. At first, Penzias and Wilson didn't know what it is. (Adapted from NASA image).

One of the first hypotheses that Penzias and Wilson had for the cosmic microwave background was that it was due to pigeons making nests in their antenna and covering the inside with "a dielectric substance," as they noted. They got rid of the dielectric substance, and of the pigeons, but the microwave background persisted.

Penzias and Wilson later won the Nobel prize. Arno Penzias became a chauffeur-driven vice-president of AT&T. See, work hard and it'll pay off.

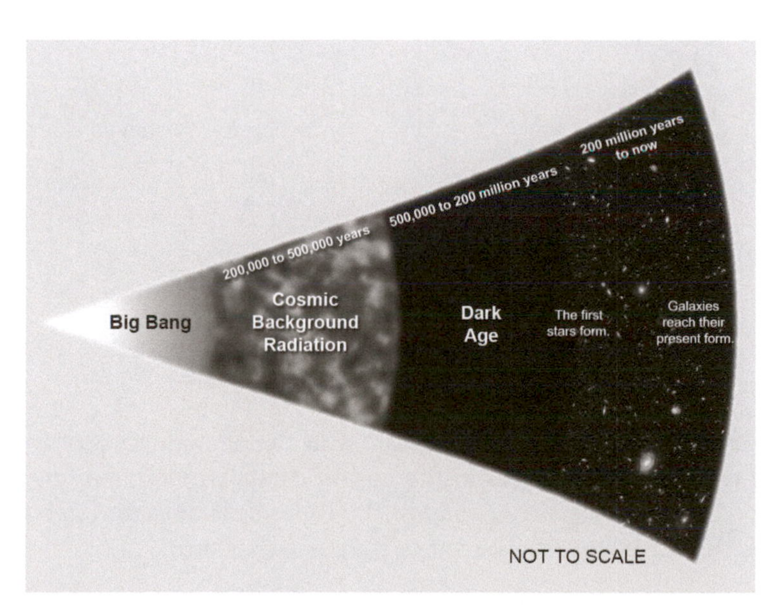

► **Figure 29-10** Recall *look-back time*. Because it takes time for light to get around, when one looks deeply into the Universe, it's like looking into a time machine. What does one see, when one looks back all the way? One sees the origin of the Universe. The Cosmic Background Radiation is the first light in the Universe: the light given off when the cosmic fireball first cooled enough to become transparent to light. In other words, **the Cosmic Background Radiation is** *an actual picture of the Big Bang!* (Image courtesy of Frederick Ringwald, using NASA images)

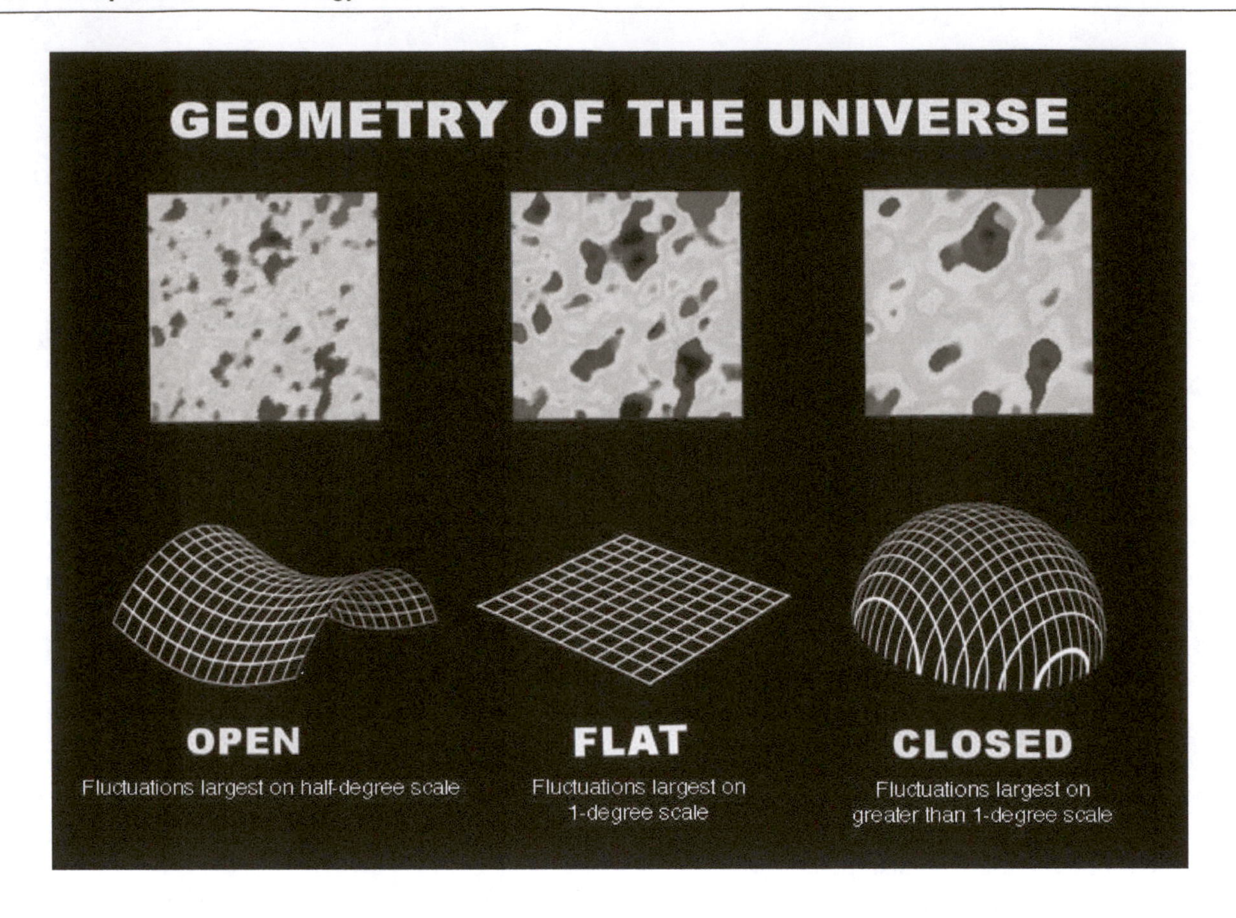

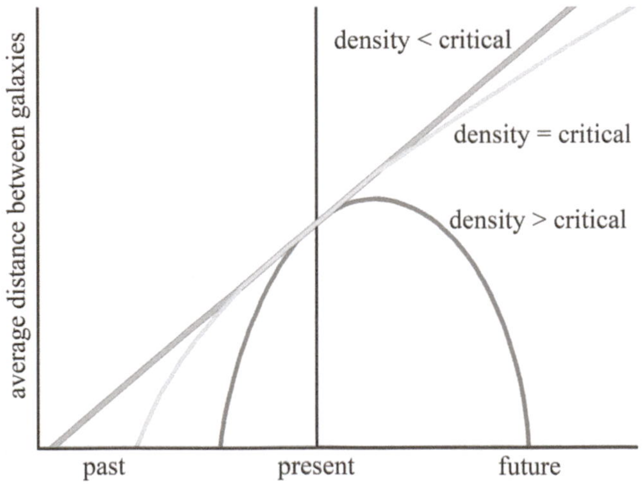

► **Figure 29-11** *Top:* Measuring the amount of variation in the temperature of the Cosmic Background Radiation can show the conditions of the gas in the cosmic fireball, much like how geologists use sound waves to tell what's inside Earth. (Source: NASA) *Bottom:* The Universe is observed to be within 2% of the density it needs to keep expanding forever. (Image courtesy of Frederick Ringwald)

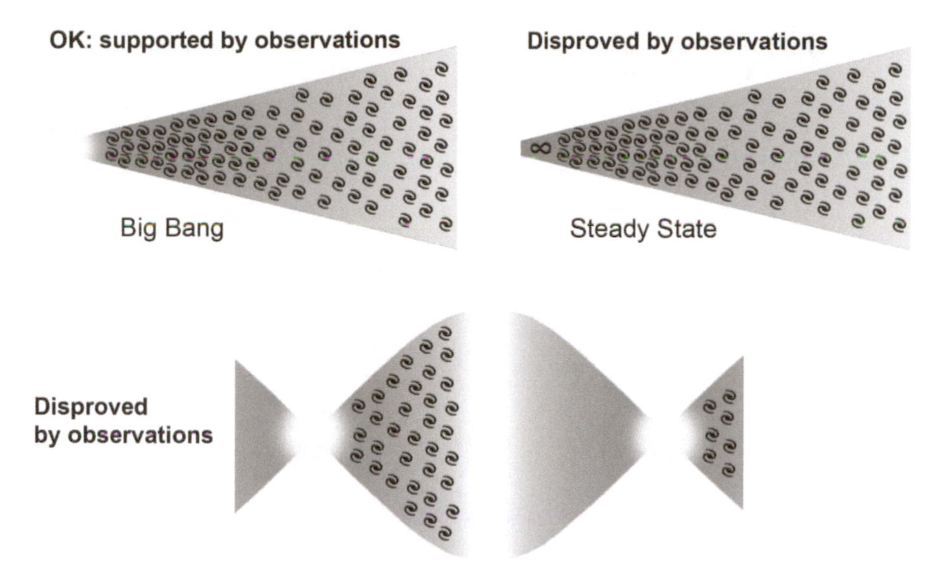

▶ **Figure 29-12** The discovery of the Cosmic Background Radiation supports the Big Bang theory, which predicted it. It disproved the Steady State theory, which did not predict it. The Oscillating Universe is the idea that the expansion of the Universe will eventually stop, because of the gravity of all the mass in the Universe. It will then collapse, billions of years in the future, in a "Big Crunch" or "Gnab Gib" (which is "Big Bang" spelled backwards). The Oscillating Universe has also been disproved, since precise observations of the Cosmic Background Radiation show it is within 2% of the density the Universe needs to keep expanding forever. (Image courtesy of Frederick Ringwald)

▶ **Figure 29-13** The Olbers paradox is that if the Universe were infinitely large and infinitely old, the sky at night shouldn't be dark. It should be as bright as the Sun, since no matter in which direction one looked, sooner or later there should be a star. One can't see the forest, because there are too many trees in the way, as in this Louisiana swamp. The solution to the paradox is that the Universe isn't infinitely old. The Big Bang theory predicts it should have a finite age. (Image courtesy of Frederick Ringwald)

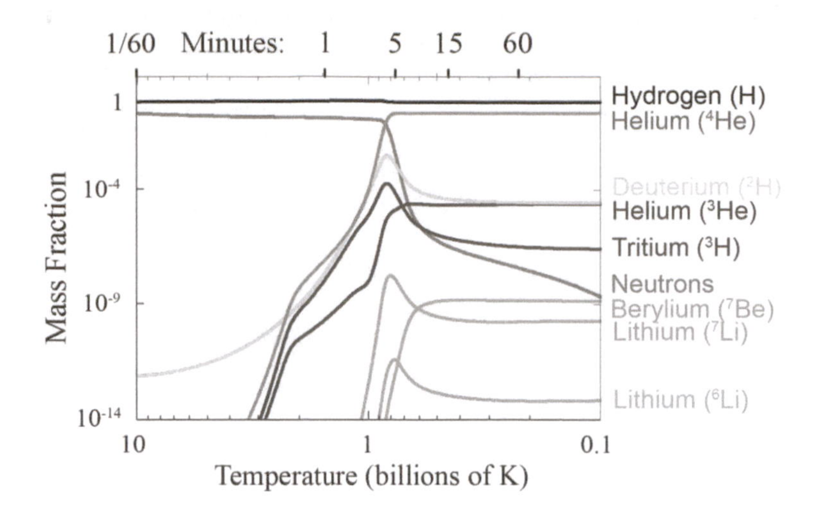

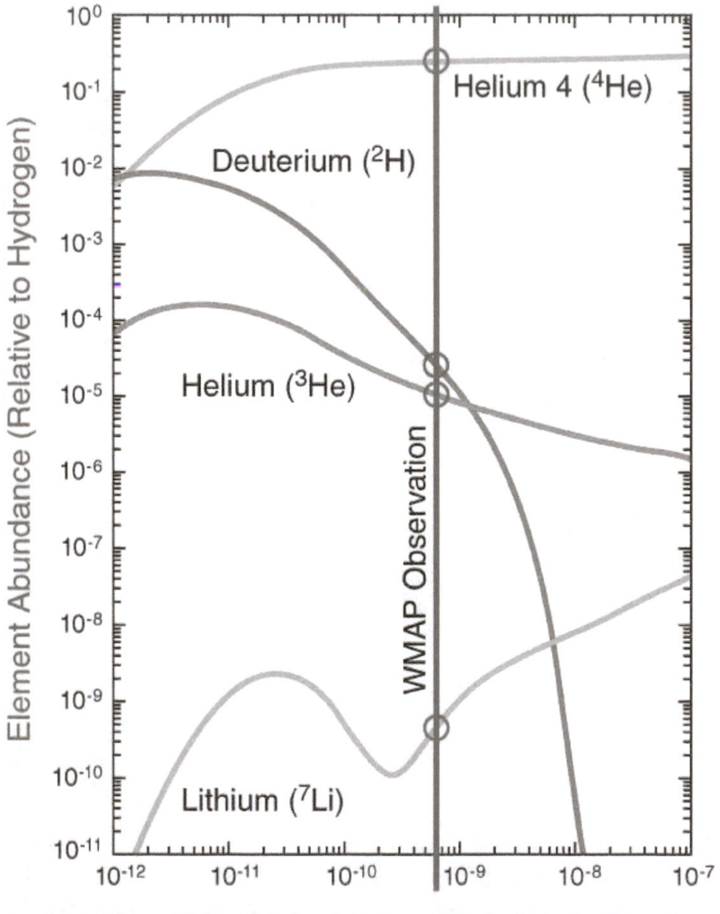

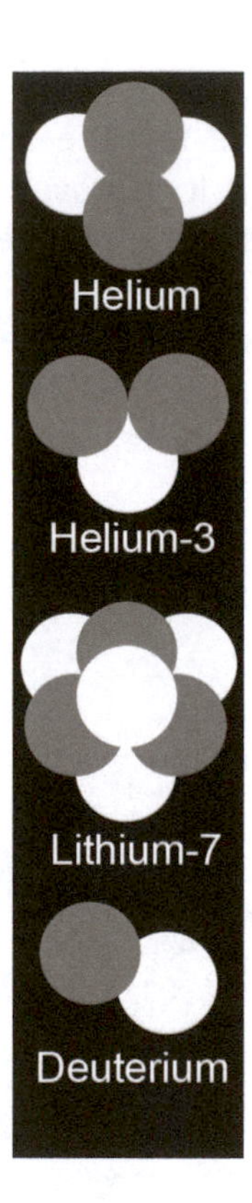

► **Figure 29-14** What other evidence did the Big Bang leave? During *the first three minutes* of the existence of the Universe, the whole Universe was as hot and dense as the interior of a star. The same nuclear reactions that occur in stars were happening *everywhere*. Today, we observe remnants of this: the *abundances of helium and other light elements*. *Top:* Big Bang Nucleosynthesis Predictions (Image courtesy of Scott Burles, Kenneth Nollett, and Michael Turner and redrawn by the author) *Bottom:* Observations of light-element abundances agree with theoretical predictions and with each other. (Source: NASA/WMAP Science Team)

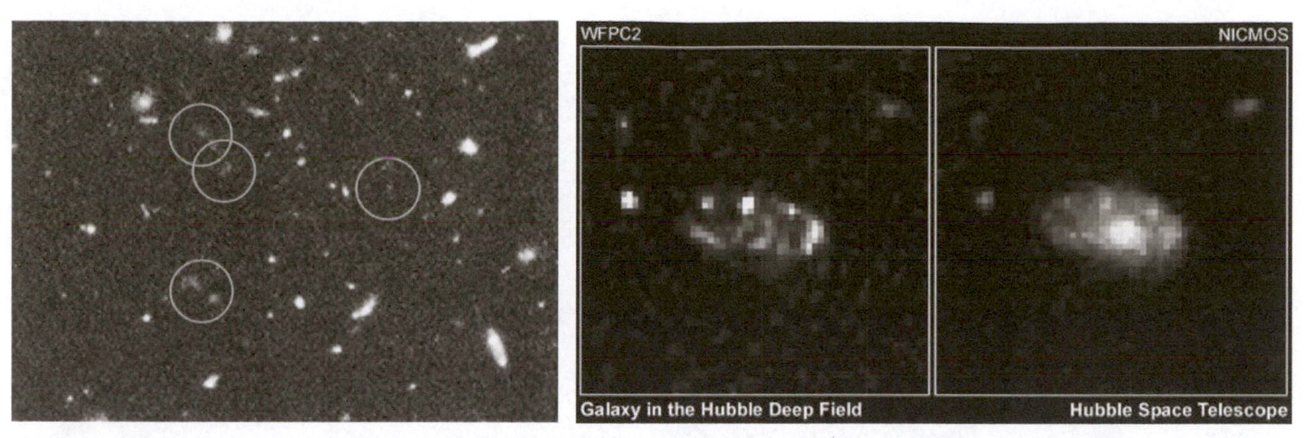

► **Figure 29-15** These examples of galaxy evolution further discredit the Steady State theory since it predicts that there should be no overall evolution of galaxies. *Left:* Because of look-back time, the little faint red protogalaxies shown here are in their infancy, as they were during the first billion years of the history of the Universe. (Source: NASA/STScI) *Right:* Galaxies in the early Universe had more hot, short-lived blue stars (OB types), after counting for redshift. (Source: NASA/STScI)

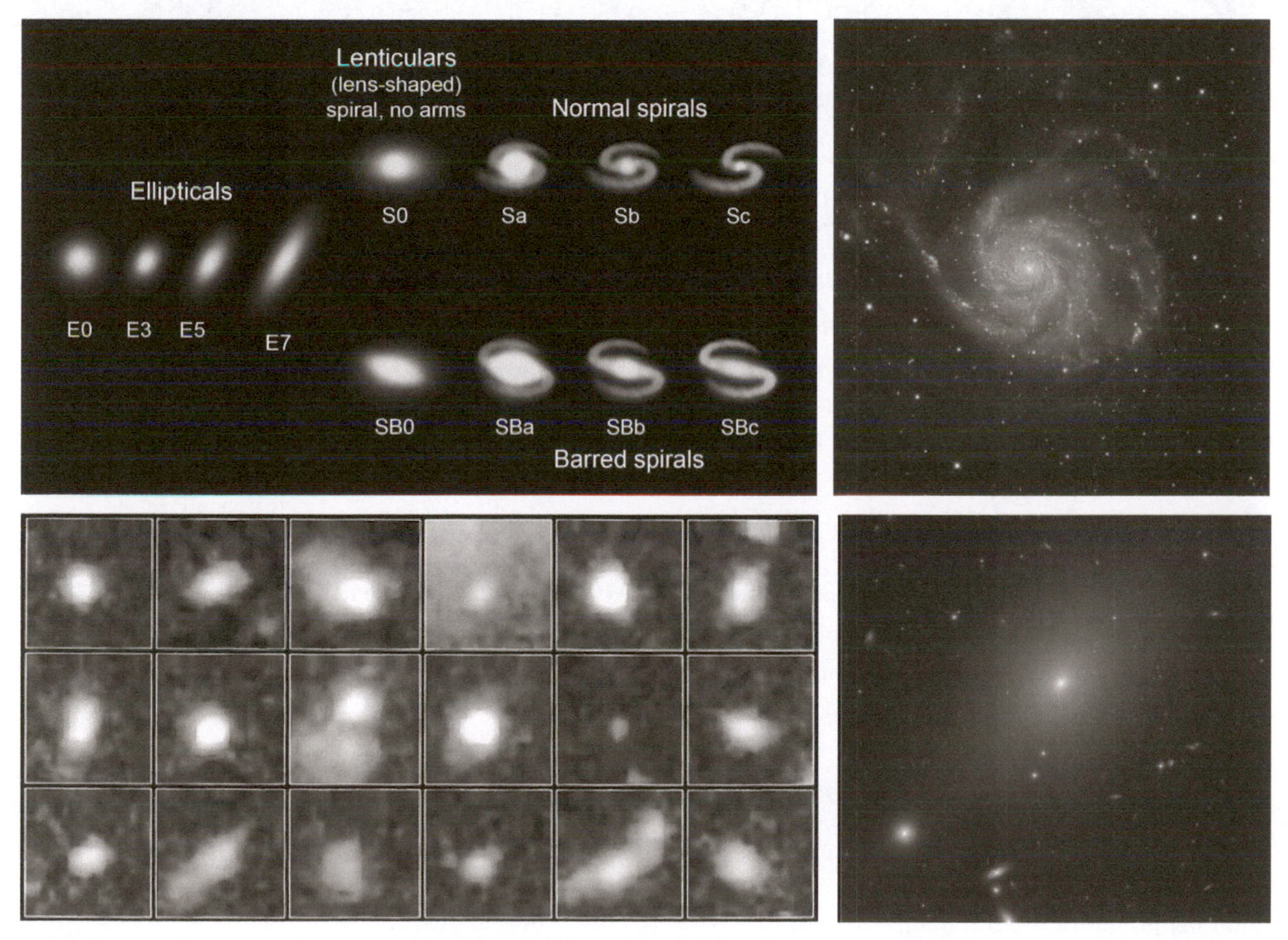

► **Figure 29-16** *Top left:* Edwin Hubble's "tuning fork" classification of galaxies has been explained by observations by *Hubble Space Telescope*. It turns out to be an evolutionary sequence: the earliest galaxies are irregulars, which merge together into spirals like the Milky Way, which merge into giant elliptical galaxies. Galaxies now are more highly clustered than in the past, because of gravity. (Image courtesy of Frederick Ringwald) *Bottom left:* Primordial irregular galaxies (Source: NASA/STScI) *Top right:* A spiral galaxy, similar to the Milky Way Galaxy (Image courtesy of Greg Morgan) *Bottom right:* A giant elliptical galaxy eating spiral galaxies. (Source: NASA/STScI)

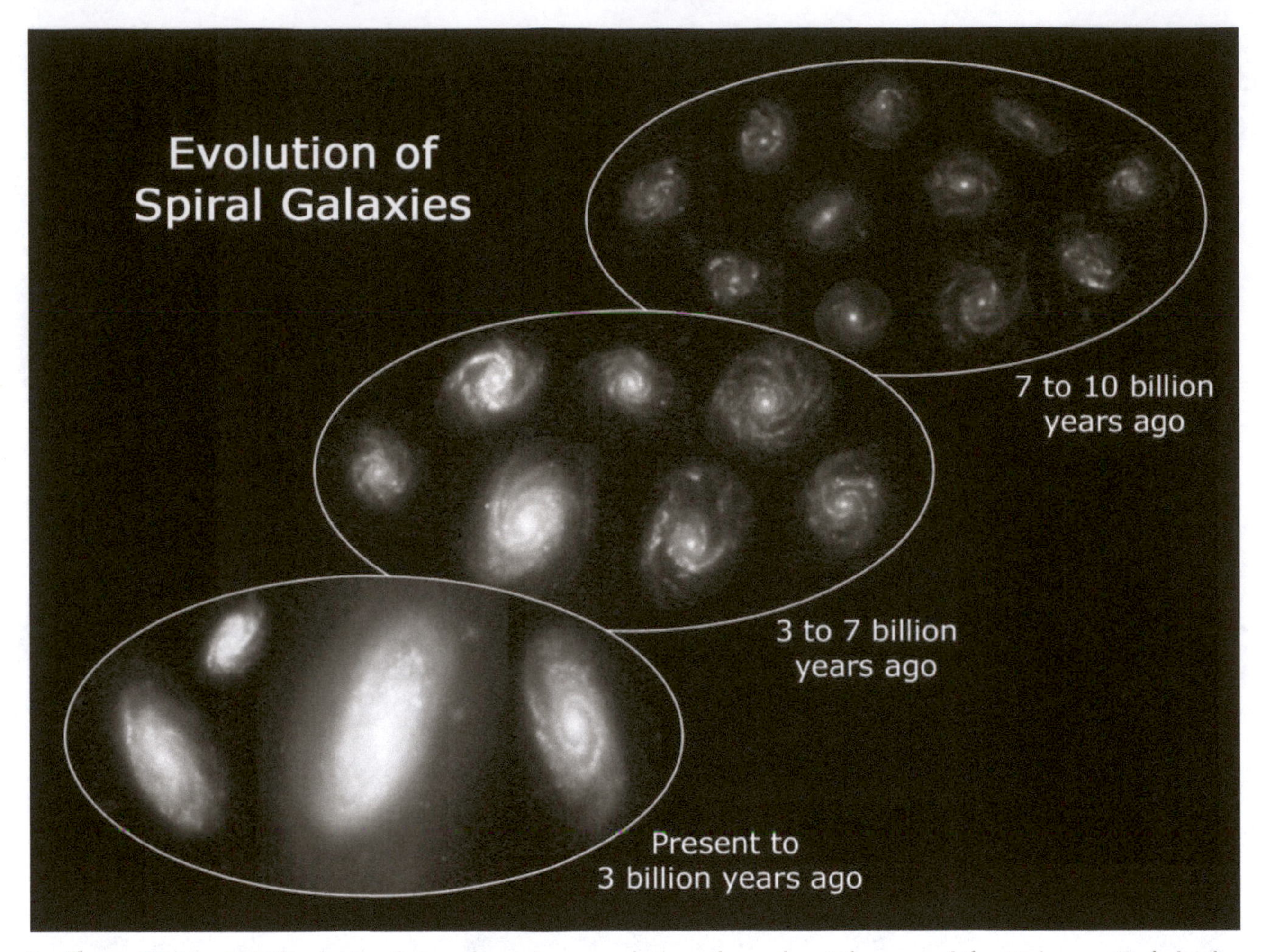

► **Figure 29-17** Spiral galaxies themselves show evolution, throughout the age of the Universe. Early in the Universe, they are much more active, with bright nuclei from stars falling into their centers. Now, they are observed to be much more relaxed, most stars having settled into stable orbits. (Source: NASA/STScI/Frank Summers)

Where did the energy of the Big Bang come from? Doesn't the Big Bang violate the First Law of Thermodynamics, which says that energy can't just appear from nowhere?

Not necessarily: the energy of the observed expansion of the Universe just balances the energy of the gravity holding the Universe together, within 2%. The *net* energy of the Universe may therefore be *zero*.

Where is the center of the Universe?

There is no center. With the whole Universe expanding, from any galaxy, it looks like it is in the center, because the other galaxies are moving away from it.

What came before the Big Bang?

We don't know: we have no information from this early time.

What is outside the Observable Universe?

We don't know: again, we have no information, since we can't see "outside" it.
The speed of light is finite, so it limits how far away we can see.

Remember also look-back time: the speed of light limits how far back in time we can see.

Can the Universe be infinite in size?

Maybe: its mass is spread very evenly, on its largest scale.

If the Universe is expanding, what is it expanding into?

The short answer is that *we don't know.*

A longer answer may be that, in a way similar to how a two-dimensional surface of a balloon can expand into a three-dimensional room, our three-dimensional Universe (plus time) may be expanding into a higher-dimensional space. Higher dimensions do get a lot of attention from theorists these days. They may or may not exist: we currently have no observations or experimental evidence that they do.

We can't currently answer *everything.* This does not mean that we don't know *anything.*

We see that the Universe is expanding. We also observe the cosmic background radiation, which is the light left over from its hot, dense origin. We can also see other traces of its origin, and we can see the galaxies in it evolving.

Why are we here?...

from: Carroll, B.W., and Ostlie, D.A., 1996, *An Introduction to Modern Astrophysics,* Addison-Wesley Pub. Comp. Inc (A.K.A "BOB": the Big Orange Book)

In our astrophysics class, a student once asked, "Why are we here?"

The answer was as amazing to us as it was to the class:

> We are here because, more than ten billion years ago, the Universe borrowed energy from the vacuum to create vast amounts of matter and antimatter in nearly equal numbers. Most of it annihilated and filled the Universe with photons. Less than one part per billion survived to form protons and neutrons, and then the hydrogen and helium that makes up most everything there is. Some of this hydrogen and helium collapsed to make the first generation of massive stars, which produced the first batch of heavy elements in their central nuclear fires. These stars exploded and enriched the interstellar clouds that would form the next generation of stars. Finally, about five billion years ago, one particular cloud in one particular galaxy collapsed to form our Sun and its planetary system. Life arose on the third planet, based on the hydrogen, carbon, nitrogen, oxygen, and other elements found in the protostellar cloud. The development of life transformed Earth's atmosphere and allowed small furry mammals to take center stage. Humans evolved and moved out of Africa to conquer the world with their new knowledge of tools, language, and agriculture. After raising food on the land, your ancestors, your parents, and then you consumed this food and breathed the air. Your own body is a collection of the atoms that were created billions of years earlier in the interior of stars, the fraction of a fraction of a percent of normal matter that escaped annihilation in the first microsecond of the Universe. Your life and everything in the world around you is intimately tied to countless aspects of modern astrophysics.
>
> (If you think this should be "How did we come to be here," rather than "Why are we here," consider that at least we got this answer by *observing* the Universe around us, and by thinking about the observations *rationally.* It wasn't done entirely by making up stories.)

Chapter 30
The Deep Universe

Quasars

Quasars are ***the most energetic things*** *known in the Universe*. One quasar can be more luminous than 10,000 normal galaxies—a whole supercluster of galaxies.

Maarten Schmidt discovered this in 1963, since he found that quasars have high redshifts. Quasars are therefore literally at the edge of the Observable Universe, and so must be extremely luminous. In the 1990s, images taken by *Hubble Space Telescope* were the first to have enough resolution to show that quasars are the centers of distant galaxies.

Quasars are the most powerful kind of *active galactic nuclei* (AGNs). Quasars and other, less powerful AGNs are galaxies that are still in the process of formation. They achieve their stupendous power by gas [and sometimes, whole stars] falling into supermassive black holes in their centers.

Galaxies are often observed to have black holes in their centers. The one in the center of the Milky Way is called Sgr A* (pronounced "Sagittarius A-star"). Astronomers have tracked stars orbiting around it, and by using Kepler's Third Law, have measured its mass at 3.6 ± 0.4 million times the Sun's mass.

Sgr A* isn't active: little gas is falling into it. It poses no danger to Earth, since the Sun and the other stars orbit it stably, much as the planets in the Solar System orbit the Sun. Sgr A* is also 27,400 ± 1,300 light-years from Earth.

Seyfert galaxies are milder, less luminous active galaxies, in which the galaxy is easily visible, but still has a bright nucleus. These are named after Carl Seyfert, who discovered them in 1943.

► **Figure 30-1** *Top:* Maarten Schmidt discovered the redshifts of quasars in 1963. He found that they had redshifts so high, Hubble's law implied that quasars are so far away, they are literally at the edges of the Observable Universe. Quasars are therefore the most energetic things known in the Universe. One quasar is as luminous as 10,000 galaxies. (James P. Blair National Geographic Creative) *Bottom:* Quasars were mysterious for many years, because from ordinary telescopes, they don't look like much. The little blue point of light in the middle is a quasar, which is 2.4 billion light-years away. (Image courtesy of Frederick Ringwald)

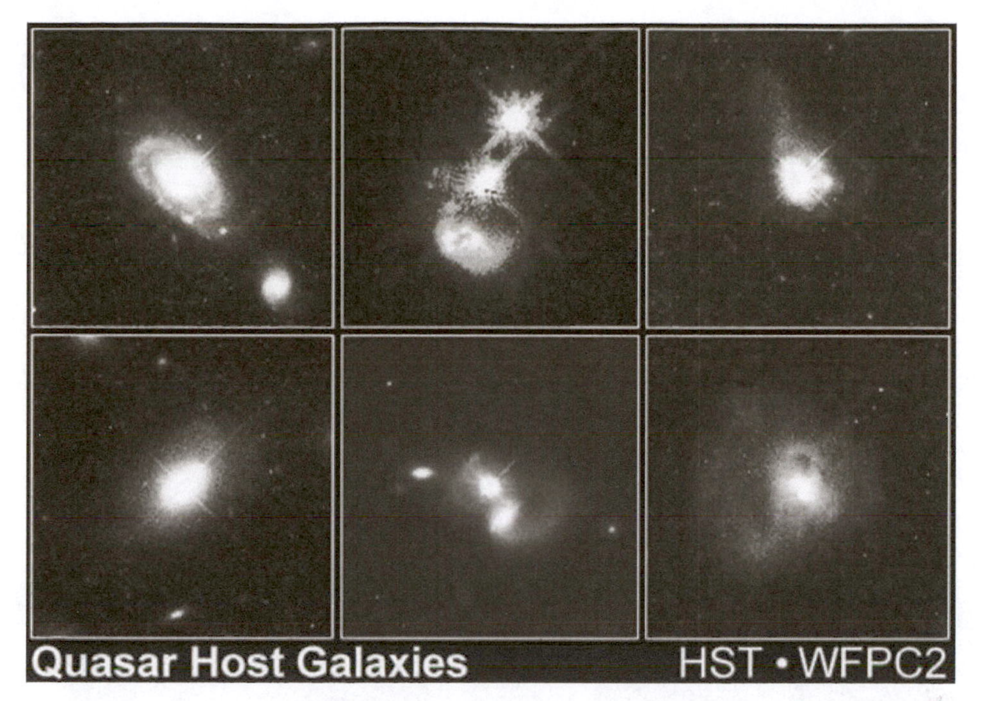

► **Figure 30-2** *Hubble Space Telescope* finally showed what quasars are, in 1995. Quasars are active galaxies, in which gas and whole stars are falling into black holes in the centers. This makes the central engines of quasars 10,000 times more luminous than the host galaxy, which can easily be lost in the glare of telescopes with less resolution than *Hubble*. (Source: NASA/STScI)

► **Figure 30-3** Whenever I show a detailed picture like this of the center of any galaxy, students ask, "What's in the center?" Would you believe it's a black hole? There is now evidence that nearly all galaxies have supermassive black holes in their centers. In quasars and other active galactic nuclei, gas and stars are falling into these central black holes. In normal galaxies, there is much less activity, with the orbits of their stars having become more relaxed. (Source: NASA/STScI)

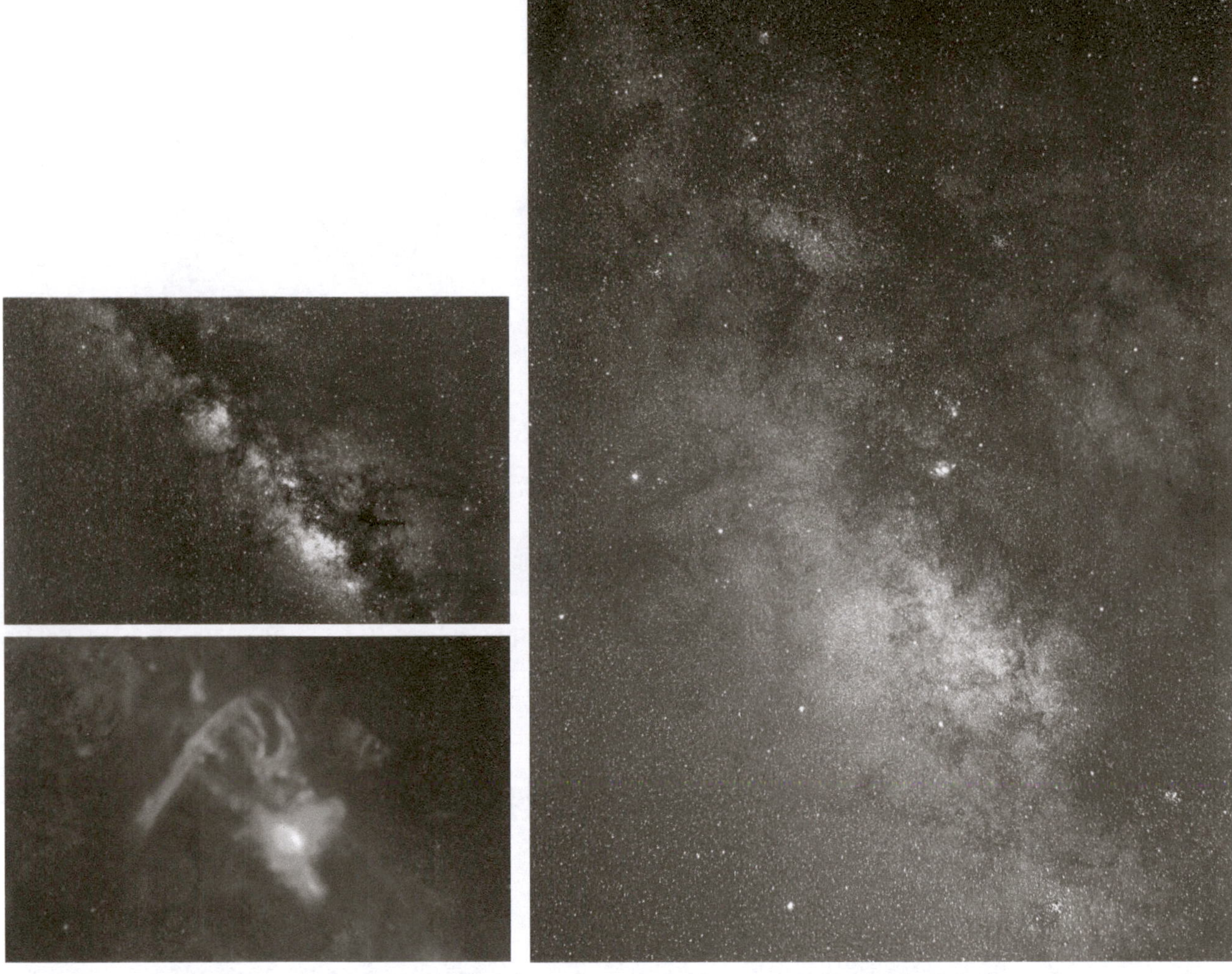

▶ **Figure 30-4** Sgr A* (pronounced "Sagittarius A-star") is the name of the black hole at the center of the Milky Way Galaxy. *Top left and right:* Visible-light images of the center of the Milky Way Galaxy do not show Sgr A*, since the Galactic Center is obscured by dark clouds of dust. (Images courtesy of Frederick Ringwald) *Bottom left:* A radio image of the Galactic Center shows Sgr A*, which emits radio and infrared radiation that passes through the dust. (Source: NRAO/AUI/NSF) Sgr A* is known to be a black hole with a mass of 3.6 ± 0.4 million times the Sun's mass because astronomers track stars that orbit Sgr A*. By Kepler's Third Law, if one can track an object orbiting a more massive object, it tells how strong the gravity is. It can therefore show the mass of the more massive object.

► **Figure 30-5** Sometimes, when the geometry and conditions are just right, a black hole in a galaxy center can be seen directly. This is such a case: the disk of gas around the black hole is visible at right. Yes, no kidding, this is really a picture of a black hole. The black hole shoots jets into space, shown at left. No one understands how it does this. (Source: NASA/STScl)

► **Figure 30-6** *Left:* A visible-light image of the nearest active galaxy, called Centaurus A, shows dust lanes obscuring its nucleus, because it is an elliptical galaxy eating a dusty spiral galaxy. (Image courtesy of Frederick Ringwald) ***Right top left:*** An infrared image of Cen A shows its central black hole, bright from the gas from the spiral galaxy it is consuming. (Source: NASA/STScI) ***Right top right:*** A radio image of Cen A shows jets spewing from the black hole. Again, how these work no one knows. (Source: NRAO/AUI/VLA/J. Burns/R. P. Rice) ***Bottom right:*** An X-ray image also shows a jet of million-degree gas shooting from the center of Cen A. (Source: NASA/CXC/SAO/R. Kraft et al.)

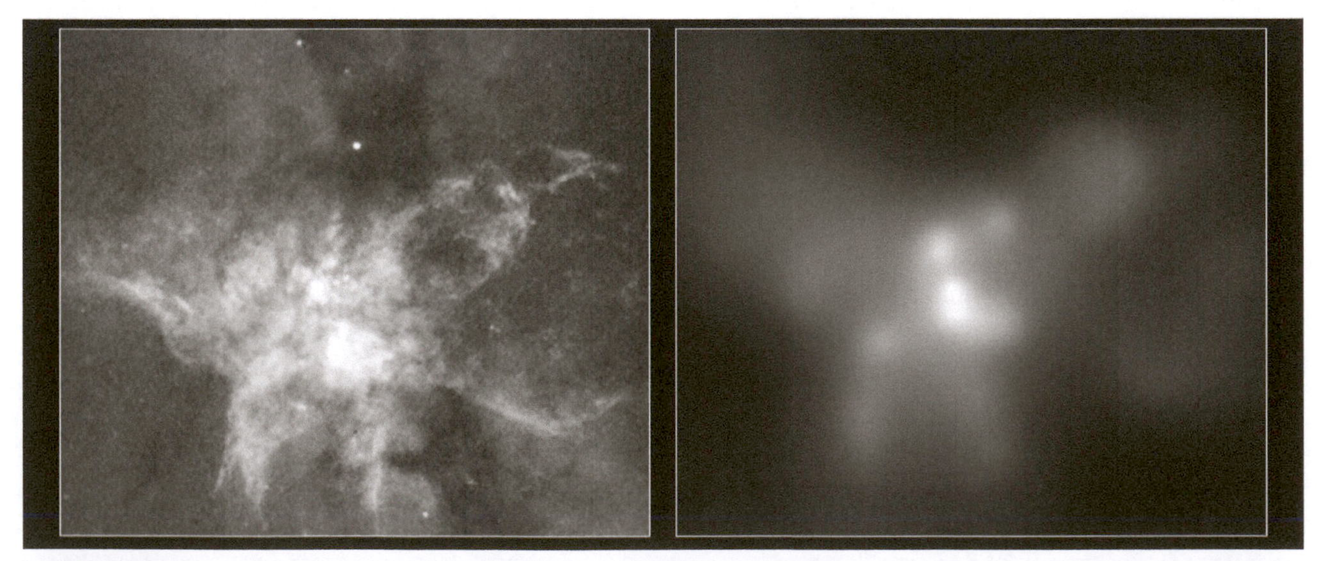

► **Figure 30-7** *Left:* Two galaxies are colliding and clearly making a mess in this visible-light image of NGC 6240. (Source: NASA/CXC/STScI) ***Right:*** This X-ray image of NGC 6240 shows two strong sources of X-rays. They are from the hot gas going down the two supermassive black holes, which used to be in the centers of the two galaxies that collided. (Source: NASA/CXC/STScI)

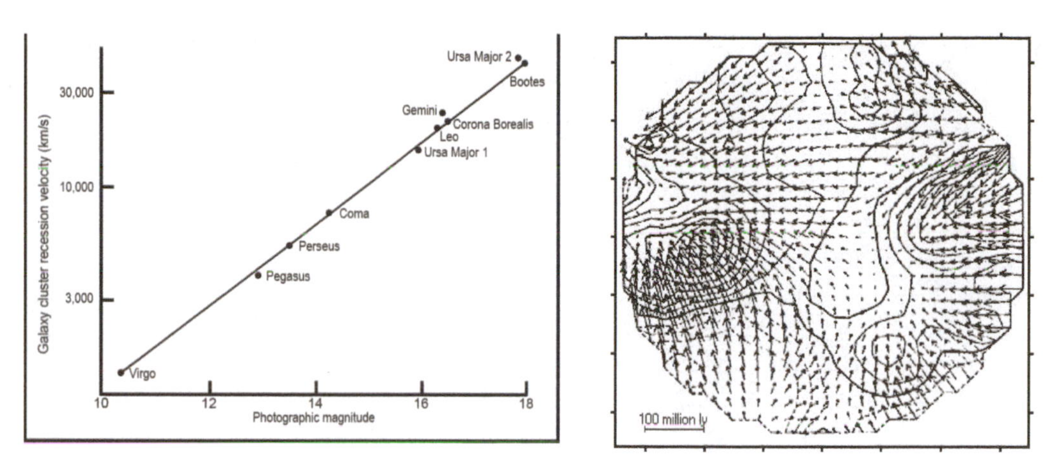

► **Figure 30-8** How can galaxies collide, in an expanding Universe? Despite the expansion of the Universe, galaxies still can clump together because of gravity. *Left:* Hubble's law is an *average* trend. (Image courtesy of Frederick Ringwald) *Right:* There can be local currents in the overall flow, much like turbulence in a river. (Dekel, A. 1994, ARA&A, 32, 371 Avishai Dekel)

► **Figure 30-9** When galaxies collide, fireworks happen. Interacting galaxies take a wide variety of shapes. At upper right is M82, a starburst galaxy. It is having a burst of star formation because it was stirred up by a collision with another galaxy. If Earth were in M82, there would be hundreds of stars in the sky bright enough to be seen in the daytime: it probably would never get very dark at night. (Images courtesy of Frederick Ringwald at Mount Laguna Observatory)

Gamma-Ray Bursts

Gamma-ray bursts are ***the most powerful things** known in the Universe.* Power is how fast energy is used, so gamma-ray bursts are more powerful than quasars, but they only last a few seconds: they're explosions. One gamma-ray burst was brighter than the whole rest of the Universe, in gamma rays, for about two seconds.

Gamma ray bursts are flashes of high-energy gamma rays. They were discovered in 1967 by U.S. Department of Defense (DoD) spacecraft, which were trying to catch the Soviet Union cheating on arms-control treaties. The Soviets weren't testing nuclear weapons in space: gamma-ray bursts are a natural phenomenon, from outside the Solar System.

The DoD tried to keep this secret, but couldn't. This casts doubt on the idea that the DoD knows that UFOs are extraterrestrial spacecraft, and has kept it secret for many years.

Gamma-ray bursts are smoothly distributed over the sky. They therefore must be *very* deep in space. If they were in the Solar System, they'd cluster around the ecliptic; if they were in the Milky Way, they'd cluster around the Galactic plane.

A typical gamma-ray burst has the energy of a supernova. This energy is liberated in a few seconds, and is only gamma rays. What could do that?

1. Long-duration gamma-ray bursts, which last more than two seconds, have been explained as "hypernovae," which are also known as "collapsars." In a hypernova, a massive star collapses into a black hole, during a supernova explosion. The gamma rays flash from a dense disk that forms around the black hole for only about a second, before the supernova obliterates it. Observations support this: many gamma-ray bursts come from galaxies in which supernovae are seen.

2. Short-duration gamma-ray bursts, which last less than two seconds, are still unexplained. Causes under investigation include colliding or merging neutron stars, or primordial black holes, which were formed during the Big Bang and are evaporating in puffs of Hawking radiation. Observations to confirm either model are lacking, however. Nature is not revealing its secrets easily here.

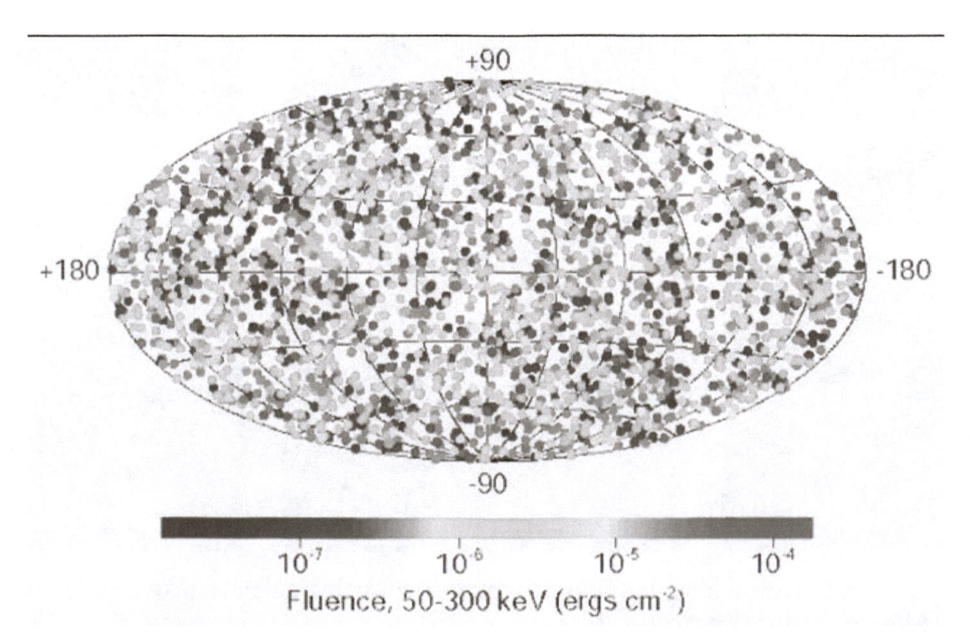

► **Figure 30-10** **Gamma-ray bursts** are the most powerful explosions known. They are the brightest things known, but they last for less than 2 seconds. They are spread out very evenly over the sky, which shows that they must be very far away, farther even than quasars. (Source: NASA/CGRO)

► **Figure 30-11** Gamma-ray bursts were mysterious until 1997, when improved instruments showed a link with supernovae. Since then, the collapsar model has successfully explained many of the observations. In a collapsar, a massive star collapses to make a supernova. Just as it does, a black hole forms in its center, and a disk of dense gas from the star around it detonates in a nuclear explosion, shooting two jets of gamma rays perpendicular to the disk. We observe only gamma-ray bursts from jets that are pointed right at us, but they are so bright, they can be detected across the Universe. The collapsar model explains many long-duration gamma-ray bursts, which last more than two seconds. It does not explain short-duration gamma-ray bursts. Nature is not letting us off the hook here. (Source: NASA/SkyWorks Digital)

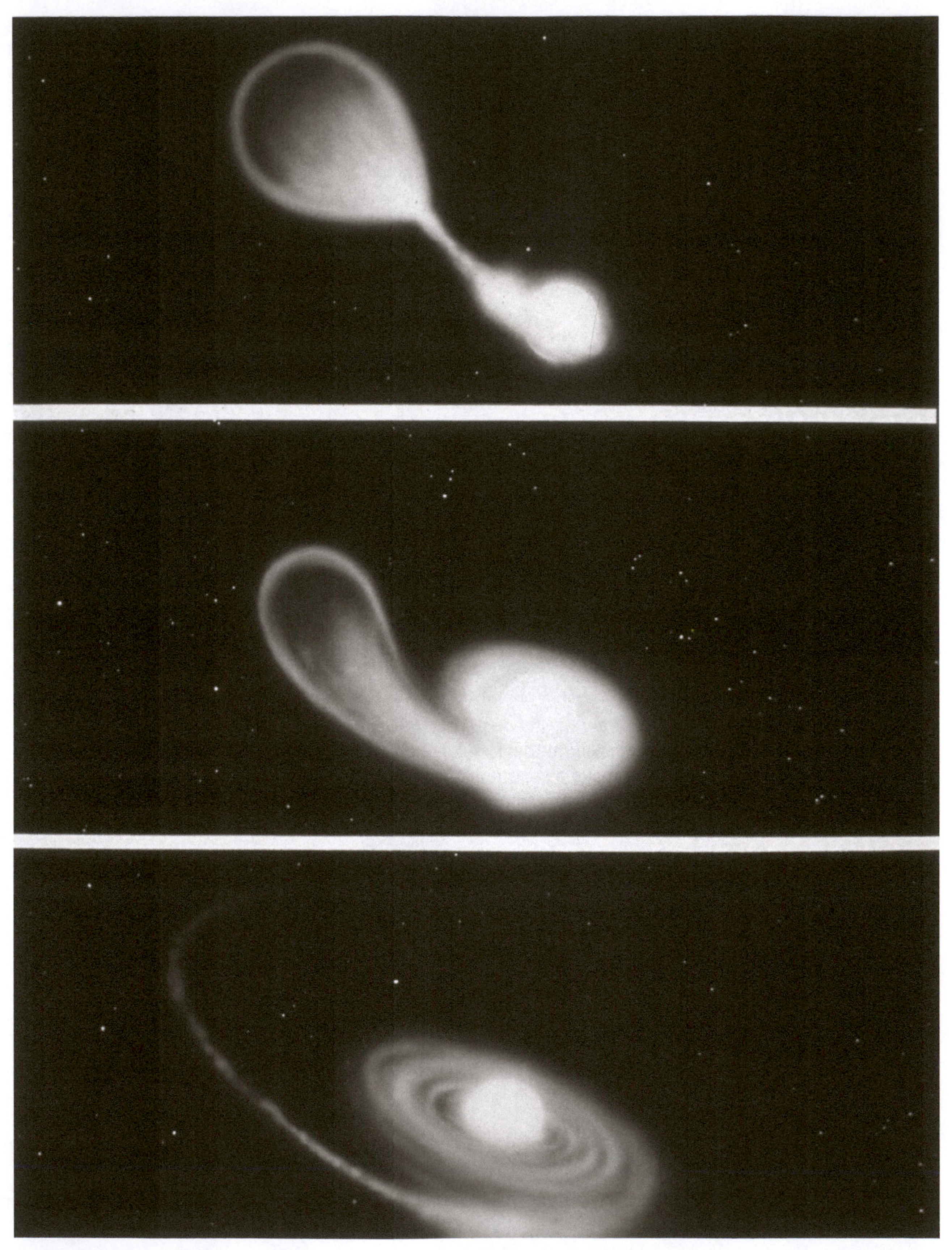

▶ **Figure 30-12** Merging neutron stars have been suggested as a possible way to make short-duration gamma-ray bursts. Observations to confirm this are still lacking, however. (Source: NASA/STScI/Dana Berry)

Dark Matter

In 1930, Fritz Zwicky discovered clusters of galaxies. He realized that for these clusters to hold together by gravity, they needed to have hundreds of times more mass than the galaxies we see. He was ignored, perhaps partly because of his aggressive personality.

In the 1970s, Vera Rubin found that the outer parts of spiral galaxies rotate so fast, they should fly apart. In order to hold together, in the way we see, galaxies would need to have hundreds of times more mass than we can see. She was able to convince many astronomers because the graphs she plotted showed this effect clearly.

In the mid-'80s, both Zwicky and Rubin were vindicated. By then, too much high-quality data had accumulated that revealed the "missing mass." More properly, this should be called "dark matter," since the problem was not that the mass was missing: the problem was that the mass was apparently there, but couldn't be seen.

Apparently, over 90% of the Universe is in a form that cannot be seen. We might be made of star stuff, but, profoundly, the Universe apparently is *not*. Dark matter could be anything: black holes, burnt-out white dwarf stars, and back issues of the *Astrophysical Journal* have all been suggested. However, dark matter *can't* be any kind of "ordinary" matter composed of atoms, either now or in the past. It must be something totally unknown. It certainly isn't the same as the "brown organic tarry gunk" found throughout space.

► **Figure 30-13** *Left:* In 1933, Fritz Zwicky found evidence for dark matter in galaxy clusters. (Image from the Archives of the California Institute of Technology) ***Right:*** Zwicky found that in order to hold together by gravity, which they clearly do, clusters of galaxies need to have five times more mass than can be seen. This unseen mass is called "dark matter," not very imaginatively. (Source: NASA/STScI)

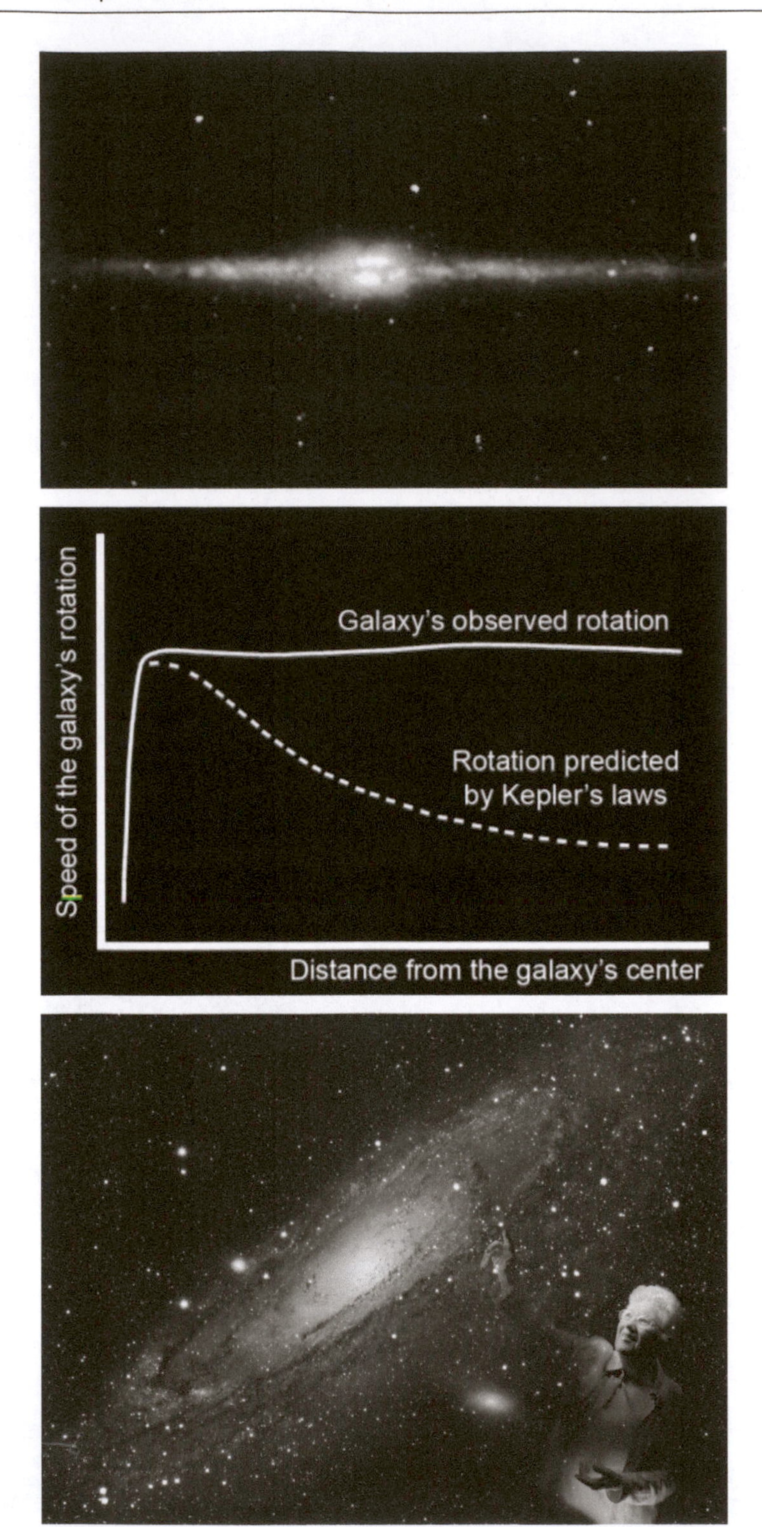

▶ **Figure 30-14** *Top:* In 1976, Vera Rubin found further evidence for dark matter with her observations of galaxy rotation. (Source: NASA/GSFC) *Middle:* One reason that Vera Rubin's results were taken seriously immediately was that she made graphs that clearly showed that something surprising was happening. Specifically, the outer parts of galaxies were rotating too rapidly to be held together by the gravity of the mass that can be seen. This implied that galaxies have invisible, massive halos. (Image courtesy of Frederick Ringwald) *Bottom:* As she said in 2002, "I said in 1980 that we'd know what dark matter was in ten years, that the particle physicists would tell us." So far, it still hasn't happened: the nature of dark matter is still a mystery. (© Peter Ginter/Science Faction/Corbis)

► **Figure 30-15** *Left:* Even though dark matter can't be seen directly, gravitational lensing can map where dark matter is. (Source: NASA/STScI) *Right:* This map of dark matter shows that it has structure, like ordinary matter. It just doesn't emit any light and apparently interacts with other matter only by its gravity. (Source: NASA/STScI)

Dark Energy

In 1998, observations of distant supernovae surprised everyone, by showing that the expansion of the Universe appears to be *accelerating* (getting faster).

So far, explanations have been sketchy. One *tentative* explanation is that the Universe is full of an unknown type of energy, called dark energy. It doesn't appear to have anything to do with dark matter, even though one might expect it would, since matter and energy are equivalent, with $E = mc^2$. The nature of dark energy is still a complete mystery.

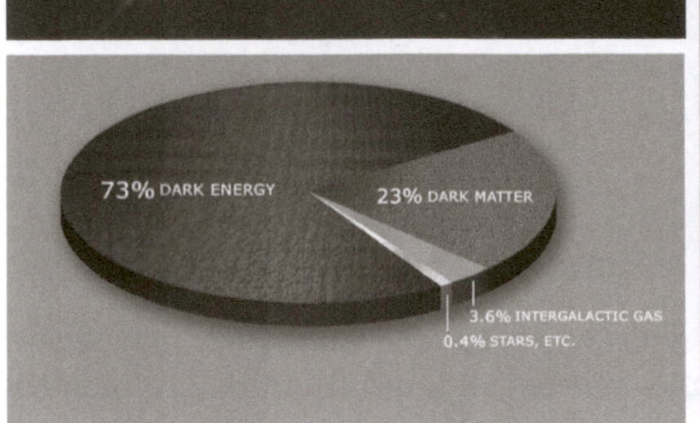

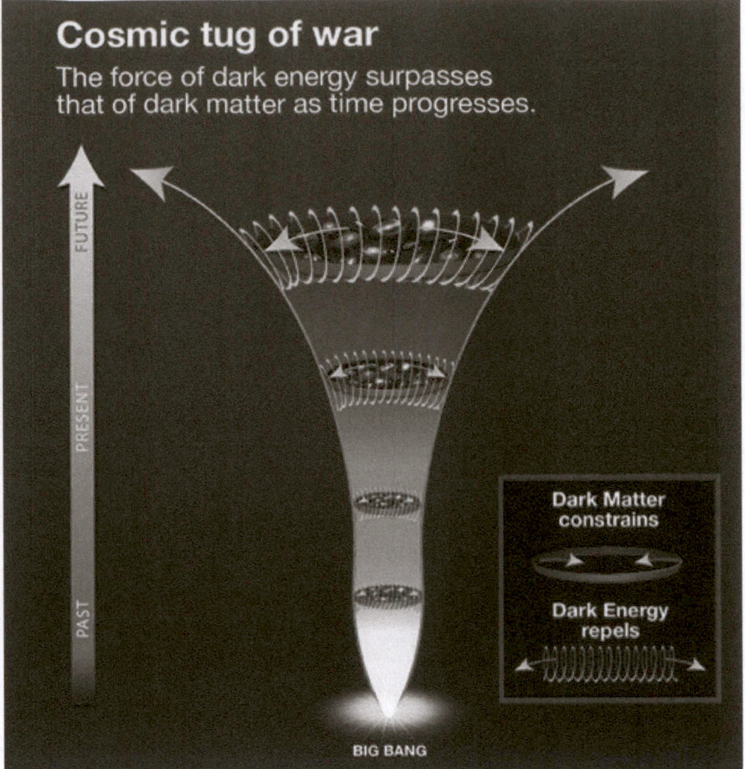

► **Figure 30-16** *Top left:* Observations of distant supernovae (to lower left of the galaxy) show that the expansion of the Universe is acceleration. This may be evidence for dark energy, a mysterious form of energy pervading the Universe. (Source: NASA/STScl) ***Bottom left:*** Apparently, 73% of the Universe is dark energy, and 23% is dark matter. Less than 4% is matter made of atoms, like we are used to. In other words, over 96% of the Universe is in two different forms that we have no understanding of whatsoever. Reality as we understand it is like whitecaps on the waves of a dark ocean. (Source: NASA) ***Right:*** Dark matter is apparently not related to dark energy, since they act very differently. Dark matter makes matter clump together; dark energy makes the Universe spread apart. (Source: NASA/STScl)

Chapter 31
Life Beyond Earth

"Extraordinary claims require extraordinary evidence."

–attributed to Carl Sagan

"It's important to keep an open mind, but not so far open that your brains fall out."

–Jim Oberg

"Nothing is too wonderful to be true, if it be consistent with the laws of nature."

–Michael Faraday

Finding life beyond Earth is really two separate problems:

1. Finding intelligent life.
2. Finding simple life, such as bacteria.

Intelligent Life Beyond Earth

Four spacecraft (*Pioneer 10* and *11*, and *Voyager 1* and *2*) have left the Solar System. All are bearing plaques, in case they are ever found. This is an unlikely way of making contact: the stars are so distant, it will take the spacecraft millions of years to get there.

The Search for Extraterrestrial Intelligence (SETI) www.seti.org

"Extra" means outside, and "Terra" means Earth. An extra-terrestrial is a being from outside Earth. It's a friendlier term than "space alien."

In the scientific field of SETI, astronomers use radio telescopes to listen for faint radio signals from extraterrestrial civilizations. A fictional dramatization of SETI was the 1997 film *Contact*, starring Jodie Foster. In 1961, Frank Drake first estimated the likelihood of success with the following equation:

The Drake Equation: $N = R_* \times f_p \times n_e \times f_l \times f_i \times f_c \times L$,

where: N = the number of life forms in the Galaxy whose signals that humans may be able to detect,

and: R_* = the rate at which Sun-like stars form in the Galaxy,
f_p = the fraction of these stars that have planets,
n_e = the average number of planets per such system that are suitable for life,
f_l = the fraction of those planets on which life develops,
f_i = the fraction of those life forms that are intelligent,
f_c = the fraction of those intelligent species that develop the technology for interstellar communication (e.g., radio telescopes; maybe also lasers),
L = the average lifetime of a technologically advanced civilization (actually, how long they broadcast radio signals into space).

Don't memorize these terms, but do understand what the equation means.

One problem with the Drake equation is that although it looks scientific because it's mathematical, little or nothing is known about many of these quantities, particularly the ones further down the list.

For example, what *is* the average lifetime of such a technologically advanced civilization? How do we determine this without finding the civilizations first? The Drake equation therefore lacks the predictive power that any scientific theory should have. Another problem with the Drake equation is that it has many assumptions built into it. What if the Universe is teeming with civilizations like the ancient Chinese or the ancient Greeks, who without question were advanced, but never got around to inventing radio? What if astronomers find signals from something utterly alien, say, an interplanetary whale emitting radio songs naturally for its own amusement? Does this count as "intelligent," or even as a "civilization"? Although mathematics may be a universal basis for communication, what if aliens don't use symbol-based language at all?

"The probability of success is difficult to estimate; but if we never search, the chance of success is zero."

–Giuseppe Cocconi and Philip Morrison (1959, *Nature*)

"Two possibilities exist: *either we are alone in the Universe or we are not*. Both are equally terrifying."

–Arthur C. Clarke

"Isolationism is neither a practical policy on the national or cosmic scale. And when the first contact with the outer universe is made, one would like to think that [Humankind] played an active and not merely a passive role–that we were the discoverers, not the discovered."

–Arthur C. Clarke

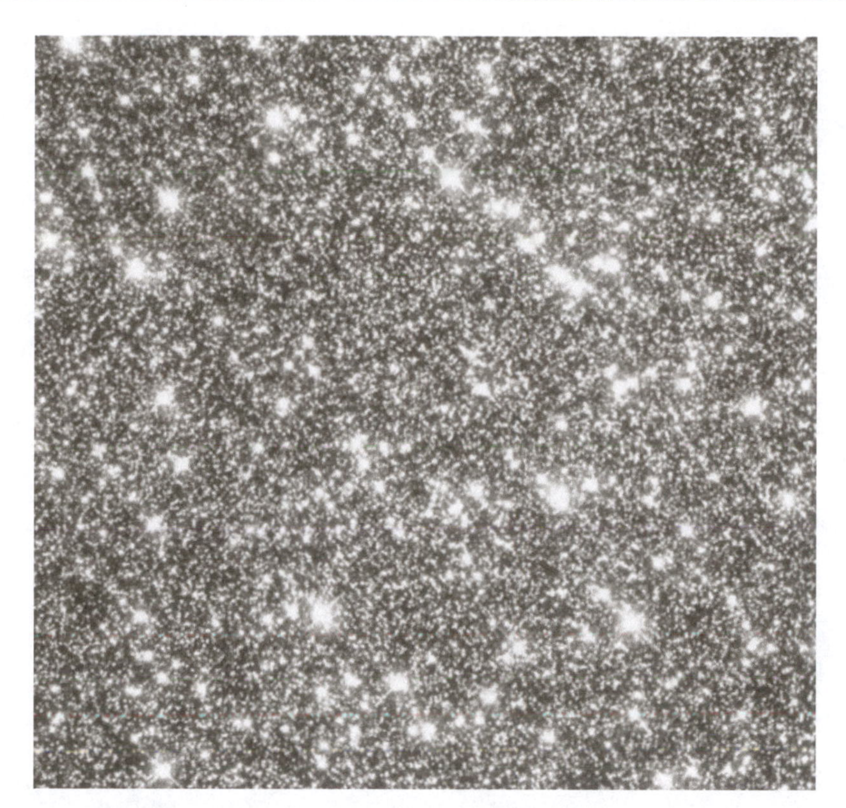

► **Figure 31-1** "Is it reasonable to suppose that in a large field, that only one shaft of wheat should grow, and in an infinite Universe, to have only one living world?"—Metrodorus, circa 500 B.C. (Source: NASA/STScI)

► **Figure 31-2** *Pioneer 10* flew by Jupiter in 1973 and crossed the orbit of Neptune in 1983. *Pioneer 10* is now drifting out into the Milky Way Galaxy. From that distance, the Solar System looks like a bright star, the Sun, which is orbited by faint points of light, the planets of the Solar System. (Source: NASA/Don Davis)

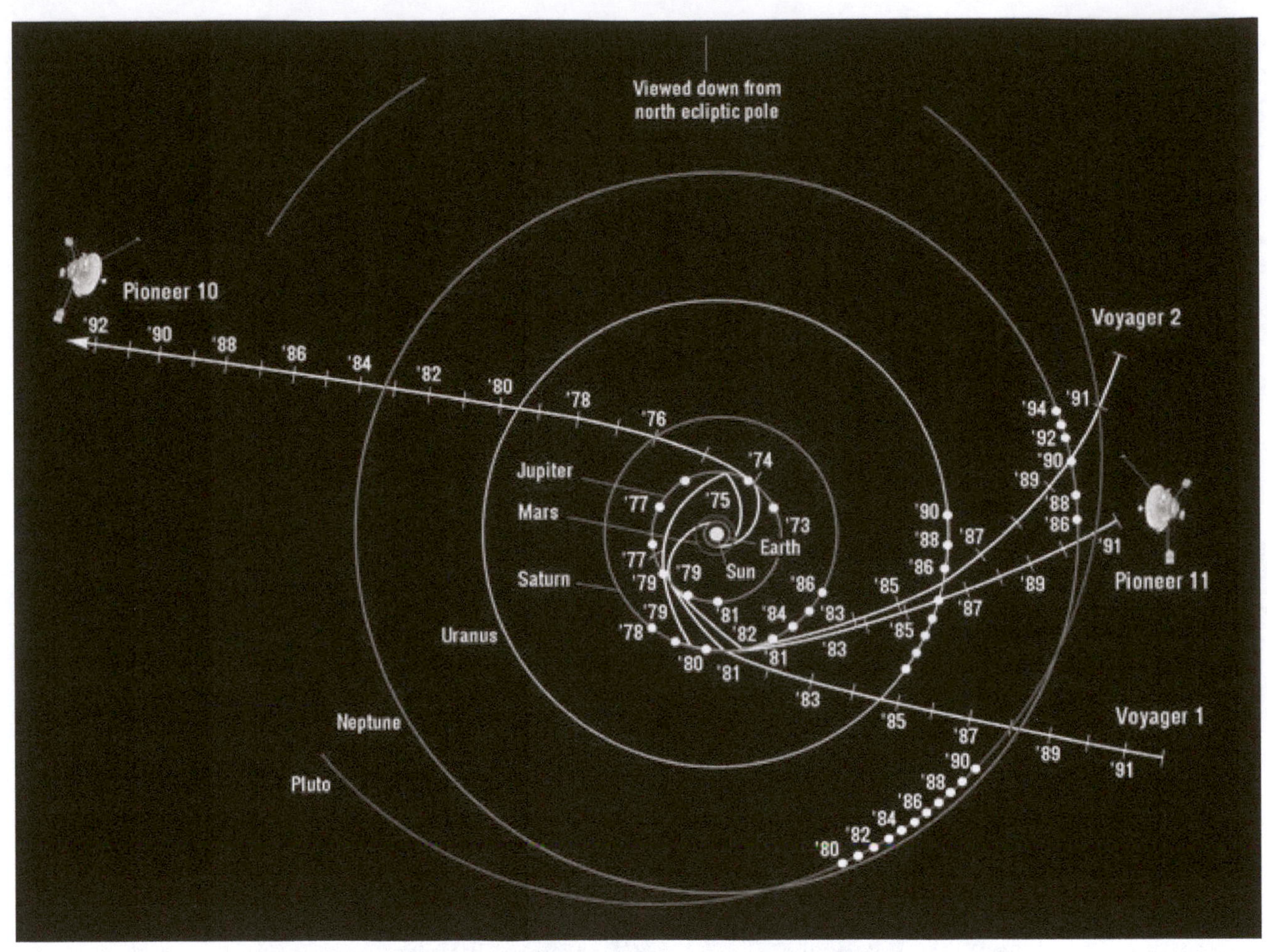

▶ **Figure 31-3** The first starships: four spacecraft (*Pioneer 10* and *11* and *Voyager 1* and *2*), now discarded, have drifted out of the Solar System. A fifth spacecraft, named *New Horizons*, will also leave the Solar System after it finishes its flyby of Pluto in 2015. It will still take all of them thousands of years to reach the next stars. (Source: NASA)

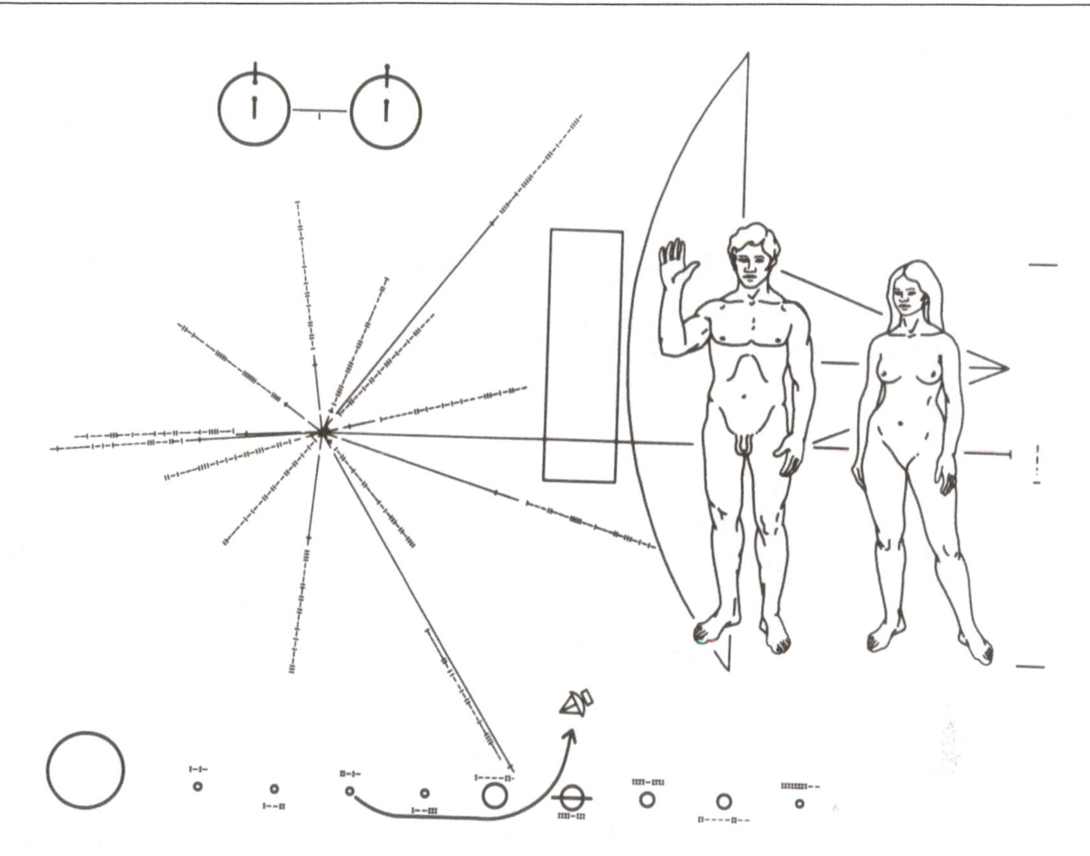

▶ **Figure 31-4** Carl Sagan made sure that the *Pioneers 10* and *11* spacecraft have plaques like this one—just in case anyone (or anything) ever finds them. The *Voyager 1* and *2* spacecraft have records of the sounds of Earth. The problem with this approach is that they are just little spacecraft, not much bigger than cars, adrift in the vastness of the space between the stars. (Source: NASA)

▶ **Figure 31-5** Some years ago, NASA sponsored a conference to discuss how to build starships that could fly to a nearby star within a few decades, with technology available now. Shown here is a solar sail, which is a piece of aluminum foil that rides the pressure of sunlight. A similar light sail that was pushed by a laser is a starship design that could be built and flown now. (Source: NASA/MSFC)

► **Figure 31-6** SETI is the Search for Extraterrestrial Intelligence. A *search* means one *hasn't* found it, yet! *Left:* Frank Drake founded the field of SETI in 1960, with his "Project Ozma" (named after the fantasy novel *The Wizard of Oz*, by L. Frank Baum). (© Roger Ressmeyer/CORBIS) As Dr. Drake says, "It's too late to worry about giving ourselves away. The deed is done. And repeated daily with every television transmission, every military radar signal, every spacecraft command…" *Right:* In Project Ozma, Frank Drake used the National Radio Astronomy Observatory's 140-foot radio telescope at Greenbank, West Virginia (shown above), to listen for radio signals of intelligent origin. Project Ozma, and many similar searches carried out since, did not find any. It is a big Universe, however: recent calculations have shown that humans shouldn't get discouraged even if we spend over a century and still find nothing. (Larry Mulvehill/Science Source)

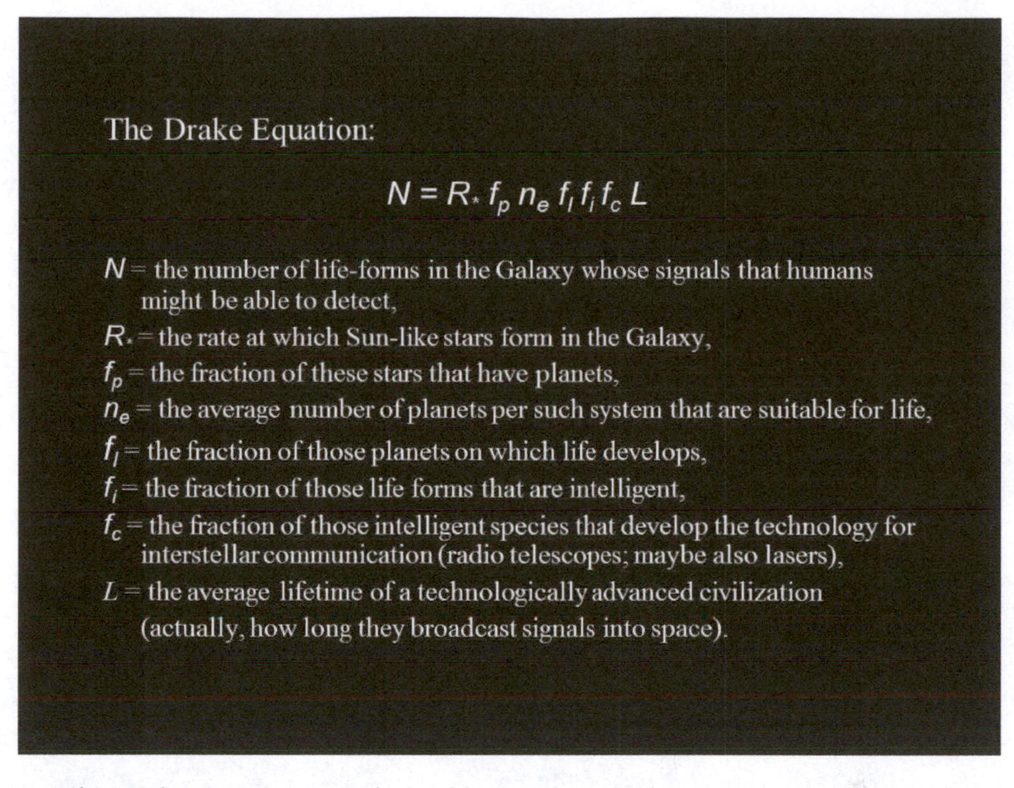

► **Figure 31-7** The Drake equation was devised by Frank Drake in 1961. It attempts to estimate the number of intelligent life-forms with which we could communicate across interstellar distances. (Image courtesy of Frederick Ringwald)

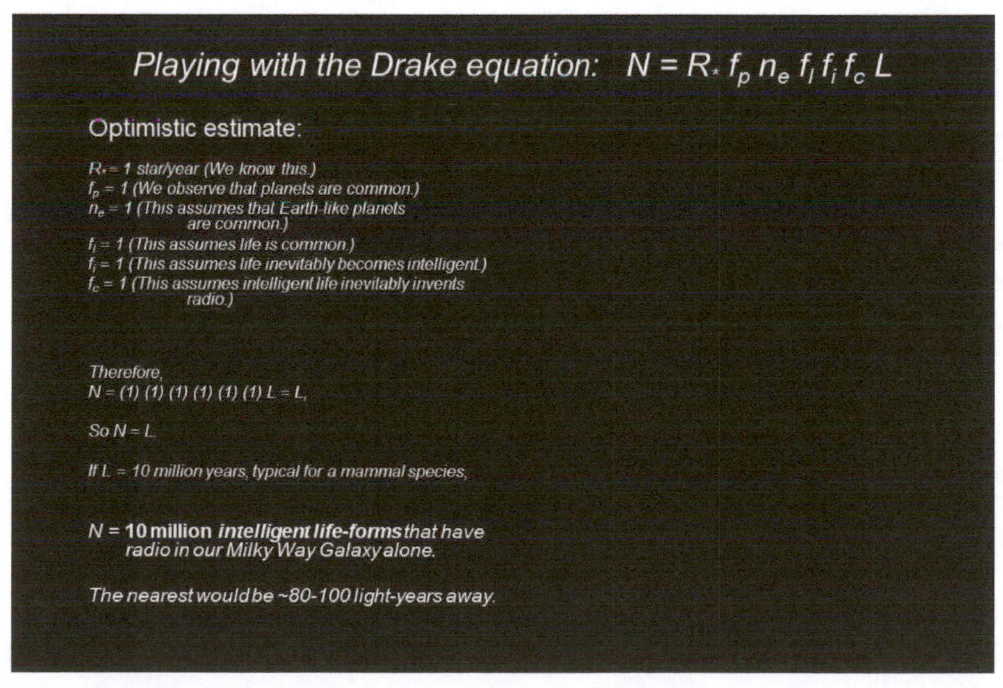

► **Figure 31-8** Many of the factors in the Drake equation are not well known. Some estimates by the author are shown here. If the nearest intelligent life-forms really are 80–100 light-years away, they might just be picking up our earliest radio signals now. We might therefore expect a response as soon as 80–100 years from now. (Image courtesy of Frederick Ringwald)

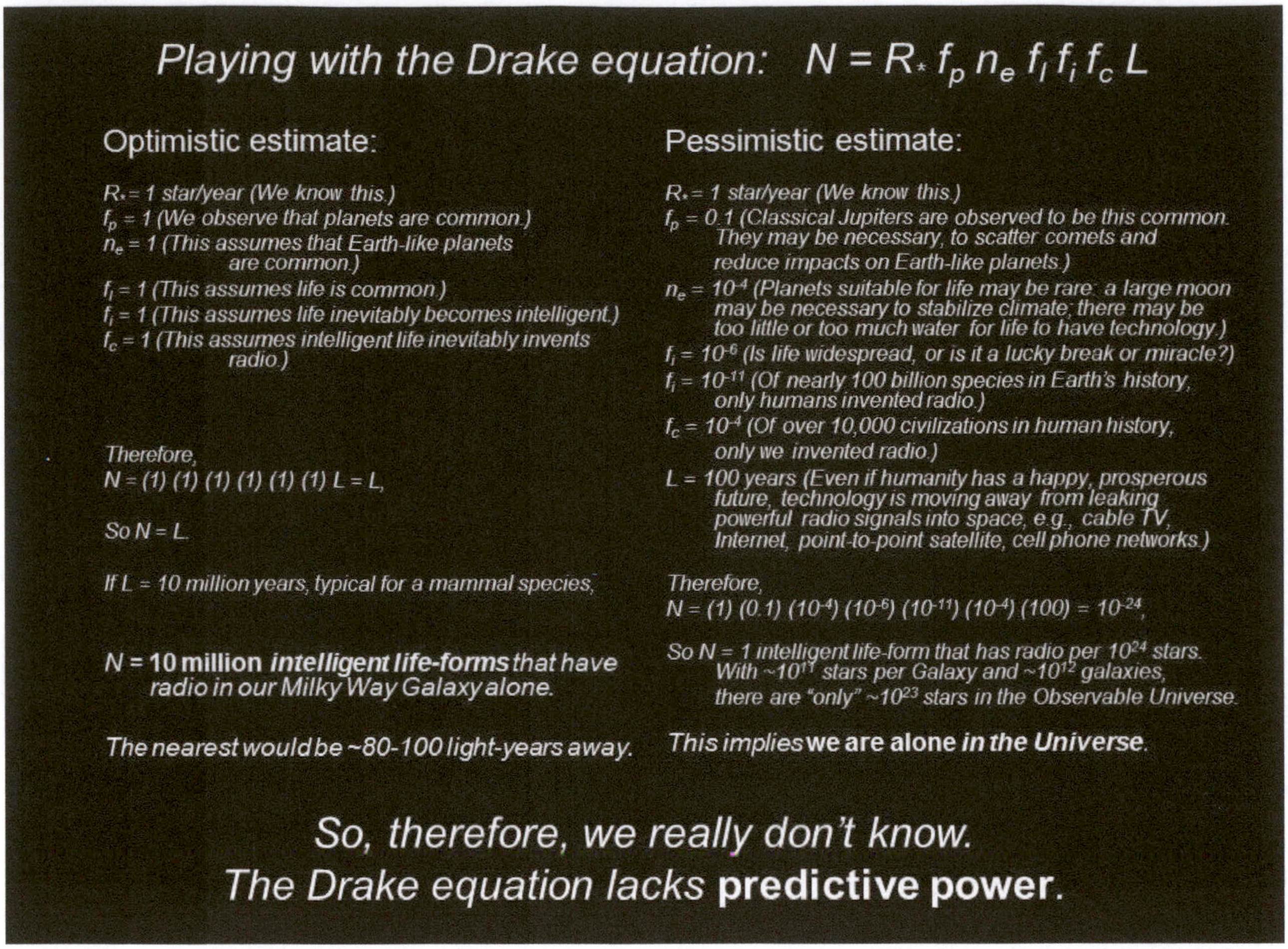

Playing with the Drake equation: $N = R_* f_p n_e f_l f_i f_c L$

Optimistic estimate:

R_* = 1 star/year (We know this.)
f_p = 1 (We observe that planets are common.)
n_e = 1 (This assumes that Earth-like planets are common.)
f_l = 1 (This assumes life is common.)
f_i = 1 (This assumes life inevitably becomes intelligent.)
f_c = 1 (This assumes intelligent life inevitably invents radio.)

Therefore,
$N = (1)(1)(1)(1)(1)(1) L = L$,

So $N = L$.

If L = 10 million years, typical for a mammal species,

N = 10 million *intelligent life-forms* *that have radio in our Milky Way Galaxy alone.*

The nearest would be ~80-100 light-years away.

Pessimistic estimate:

R_* = 1 star/year (We know this.)
f_p = 0.1 (Classical Jupiters are observed to be this common. They may be necessary, to scatter comets and reduce impacts on Earth-like planets.)
$n_e = 10^{-4}$ (Planets suitable for life may be rare: a large moon may be necessary to stabilize climate; there may be too little or too much water for life to have technology.)
$f_l = 10^{-6}$ (Is life widespread, or is it a lucky break or miracle?)
$f_i = 10^{-11}$ (Of nearly 100 billion species in Earth's history, only humans invented radio.)
$f_c = 10^{-4}$ (Of over 10,000 civilizations in human history, only we invented radio.)
L = 100 years (Even if humanity has a happy, prosperous future, technology is moving away from leaking powerful radio signals into space, e.g., cable TV, Internet, point-to-point satellite, cell phone networks.)

Therefore,
$N = (1)(0.1)(10^{-4})(10^{-6})(10^{-11})(10^{-4})(100) = 10^{-24}$,

So N = 1 intelligent life-form that has radio per 10^{24} stars. With ~10^{11} stars per Galaxy and ~10^{12} galaxies, there are "only" ~10^{23} stars in the Observable Universe.

This implies ***we are alone in the Universe.***

So, therefore, we really don't know.
The Drake equation lacks **predictive power.**

► **Figure 31-9** A problem with the Drake equation is that, again, many of the factors in the equation are not well known. Another problem is that the Drake equation has many assumptions built into it, many of which assume intelligent life-forms essentially just like us, whereas real ones may be different in many ways. This is a less-optimistic estimate that attempts to take into account various complications in the emergence of intelligent life. These complications come with their own assumptions too, of course.

The result of this less optimistic estimate is that intelligent life is not common throughout the Universe. Indeed, it may be so rare, we are alone in the Universe. So, therefore, we really don't know. A major problem with the Drake equation is that it lacks predictive power. Recall that, to be considered scientific, an idea must make predictions that can be tested by experiment, or at least by critical observations. (Image courtesy of Frederick Ringwald)

▶ **Figure 31-10** The SETI Institute (www.seti.org) is in Mountain View, CA. The 1997 film *Contact* (based on a science-fiction novel by Carl Sagan) is (loosely) based on them. For example, after having been made fun of by a U.S. Senator for wasting tax dollars, the SETI Institute had to turn to private funding. Since it is private funding, and since they now have their own radio telescope, if they have competent bankers, the search could go on indefinitely. How can one possibly not wish them luck? Jill Tarter, who founded the SETI Institute, says, "We don't use the headphones." This is because SETI is an exercise in picking out faint signals in a noisy Universe, referred to as "the cosmic haystack." Computers therefore are essential for searching for signals from billions of stars and wavelengths. (© Roger Ressmeyer/CORBIS)

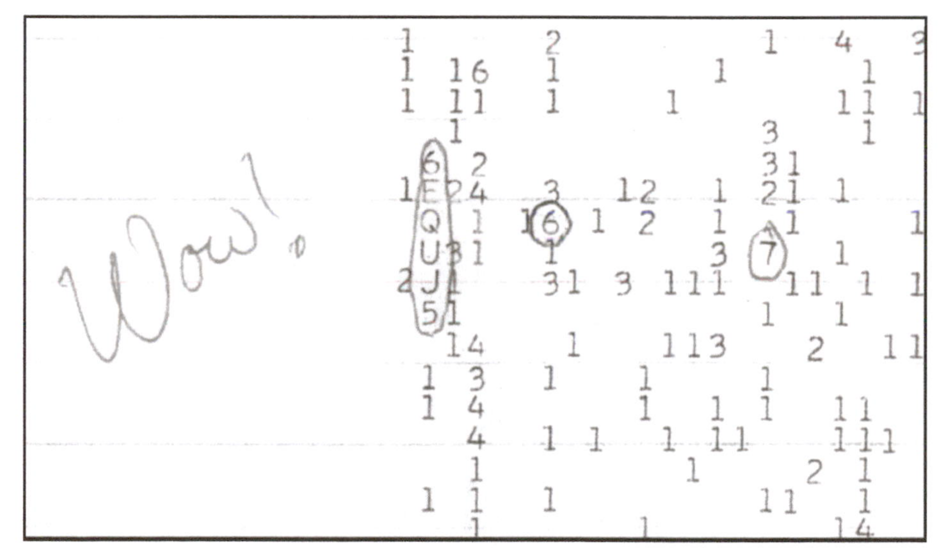

▶ **Figure 31-11** Candidate signals have turned up in SETI. Shortly before midnight on August 15, 1977, Ohio State's "Big Ear" radio telescope picked up a powerful, narrow-band signal that appeared to be exactly what an alien broadcast would look like. Jerry Ehman, the astronomer who was watching the telescope that night, was so surprised that he wrote "Wow!" on the printout. This is why this is called the "Wow" signal. The "Wow!" signal lasted for the 72 seconds that Big Ear observed it. It has never been detected again, despite repeated searches of that part of the sky, near the center of the Milky Way Galaxy.

Notice that on the very next day, August 16, 1977, Elvis Presley died. Four days later, on August 20, 1977, the *Voyager 2* spacecraft was launched for the Outer Solar System. This illustrates a problem with SETI: "the giggle factor." Politicians especially have a history of making fun of SETI, despite the potential for a successful search being among the most influential findings in the history of science. This is why much SETI research today is privately funded. (Public domain image by Jerry R. Ehman)

The Fermi Paradox, or "Where Are They?"

This is another problem with the SETI.

Over lunch sometime in 1950, Enrico Fermi and three other scientists noted that:

1. The Universe is very old, billions of years older than Earth.
2. Sun-like stars are common: they make up 5–10% of stars.
3. Planets appear to be common around Sun-like stars. (We still haven't found any Earth-like planets, but our instruments can't, yet).
4. The commonest elements in living matter (hydrogen, carbon, nitrogen, oxygen) are also the commonest in the Universe (except helium, but it's chemically inert).
5. Conditions hospitable for life appear to be common. For example, water, complex organic substances, and planets around Sun-like stars all are common.
6. In the one case we know, Earth, life developed very early in the history of the planet, seemingly as soon as Earth was capable of supporting life.
7. We are intelligent (or at least we like to think so), and we are a relatively young species of life on Earth.

We might summarize (2) through (7) and say that we might expect life and intelligence to be common throughout the Universe. When coupled with (1), we might expect most intelligent life in the Universe to be much older than we are.

SO: Where are they? Why do we not see extra-terrestrials *here*?

Just about any possible solution to this is mind-boggling. Space may be very big, but aliens with advanced technology have had billions of years to get here.

The SETI Institute and several other groups around the world continue their searches, often with private funding. How can we possibly not wish them luck?

► **Figure 31-12** Enrico Fermi posed the Fermi Paradox: If intelligent life in the Universe is so common, and the Universe is so old, then "Where are they?" (Source: U.S. Department of Energy)

► **Figure 31-13** There is no good evidence for alien encounters at any time in history. Even if aliens had visited Earth a million times randomly over time, on average the last visit would have been 4,600 years ago. (Image courtesy of Dr. Greg Morgan and digitally processed by Frederick Ringwald)

Searching for Simple Life Beyond Earth

A cynic might say, "Who cares about finding bacteria?" Bacteria might seem simple, but they really aren't. Finding bacteria in space would show that life is not unique to Earth. It would encourage scientists to keep looking for intelligent life. Even the simplest life is quite complex, and still not completely understood. Understanding it better would help understanding of all life, potentially useful for understanding cancer and other disease.

In 1995, two paleontologists found bacteria spores in a bee preserved in amber that was 30 million years old. These bacteria were in deep hibernation. (The technical term is *stasis*.) Upon placement in a nutrient-rich culture, they became active again: in other words, came back to life.

In 2000, scientists found bacteria in stasis in salt crystals that are 250 million years old. They were able to culture these bacteria back to life, too.

- Evidence for life in a Martian meteorite (ALH 84001) was claimed by NASA in 1996. This claim is still *inconclusive*. All the evidence is indirect: no actual organisms that are alive now have been found. There were also no claimed remnants of life that couldn't possibly have been anything else.
- The *Viking 1* and 2 spacecraft that landed on Mars (in 1976) scooped up some Martian soil and made chemical tests of it, specifically in order to search for life. The results of these tests were *inconclusive*.

  A problem with these tests is that they literally only scratched the surface of Mars, and the surface of Mars is quite hostile to life. Mars only has a thin atmosphere, with no ozone layer to protect it from high-energy ultraviolet radiation from the Sun.

  (A claim that bacteria were brought back by the *Apollo 12* mission after 30 months in equipment on the Moon was discredited in 2011. Examination of archival footage showed that the clean room that made this claim wasn't clean, by today's standards.)

What about life underground? There is abundant life deep inside Earth, where the Sun never shines. It lives off heat and chemicals coming from inside Earth.

- Deep-sea vents in Earth's oceans are also full of life. Some of the most primitive life forms known on Earth are found there. Did life on Earth originate there?

Does this mean that Europa, Ganymede, and Callisto (the large, icy moons of Jupiter) are likely places to find life? All might have liquid water, under their icy crusts. Enceladus, an icy moon of Saturn, has geysers of liquid water that erupt from just under its icy crust.

- What about life on planets around other stars (exoplanets)? NASA is designing spacecraft to get spectra of the atmospheres of exoplanets. The idea is to search for spectral lines of oxygen (O_2), methane (CH_4) or other gases that are out of chemical equilibrium, a sign of life. Earth's atmosphere wouldn't have nearly as much oxygen or methane as it does if it weren't for life on Earth, which replenishes these gases every day.

In 2004, methane was discovered in the atmosphere of Mars. It's out of equilibrium with other gases in the atmosphere of Mars. Might it be from Martian bacteria?

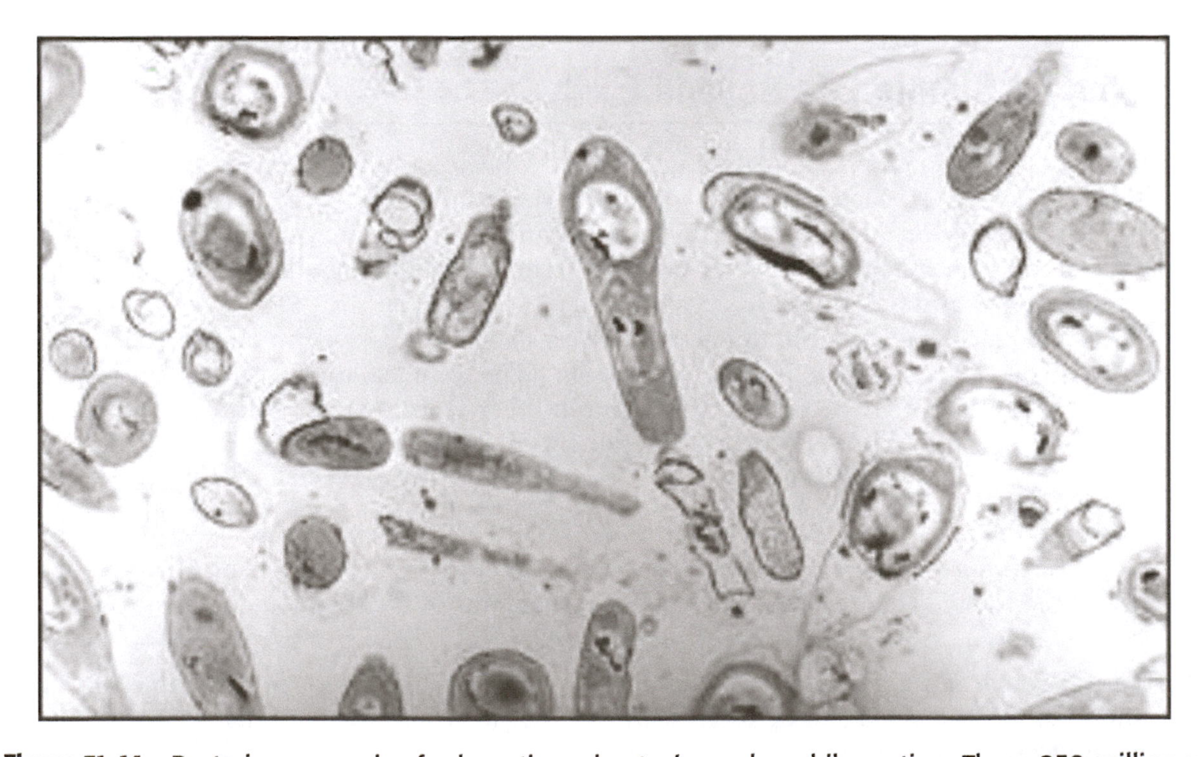

► **Figure 31-14** **Bacteria can survive for long times in *stasis*, or deep hibernation. These 250 million year-old bacteria were preserved in a salt crystal, and were cultured back to life when put in a warm, wet environment. Because life can do this, life carried in meteorites has the ability to travel across space. It's not necessary to wait until starships are invented.** (250 Million Year Old Halophilic Bacterium From a Salt Crystal. Copyright Russell H. Vreeland. Reprinted with permission)

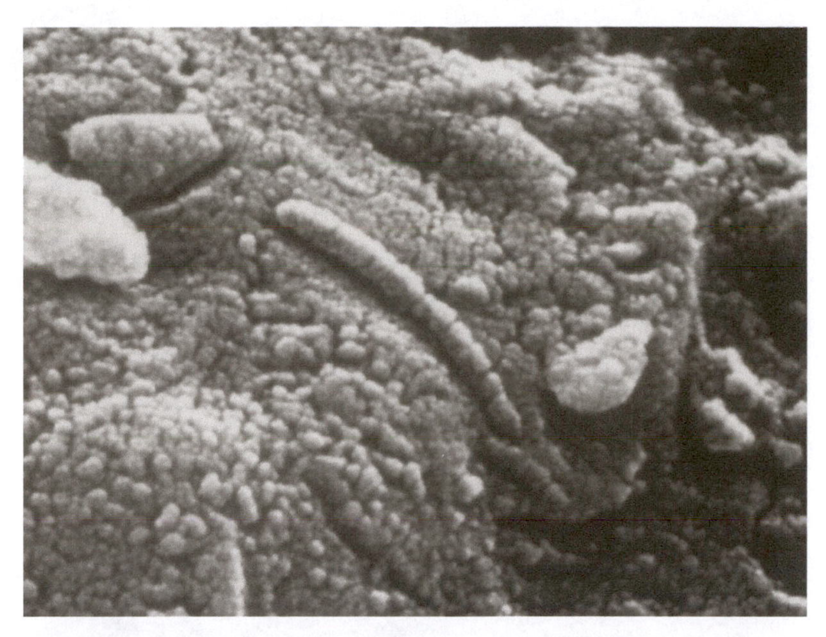

▶ **Figure 31-15** In 1996, NASA announced it had found evidence of life in a meteorite from Mars. NASA claimed that these long, thin things are nanobacteria, which are bacteria unlike most bacteria, in that they are much smaller. Many other scientists are skeptical of this claim, however, because of a lack of other evidence that is compelling, meaning that it could not be explained in any other way. As Carl Sagan liked to say, "Extraordinary claims require extraordinary evidence." Also, as Sigmund Freud noted, "Sometimes a cigar is only a cigar." (Source: NASA)

the ONION

Mean Scientists Dash Hopes Of Life On Mars

April 9, 2003 | Issue 39•13

PASADENA, CA—A team of cold-hearted, killjoy scientists at NASA's Jet Propulsion Laboratory callously announced Monday that the likelihood of complex life on Mars is "extraordinarily low," dashing the hopes of the public just like that.

▶ **Figure 31-16** This is a joke from a humor magazine I like, *The Onion*. It shows that scientists aren't being "mean," or "stubborn," or "difficult," or "closed-minded," or, worst of all, "unscientific" when they are skeptical of claims of life from Mars or elsewhere in Outer Space. They are being perfectly scientific, since extraordinary claims require extraordinary evidence, and so far, the evidence hasn't been compelling. Remember, we've been burned on this question before, by Percival Lowell and the canals he claimed he could see on Mars (which spacecraft later showed are not there). Actually, no one would be happier than most scientists would be, if we had *compelling* evidence. If a Martian lifeform came up and bit me, I'd change my mind, and I'd acknowledge that it existed. (Reprinted with permission of The Onion., Copyright © 2013, by Onion, Inc. www.theonion.com)

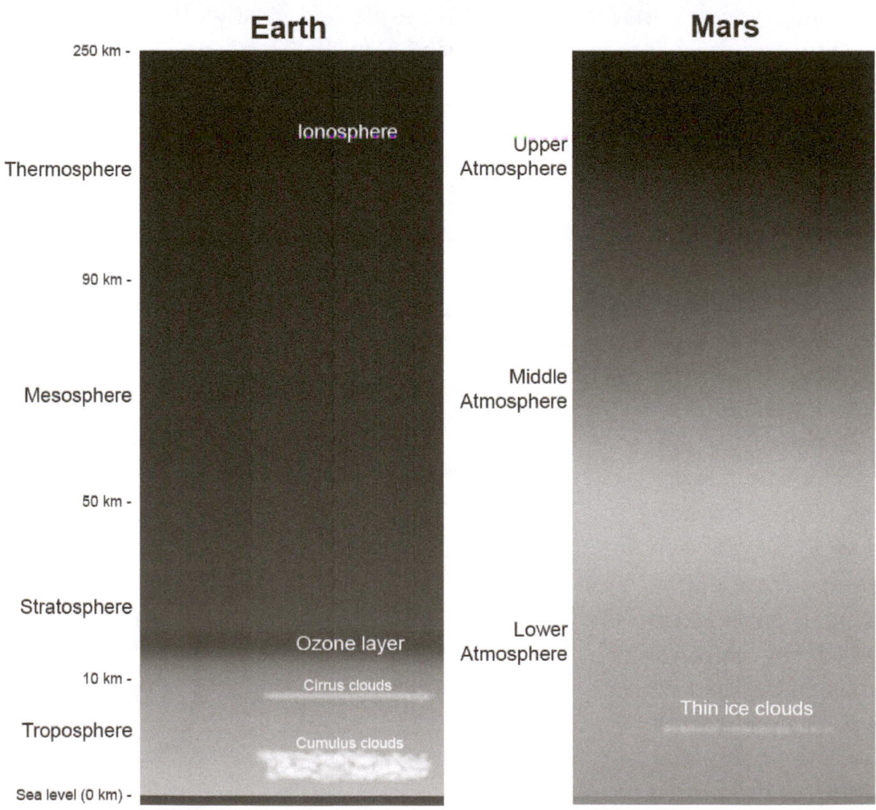

▶ **Figure 31-17** *Top:* The *Viking 1* and *2* landers searched for life on Mars in 1976. The results were inconclusive. (They didn't find any.) (Source: NASA) ***Bottom:*** A problem with searching for life on Mars is that, with no ozone layer, the surface of Mars is sterilized by high-energy ultraviolet radiation from the Sun. Might there still be life deep inside Mars? There are bacteria deep inside Earth. (Image courtesy of Frederick Ringwald)

▶ **Figure 31-18** There are no large life-forms that are obvious to the eye on Mars. (Source: NASA)

▶ **Figure 31-19** That there is no life obvious to the eye on Mars doesn't prove that there is no life on Mars. This is one of the Dry Valleys of Antarctica. They are some of the coldest and driest extreme deserts on Earth. There is no life obvious to the eye, but bacteria do live inside the rocks here. (NSF/Josh Landis)

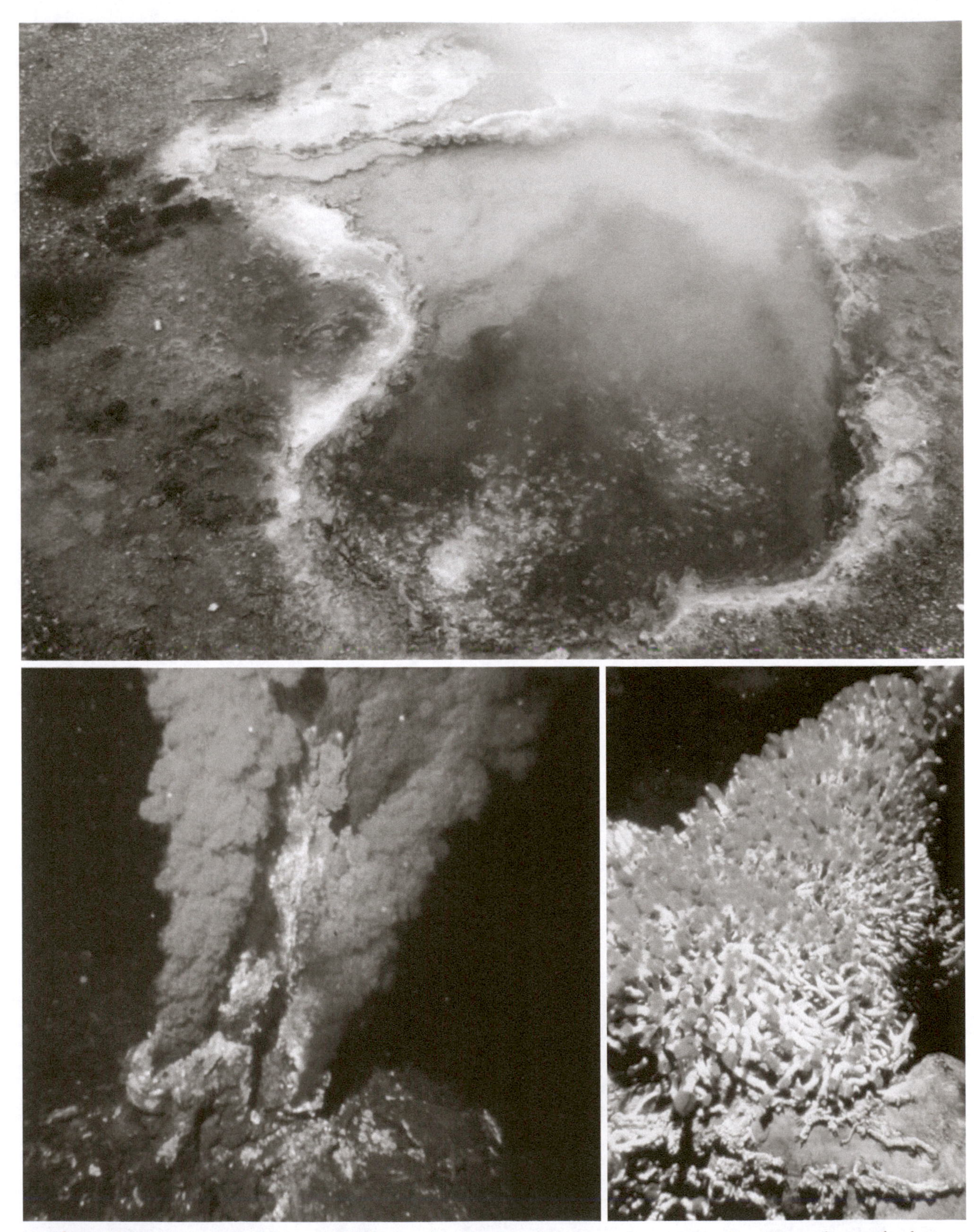

► **Figure 31-20** Extremophiles are life-forms that live in extreme environments. *Top:* Bacteria are the brown mats living in boiling water in geyser pools in Yellowstone National Park. (Image courtesy of Frederick Ringwald) *Bottom:* Bacteria also live deep inside Earth, and complex ecosystems of tube worms live in boiling-hot water around volcanic vents in the ocean floor. (Source: NOAA)

► **Figure 31-21** Life may exist beneath the icy surfaces of Europa, a large moon of Jupiter that has liquid water, heat (from Jupiter), and organic chemistry. What more could one want for life? If life does exist inside Europa, can it exist inside the many other icy moons and other bodies (including Pluto) in the Outer Solar System? (Source: NASA)

► **Figure 31-22** Spacecraft are planned for Europa, the icy moon of Jupiter. *Left:* An orbiter with radar that could see through the ice could prove whether oceans really do exist under the ice. If they do, they could conceivably find beds of coral or seaweed, if any exist. (Source: NASA) *Right:* A hydrobot, similar to ones used in Antarctica, might melt its way through the ice. (Source: NASA)

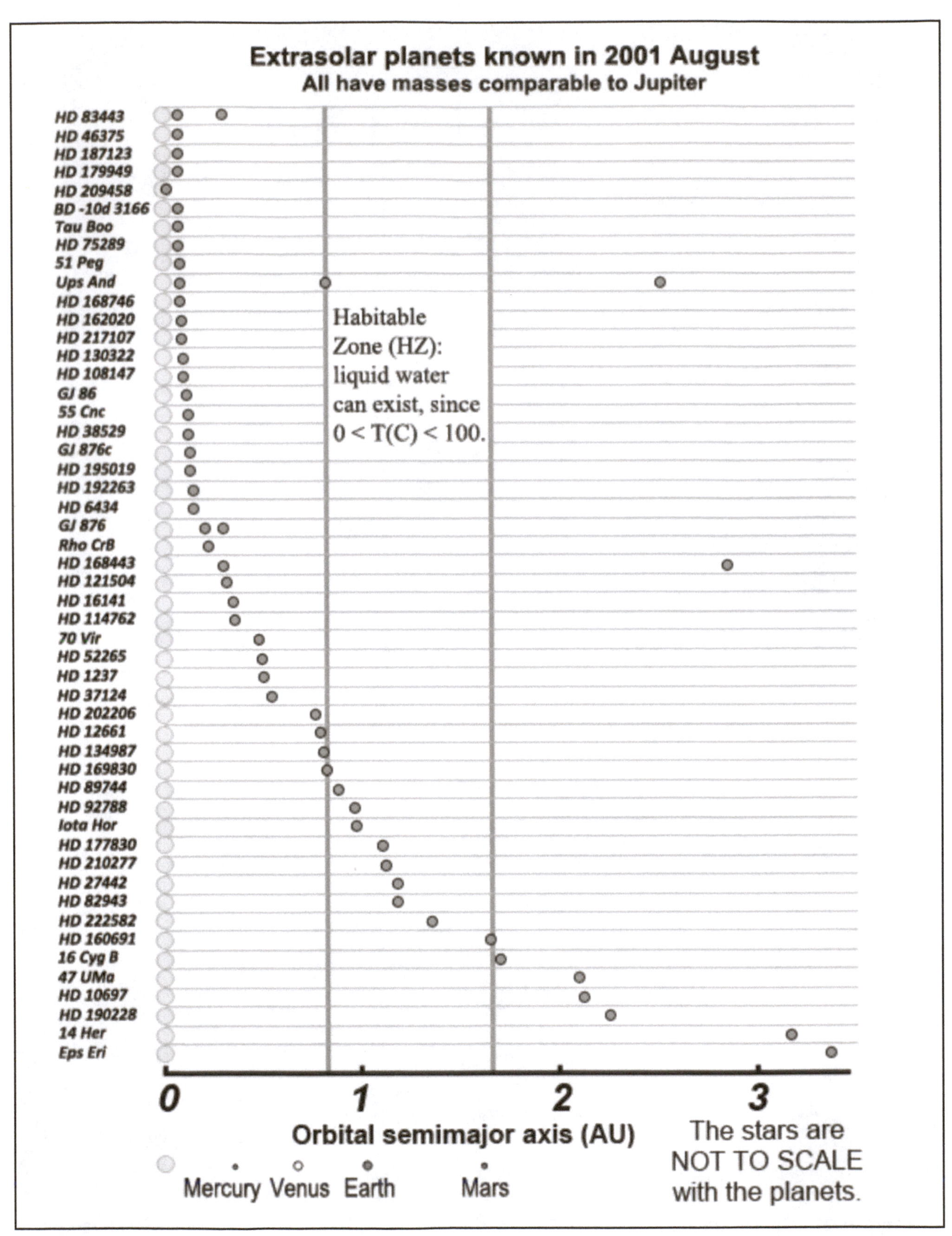

► **Figure 31-23** Exoplanets are planets that orbit other stars. Hundreds are now known. Several of these are Earth-sized and in the Habitable Zones of stars similar to the Sun. The Habitable Zones are where temperatures are between 0 and 100°C, where liquid water could exist. (Image courtesy of Frederick Ringwald)

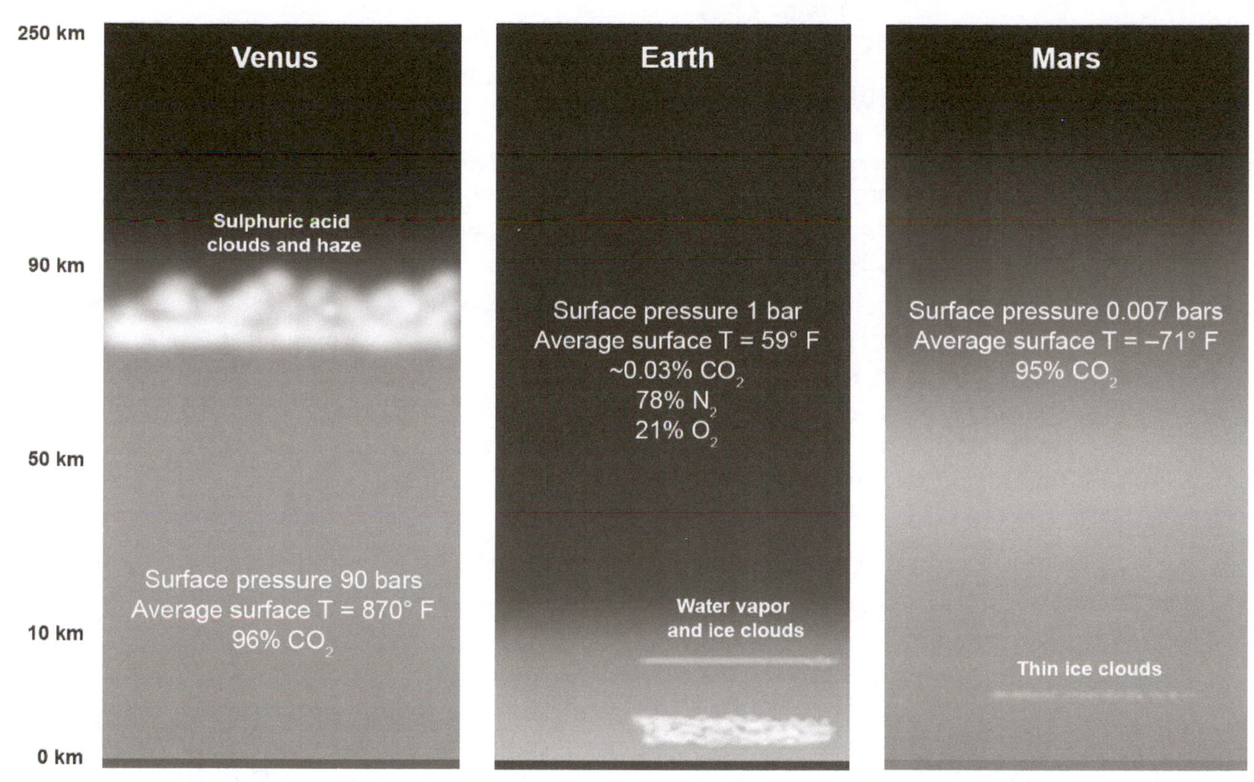

▶ **Figure 31-24** Unlike Venus and Mars, Earth's atmosphere is full of oxygen, *because of life*. It may be possible to detect life on planets and even quite distant exoplanets by getting spectra of their atmospheres, and looking for spectral lines that come from oxygen (O_2), methane (CH_4), and other gases that are produced by life, or are otherwise out of chemical equilibrium with other gases in the planets' atmospheres. (Image courtesy of Frederick Ringwald)

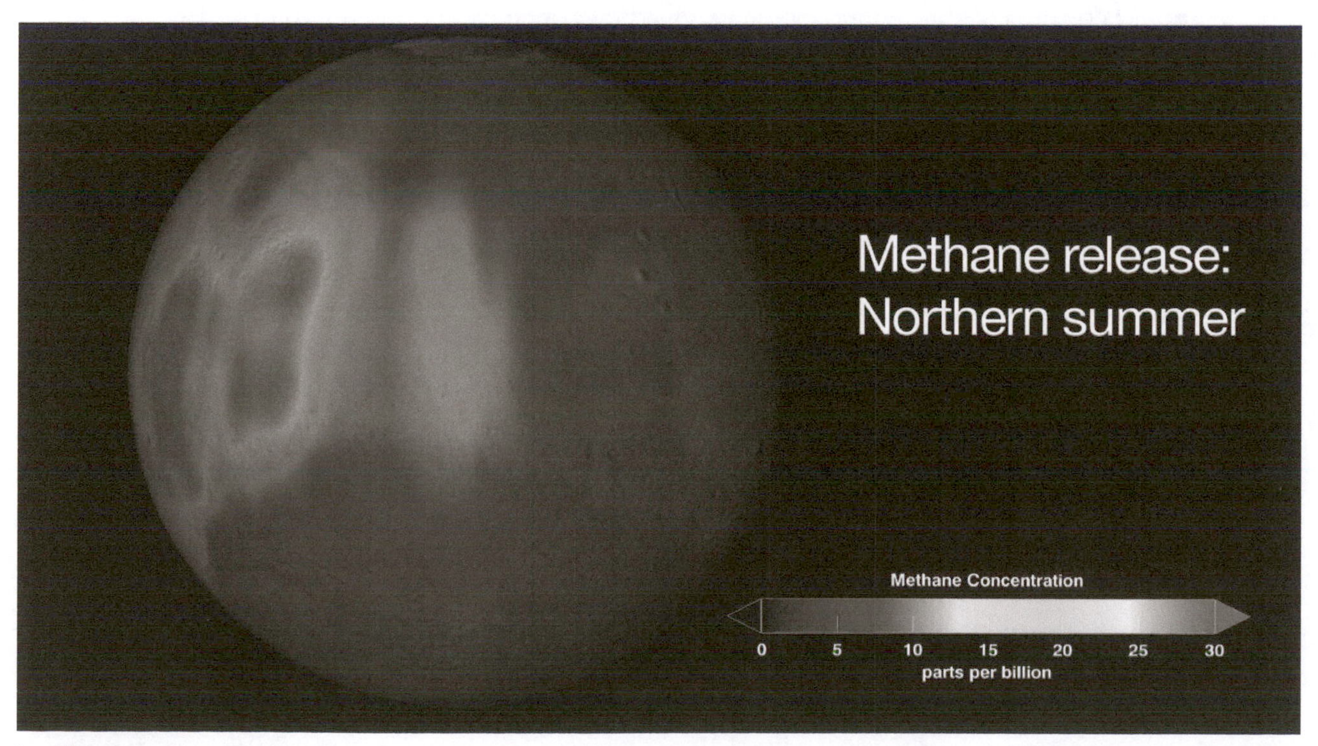

▶ **Figure 31-25** To find life, search a planet's atmosphere for oxygen or other out-of-equilibrium gases. Is the methane in the atmosphere of Mars from life? (Source: NASA)

What any discussion of "Life Beyond Earth" is almost certainly *not* about:

Unidentified Flying Objects (UFOs), or "flying saucers."

1. So far, the only evidence we have is someone's say-so. No clear, unambiguous physical evidence has ever been found. Typical physical evidence consists mainly of broken tree branches and burned spots on the ground. Have you ever seen a clear, well-focused, well-documented video of an alien spacecraft?
2. UFO means Unidentified Flying Object. Why does everyone assume this must necessarily mean "alien spacecraft"? Many false alarms have been well documented: the planet Venus is the most widely reported UFO, because it's so bright. (Former President Jimmy Carter called the Georgia State Police on it, when he was Governor of Georgia.) There are many unfamiliar phenomena in the sky, including sundogs, haloes, glories, and mirages. How many are truly *unidentifiable*, and not just *unidentified*?
3. Perhaps most damning is that people who know the sky well, including over 6,000 professional astronomers and 200,000 amateur astronomers in the United States alone, almost never see or report UFOs.
4. Even the term *flying saucer* is a misnomer. "Flying saucer" was coined in 1947, when a pilot, Kenneth Arnold, reported seeing nine objects flying in a "V" formation over Mount Rainier. Arnold told a reporter that they flew erratically, fluttering and tipping their wings, like "a saucer if you skip it across water." Although the objects Arnold reported were not saucer-shaped, the press distorted his words into the phrase "flying saucer." Shortly after the story appeared in newspapers, many people reported seeing "flying saucers." A likely rational explanation of this is the power of suggestion.
5. An especially unlikely explanation is that conspiracy groups, particularly within the U.S. government, are keeping this evidence "secret." It is strange, then, that we often see the government unable to keep secrets about even the most obvious things, such as embarrassing things about certain U.S. Presidents, or important things such as nuclear bomb designs.

   A finding of this magnitude *couldn't* be kept secret. It *wouldn't* be kept secret, either. It's hard to believe that any scientist, government, or journalist would hesitate at the chance for fame.
6. The claim that contact with aliens is being kept secret because humans would panic, with "rioting in the streets," is unlikely. A 1997 Gallup poll and a 2002 Roper poll showed that half the American public (incorrectly) believes that we've made contact already. Furthermore, two claims of detections have been made in public: (1) Jocelyn Bell's discovery in 1967 of LGMs ("Little Green Men"), which turned out to be pulsars, and (2) Nikolai Kardashev's claim in 1963 of a periodic signal from the radio source CTA-102, which turned out to be a quasar (and not periodic). CTA-102 was so widely publicized that the Byrds, a famous rock 'n' roll band, recorded a song about it. In neither case was there any rioting in the streets.

► **Figure 31-26** Typical physical evidence of UFOs is broken tree branches and burned spots on the ground. There has never been anything compelling, which could not be explained in any other way, such as a ray-gun that could vaporize a person dropped by a flying saucer occupant. (If you come across such a thing, please let me know: I have a list of people I'd like to vaporize.) (Image courtesy of Frederick Ringwald)

► **Figure 31-27** This image was carefully investigated, and was identified positively as a hoax. They do happen. (Source: CIA)

▶ **Figure 31-28** This is a lens flare, which is a reflection inside the camera's lens. (Image courtesy of Frederick Ringwald)

▶ **Figure 31-29** The planet Venus is seen here to the upper left of the Moon. It is the brightest object in the sky, other than the Sun and Moon. It is therefore unsurprising that Venus is the most commonly reported UFO. To the unaided eye, it looks like a very bright star. Air-traffic controllers often give it clearance to land, because they think it is an oncoming airplane; pilots from both the U.S. Navy and U.S. Air Force have chased it, and fired at it. (Image courtesy of Frederick Ringwald)

► **Figure 31-30** Not all UFOs are weather balloons, it is true, but some are. These are weather balloons. The famous "Incident at Roswell" in 1947, while claimed by the U.S. Air Force to be a weather balloon, was actually the crash of a top-secret spy balloon, called Project Mogul. (*Top left:* Source: NASA Dryden Flight Research Center; *Bottom left:* NASA/GSFC; *Top right:* Source: NOAA; *Bottom right:* Source: NOAA)

▶ **Figure 31-31** Military flares like these explain one famous UFO sighting, the "Phoenix Lights" of March 13, 1997. The Phoenix lights were seen by many people and were even recorded on videotape. When I lived in Tucson, I saw A-10s and F-16s flying out of Davis-Monthan Air Force Base every day. Never once did I see an alien spacecraft. (*Left:* Public domain image by Staff Sgt Justin Weaver, USAF; *Right:* Source: DoD)

▶ **Figure 31-32** This is a mirage. In a mirage, the air immediately above a hot surface is cooler than the surface, in such a way to become reflective to light. To someone dying of thirst in the desert, this may look like water; although to someone dying of thirst in the desert, lots of things start to look like water. (Image courtesy of Frederick Ringwald)

▶ **Figure 31-33** These are lenticular clouds, which are disturbingly often mistaken for flying saucers. ("Lenticular" means lens-shaped.) (Images courtesy of Frederick Ringwald)

▶ **Figure 31-34** This is the trail from a rocket launch. Since the launch was just after sunset, the trail reflects the colors of the sunset, just over the horizon. (Used by permission of the photographer, James W. Young)

► **Figure 31-35** This is a solar halo, with sundogs (also called "mock Suns") on either side, and a parhelion above. It is caused by scattering of sunlight by ice crystals in Earth's atmosphere. A similar display may have been reported by the Prophet Ezekiel in the Old Testament (Ezekiel 1:3-17), as a fiery wheel with the faces of four angels. (Courtesy, Peter Rosen)

► **Figure 31-36** These are crepuscular rays, or "the rays of Buddha," from shadows and perspective. Knowing this doesn't make them any less beautiful, does it? (Image courtesy of Frederick Ringwald)

Chapter

32

(Some of) The Most Influential Scientific Findings of All Time

One way to start scientists arguing is to suggest which are the most influential scientific findings of all time. Despite the risk, doing this is useful for summarizing all of science on one page. It shows that the most influential findings aren't necessarily those with the most immediate practical applications, although these do come, in time. They are the findings that change how we see the world around us. My picks are:

10. The electromagnetic theory of James Clerk Maxwell (1865) shows that electricity, magnetism, and light are all forms of the same phenomenon. This gave us radio and TV: so much for it being "just a theory."
9. Thermodynamics, the study of energy (Maxwell, Kelvin, and others, 1800s), has wide generality, dealing with ideas both practical (such as how to fuel cars) and abstruse (such as black holes and the arrow of time).
8. Living things are observed to change over time, or evolve, by natural selection (Charles Darwin, 1859). This is not "just a theory," one can see it happen:
   a. My favorite dog is a Golden Retriever/Basset Hound mix. He has a Golden Retriever's fur, and a Basset's long body and short legs. If this can happen in one generation, imagine what can happen in a million years.
   b. The fossil record shows what can happen over millions of years. It includes many transitional forms, such as walking fish, feathered dinosaurs, *Archaeopteryx*, *Eohippus, Ambulocetus*, *Homo Habilis-Ergaster-Erectus*.
   c. Peppered moths were white in the 1700s. They became black when coal was used. They're now white.
   d. Pesticide-resistant insects, herbicide-resistant superweeds, antibiotic-resistant bacteria, new diseases such as MRSA, SARS, and new strains of flu.
   e. Experiments with fruit flies (*Drosophila*) and clam worms (*Nereis acuminata*) have shown directly that they can change into different species.
   f. Vestigial organs are still present, such as leg bones in snakes and whales.
   g. Homology: similarities due to common ancestry, such as bones in vertebrates, and protein structure.
   h. We can trace specific genes in specific DNA sequences back in time. Humans share over 98% of their genes with chimpanzees. This doesn't mean you're a chimp, however!

7. The heredity of living things is carried by genes (Gregor Mendel, 1865). The genetic information is carried by DNA, a molecule present in all human cells (James Watson and Francis Crick, 1954).
6. Living things are composed of cells (Robert Hooke, 1665). That single-celled microscopic organisms, or germs, can cause disease is called "the germ theory of disease" (Louis Pasteur, 1865).
5. Plate tectonics: Earth's surface is broken into moving plates, which unifies geology (mountain building, continents, earthquakes). Proposed by Alfred Wegener in 1912, this was not accepted until after 1965.
4. The Copernican Principle: the discovery that we are not the center of the Universe, beginning with Copernicus (1543) and continuing with Edwin Hubble's discovery of the expansion of the Universe (1929).
3. The nuclear model of the atom, based on the discovery of chemical elements by Antoine Lavoisier (1776), the realization that atoms exist by John Dalton (1800), the discovery of the atomic nucleus by Ernest Rutherford (1911), and the theory of atomic structure by Niels Bohr (1913). This would become quantum mechanics (1926), which would explain the periodic table and the chemical bond (Linus Pauling, 1954).
2. The Principle of Relativity (Albert Einstein, 1905, 1916): that all physical law depends on the observer. This shows that matter and energy are equivalent, and explains gravity as curvature of space and time.
1. Newton's Laws of Motion and Gravity (1687), with contributions by Galileo, Johannes Kepler, and Tycho Brahe. This includes Newton's realization of the Universality of Physical Law: that the Universe follows orderly laws, and that we humans can learn them, by careful observation and rational thought.

▶ **Figure 32-1** "There can be no thought of finishing, for 'aiming at the stars,' both literally and figuratively, is a problem to occupy generations, so that no matter how much progress one makes, there is always the thrill of just beginning."—Robert Goddard, 1932, writing to his hero, H. G. Wells (Source: NASA/Esther C. Kisk Goddard)

▶ **Figure 32-2** To be considered scientific, any theory must make *predictions,* which can be tested by experiment. James Clerk Maxwell's electromagnetic theory (1865) predicted the existence of radiation that the unaided eye can't see. This gave us radio, television, and X-rays. So much for it being "just a theory." (Image © Nicku, 2013. Used under license from Shutterstock, Inc.)

▶ **Figure 32-3** Maxwell's electromagnetic theory predicted the existence of wavelengths of light that unaided human eyes can't see: *the electromagnetic spectrum*. This is perhaps the biggest advance in astronomy in the past two generations. It used to be that astronomers could detect only the visible (optical) light that our eyes can see: it was like watching television where one can get only one channel. Now, we get *all* the channels, and we learn about the Universe much faster because of it. (Source: NASA)

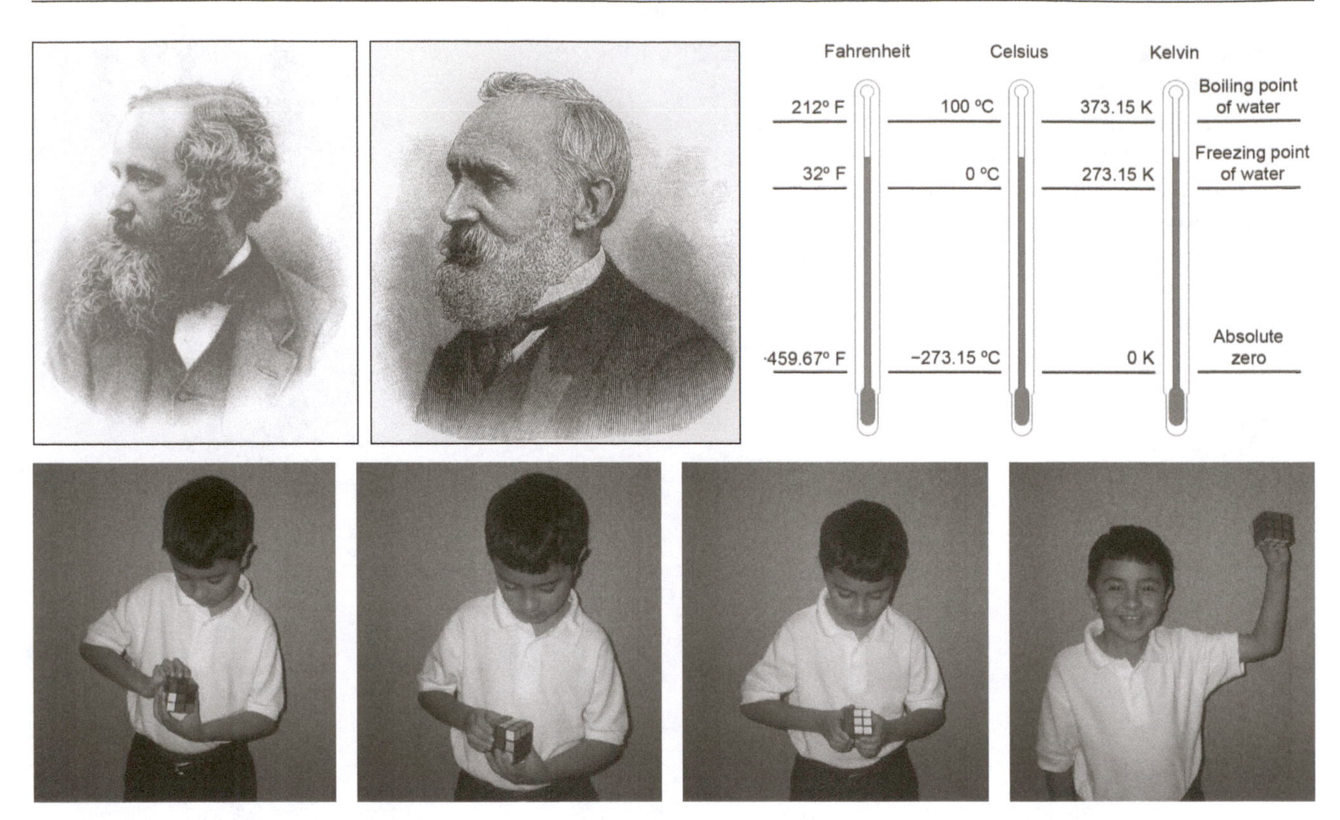

► **Figure 32-4** Thermodynamics is the study of energy. It was a team effort, a characteristic of science today: Maxwell, Kelvin, and others contributed to it. *Top left:* Maxwell (Image © Nicku, 2013. Used under license from Shutterstock, Inc.)

Top center: Kelvin (Image © Nicku, 2013. Used under license from Shutterstock, Inc.)

Top right: Temperature scales (Image courtesy of Frederick Ringwald)

Bottom: At first, thermodynamics was a practical exploration of how to make steam engines more efficient. The laws of thermodynamics discovered by this turned out to be among the most general laws known about the Universe: among other things, they show that nature has a built-in direction to it—"the arrow of time."

This generality is an example of what scientists like: a small number of ideas, sometimes called laws, which can be used to understand so many different phenomena that, out of context, would be difficult to understand. (Images courtesy of Frederick Ringwald)

► **Figure 32-5** Charles Darwin's idea of evolution isn't "just a theory"; you can see it happen. My favorite example is Howard. (Bottom; image courtesy of Kevin S. Swaney). Howard is a cross between a Golden Retriever (top left; Image © USBFCO, 2013. Used under license from Shutterstock, Inc.) and a Bassett Hound (top right; Image © cynoclub, 2013. Used under license from Shutterstock, Inc.). If this can happen in one generation, imagine what can happen in millions of years. That was Darwin's main argument in his book, *The Origin of Species*, which starts with a long chapter on the horticulture of gooseberries. Darwin reasoned that, if humans can selectively breed plants and animals, nature can do it too, using natural selection.

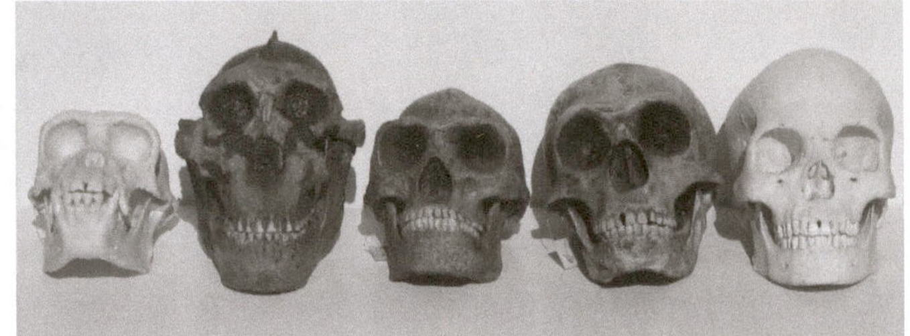

► **Figure 32-6** Other evidence of evolution includes *transitional forms* (formerly known as missing links). Transitional forms include: *Left: Archaeopteryx* (an early bird with teeth and fingers) and feathered dinosaurs (Image courtesy of Frederick Ringwald) *Top right: Eohippus* (the dawn horse) (Public domain image from the American Museum of Natural History—copyright expired, 1919) *Bottom right:* Skulls, from left to right, of a chimpanzee, a *Paranthropus Boisei* (a gorilla-like vegetarian that walked on two legs), a *Homo Ergaster* (who had stone tools and fire, but maybe not articulate speech, although I'm sure they howled at each other a lot), a *Homo Neanderthalensis* (which was to you what a Saber-toothed cat is to a house cat), and a *Homo Sapiens.* (Image courtesy of Frederick Ringwald)

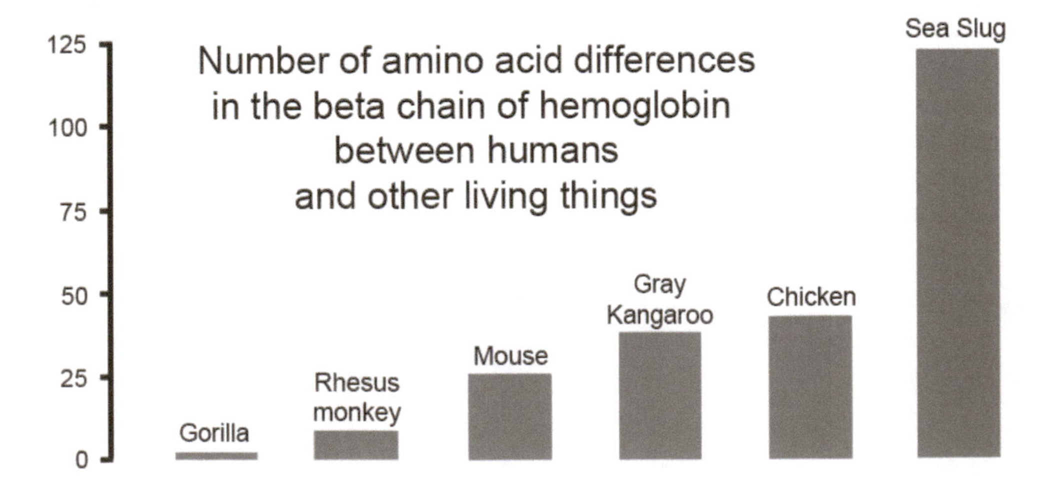

► **Figure 32-7** Further evidence for evolution is *homology*, which is similarity due to common ancestry. A human arm, a dog's leg, a bird's wing, and a whale's flipper are all used differently, and yet they have the same bones in the same pattern. This happens on the molecular level, too. The closer the relation, the more similar the proteins and the DNA. The amino acids in our proteins are similar to those of gorillas, but less so of sea slugs. (Image courtesy of Frederick Ringwald)

► **Figure 32-8** Examples of evolution before your very eyes include color changes in peppered moths. Peppered moths come in two varieties: black and white. *Left:* In the 1700s and today, the black ones were less common, since the white ones are camouflaged against trees and so can't easily be seen by predators. (Michael W.Tweedie/Science Photo Library) *Right:* In the 1800s, when soot from burning coal was common, the white ones became uncommon, because they were no longer camouflaged against the black Soot. The black ones became common, because they now were camouflaged. Evolution isn't "just a theory": you can see it happen. (Michael W.Tweedie/Science Photo Library)

► **Figure 32-9** In 1954, James Watson (left) and Francis Crick (right) found that DNA carries the genetic information of living things. "You can just reckon that any powerful technique you invent, even if it is beneficial, is going to have effects much wider than you think, and it's going to have some disadvantages."—Francis Crick

Manipulating DNA, also called genetic engineering, is potentially a very powerful way to cure disease, especially inherited disease. Nevertheless, Crick's partner, James Watson, has a long history of saying stupid things about it, such as "People say it would be terrible if we made all girls pretty. I think it would be great."

Pretty to whom, you may wonder? Pretty to James Watson, of course. In other words, if you don't conform to the standards of beauty set by James Watson, you may expect trouble from this.

This illustrates the value of a course like this, on science but not for science majors. Science is too important to be left to the scientists. People like you have to take the consequences: therefore, a well-informed public needs to have a say in the practice of science. (A. Barrington Brown/Science Photo Library)

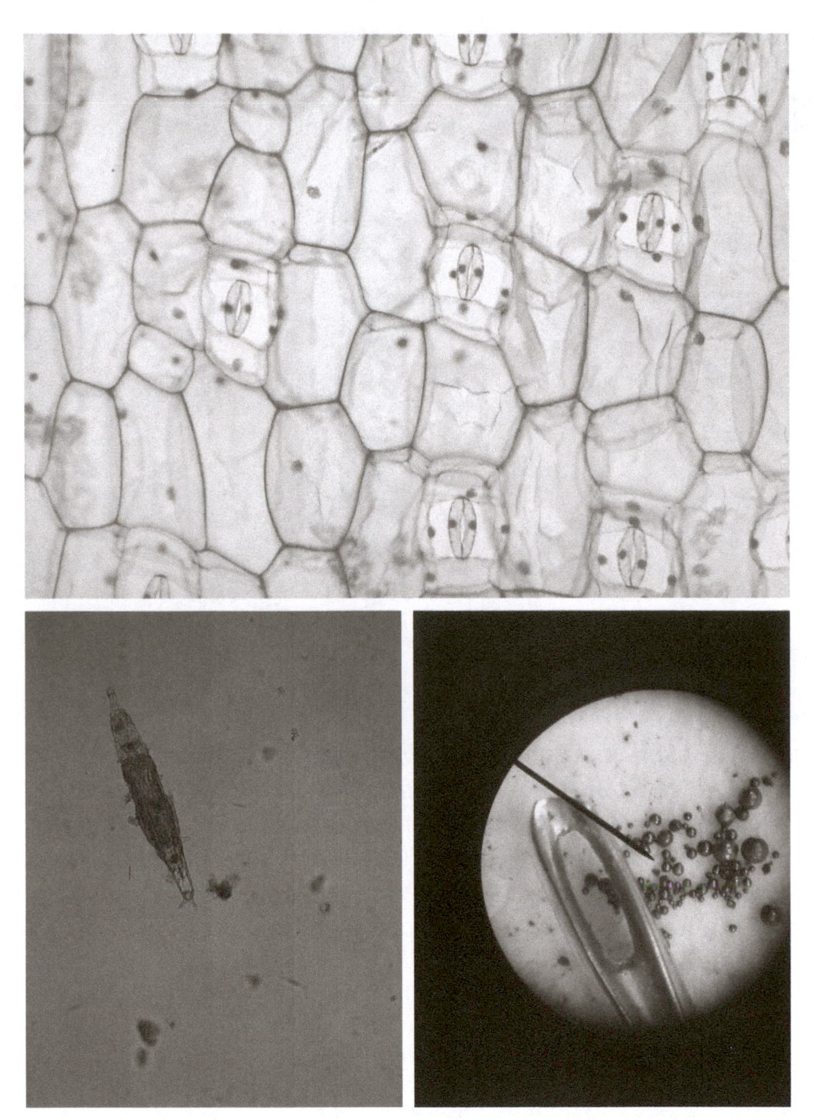

► **Figure 32-10** It's fun to play with a microscope! All the considerations mentioned earlier about buying telescopes for children apply to microscopes, too. There are many junky ones, but it's possible to buy a good, digital one for $200. *Top:* A microscopic view of the leaf surface of spiderwort. (Image © Jubal Harshaw, 2013. Used under license from Shutterstock, Inc.) ***Bottom Left:*** A rotifer, a multi-celled animal (Public domain) ***Bottom Right:*** Micrometeorites from rain with the eye of a needle (Image courtesy of Brian Bellis)

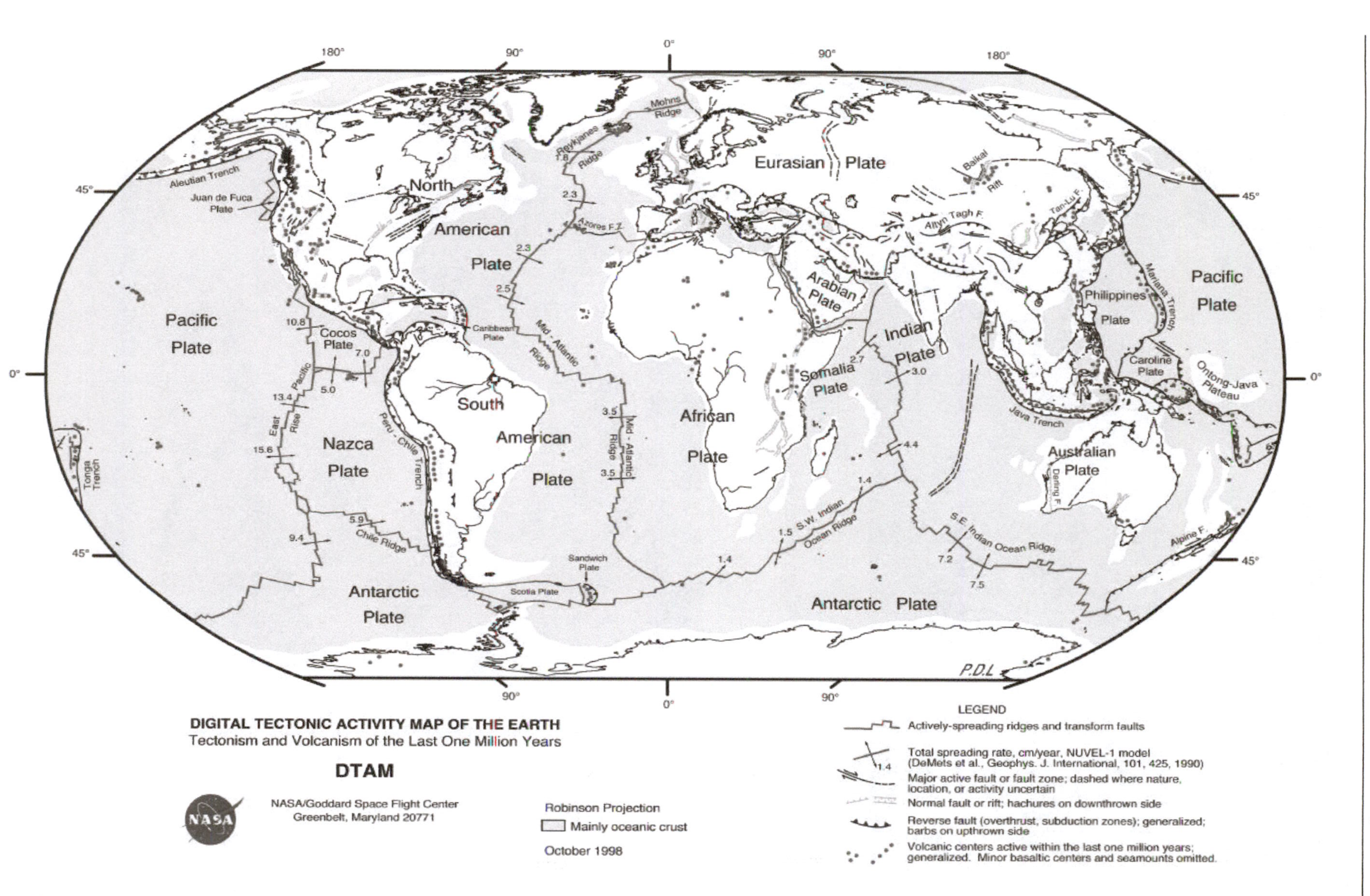

▶ **Figure 32-11** Plate tectonics is the idea that Earth's surface is broken into 15 or so rigid plates, which move because of heat bubbling up from Earth's interior. Plate tectonics is another example of what scientists like: *generality*. The relatively simple idea of plate tectonics explains so much, such as mountain building (when two plates push together), earthquakes (when two plates slide along each other), and continental drift and seafloor spreading (when two plates pull apart). (Source: NASA)

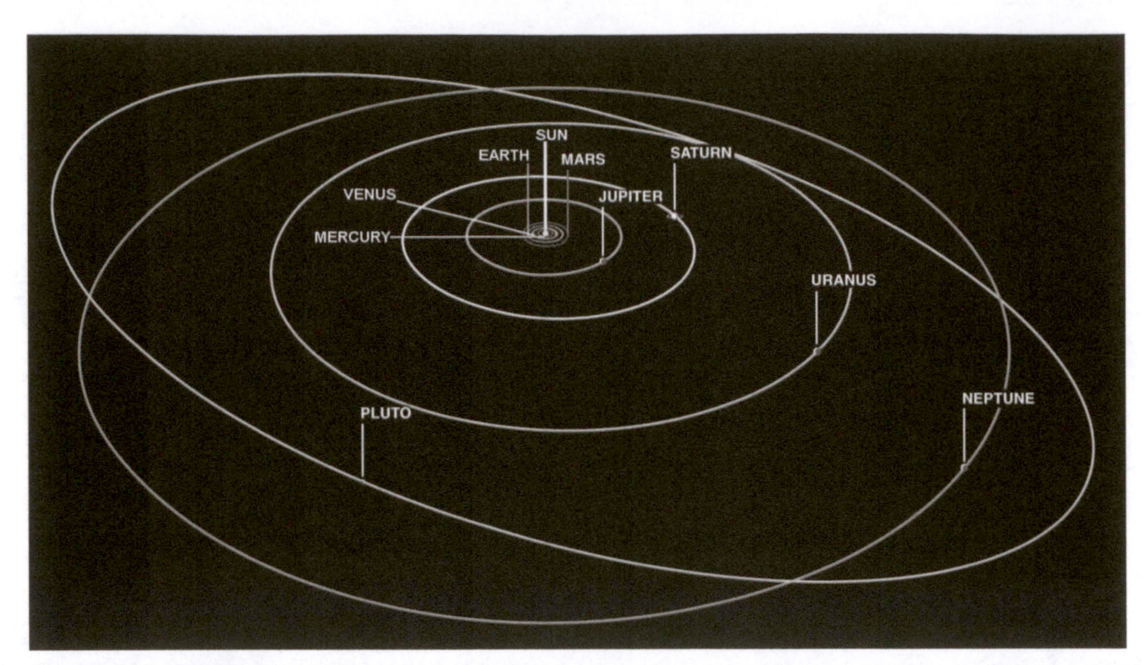

▶ **Figure 32-12** "Perhaps no discovery or opinion ever produced a greater effect on the human spirit than did the teaching of Copernicus. No sooner was the Earth recognized as being round and self-contained, than it was obliged to relinquish the colossal privilege of being the center of the Universe."—Johann Wolfgang von Goethe. (Source: NASA/LPI)

▶ **Figure 32-13** This is an image by a spacecraft through the rings of Saturn. The arrow points to a pale blue dot, seen through the rings of Saturn. That pale, blue dot is Earth. *That pale blue dot* is you. Don't forget this, in your dealings with others. (Source: NASA/JPL)

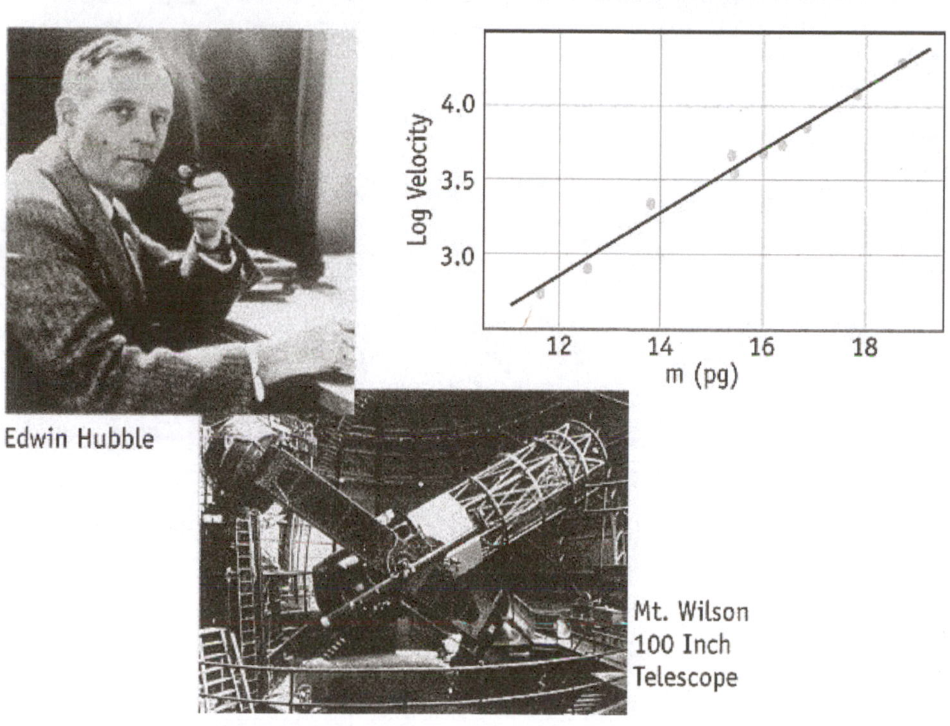

► **Figure 32-14** The idea of Copernicus that Earth is not the center of the Universe continued with Edwin Hubble, who discovered that the Universe is expanding, and therefore has no center. (Source: NASA)

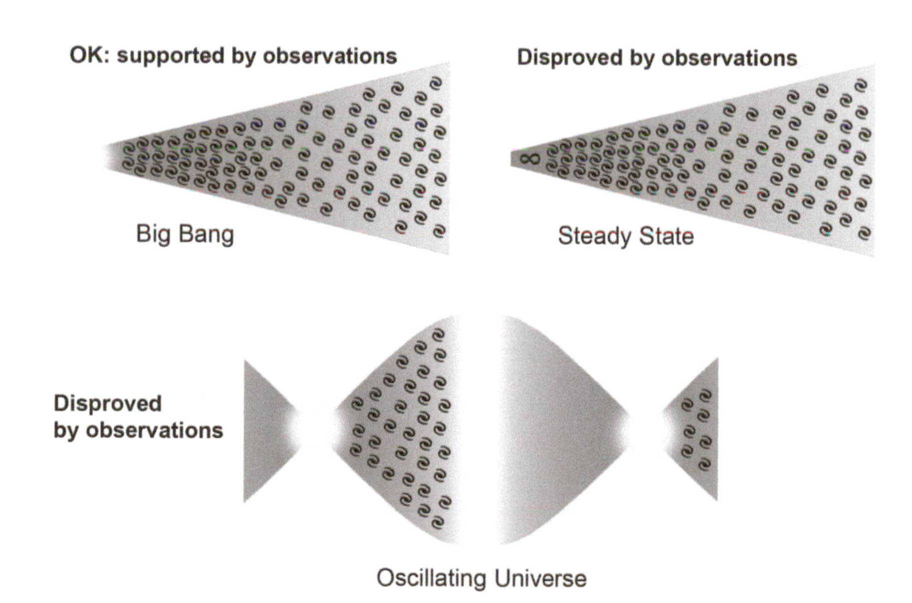

► **Figure 32-15** The Universe is observed to be expanding. Careful measurements of this expansion imply that, 13.8 billion years ago, the Universe began in a hot, dense state, widely referred to as the Big Bang. The Big Bang theory made the prediction that some heat would be left over as the Cosmic Background Radiation, and it has been found. The Steady State theory was disproved by failing to predict the Cosmic Background Radiation. The Oscillating Universe, the idea that the expansion of the Universe will eventually stop and contract because of the gravity of the mass in the Universe, has also been disproved, since apparently the Universe will expand forever. (Image courtesy of Frederick Ringwald)

► **Figure 32-16** Recall the idea of look-back time. What does one see, if one looks back all the way? One sees the origin of the Universe, shown here. This is an actual picture of the Big Bang. (Adapted from NASA image.)

► **Figure 32-17** The Rutherford-Bohr model of the atom. (Image courtesy of Frederick Ringwald)

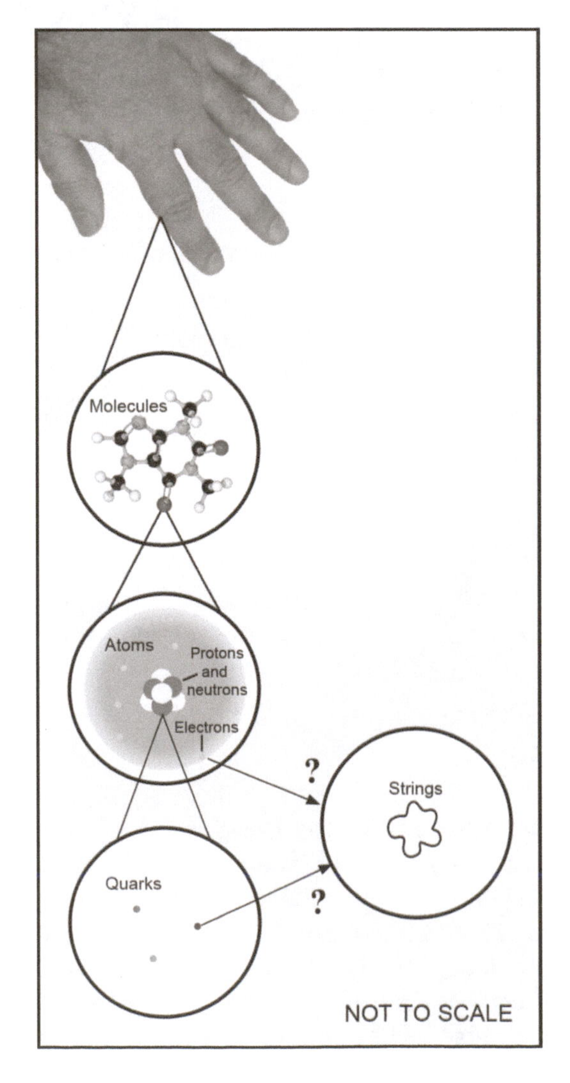

▶ **Figure 32-18** "One thing I have learned in a long life: that all our science, measured against reality, is primitive and childlike—and yet it is the most precious thing we have."—Albert Einstein (Source: Library of Congress)

▶ **Figure 32-19** "Human history becomes more and more a race between education and catastrophe." —H. G. Wells

"If knowledge can create problems, it is not through ignorance that we can solve them."—Isaac Asimov (Left: Source: DOE; Right: Source: NASA)

▶ **Figure 32-20** Earth's atmosphere extends only about 100 km (60 miles) above the surface of Earth. Above this is empty space. More than one astronaut has expressed surprise while looking down on it from an orbiting spacecraft.

"Obviously this was not the ocean of air I had been told it was so many times in my life. I was terrified by its fragile appearance."—Astronaut Ulf Merbold (Source: NASA/International Space Station 7 crew)

▶ **Figure 32-21** "If I have seen farther, it is because I have stood on the shoulders of giants."—Isaac Newton

This acknowledged the contributions of Copernicus, Tycho, Kepler, and especially Galileo. (Image © Georgios Kollidas, 2013. Used under license from Shutterstock, Inc.)

► **Figure 32-22** "I do not know what I may appear to the world; but to myself I seem to have been only like a boy playing on the sea-shore, and diverting myself in now and then finding a smoother pebble or a prettier shell than ordinary, whilst the great ocean of truth lay all undiscovered before me."—Isaac Newton (Image courtesy of Frederick Ringwald)

The Take-Home Message

The Universe may be very, very big,
but *we can understand it*,
by careful observation, rational thought,
and testing ideas by experiment
(or in other words, scientific method).

You are more than an ape.
You are certainly *much* more than a machine!
Act accordingly, with your human facility of reason.

Insist on independent verification of the facts.
Beware of anything based solely on anyone's say-so.
You're entitled to your opinion. You're not entitled to make up your own facts.

If you don't know the answer, it's OK to admit you don't know.
This is really the only way to learn anything new.
This is especially important whenever dealing with children.

It's OK to admit you were wrong, or it should be, as astronomers did with Pluto
(and look at all the nonsense we took for it).

Much of reality is complex, and requires careful, rational thought.
Beware of anyone who says "it's just that simple,"
or "what could be more clear," or "it's just a theory,"
or "it's common sense," or that they have unique access to "absolute truth."

> "Most people would rather die than think; in fact, they do so… The whole problem with the world is that fools and fanatics are always so certain of themselves, but wiser people so full of doubts."
>
> –Bertrand Russell

Beware of *anyone* who says they know it all.
I observe that they never do.

Bibliography and Additional Resources

Science Fiction

"[Science fiction] consists of stories that extrapolate known science to ask 'What if?'"

–Siney Perkowitz, in *Hollywood Science*

Douglas N. Adams, The Hitchhikers Guide to the Galaxy (Del Rey), 1979.
Douglas N. Adams, The Restaurant at the End of the Universe (Harmony), 1982.
Ray Bradbury, The Martian Chronicles (Bantam Books), 1951. (Sublimely beautiful science fiction, even though the real Mars isn't like that.)
Arthur C. Clarke, Childhood's End (Ballantine Books), 1953.
Fred Hoyle, The Black Cloud (Buccaneer), 1992.
Aldous Huxley, Brave New World (Harpercollins), 1995.
George Orwell, Animal Farm (New American Library), 1974.
George Orwell, 1984 (New American Library), 1961.
Robert Silverberg (ed.), The Science Fiction Hall of Fame, Vol. One, 1929–1964 (Orb Books), 2005.
George Gaylord Simpson, The Dechronization of Sam Macgruder (St. Martin's Griffin), 1997.
Robert Louis Stevenson, The Strange Case of Doctor Jekyll and Mister Hyde (Longmans, Green & co.), 1886.
Kurt Vonnegut, Jr., Player Piano (Delacorte), 1952.
H. G. Wells, The Time Machine (William Heinemann), 1895.
H. G. Wells, The War of the Worlds (William Heinemann), 1898.

Space Science and Speculations

J. Kelly Beatty, Carolyn Collins Petersen, & Andrew Chaikin, The New Solar System, 4th ed. (Cambridge University Press), 1998.
Andrew Chaikin, A Man on the Moon (Penguin), 1994.
Oriana Fallaci, If the Sun Dies (Antheneum), 1967.
Richard Feynman, The Feynman Lectures on Physics: The Definitive and Extended Edition (Addison-Wesley), 2005.
Robert Zimmerman, Genesis: The Story of Apollo 8 (Four Walls Eight Windows), 1998.
Robert Zubrin, The Case for Mars (Free Press), 1997.

Graphics

Darrell Huff, How to Lie with Statistics (Norton), 1993.
Edward R. Tufte, Beautiful Evidence (Graphics Press), 2006.
Edward R. Tufte, Envisioning Information (Graphics Press), 1990.
Edward R. Tufte, The Visual Display of Quantitative Information 2nd ed. (Graphics Press), 2001.
Edward R. Tufte, Visual Explanations (Graphics press), 1997.

Resources for Skeptics

George O. Abell and Barry Singer (eds.), Science and the Paranormal (Scribner's), 1981.
A. K. Dewdney, Yes, We Have No Neutrons (Wiley), 1998.
Martin Gardner, Fads and Fallacies in the Name of Science (Dover), 1957.
(This is a thorough, early exposé of older nonsense, such as by Charles Fort or Edgar Cayce, but it has an angry, ridiculing tone. Remember, "Ridicule is not part of the scientific method" – J. Allen Hynek)
Terence Hines, Pseudoscience and the Paranormal, 2nd ed. (Prometheus Books), 2003.
(This is probably the most thorough, best researched, and most scientific of the anti-pseudoscience books: highly recommended!)
Charles Mackay, Extraordinary Popular Delusions and the Madness of Crowds, 1841.
(This may be the best discussion ever of witch burning.)
Phil Plait, Bad Astronomy: Misconceptions and Misuses Revealed, from Astrology to the Moon Landing "Hoax" (Wiley), 2002.
James Randi, Flim-Flam! (Prometheus Books), 1982.
(This is a killer exposé by a master magician, but it has a very angry tone.)
Carl Sagan, The Demon-Haunted World: Science as a Candle In the Dark (Ballantine), 1997.
Michael Shermer, Why People Believe Weird Things (Freeman), 1997.
Max Shulman, Love is a Fallacy (short story) (quarterly magazine)
Skeptical Inquirer

Resources on the Age of Earth

Badash, L. 1989, Scientific American, Volume 261, Number 2 (August issue), p. 90.
Barrell, J. 1917, Rhythms and the Measurements of Geologic Time: Bulletin of the Geological Society of America, v. 28, p. 745.
Bonanno, A. et al. 2002, Astronomy & Astrophysics, v. 390, p. 1115.
Chaboyer B. 1999, in A. Heck, F. Caputo (eds.), Post-Hipparcos Cosmic Candles, Astrophysics and Space Science, v. 237, p. 111.
Chincarini, G. 2005, European Southern Observatory Press Release 22/05.
Cole, G. H. A., and Woolfson, M. M. 2002, Planetary Science, Topic B (Geochronology), p. 202.
Dalrymple, G. B. 1991, The Age of Earth, Stanford University Press.
Dauphas, N. 2005, Nature, v. 435, p. 1203.
Frebel, A. et al. 2007, The Astrophysical Journal Letters, v. 660, p. L117.
Freedman, W. et al. 2001, The Astrophysical Journal, v. 553, p. 47.
Gore, Pamela J. W. 1999, http://www.dc.peachnet.edu/~pgore/geology/geo102/age.htm
Hansen, B. et al. 2002, The Astrophysical Journal, v. 574, p. 155.
Hinshaw, G. et al. 2009, The Astrophysical Journal Supplement, 180, 225
Illingworth, G., and Faber, S. 2004, University of California News Article 6181 (2004-03-09).
Norman, M. 2004, Planetary Science Research Discoveries: The Oldest Moon Rocks, http://www.psrd.hawaii.edu/April04/lunarAnorthosites.html
Tanford, C., and Reynolds, J. 1993, The Scientific Traveler, John Wiley and Sons.
Tegmark, M. et al. 2004, Physical Review Letters D, v. 69, p. 103501.
Zimmer, C. 2001, National Geographic, v. 200, no. 3, p. 78 (September issue).

Other books that helped the author write this book

Joseph P. Allen, Entering Space (Stewart, Tabori, & Chang), 1986.
Lawrence H. Aller, Atoms, Stars, and Nebulae, Revised Edition (Harvard), 1971.
David Alt, Roadside Geology of Northern and Central California (Mountain Press), 2000.
Roy Chapman Andrews, All About Dinosaurs (E. M. Hale), 1953.
Roy Chapman Andrews, Nature's Ways (Crown), 1951.
Isaac Asimov, Adding a Dimension (Doubleday), 1964.
Isaac Asimov, View from a Height (Discus), 1964.
Alan Bean, Apollo: An Eyewitness Account (Greenwich Workshop), 1998.
Otto Bettmann, The Good Old Days: They Were Terrible! (Random House), 1974.
John Billingham (ed.), Life in the Universe (MIT Press), 1981.
Robert Burnham Jr., Burnham's Celestial Handbook, Volumes 1–3, (Dover), 1978.
Edgar M. Cortright (ed.), Exploring Space with a Camera (NASA SP-168), 1968.
Charles Darwin, The Origin of Species (Modern Library), 1993, first edition 1859.
Ingri D'Aulaire and Edgar Parin D'Aulaires, D'Aulaires Book of Greek Myths (Doubleday), 1962.
Ingri D'Aulaire and Edgar Parin D'Aulaire, Norse Gods and Giants (Doubleday), 1967.
Lowell Dingus, Next of Kin: Great Fossils at the American Museum of Natural History (Rizzoli International), 1996.
Rene Dubos, Henry Margenau, and C. P. Snow (eds.), LIFE Science Library (Time-Life), 1963–1967.
Betty Edwards, The New Drawing on the Right Side of the Brain (Tarcher), 1999.
Peter Feibelman, A Ph. D. Is Not Enough! (Basic Books), 1993.
Richard Feynman, The Character of Physical Law (Modern Library), 1994.
Roy A. Gallant, many of his books!
Galileo Galilei, The Starry Messenger (U. of Chicago Press), 1989.
George Gamow, Mr. Tompkins in Paperback (Cambridge University Press), 1993.
Mark A. Garlick, Astronomy: A Visual Guide (Firefly), 2004.
Richard L. Gregory, Eye and Brain, 5th ed. (Princeton), 1997.
William K. Hartmann, Moons & Planets, 5th ed. (Cambridge University Press), 2004.
Martin Harwitt, Astrophysical Concepts (Wiley), 1973.
Steven Hawking, A Brief History of Time (Bantam), 1988.
S. I. Hayakawa, Language in Thought and Action, 3rd ed. (Harcourt Brace Jovanovich), 1972.
T. L. Heath (transl.), Euclid's Elements (Green Lion), 2002.
Grant Heiken, David Vaniman, and Bevan M. French, Lunar Sourcebook: A User's Guide to the Moon (Cambridge University Press), 1991.
Douglas R. Hofstadter, Gödel, Escher, Bach (Vintage Books), 1980.
Rolfe Humphries, The Metamorphoses of Ovid (Indiana University Press), 1955.
Bruce Jakosky, The Search for Life on Other Planets (Cambridge University Press), 1998.
Fred H. Knelman (ed.), 1984 and All That: Modern Science, Social Change, and Human Values (Wadsworth), 1971.
Michael Light, Full Moon (Knopf), 1999.
James Loewen, Lies My Teacher Told Me (New Press), 1995.
Spencer George Lucas, Dinosaurs: the Textbook (McGraw-Hill), 2005.
Peter B. Medawar, Advice to a Young Scientist (Basic Books), 1981.
Michael A. G. Michaud, Contact with Alien Civilizations (Copernicus Books), 2007.
Marcel Minnaert, Light and Color in the Outdoors (Springer), 1995.
Richard A. Muller, Physics for Future Presidents (Norton), 2008.
Henning Nelms, Thinking with a Pencil (Ten Speed Press), 1957.

Arthur P. Norton, Norton's Star Atlas and Telescopic Handbook (Sky Publishing), 1969.
Sidney Perkowitz, Hollywood Science (Columbia), 2007.
Robert Pirsing, Zen and the Art of Motorcycle Maintenance (Bantam), 1974.
Neil Postman, Amusing Ourselves to Death (Penguin), 1985.
Neil Postman, The End of Education (Vintage), 1996.
Neil Postman, Technopoly (Vintage), 1993.
William L. Ramsey, Raymond A. Burckley, Clifford R. Phillips, and Frank M. Watenpaugh, Modern Earth Science (Holt, Rinehart, and Winston), 1969.
Roger Ressmeyer, Space Places (Collins), 1990.
H. A. Rey, The Stars: A New Way to See Them (Houghton Mifflin), 2008, first edition 1953.
Tom Rogers, Insultingly Stupid Movie Physics (Sourcebooks Hysteria), 2007.
Bertrand Russell, A History of Western Philosophy (Touchstone), 1945.
Carl Sagan, Cosmos (Random House), 1980.
Carl Sagan, Pale Blue Dot: A Vision of the Human Future in Space (Ballantine), 1994.
Walter Scheider, A Serious But Not Ponderous Guide to Nuclear Energy (Cavendish Press), 2001.
Frank H. Shu, The Physical Universe (University Science Books), 1982.
Andrei Sokolov, Ron Miller, Vitaly Myagkov, and William K. Hartmann, In the Stream of Stars (Workman), 1990.
Paul D. Spudis, The Once and Future Moon (Smithsonian), 1996.
W. Strunk Jr. and E. B. White, The Elements of Style 4th ed. (Longman), 2000.
Clifford E. Swartz, Used Math: For the First Two Years of College Science, 2nd ed. (American Association of Physics Teachers), 1993.
Charles Tanford and Jacqueline Reynolds, The Scientific Traveler (Wiley), 1992.
Edwin F. Taylor and John Archibald Wheeler, Spacetime Physics, 2nd ed. (Freeman), 1992.
Edward K. Thompson et al., The World We Live In (Time), 1955.
Kip Thorne, Black Holes and Warped Spacetime (Norton), 1995.
Alvin Toffler, Future Shock (Random House), 1970.
Albrecht Unsöld, The New Cosmos (Longmans), 1969.
Alan Walker and Pat Shipman, The Wisdom of the Bones (Knopf), 1996.
Steven Weinberg, The First Three Minutes: A Modern View of the Origin of the Universe (Basic Books), 1977.
Tom Weller, Science Made Stupid (humor) (Mariner Books), 1985.
Any of the Golden Guides series
Any of the Peterson's Field Guides
(video) Infrared: More Than Your Eyes Can See (NASA), 2002.
(video) Powers of Ten (The Films of Charles and Ray Eames, Volume 1, narrated by Phillip Morrison, 1968.
(video) A Private Universe (Harvard-Smithsonian Center for Astrophysics), 1987.
(video) Walking with Dinosaurs series (BBC), 1999.
(video) Walking with Prehistoric Beasts series (BBC), 2001.

Online Resources

Jonathan's Space Report (http://www.planet4589.org/space/jsr/jsr.html)
Space.com
Mr. Eclipse website (www.mreclipse.com)